Multivariate Statistical Analysis

STATISTICS: Textbooks and Monographs

D. B. Owen
Founding Editor, 1972–1991

Associate Editors

Statistical Computing/
Nonparametric Statistics
Professor William R. Schucany
Southern Methodist University

Multivariate Analysis
Professor Anant M. Kshirsagar
University of Michigan

Probability
Professor Marcel F. Neuts
University of Arizona

Quality Control/Reliability
Professor Edward G. Schilling
Rochester Institute of Technology

Editorial Board

Applied Probability
Dr. Paul R. Garvey
The MITRE Corporation

Statistical Distributions
Professor N. Balakrishnan
McMaster University

Economic Statistics
Professor David E. A. Giles
University of Victoria

Statistical Process Improvement
Professor G. Geoffrey Vining
Virginia Polytechnic Institute

Experimental Designs
Mr. Thomas B. Barker
Rochester Institute of Technology

Stochastic Processes
Professor V. Lakshmikantham
Florida Institute of Technology

Multivariate Analysis
Professor Subir Ghosh
University of California–Riverside

Survey Sampling
Professor Lynne Stokes
Southern Methodist University

Time Series
Sastry G. Pantula
North Carolina State University

1. The Generalized Jackknife Statistic, *H. L. Gray and W. R. Schucany*
2. Multivariate Analysis, *Anant M. Kshirsagar*
3. Statistics and Society, *Walter T. Federer*
4. Multivariate Analysis: A Selected and Abstracted Bibliography, 1957–1972, *Kocherlakota Subrahmaniam and Kathleen Subrahmaniam*
5. Design of Experiments: A Realistic Approach, *Virgil L. Anderson and Robert A. McLean*
6. Statistical and Mathematical Aspects of Pollution Problems, *John W. Pratt*
7. Introduction to Probability and Statistics (in two parts), Part I: Probability; Part II: Statistics, *Narayan C. Giri*
8. Statistical Theory of the Analysis of Experimental Designs, *J. Ogawa*
9. Statistical Techniques in Simulation (in two parts), *Jack P. C. Kleijnen*
10. Data Quality Control and Editing, *Joseph I. Naus*
11. Cost of Living Index Numbers: Practice, Precision, and Theory, *Kali S. Banerjee*
12. Weighing Designs: For Chemistry, Medicine, Economics, Operations Research, Statistics, *Kali S. Banerjee*
13. The Search for Oil: Some Statistical Methods and Techniques, *edited by D. B. Owen*
14. Sample Size Choice: Charts for Experiments with Linear Models, *Robert E. Odeh and Martin Fox*
15. Statistical Methods for Engineers and Scientists, *Robert M. Bethea, Benjamin S. Duran, and Thomas L. Boullion*
16. Statistical Quality Control Methods, *Irving W. Burr*
17. On the History of Statistics and Probability, *edited by D. B. Owen*
18. Econometrics, *Peter Schmidt*
19. Sufficient Statistics: Selected Contributions, *Vasant S. Huzurbazar (edited by Anant M. Kshirsagar)*
20. Handbook of Statistical Distributions, *Jagdish K. Patel, C. H. Kapadia, and D. B. Owen*
21. Case Studies in Sample Design, *A. C. Rosander*
22. Pocket Book of Statistical Tables, *compiled by R. E. Odeh, D. B. Owen, Z. W. Birnbaum, and L. Fisher*
23. The Information in Contingency Tables, *D. V. Gokhale and Solomon Kullback*
24. Statistical Analysis of Reliability and Life-Testing Models: Theory and Methods, *Lee J. Bain*
25. Elementary Statistical Quality Control, *Irving W. Burr*
26. An Introduction to Probability and Statistics Using BASIC, *Richard A. Groeneveld*
27. Basic Applied Statistics, *B. L. Raktoe and J. J. Hubert*
28. A Primer in Probability, *Kathleen Subrahmaniam*
29. Random Processes: A First Look, *R. Syski*
30. Regression Methods: A Tool for Data Analysis, *Rudolf J. Freund and Paul D. Minton*
31. Randomization Tests, *Eugene S. Edgington*
32. Tables for Normal Tolerance Limits, Sampling Plans and Screening, *Robert E. Odeh and D. B. Owen*
33. Statistical Computing, *William J. Kennedy, Jr., and James E. Gentle*
34. Regression Analysis and Its Application: A Data-Oriented Approach, *Richard F. Gunst and Robert L. Mason*
35. Scientific Strategies to Save Your Life, *I. D. J. Bross*
36. Statistics in the Pharmaceutical Industry, *edited by C. Ralph Buncher and Jia-Yeong Tsay*
37. Sampling from a Finite Population, *J. Hajek*
38. Statistical Modeling Techniques, *S. S. Shapiro and A. J. Gross*
39. Statistical Theory and Inference in Research, *T. A. Bancroft and C.-P. Han*
40. Handbook of the Normal Distribution, *Jagdish K. Patel and Campbell B. Read*
41. Recent Advances in Regression Methods, *Hrishikesh D. Vinod and Aman Ullah*
42. Acceptance Sampling in Quality Control, *Edward G. Schilling*
43. The Randomized Clinical Trial and Therapeutic Decisions, *edited by Niels Tygstrup, John M Lachin, and Erik Juhl*

44. Regression Analysis of Survival Data in Cancer Chemotherapy, *Walter H. Carter, Jr., Galen L. Wampler, and Donald M. Stablein*
45. A Course in Linear Models, *Anant M. Kshirsagar*
46. Clinical Trials: Issues and Approaches, *edited by Stanley H. Shapiro and Thomas H. Louis*
47. Statistical Analysis of DNA Sequence Data, *edited by B. S. Weir*
48. Nonlinear Regression Modeling: A Unified Practical Approach, *David A. Ratkowsky*
49. Attribute Sampling Plans, Tables of Tests and Confidence Limits for Proportions, *Robert E. Odeh and D. B. Owen*
50. Experimental Design, Statistical Models, and Genetic Statistics, *edited by Klaus Hinkelmann*
51. Statistical Methods for Cancer Studies, *edited by Richard G. Cornell*
52. Practical Statistical Sampling for Auditors, *Arthur J. Wilburn*
53. Statistical Methods for Cancer Studies, *edited by Edward J. Wegman and James G. Smith*
54. Self-Organizing Methods in Modeling: GMDH Type Algorithms, *edited by Stanley J. Farlow*
55. Applied Factorial and Fractional Designs, *Robert A. McLean and Virgil L. Anderson*
56. Design of Experiments: Ranking and Selection, *edited by Thomas J. Santner and Ajit C. Tamhane*
57. Statistical Methods for Engineers and Scientists: Second Edition, Revised and Expanded, *Robert M. Bethea, Benjamin S. Duran, and Thomas L. Boullion*
58. Ensemble Modeling: Inference from Small-Scale Properties to Large-Scale Systems, *Alan E. Gelfand and Crayton C. Walker*
59. Computer Modeling for Business and Industry, *Bruce L. Bowerman and Richard T. O'Connell*
60. Bayesian Analysis of Linear Models, *Lyle D. Broemeling*
61. Methodological Issues for Health Care Surveys, *Brenda Cox and Steven Cohen*
62. Applied Regression Analysis and Experimental Design, *Richard J. Brook and Gregory C. Arnold*
63. Statpal: A Statistical Package for Microcomputers—PC-DOS Version for the IBM PC and Compatibles, *Bruce J. Chalmer and David G. Whitmore*
64. Statpal: A Statistical Package for Microcomputers—Apple Version for the II, II+, and IIe, *David G. Whitmore and Bruce J. Chalmer*
65. Nonparametric Statistical Inference: Second Edition, Revised and Expanded, *Jean Dickinson Gibbons*
66. Design and Analysis of Experiments, *Roger G. Petersen*
67. Statistical Methods for Pharmaceutical Research Planning, *Sten W. Bergman and John C. Gittins*
68. Goodness-of-Fit Techniques, *edited by Ralph B. D'Agostino and Michael A. Stephens*
69. Statistical Methods in Discrimination Litigation, *edited by D. H. Kaye and Mikel Aickin*
70. Truncated and Censored Samples from Normal Populations, *Helmut Schneider*
71. Robust Inference, *M. L. Tiku, W. Y. Tan, and N. Balakrishnan*
72. Statistical Image Processing and Graphics, *edited by Edward J. Wegman and Douglas J. DePriest*
73. Assignment Methods in Combinatorial Data Analysis, *Lawrence J. Hubert*
74. Econometrics and Structural Change, *Lyle D. Broemeling and Hiroki Tsurumi*
75. Multivariate Interpretation of Clinical Laboratory Data, *Adelin Albert and Eugene K. Harris*
76. Statistical Tools for Simulation Practitioners, *Jack P. C. Kleijnen*
77. Randomization Tests: Second Edition, *Eugene S. Edgington*
78. A Folio of Distributions: A Collection of Theoretical Quantile-Quantile Plots, *Edward B. Fowlkes*
79. Applied Categorical Data Analysis, *Daniel H. Freeman, Jr.*
80. Seemingly Unrelated Regression Equations Models: Estimation and Inference, *Virendra K. Srivastava and David E. A. Giles*

81. Response Surfaces: Designs and Analyses, *Andre I. Khuri and John A. Cornell*
82. Nonlinear Parameter Estimation: An Integrated System in BASIC, *John C. Nash and Mary Walker-Smith*
83. Cancer Modeling, *edited by James R. Thompson and Barry W. Brown*
84. Mixture Models: Inference and Applications to Clustering, *Geoffrey J. McLachlan and Kaye E. Basford*
85. Randomized Response: Theory and Techniques, *Arijit Chaudhuri and Rahul Mukerjee*
86. Biopharmaceutical Statistics for Drug Development, *edited by Karl E. Peace*
87. Parts per Million Values for Estimating Quality Levels, *Robert E. Odeh and D. B. Owen*
88. Lognormal Distributions: Theory and Applications, *edited by Edwin L. Crow and Kunio Shimizu*
89. Properties of Estimators for the Gamma Distribution, *K. O. Bowman and L. R. Shenton*
90. Spline Smoothing and Nonparametric Regression, *Randall L. Eubank*
91. Linear Least Squares Computations, *R. W. Farebrother*
92. Exploring Statistics, *Damaraju Raghavarao*
93. Applied Time Series Analysis for Business and Economic Forecasting, *Sufi M. Nazem*
94. Bayesian Analysis of Time Series and Dynamic Models, *edited by James C. Spall*
95. The Inverse Gaussian Distribution: Theory, Methodology, and Applications, *Raj S. Chhikara and J. Leroy Folks*
96. Parameter Estimation in Reliability and Life Span Models, *A. Clifford Cohen and Betty Jones Whitten*
97. Pooled Cross-Sectional and Time Series Data Analysis, *Terry E. Dielman*
98. Random Processes: A First Look, Second Edition, Revised and Expanded, *R. Syski*
99. Generalized Poisson Distributions: Properties and Applications, *P. C. Consul*
100. Nonlinear L_p-Norm Estimation, *Rene Gonin and Arthur H. Money*
101. Model Discrimination for Nonlinear Regression Models, *Dale S. Borowiak*
102. Applied Regression Analysis in Econometrics, *Howard E. Doran*
103. Continued Fractions in Statistical Applications, *K. O. Bowman and L. R. Shenton*
104. Statistical Methodology in the Pharmaceutical Sciences, *Donald A. Berry*
105. Experimental Design in Biotechnology, *Perry D. Haaland*
106. Statistical Issues in Drug Research and Development, *edited by Karl E. Peace*
107. Handbook of Nonlinear Regression Models, *David A. Ratkowsky*
108. Robust Regression: Analysis and Applications, *edited by Kenneth D. Lawrence and Jeffrey L. Arthur*
109. Statistical Design and Analysis of Industrial Experiments, *edited by Subir Ghosh*
110. U-Statistics: Theory and Practice, *A. J. Lee*
111. A Primer in Probability: Second Edition, Revised and Expanded, *Kathleen Subrahmaniam*
112. Data Quality Control: Theory and Pragmatics, *edited by Gunar E. Liepins and V. R. R. Uppuluri*
113. Engineering Quality by Design: Interpreting the Taguchi Approach, *Thomas B. Barker*
114. Survivorship Analysis for Clinical Studies, *Eugene K. Harris and Adelin Albert*
115. Statistical Analysis of Reliability and Life-Testing Models: Second Edition, *Lee J. Bain and Max Engelhardt*
116. Stochastic Models of Carcinogenesis, *Wai-Yuan Tan*
117. Statistics and Society: Data Collection and Interpretation, Second Edition, Revised and Expanded, *Walter T. Federer*
118. Handbook of Sequential Analysis, *B. K. Ghosh and P. K. Sen*
119. Truncated and Censored Samples: Theory and Applications, *A. Clifford Cohen*
120. Survey Sampling Principles, *E. K. Foreman*
121. Applied Engineering Statistics, *Robert M. Bethea and R. Russell Rhinehart*
122. Sample Size Choice: Charts for Experiments with Linear Models: Second Edition, *Robert E. Odeh and Martin Fox*
123. Handbook of the Logistic Distribution, *edited by N. Balakrishnan*
124. Fundamentals of Biostatistical Inference, *Chap T. Le*
125. Correspondence Analysis Handbook, *J.-P. Benzécri*

126. Quadratic Forms in Random Variables: Theory and Applications, *A. M. Mathai and Serge B. Provost*
127. Confidence Intervals on Variance Components, *Richard K. Burdick and Franklin A. Graybill*
128. Biopharmaceutical Sequential Statistical Applications, *edited by Karl E. Peace*
129. Item Response Theory: Parameter Estimation Techniques, *Frank B. Baker*
130. Survey Sampling: Theory and Methods, *Arijit Chaudhuri and Horst Stenger*
131. Nonparametric Statistical Inference: Third Edition, Revised and Expanded, *Jean Dickinson Gibbons and Subhabrata Chakraborti*
132. Bivariate Discrete Distribution, *Subrahmaniam Kocherlakota and Kathleen Kocherlakota*
133. Design and Analysis of Bioavailability and Bioequivalence Studies, *Shein-Chung Chow and Jen-pei Liu*
134. Multiple Comparisons, Selection, and Applications in Biometry, *edited by Fred M. Hoppe*
135. Cross-Over Experiments: Design, Analysis, and Application, *David A. Ratkowsky, Marc A. Evans, and J. Richard Alldredge*
136. Introduction to Probability and Statistics: Second Edition, Revised and Expanded, *Narayan C. Giri*
137. Applied Analysis of Variance in Behavioral Science, *edited by Lynne K. Edwards*
138. Drug Safety Assessment in Clinical Trials, *edited by Gene S. Gilbert*
139. Design of Experiments: A No-Name Approach, *Thomas J. Lorenzen and Virgil L. Anderson*
140. Statistics in the Pharmaceutical Industry: Second Edition, Revised and Expanded, *edited by C. Ralph Buncher and Jia-Yeong Tsay*
141. Advanced Linear Models: Theory and Applications, *Song-Gui Wang and Shein-Chung Chow*
142. Multistage Selection and Ranking Procedures: Second-Order Asymptotics, *Nitis Mukhopadhyay and Tumulesh K. S. Solanky*
143. Statistical Design and Analysis in Pharmaceutical Science: Validation, Process Controls, and Stability, *Shein-Chung Chow and Jen-pei Liu*
144. Statistical Methods for Engineers and Scientists: Third Edition, Revised and Expanded, *Robert M. Bethea, Benjamin S. Duran, and Thomas L. Boullion*
145. Growth Curves, *Anant M. Kshirsagar and William Boyce Smith*
146. Statistical Bases of Reference Values in Laboratory Medicine, *Eugene K. Harris and James C. Boyd*
147. Randomization Tests: Third Edition, Revised and Expanded, *Eugene S. Edgington*
148. Practical Sampling Techniques: Second Edition, Revised and Expanded, *Ranjan K. Som*
149. Multivariate Statistical Analysis, *Narayan C. Giri*
150. Handbook of the Normal Distribution: Second Edition, Revised and Expanded, *Jagdish K. Patel and Campbell B. Read*
151. Bayesian Biostatistics, *edited by Donald A. Berry and Dalene K. Stangl*
152. Response Surfaces: Designs and Analyses, Second Edition, Revised and Expanded, *André I. Khuri and John A. Cornell*
153. Statistics of Quality, *edited by Subir Ghosh, William R. Schucany, and William B. Smith*
154. Linear and Nonlinear Models for the Analysis of Repeated Measurements, *Edward F. Vonesh and Vernon M. Chinchilli*
155. Handbook of Applied Economic Statistics, *Aman Ullah and David E. A. Giles*
156. Improving Efficiency by Shrinkage: The James-Stein and Ridge Regression Estimators, *Marvin H. J. Gruber*
157. Nonparametric Regression and Spline Smoothing: Second Edition, *Randall L. Eubank*
158. Asymptotics, Nonparametrics, and Time Series, *edited by Subir Ghosh*
159. Multivariate Analysis, Design of Experiments, and Survey Sampling, *edited by Subir Ghosh*

160. Statistical Process Monitoring and Control, *edited by Sung H. Park and G. Geoffrey Vining*
161. Statistics for the 21st Century: Methodologies for Applications of the Future, *edited by C. R. Rao and Gábor J. Székely*
162. Probability and Statistical Inference, *Nitis Mukhopadhyay*
163. Handbook of Stochastic Analysis and Applications, *edited by D. Kannan and V. Lakshmikantham*
164. Testing for Normality, *Henry C. Thode, Jr.*
165. Handbook of Applied Econometrics and Statistical Inference, *edited by Aman Ullah, Alan T. K. Wan, and Anoop Chaturvedi*
166. Visualizing Statistical Models and Concepts, *R. W. Farebrother*
167. Financial and Actuarial Statistics: An Introduction, *Dale S. Borowiak*
168. Nonparametric Statistical Inference: Fourth Edition, Revised and Expanded, *Jean Dickinson Gibbons and Subhabrata Chakraborti*
169. Computer-Aided Econometrics, *edited by David E. A. Giles*
170. The EM Algorithm and Related Statistical Models, *edited by Michiko Watanabe and Kazunori Yamaguchi*
171. Multivariate Statistical Analysis: Second Edition, Revised and Expanded, *Narayan C. Giri*
172. Computational Methods in Statistics and Econometrics, *Hisashi Tanizaki*

Additional Volumes in Preparation

Multivariate Statistical Analysis
Second Edition, Revised and Expanded

Narayan C. Giri
University of Montreal
Montreal, Quebec, Canada

MARCEL DEKKER, INC.　　　　NEW YORK • BASEL

Although great care has been taken to provide accurate and current information, neither the author(s) nor the publisher, nor anyone else associated with this publication, shall be liable for any loss, damage, or liability directly or indirectly caused or alleged to be caused by this book. The material contained herein is not intended to provide specific advice or recommendations for any specific situation.

Trademark notice: Product or corporate names may be trademarks or registered trademarks and are used only for identification and explanation without intent to infringe.

Library of Congress Cataloging-in-Publication Data
A catalog record for this book is available from the Library of Congress.

ISBN: 0-8247-4713-5

This book is printed on acid-free paper.

Headquarters
Marcel Dekker, Inc., 270 Madison Avenue, New York, NY 10016, U.S.A.
tel: 212-696-9000; fax: 212-685-4540

Distribution and Customer Service
Marcel Dekker, Inc., Cimarron Road, Monticello, New York 12701, U.S.A.
tel: 800-228-1160; fax: 845-796-1772

Eastern Hemisphere Distribution
Marcel Dekker AG, Hutgasse 4, Postfach 812, CH-4001 Basel, Switzerland
tel: 41-61-260-6300; fax: 41-61-260-6333

World Wide Web
http://www.dekker.com

The publisher offers discounts on this book when ordered in bulk quantities. For more information, write to Special Sales/Professional Marketing at the headquarters address above.

Copyright © 2004 by Marcel Dekker, Inc. All Rights Reserved.

Neither this book nor any part may be reproduced or transmitted in any form or by any means, electronic or mechanical, including photocopying, microfilming, and recording, or by any information storage and retrieval system, without permission in writing from the publisher.

Current printing (last digit):

10 9 8 7 6 5 4 3 2 1

PRINTED IN THE UNITED STATES OF AMERICA

To Nilima, Nabanita, and Nandan

Preface to the Second Edition

As in the first edition the aim has been to provide an up-to-date presentation of both the theoretical and applied aspects of multivariate analysis using the invariance approach for readers with a basic knowledge of mathematics and statistics at the undergraduate level. This new edition updates the original book by adding new results, examples, problems, and references. The following new subsections are added. Section 4.3 deals with the symmetric distributions: its properties and characterization. Section 4.3.6 treats elliptically symmetric distributions (multivariate) and Section 4.3.7 considers the singular symmetrical distribution. Regression and correlations in symmetrical distributions are discussed in Section 4.5.1. The redundancy index is included in Section 4.7. In Section 5.3.7 we treat the problem of estimation of covariance matrices and the equivariant estimation under curved model of mean, and covariance matrix is treated in Section 5.4. Basic distributions in symmetrical distributions are given in Section 6.12. Tests of mean against one-sided alternatives are given in Section 7.3.1. Section 8.5.2 treats multiple correlation with partial information and Section 8.1 deals with tests with missing data. In Section 9.5 we discuss the relationship between discriminant analysis and cluster analysis.

A new Appendix A dealing with tables of chi-square adjustments to the Wilks' criterion U (Schatkoff, M. (1966), *Biometrika*, pp. 347–358, and Pillai, K.C.S. and Gupta, A.K. (1969), *Biometrika*, pp. 109–118) is added. Appendix B lists the publications of the author.

In preparing this volume I have tried to incorporate various comments of reviewers of the first edition and colleagues who have used it. The comments of my own students and my long experience in teaching the subject have also been utilized in preparing the Second Edition.

Narayan C. Giri

Preface to the First Edition

This book is an up-to-date presentation of both theoretical and applied aspects of multivariate analysis using the invariance approach. It is written for readers with knowledge of mathematics and statistics at the undergraduate level. Various concepts are explained with live data from applied areas. In conformity with the general nature of introductory textbooks, we have tried to include many examples and motivations relevant to specific topics. The material presented here is developed from the subjects included in my earlier books on multivariate statistical inference. My long experience teaching multivariate statistical analysis courses in several universities and the comments of my students have also been utilized in writing this volume.

Invariance is the mathematical term for symmetry with respect to a certain group of transformations. As in other branches of mathematics the notion of invariance in statistical inference is an old one. The unpublished work of Hunt and Stein toward the end of World War II has given very strong support to the applicability and meaningfulness of this notion in the framework of the general class of statistical tests. It is now established as a very powerful tool for proving the optimality of many statistical test procedures. It is a generally accepted principle that if a problem with a unique solution is invariant under a certain transformation, then the solution should be invariant under that transformation. Another compelling reason for discussing multivariate analysis through invariance is that most of the commonly used test procedures are likelihood ratio tests. Under a mild restriction on the parametric space and the probability density functions under consideration, the likelihood ratio tests are almost invariant.

Invariant tests depend on the observations only through maximal invariant. To find optimal invariant tests we need to find the explicit form of the maximal invariant statistic and its distribution. In many testing problems it is not always convenient to find the explicit form of the maximal invariant. Stein (1956) gave a representation of the ratio of probability densities of a maximal invariant by integrating with respect to a invariant measure on the group of transformations leaving the problem invariant. Stein did not give explicitly the conditions under which his representation is valid. Subsequently many workers gave sufficient conditions for the validity of his representation. Spherically and elliptically symmetric distributions form an important family of nonnormal symmetric distributions of which the multivariate normal distribution is a member. This family is becoming increasingly important in robustness studies where the aim is to determine how sensitive the commonly used multivariate methods are to the multivariate normality assumption. Chapter 1 contains some special results regarding characteristic roots and vectors, and partitioned submatrices of real and complex matrices. It also contains some special results on determinants and matrix derivatives and some special theorems on real and complex matrices.

Chapter 2 deals with the theory of groups and related results that are useful for the development of invariant statistical test procedures. It also contains results on Jacobians of some important transformations that are used in multivariate sampling distributions.

Chapter 3 is devoted to basic notions of multivariate distributions and the principle of invariance in statistical inference. The interrelationship between invariance and sufficiency, invariance and unbiasedness, invariance and optimal tests, and invariance and most stringent tests are examined. This chapter also includes the Stein representation theorem, Hunt and Stein theorem, and robustness studies of statistical tests.

Chapter 4 deals with multivariate normal distributions by means of the probability density function and a simple characterization. The second approach simplifies multivariate theory and allows suitable generalization from univariate theory without further analysis. This chapter also contains some characterizations of the real multivariate normal distribution, concentration ellipsoid and axes, regression, multiple and partial correlation, and cumulants and kurtosis. It also deals with analogous results for the complex multivariate normal distribution, and elliptically and spherically symmetric distributions. Results on vec operator and tensor product are also included here.

Maximum likelihood estimators of the parameters of the multivariate normal, the multivariate complex normal, the elliptically and spherically symmetric distributions and their optimal properties are the main subject matter of Chapter 5. The James–Stein estimator, the positive part of the James–Stein estimator,

Preface to the First Edition

unbiased estimation of risk, smoother shrinkage estimation of mean with known and unknown covariance matrix are considered here.

Chapter 6 contains a systematic derivation of basic multivariate sampling distributions for the multivariate normal case, the complex multivariate normal case, and the case of symmetric distributions.

Chapter 7 deals with tests and confidence regions of mean vectors of multivariate normal populations with known and unknown covariance matrices and their optimal properties, tests of hypotheses concerning the subvectors of μ in multivariate normal, tests of mean in multivariate complex normal and symmetric distributions, and the robustness of the T^2-test in the family of elliptically symmetric distributions.

Chapter 8 is devoted to a systematic derivation of tests concerning covariance matrices and mean vectors, the sphericity test, tests of independence, the R^2-test, a special problem in a test of independence, MANOVA, GMANOVA, extended GMANOVA, equality of covariance matrice in multivariate normal populations and their extensions to complex multivariate normal, and the study of robustness in the family of elliptically symmetric distributions.

Chapter 9 contains a modern treatment of discriminant analysis. A brief history of discriminant analysis is also included here.

Chapter 10 deals with several aspects of principal component analysis in multivariate normal populations.

Factor analysis is treated in Chapter 11 and various aspects of canonical correlation analysis are treated in Chapter 12.

I believe that it would be appropriate to spread the materials over two three-hour one-semester basic courses on multivariate analysis for statistics graduate students or one three-hour one-semester course for graduate students in nonstatistic majors by proper selection of materials according to need.

Narayan C. Giri

Contents

Preface to the Second Edition   *v*
Preface to the First Edition   *vii*

1   VECTOR AND MATRIX ALGEBRA   1
    1.0   Introduction   1
    1.1   Vectors   1
    1.2   Matrices   4
    1.3   Rank and Trace of a Matrix   7
    1.4   Quadratic Forms and Positive Definite Matrix   7
    1.5   Characteristic Roots and Vectors   8
    1.6   Partitioned Matrix   16
    1.7   Some Special Theorems on Matrix Derivatives   21
    1.8   Complex Matrices   24
        Exercises   25
        References   27

2   GROUPS, JACOBIAN OF SOME TRANSFORMATIONS, FUNCTIONS AND SPACES   29
    2.0   Introduction   29
    2.1   Groups   29
    2.2   Some Examples of Groups   30
    2.3   Quotient Group, Homomorphism, Isomorphism   31
    2.4   Jacobian of Some Transformations   33
    2.5   Functions and Spaces   38
        References   39

3	**MULTIVARIATE DISTRIBUTIONS AND INVARIANCE**		41
	3.0 Introduction		41
	3.1 Multivariate Distributions		41
	3.2 Invariance in Statistical Testing of Hypotheses		44
	3.3 Almost Invariance and Invariance		49
	3.4 Sufficiency and Invariance		55
	3.5 Unbiasedness and Invariance		56
	3.6 Invariance and Optimum Tests		57
	3.7 Most Stringent Tests and Invariance		58
	3.8 Locally Best and Uniformly Most Powerful Invariant Tests		58
	3.9 Ratio of Distributions of Maximal Invariant, Stein's Theorem		59
	3.10 Derivation of Locally Best Invariant Tests (LBI)		61
	Exercises		63
	References		65
4	**PROPERTIES OF MULTIVARIATE DISTRIBUTIONS**		69
	4.0 Introduction		69
	4.1 Multivariate Normal Distribution (Classical Approach)		70
	4.2 Complex Multivariate Normal Distribution		84
	4.3 Symmetric Distribution: Its Properties and Characterizations		91
	4.4 Concentration Ellipsoid and Axes (Multivariate Normal)		110
	4.5 Regression, Multiple and Partial Correlation		112
	4.6 Cumulants and Kurtosis		118
	4.7 The Redundancy Index		120
	Exercises		120
	References		127
5	**ESTIMATORS OF PARAMETERS AND THEIR FUNCTIONS**		131
	5.0 Introduction		131
	5.1 Maximum Likelihood Estimators of μ, Σ in $N_p(\mu, \Sigma)$		132
	5.2 Classical Properties of Maximum Likelihood Estimators		141
	5.3 Bayes, Minimax, and Admissible Characters		151
	5.4 Equivariant Estimation Under Curved Models		184
	Exercises		202
	References		206
6	**BASIC MULTIVARIATE SAMPLING DISTRIBUTIONS**		211
	6.0 Introduction		211
	6.1 Noncentral Chi-Square, Student's t-, F-Distributions		211
	6.2 Distribution of Quadratic Forms		213
	6.3 The Wishart Distribution		218
	6.4 Properties of the Wishart Distribution		224
	6.5 The Noncentral Wishart Distribution		231

	6.6	Generalized Variance	232
	6.7	Distribution of the Bartlett Decomposition (Rectangular Coordinates)	233
	6.8	Distribution of Hotelling's T^2	234
	6.9	Multiple and Partial Correlation Coefficients	241
	6.10	Distribution of Multiple Partial Correlation Coefficients	245
	6.11	Basic Distributions in Multivariate Complex Normal	248
	6.12	Basic Distributions in Symmetrical Distributions	250
		Exercises	258
		References	264
7	**TESTS OF HYPOTHESES OF MEAN VECTORS**		269
	7.0	Introduction	269
	7.1	Tests: Known Covariances	270
	7.2	Tests: Unknown Covariances	272
	7.3	Tests of Subvectors of μ in Multivariate Normal	299
	7.4	Tests of Mean Vector in Complex Normal	307
	7.5	Tests of Means in Symmetric Distributions	309
		Exercises	317
		References	320
8	**TESTS CONCERNING COVARIANCE MATRICES AND MEAN VECTORS**		325
	8.0	Introduction	325
	8.1	Hypothesis: A Covariance Matrix Is Unknown	326
	8.2	The Sphericity Test	337
	8.3	Tests of Independence and the R^2-Test	342
	8.4	Admissibility of the Test of Independence and the R^2-Test	349
	8.5	Minimax Character of the R^2-Test	353
	8.6	Multivariate General Linear Hypothesis	369
	8.7	Equality of Several Covariance Matrices	389
	8.8	Complex Analog of R^2-Test	406
	8.9	Tests of Scale Matrices in $E_p(\mu, \Sigma)$	407
	8.10	Tests with Missing Data	412
		Exercises	423
		References	427
9	**DISCRIMINANT ANALYSIS**		435
	9.0	Introduction	435
	9.1	Examples	437
	9.2	Formulation of the Problem of Discriminant Analysis	438
	9.3	Classification into One of Two Multivariate Normals	444
	9.4	Classification into More than Two Multivariate Normals	468

	9.5	Concluding Remarks	473
	9.6	Discriminant Analysis and Cluster Analysis	473
		Exercises	474
		References	477
10	**PRINCIPAL COMPONENTS**	483
	10.0	Introduction	483
	10.1	Principal Components	483
	10.2	Population Principal Components	485
	10.3	Sample Principal Components	490
	10.4	Example	492
	10.5	Distribution of Characteristic Roots	495
	10.6	Testing in Principal Components	498
		Exercises	501
		References	502
11	**CANONICAL CORRELATIONS**	505
	11.0	Introduction	505
	11.1	Population Canonical Correlations	506
	11.2	Sample Canonical Correlations	510
	11.3	Tests of Hypotheses	511
		Exercises	514
		References	515
12	**FACTOR ANALYSIS**	517
	12.0	Introduction	517
	12.1	Orthogonal Factor Model	518
	12.2	Oblique Factor Model	519
	12.3	Estimation of Factor Loadings	519
	12.4	Tests of Hypothesis in Factor Models	524
	12.5	Time Series	525
		Exercises	526
		References	526
13	**BIBLIOGRAPHY OF RELATED RECENT PUBLICATIONS**	529
Appendix A	**TABLES FOR THE CHI-SQUARE ADJUSTMENT FACTOR**	531
Appendix B	**PUBLICATIONS OF THE AUTHOR**	543
Author Index		*551*
Subject Index		*555*

Multivariate Statistical Analysis

1
Vector and Matrix Algebra

1.0. INTRODUCTION

The study of multivariate analysis requires knowledge of vector and matrix algebra, some basic results of which are considered in this chapter. Some of these results are stated herein without proof; proofs can be obtained from Besilevsky (1983), Giri (1993), Graybill (1969), Maclane and Birkoff (1967), Markus and Mine (1967), Perlis (1952), Rao (1973), or any textbook on matrix algebra.

1.1. VECTORS

A vector is an ordered p-tuple $x_1, \ldots, x_p$ and is written as

$$x = \begin{pmatrix} x_1 \\ \vdots \\ x_p \end{pmatrix}.$$

Actually it is called a p-dimensional column vector. For brevity we shall simply call it a p-vector or a vector. The transpose of x is given by $x' = (x_1, \ldots, x_p)$. If all components of a vector are zero, it is called the null vector 0. Geometrically a p-vector represents a point $A = (x_1, \ldots, x_p)$ or the directed line segment $\overrightarrow{0A}$ with the point A in the p-dimensional Euclidean space E^p. The set of all p-vectors is denoted by V^p. Obviously $V^p = E^p$ if all components of the vectors are real numbers. For any two vectors $x = (x_1, \ldots, x_p)'$ and $y = (y_1, \ldots, y_p)'$ we define the vector sum $x + y = (x_1 + y_1, \ldots, x_p + y_p)'$ and scalar multiplication by a constant a by

$$ax = (ax_1, \ldots, ax_p)'.$$

Obviously vector addition is an associative and commutative operation, i.e., $x + y = y + x$, $(x + y) + z = x + (y + z)$ where $z = (z_1, \ldots, z_p)'$, and scalar multiplication is a distributive operation, i.e., for constants a, b, $(a + b)x = ax + bx$. For $x, y \in V^p$, $x + y$ and ax also belong to V^p. Furthermore, for scalar constants a, b, $a(x + y) = ax + ay$ and $a(bx) = b(ax) = abx$.

The quantity $x'y = y'x = \sum_1^p x_i y_i$ is called the dot product of two vectors x, y in V^p. The dot product of a vector $x = (x_1, \ldots, x_p)'$ with itself is denoted by $\|x\|^2 = x'x$, where $\|x\|$ is called the norm of x. Some geometrical significances of the norm are

1. $\|x\|^2$ is the square of the distance of the point x from the origin in E^p,
2. the square of the distance between two points $(x_1, \ldots, x_p), (y_1, \ldots, y_p)$ is given by $\|x - y\|^2$,
3. the angle θ between two vectors x, y is given by $\cos \theta = (x/\|x\|)'(y/\|y\|)$.

Definition 1.1.1. *Orthogonal vectors.* Two vectors x, y in V^p are said to be orthogonal to each other if and only if $x'y = y'x = 0$. A set of vectors in V^p is orthogonal if the vectors are pairwise orthogonal.

Geometrically two vectors x, y are orthogonal if and only if the angle between them is $90°$. An orthogonal vector x is called an orthonormal vector if $\|x\|^2 = 1$.

Definition 1.1.2. *Projection of a vector.* The projection of a vector x on $y (\neq 0)$, both belonging to V^p, is given by $\|y\|^{-2}(x'y)y$. (See Fig. 1.1.)

If $\vec{OA} = x$, $\vec{OB} = y$, and P is the foot of the perpendicular from the point A on OB, then $\vec{OP} = \|y\|^{-2}(x'y)y$ where 0 is the origin of E^p. For two orthogonal vectors x, y the projection of x on y is zero.

Definition 1.1.3. A set of vectors $\alpha_1, \ldots, \alpha_k$ in V^p is said to be linearly independent if none of the vectors can be expressed as a linear combination of the others.

Thus if $\alpha_1, \ldots, \alpha_k$ are linearly independent, then there does not exist a set of scalar constants $c_1, \ldots, c_k$ not all zero such that $c_1 \alpha_1 + \cdots + c_k \alpha_k = 0$. It may be verified that a set of orthogonal vectors in V^p is linearly independent.

Vector and Matrix Algebra

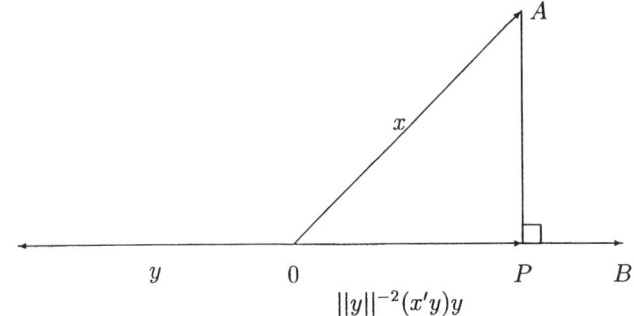

Figure 1.1. Projection of x on y

Definition 1.1.4. *Vector space spanned by a set of vectors.* Let $\alpha_1, \ldots, \alpha_k$ be a set of k vectors in V^p. Then the vector space V spanned by $\alpha_1, \ldots, \alpha_k$ is the set of all vectors which can be expressed as linear combinations of $\alpha_1, \ldots, \alpha_k$ and the null vector 0.

Thus if $\alpha, \beta \in V$, then for scalar constants $a, b, a\alpha + b\beta$ and $a\alpha$ also belong to V. Furthermore, since $\alpha_1, \ldots, \alpha_k$ belong to V^p, any linear combination of $\alpha_1, \ldots, \alpha_k$ also belongs to V^p and hence $V \subset V^p$. So V is a linear subspace of V^p.

Definition 1.1.5. *Basis of a vector space.* A basis of a vector space V is a set of linearly independent vectors which span V.

In V^p the unit vectors $\epsilon_1 = (1, 0, \ldots, 0)'$, $\epsilon_2 = (0, 1, 0, \ldots, 0)', \ldots, \epsilon_p = (0, \ldots, 0, 1)'$ form a basis of V^p. If A and B are two disjoint linear subspaces of V^p such that $A \cup B = V^p$ then A and B are complementary subspaces.

Theorem 1.1.1. *Every vector space V has a basis and two bases of V have the same number of elements.*

Theorem 1.1.2. *Let the vector space V be spanned by the vectors $\alpha_1, \ldots, \alpha_k$. Any element $\alpha \in V$ can be uniquely expressed as $\alpha = \sum_1^k c_i \alpha_i$ for scalar constants $c_1, \ldots, c_k$, not all zero, if and only if $\alpha_1, \ldots, \alpha_k$ is a basis of V.*

Definition 1.1.6. *Coordinates of a vector.* If $\alpha_1, \ldots, \alpha_k$ is a basis of a vector space V and if $\alpha \in V$ is uniquely expressed as $\alpha = \sum_1^k c_i \alpha_i$ for scalar constants $c_1, \ldots, c_k$, then the coefficient c_i of the vector α_i is called the ith coordinate of α with respect to the basis $\alpha_1, \ldots, \alpha_k$.

Definition 1.1.7. *Rank of a vector space.* The number of vectors in a basis of a vector space V is called the rank or the dimension of V.

1.2. MATRICES

Definition 1.2.1. *Matrix.* A real matrix A is an ordered rectangular array of elements a_{ij} (reals)

$$A = \begin{pmatrix} a_{11} & \cdots & a_{1q} \\ \vdots & & \vdots \\ a_{p1} & \cdots & a_{pq} \end{pmatrix} \quad (1.1)$$

and is written as $A_{p \times q} = (a_{ij})$.

A matrix with p rows and q columns is called a matrix of dimension $p \times q$ (p by q), the number of rows always being listed first. If $p = q$, we call it a square matrix of dimension p.

A p-dimensional column vector is a matrix of dimension $p \times 1$. Two matrices of the same dimension $A_{p \times q}, B_{p \times q}$ are said to be equal (written as $A = B$) if $a_{ij} = b_{ij}$ for $i = 1, \ldots, p, j = 1, \ldots, q$. If all $a_{ij} = 0$, then A is called a null matrix and is denoted 0. The transpose of a $p \times q$ matrix A is a $q \times p$ matrix A':

$$A' = \begin{pmatrix} a_{11} & \cdots & a_{p1} \\ \vdots & & \vdots \\ a_{1q} & \cdots & a_{pq} \end{pmatrix} \quad (1.2)$$

and is obtained by interchanging the rows and columns of A. Obviously $(A')' = A$. A square matrix A is said to be symmetric if $A = A'$ and is skew symmetric if $A = -A'$. The diagonal elements of a skew symmetric matrix are zero. In what follows we shall use the notation "A of dimension $p \times q$" instead of $A_{p \times q}$.

For any two matrices $A = (a_{ij})$ and $B = (b_{ij})$ of the same dimension $p \times q$ we define the matrix sum $A + B$ as a matrix $(a_{ij} + b_{ij})$ of dimension $p \times q$. The matrix $A - B$ is to be understood in the same sense as $A + B$ where the plus (+) is replaced by the minus (−) sign. Clearly $(A + B)' = A' + B', A + B = B + A$, and for any three matrices $A, B, C, (A + B) + C = A + (B + C)$. Thus the operation matrix sum is commutative and associative.

For any matrix $A = (a_{ij})$ and a scalar constant c, the scalar product cA is defined by $cA = Ac = (ca_{ij})$. Obviously $(cA)' = cA'$, so scalar product is a distributive operation.

Vector and Matrix Algebra

The matrix product of two matrices $A_{p\times q} = (a_{ij})$ and $B_{q\times r} = (b_{ij})$ is a matrix $C_{p\times r} = AB = (c_{ij})$ where

$$c_{ij} = \sum_{k=1}^{q} a_{ik}b_{kj}, \quad i = 1,\ldots,p, j = 1,\ldots,r. \tag{1.3}$$

The product AB is defined if the number of columns of A is equal to the number of rows of B and in general $AB \neq BA$. Furthermore $(AB)' = B'A'$. The matrix product is distributive and associative provided the products are defined, i.e., for any three matrices A, B, C,

1. $A(B+C) = AB + AC$ (distributive),
2. $(AB)C = A(BC)$ (associative).

Definition 1.2.2. *Diagonal matrix.* A square matrix A is said to be a diagonal matrix if all its off-diagonal elements are zero.

Definition 1.2.3. *Identity matrix.* A diagonal matrix whose diagonal elements are unity is called an identity matrix and is denoted by I.

For any square matrix A, $AI = IA = A$.

Definition 1.2.4. *Triangular matrix.* A square matrix $A = (a_{ij})$ with $a_{ij} = 0, j < i$, is called an upper triangular matrix. If $a_{ij} = 0$ for $j > i$, then A is called a lower triangular matrix.

Definition 1.2.5. *Orthogonal matrix.* A square matrix A is said to be orthogonal if $AA' = A'A = I$.

Associated with any square matrix $A = (a_{ij})$ of dimension $p \times p$ is a unique scalar quantity $|A|$, or det A, called the determinant of A which is defined by

$$|A| = \sum_{\pi} \delta(\pi) a_{1\pi(1)} a_{2\pi(2)} \cdots a_{p\pi(p)}, \tag{1.4}$$

where π runs over all $p!$ permutations of columns subscripts $(1, 2, \ldots, p)$ and $\delta(\pi) = 1$ if the number of inversions in $\pi(1), \ldots, \pi(p)$ from the standard order $1, \ldots, p$ is even and $\delta(\pi) = -1$ if the number of such inversions is odd. The number of inversions in a particular permutation is the total number of times in which an element is followed by numbers which would ordinarily precede it in the standard order $1, 2, \ldots, p$. From Chapter 3 on we shall consistently use the symbol det A for the determinant and reserve $\|$ for the absolute value symbol.

Definition 1.2.6. *Minor and cofactor.* For any square matrix $A = (a_{ij})$ of dimension $p \times p$, the minor of the element a_{ij} is the determinant of the matrix formed by deleting the ith row and the jth column of A. The quantity $(-1)^{i+j} \times$ the minor of a_{ij} is called the cofactor of a_{ij} and is symbolically denoted by A_{ij}.

The determinant of a submatrix (of A) of dimension $i \times i$ whose diagonal elements are also the diagonal elements of A is called a principal minor of order i. The set of leading principal minors is a set of p principal minors of orders $1, 2, \ldots, p$, respectively, such that the matrix of principal minor of order i is a submatrix of the matrix of the principal minor of order $i+1, i = 1, \ldots, p$. It is easy to verify that for any square matrix $A = (a_{ij})$ of dimension $p \times p$

$$|A| = \sum_{j=1}^{p} a_{ij}A_{ij} = \sum_{i=1}^{p} a_{ij}A_{ij}, \tag{1.5}$$

and for $j \neq j', i \neq i'$,

$$\sum_{i=1}^{p} a_{ij}A_{ij'} = \sum_{j=1}^{p} a_{ij}A_{i'j} = 0. \tag{1.6}$$

Furthermore, if A is symmetric, then $A_{ij} = A_{ji}$ for all i, j. For a triangular or a diagonal matrix A of dimension $p \times p$ with diagonal elements a_{ii}, $|A| = \prod_{i=1}^{p} a_{ii}$. If any two columns or rows of A are interchanged, then $|A|$ changes its sign, and $|A| = 0$ if two columns or rows of A are equal or proportional.

Definition 1.2.7. *Nonsingular matrix.* A square matrix A is called nonsingular if $|A| \neq 0$. If $|A| = 0$, then we call it a singular matrix.

The rows and the columns of a nonsingular matrix are linearly independent. Since for any two square matrices A, B, $|AB| = |A||B|$, we conclude that the product of two nonsingular matrices is a nonsingular matrix. However, the sum of two nonsingular matrices is not necessarily a nonsingular matrix. One such trivial case is $A = -B$ where both A and B are nonsingular matrices.

Definition 1.2.8. *Inverse matrix.* The inverse of a nonsingular matrix A of dimension $p \times p$ is the unique matrix A^{-1} such that $A^{-1}A = AA^{-1} = I$.

Let A_{ij} be the cofactor of the element a_{ij} of A and

$$C = \begin{pmatrix} \frac{A_{11}}{|A|} & \cdots & \frac{A_{1p}}{|A|} \\ \vdots & & \vdots \\ \frac{A_{p1}}{|A|} & \cdots & \frac{A_{pp}}{|A|} \end{pmatrix} \tag{1.7}$$

From (1.6) and (1.7) we get $AC' = I$. Hence $A^{-1} = C'$. The inverse matrix is defined only for the nonsingular matrix and A^{-1} is symmetric if A is symmetric. Furthermore $|A^{-1}| = (|A|)^{-1}$, $(A')^{-1} = (A^{-1})'$, and $(AB)^{-1} = B^{-1}A^{-1}$.

Vector and Matrix Algebra

1.3. RANK AND TRACE OF A MATRIX

Let A be a matrix of dimension $p \times q$. Let $R(A)$ be the vector space spanned by the rows of A and let $C(A)$ be the vector space spanned by the columns of A. The space $R(A)$ is called the row space of A and its rank $r(A)$ is called the row rank of A. The space $C(A)$ is called the column space of A and its rank $c(A)$ is called the column rank of A. For any matrix A, $r(A) = c(A)$.

Definition 1.3.1. *Rank of matrix.* The common value of the row rank and the column rank is called the rank of the matrix A and is denoted by $\rho(A)$.

For any matrix A of dimension $p \times q$, $q < p$, $\rho(A)$ may vary from 0 to q. If $\rho(A) = q$, then A is called the matrix of full rank. The rank of the null matrix 0 is 0. For any two matrices A, B for which AB is defined, the columns of AB are linear combinations of the columns of A. Thus the number of linearly independent columns of AB cannot exceed the number of linearly independent columns of A. Hence $\rho(AB) \leq \rho(A)$. Similarly, considering the rows of AB we can argue that $\rho(AB) \leq \rho(B)$. Hence $\rho(AB) \leq \min(\rho(A), \rho(B))$.

Theorem 1.3.1. *If A, B, C are matrices of dimensions $p \times q, p \times p, q \times q$, respectively, then $\rho(A) = \rho(AC) = \rho(BA) = \rho(BAC)$.*

Definition 1.3.2. *Trace of a matrix.* The trace of a square matrix $A = (a_{ij})$ of dimension $p \times p$ is defined by the sum of its diagonal elements and is denoted by $\text{tr} A = \sum_{1}^{p} a_{ii}$.

Obviously $\text{tr} A = \text{tr} A'$, $\text{tr}(A + B) = \text{tr}(A) + \text{tr}(B)$. Furthermore, $\text{tr} AB = \text{tr} BA$, provided both AB and BA are defined. Hence for any orthogonal matrix θ, $\text{tr} \theta' A \theta = \text{tr} A \theta \theta' = \text{tr} A$.

1.4. QUADRATIC FORMS AND POSITIVE DEFINITE MATRIX

A quadratic form in the real variables $x_1, \ldots, x_p$, is an expression of the form $Q = \sum_{i=1}^{p} \sum_{j=1}^{p} a_{ij} x_i x_j$, where a_{ij} are real constants. Writing $x = (x_1, \ldots, x_p)'$, $A = (a_{ij})$ we can write $Q = x'Ax$. Without any loss of generality we can take the matrix A in the quadratic form Q to be a symmetric one. Since Q is a scalar quantity

$$Q = Q' = x'A'x = \frac{1}{2}(Q + Q') = x'((A + A')/2)x$$

and $\frac{1}{2}(A + A')$ is a symmetric matrix.

Definition 1.4.1. *Positive definite matrix.* A square matrix A or the associated quadratic form $x'Ax$ is called positive definite if $x'Ax > 0$ for all $x \neq 0$ and is called positive semidefinite if $x'Ax \geq 0$ for all x.

The matrix A or the associated quadratic form $x'Ax$ is negative definite or negative semidefinite if $-x'Ax$ is positive definite or positive semidefinite, respectively.

Example 1.4.1.

$$(x_1, x_2)\begin{pmatrix} 2 & 1 \\ 1 & 3 \end{pmatrix}(x_1, x_2)' = 2x_1^2 + 2x_1 x_2 + 3x_2^2 = 2\left(x_1 + \frac{1}{2}x_2\right)^2 + \frac{5}{2}x_2^2 > 0$$

for all $x_1 \neq 0, x_2 \neq 0$. Hence the matrix $\begin{pmatrix} 2 & 1 \\ 1 & 3 \end{pmatrix}$ is positive definite.

1.5. CHARACTERISTIC ROOTS AND VECTORS

The characteristic roots of a square matrix $A = (a_{ij})$ of dimension $p \times p$ are given by the roots of the characteristic equation

$$|A - \lambda I| = 0 \tag{1.8}$$

where λ is real. Obviously this is an equation of degree p in λ and thus has exactly p roots. If A is a diagonal matrix, then the diagonal elements are themselves the characteristic roots of A. In general we can write (1.8) as

$$(-\lambda)^p + (-\lambda)^{p-1} S_1 + (-\lambda)^{p-2} S_2 + \cdots + (-\lambda) S_{p-1} + |A| = 0 \tag{1.9}$$

where S_i is the sum of all principal minors of order i of A. In particular, $S_1 = \text{tr} A$. Thus the product of the characteristic roots of A is equal to $|A|$ and the sum of the characteristics roots of A is equal to tr A. The vector $x = (x_1, \ldots, x_p)'$, not identically zero, satisfying

$$(A - \lambda I)x = 0, \tag{1.10}$$

is called the characteristic vector of the matrix A, corresponding to its characteristic root λ. Clearly, if x is a characteristic vector of the matrix A corresponding to its characteristic root λ, then any scalar multiple $cx, c \neq 0$, is also a characteristic vector of A corresponding to λ. Since, for any orthogonal matrix θ of dimension $p \times p$,

$$|\theta A \theta' - \lambda I| = |\theta A \theta' - \lambda \theta \theta'| = |A - \lambda I|,$$

the characteristic roots of the matrix A remain invariant (unchanged) with respect to the transformation $A \to \theta A \theta'$.

Vector and Matrix Algebra

Theorem 1.5.1. *If A is a real symmetric matrix (of order $p \times p$), then all its characteristic roots are real.*

Proof. Let λ be a complex characteristic root of A and let $x + iy, x = (x_1, \ldots, x_p)', y = (y_1, \ldots, y_p)'$, be the characteristic vector (complex) corresponding to λ. Then from (1.10)

$$A(x + iy) = \lambda(x + iy), (x - iy)'A(x + iy) = \lambda(x'x + y'y).$$

But

$$(x - iy)'A(x + iy) = x'Ax + y'Ay.$$

Hence we conclude that λ must be real. Q.E.D.

Note The characteristic vector z corresponding to a complex characteristic root λ must be complex. Otherwise $Az = \lambda z$ will imply that a real vector is equal to a complex vector.

Theorem 1.5.2. *The characteristic vectors corresponding to distinct characteristic roots of a symmetric matrix are orthogonal.*

Proof. Let λ_1, λ_2 be two distinct characteristic roots of a symmetric (real) matrix A and let $x = (x_1, \ldots, x_p)', y = (y_1, \ldots, y_p)'$ be the characteristic vectors corresponding to λ_1, λ_2, respectively. Then

$$Ax = \lambda_1 x, Ay = \lambda_2 y.$$

So

$$y'Ax = \lambda_1 y'x, x'Ay = \lambda_2 x'y.$$

Thus

$$\lambda_1 x'y = \lambda_2 x'y.$$

Since $\lambda_1 \neq \lambda_2$ we conclude that $x'y = 0$. Q.E.D.

Let λ be a characteristic root of a symmetric positive definite matrix A and let x be the corresponding characteristic vector. Then

$$x'Ax = \lambda x'x > 0.$$

Hence we get the following Theorem.

Theorem 1.5.3. *The characteristic roots of a symmetric positive definite matrix are all positive.*

Theorem 1.5.4. *For every real symmetric matrix A, there exists an orthogonal matrix θ such that $\theta A \theta'$ is a diagonal matrix whose diagonal elements are the characteristic roots of A.*

Proof. Let $\lambda_1 \geq \lambda_2 \cdots \geq \lambda_p$ denote the characteristic roots of A including multiplicities and let x_i be the characteristic vector of A, corresponding to the characteristic root $\lambda_i, i = 1, \ldots, p$. Write

$$y_i = x_i / \|x_i\|, i = 1, \ldots, p;$$

obviously $y_1, \ldots, y_p$ are the normalized characteristic vectors of A. Suppose there exists $s(\leq p)$ orthonormal vectors $y_1, \ldots, y_s$ such that $(A - \lambda_i I) y_i = 0$, $i = 1, \ldots, s$. Denoting by A^r the product of r matrices each equal to A we get

$$A^r y_i = \lambda_i A^{r-1} y_i = \cdots = \lambda_i^r y_i, i = 1, \ldots, s.$$

Let x be orthogonal to the vector space spanned by $y_1, \ldots, y_s$. Then

$$(A^r x)' y_i = x' A^r y_i = \lambda_i^r x' y_i = 0$$

for all r including zero and $i = 1, \ldots, s$. Hence any vector belonging to the vector space spanned by the vectors $x, Ax, A^2 x, \ldots$ is orthogonal to any vector spanned by $y_1, \ldots, y_s$. Obviously not all vectors $x, Ax, A^2 x, \ldots$ are linearly independent. Let k be the smallest value of r such that for real constants $c_1, \ldots, c_k$

$$A^k x + c_1 A^{k-1} x + \cdots + c_k x = 0.$$

Factoring the left-hand side of this expression we can, for constants $u_1, \ldots, u_k$ write it as

$$\prod_{i=1}^{k} (A - u_i I) x = 0.$$

Let

$$y_{s+1} = \prod_{i=2}^{k} (A - u_i I) x.$$

Then $(A - u_1 I) y_{s+1} = 0$. In other words there exists a normalized vector y_{s+1} in the space spanned by $(x, Ax, A^2 x, \ldots)$ which is a characteristic vector of A corresponding to its root $u_1 = \lambda_{s+1}$ (say) and y_{s+1} is orthogonal to $y_1, \ldots, y_s$. Since y_1 can be chosen corresponding to any characteristic root to start with, we have proved the existence of p orthogonal vectors $y_1, \ldots, y_p$ satisfying

Vector and Matrix Algebra

$Ay_i = \lambda_i y_i, i = 1, \ldots, p$. Let θ be an orthogonal matrix of dimension $p \times p$ with y_i as its rows. Obviously then $\theta A \theta'$ is a diagonal matrix with diagonal elements $\lambda_1, \ldots, \lambda_p$.

Q.E.D.

From this theorem it follows that any positive definite quadratic form $x'Ax$ can be transformed into a diagonal form $\sum_{i=1}^{p} \lambda_i y_i^2$ where $y = (y_1, \ldots, y_p)' = \theta x$ and the orthogonal matrix θ is such that $\theta A \theta'$ is a diagonal matrix with diagonal elements $\lambda_1, \ldots, \lambda_p$ (characteristic roots of A). Note that $x'Ax = (\theta x)'(\theta A \theta')(\theta x)$. Since the characteristic roots of a positive definite matrix A are all positive, $|A| = |\theta A \theta'| = \prod_{i=1}^{p} \lambda_i > 0$.

Theorem 1.5.5. *For every positive definite matrix A there exists a nonsingular matrix C such that $A = C'C$.*

Proof. From Theorem 1.5.4 there exists an orthogonal matrix θ such that $\theta A \theta'$ is a diagonal matrix D with diagonal elements $\lambda_1, \ldots, \lambda_p$, the characteristic roots of A. Let $D^{\frac{1}{2}}$ be a diagonal matrix with diagonal elements $\lambda_1^{\frac{1}{2}}, \ldots, \lambda_p^{\frac{1}{2}}$ and let $D^{\frac{1}{2}} \theta = C$. Then $A = \theta' D \theta = C'C$ and obviously C is a nonsingular matrix.

Q.E.D.

Any positive definite quadratic form $x'Ax$ can be transformed to a diagonal form $y'y$ where $y = Cx$ and C is a nonsingular matrix such that $A = C'C$. Furthermore, given any positive definite matrix A there exists a nonsingular matrix B such that $B'AB = I (B = C^{-1})$.

Theorem 1.5.6. *If A is a positive definite matrix, then A^{-1} is also positive definite.*

Proof. Let $A = C'C$ where C is a nonsingular matrix. Then

$$x'A^{-1}x = ((C')^{-1}x)'((C')^{-1}x) > 0 \text{ for all } x \neq 0.$$

Q.E.D.

Theorem 1.5.7. *Let A be a symmetric and at least positive semidefinite matrix of dimension $p \times p$ and of rank $r \leq p$. Then A has exactly r positive characteristic roots and the remaining $p - r$ characteristic roots of A are zero.*

The proof is left to the reader.

Theorem 1.5.8. *Let A be a symmetric nonsingular matrix of dimension $p \times p$. Then there exists a nonsingular matrix C such that*

$$CAC' = \begin{pmatrix} I & 0 \\ 0 & -I \end{pmatrix}$$

where the order of I is the number of positive characteristic roots of A and that of $-I$ is the number of negative characteristic roots of A.

Proof. From theorem 1.5.4 there exists an orthogonal matrix θ such that $\theta A \theta'$ is a diagonal matrix with diagonal elements $\lambda_1, \ldots, \lambda_p$, the characteristic roots of A. Without any loss of generality let us assume that $\lambda_1 \geq \cdots \geq \lambda_q > 0 > \lambda_{q+1} \geq \cdots \geq \lambda_p$. Let D be a diagonal matrix with diagonal elements $(\lambda_1)^{-\frac{1}{2}}, \ldots, (\lambda_q)^{-\frac{1}{2}}, (-\lambda_{q+1})^{-\frac{1}{2}}, \ldots, (-\lambda_p)^{-\frac{1}{2}}$, respectively. Then

$$D\theta A \theta' D' = \begin{pmatrix} I & 0 \\ 0 & -I \end{pmatrix}$$

If A is a symmetric square matrix (order p) of rank $r(\leq p)$ then there exists a nonsingular matrix C such that

$$CAC' = \begin{pmatrix} I & 0 & 0 \\ 0 & -I & 0 \\ 0 & 0 & 0 \end{pmatrix}$$

where the order of I is the number of positive characteristic roots of A and, the order of I plus the order of $-I$ is equal to r. Q.E.D.

Theorem 1.5.9. *Let A, B be two matrices of dimensions $p \times q, q \times p$ respectively. (a) Every nonzero characteristic root of AB is also a characteristic root of BA. (b) $|I_p + AB| = |I_q + BA|$.*

Proof. (a) Let λ be a nonzero characteristic root of AB. Then $|AB - \lambda I_p| = 0$. This implies

$$\begin{vmatrix} \lambda I_p & A \\ B & I_q \end{vmatrix} = 0.$$

But we can obviously write this as

$$\begin{vmatrix} \lambda I_q & B \\ A & I_p \end{vmatrix} = 0,$$

which implies $|BA - \lambda I_q| = 0$.

Vector and Matrix Algebra

(b) Since

$$\begin{pmatrix} I_p + AB & A \\ 0 & I_q \end{pmatrix} = \begin{pmatrix} I_p & A \\ -B & I_q \end{pmatrix} \begin{pmatrix} I_p & 0 \\ B & I_q \end{pmatrix}$$

we get

$$|I_p + AB| = \begin{vmatrix} I_p & A \\ -B & I_q \end{vmatrix}$$

Similarly from

$$\begin{pmatrix} I_p & A \\ 0 & I_q + BA \end{pmatrix} = \begin{pmatrix} I_p & 0 \\ B & I_q \end{pmatrix} \begin{pmatrix} I_p & A \\ -B & I_q \end{pmatrix}$$

we get

$$|I_q + BA| = \begin{vmatrix} I_p & A \\ -B & I_q \end{vmatrix}.$$

Q.E.D.

Thus it follows from Theorem 1.5.7 that a positive semidefinite quadratic form $x'Ax$ of rank $r \leq p$ can be reduced to the diagonal form $\sum_1^r \lambda_i y_i^2$ where $\lambda_1, \ldots, \lambda_r$ are the positive characteristic roots of A and $y_1, \ldots, y_r$ are linear combinations of the components $x_1, \ldots, x_p$ of x.

Theorem 1.5.10. *If A is positive definite and B is positive semidefinite of the same dimension $p \times p$, then there exists a nonsingular matrix C such that $CAC' = I$ and CBC' is diagonal matrix with diagonal elements $\lambda_1, \ldots, \lambda_p$, the roots of the equation $|B - \lambda A| = 0$.*

Proof. Since A is positive definite, there exists a nonsingular matrix D such that $DAD' = I$. Let $DBD' = B^{*'}$. Since $B^{*'}$ is a real symmetric matrix there exists an orthogonal matrix θ such that $\theta DBD'\theta'$ is a diagonal matrix. Write $\theta D = C$, where C is a nonsingular matrix. Obviously $CAC' = I$ and CBC' is a diagonal matrix whose diagonal elements are the characteristic roots of B^*, which are, in turn, the roots of $|B - \lambda A| = 0$. Q.E.D.

Theorem 1.5.11. *Let A be a matrix of dimension $p \times q, p < q$. Then AA' is symmetric and positive semidefinite if the rank of $A < p$ and positive definite if the rank of $A = p$.*

Proof. Obviously AA' is symmetric and the rank of AA' is equal to the rank of A. Let the rank of AA' be $r(\leq p)$ Since AA' is symmetric there exists an orthogonal $p \times p$ matrix θ such that $\theta AA'\theta'$ is a diagonal matrix with nonzero diagonal elements $\lambda_1, \ldots, \lambda_r$. Let $x = (x_1, \ldots, x_p)'$, $y = \theta x$. Then

$$x'AA'x = \sum_1^r \lambda_i y_i^2 \geq 0 \text{ for all } x.$$

If $r = p$, then

$$x'AA'x = \sum_1^p \lambda_i y_i^2 > 0 \text{ for all } x \neq 0.$$

Q.E.D.

Theorem 1.5.12. *Let A be a symmetric positive definite matrix of dimension $p \times p$ and let B be a $q \times p$ matrix. Then BAB' is symmetric and at least positive semidefinite of the same rank as B.*

Proof. Since A is positive definite there exists a nonsingular matrix C such that $A = CC'$. Hence $BAB' = (BC)(BC)'$. Proceeding exactly in the same way as in Theorem 1.5.11 we get the result. Q.E.D.

Theorem 1.5.13. *Let A be a symmetric positive definite matrix and let B be a symmetric positive semidefinite matrix of the same dimension $p \times p$ and of rank $r \leq p$. Then*

1. *all roots of the equation $|B - \lambda A| = 0$ are zero if and only if $B = 0$;*
2. *all roots of $|B - \lambda A| = 0$ are unity if and only if $B = A$.*

Proof. Since A is positive definite there exists a nonsingular matrix C such that $CAC' = I$ and CBC' is a diagonal matrix whose diagonal elements are the roots of the equation $|CBC' - \lambda I| = 0$ (see Theorem 1.5.10). Since the rank of $CBC' = $ rank B, by Theorem 1.5.7, and the fact that $|CBC' - \lambda I| = 0$ implies $|B - \lambda A| = 0$ we conclude that all roots of $|B - \lambda A| = 0$ are zero if and only if the rank of B is zero, i.e., $B = 0$. Let $\lambda = 1 - u$. Then $|B - \lambda A| = |B - A + uA|$. By part (i) all roots u of $|B - A + uA| = 0$ are zero if and only if $B - A = 0$. Q.E.D.

To prove Theorem 1.5.14 we need the following Lemmas.

Vector and Matrix Algebra

Lemma 1.5.1. *Let X be a $p \times q$ matrix of rank $r \leq q \leq p$ and let U be a $r \times q$ matrix of rank r. If $X'X = U'U$ then there exists a $p \times p$ orthogonal matrix θ such that $\theta X = \binom{U}{0}$.*

Proof. Let V be the subspace spanned by the columns of X and let $V^\perp$ be the space of all vectors orthogonal to V. Let R be an orthogonal basis matrix of $V^\perp$. Obviously R is a $p \times (p-r)$ matrix. Since UU' is of rank r $(UU')^{-1}$ exists. Write

$$\theta = \begin{pmatrix} (UU')^{-1}UX' \\ R' \end{pmatrix}.$$

Since $X'R = 0$ we get

$$\theta\theta' = \begin{pmatrix} (UU')^{-1}UU'UU'(UU')^{-1} & (UU')^{-1}UX'R \\ R'XU'(UU')^{-1} & R'R \end{pmatrix}$$

$$= \begin{pmatrix} I & 0 \\ 0 & I \end{pmatrix} = I,$$

and θ is an $p \times p$ orthogonal matrix satisfying $\theta X = \binom{U}{0}$. Q.E.D.

Lemma 1.5.2. *Let X, Y be $p \times q$ matrices with $q < p$. $X'X = Y'Y$ if and only if there exists an $p \times p$ orthogonal matrix θ such that $Y = \theta X$.*

Proof. If $Y = \theta X$ then $X'X = Y'\theta\theta'Y = Y'Y$. To prove the converse let us assume that the rank of $(X) = r = \text{rank}(Y)$, $r \leq q$ and let U be a $r \times q$ matrix such that

$$U'U = X'X = Y'Y.$$

By Lemma 1.5.1 there exist $p \times p$ orthogonal matrices θ_1, θ_2 such that

$$\theta_1 X = \begin{pmatrix} U \\ 0 \end{pmatrix} = \theta_2 Y.$$

This implies that

$$Y = \theta_2' \theta_1 X = \theta_3 X$$

where θ_3 is a $p \times p$ orthogonal matrix. Q.E.D.

Theorem 1.5.14. *Let A be a $p \times q (q \leq p)$ matrix of rank q. There exist a $q \times q$ nonsingular matrix B and a $p \times p$ orthogonal matrix θ such that*

$$A = \theta \begin{pmatrix} Iq \\ 0 \end{pmatrix} B.$$

Proof. Since $A'A$ is positive definite there exists a $q \times q$ nonsingular matrix B such that $A'A = B'B$. By Lemma 1.5.1 there exists a $p \times q$ matrix $\theta_{(1)}$ such that $A = \theta_{(1)} B$ where $\theta'_{(1)} \theta_{(1)} = Iq$. Choosing $\theta_{(2)}$ a $p \times (p-q)$ matrix such that $\theta = (\theta_{(1)}, \theta_{(2)})$ is orthogonal we get

$$A = \theta_{(1)} B = \theta \begin{pmatrix} Iq \\ 0 \end{pmatrix} B.$$

Q.E.D.

1.6. PARTITIONED MATRIX

A matrix $A = (a_{ij})$ of dimension $p \times q$ is said to be partitioned into submatrices $A_{ij}, i, j = 1, 2$, if A can be written as

$$A = \begin{pmatrix} A_{11} & A_{12} \\ A_{21} & A_{22} \end{pmatrix}$$

where $A_{11} = (a_{ij}) (i = 1, \ldots, m; j = 1, \ldots, n)$; $A_{12} = (a_{ij}) (i = 1, \ldots, m; j = n+1, \ldots, q)$; $A_{21} = (a_{ij}) (i = m+1, \ldots, p; j = 1, \ldots, n)$; $A_{22} = (a_{ij}) (i = m+1, \ldots, p; j = n+1, \ldots, q)$. If two matrices A, B of the same dimension are similarly partitioned, then

$$A + B = \begin{pmatrix} A_{11} + B_{11} & A_{12} + B_{12} \\ A_{21} + B_{21} & A_{22} + B_{22} \end{pmatrix}.$$

Let the matrix A of dimension $p \times q$ be partitioned as above and let the matrix C of dimension $q \times r$ be partitioned into submatrices C_{ij} where C_{11}, C_{12} have n rows. Then

$$AC = \begin{pmatrix} A_{11}C_{11} + A_{12}C_{21} & A_{11}C_{12} + A_{12}C_{22} \\ A_{21}C_{11} + A_{22}C_{21} & A_{21}C_{12} + A_{22}C_{22} \end{pmatrix}.$$

Theorem 1.6.1. *For any square matrix*

$$A = \begin{pmatrix} A_{11} & A_{12} \\ A_{21} & A_{22} \end{pmatrix}$$

where A_{11}, A_{22} are square submatrices and A_{22} is nonsingular, $|A| = |A_{22}||A_{11} - A_{12}A_{22}^{-1}A_{21}|$.

Vector and Matrix Algebra

Proof.

$$\begin{vmatrix} A_{11} & A_{12} \\ A_{21} & A_{22} \end{vmatrix} = \begin{vmatrix} A_{11} & A_{12} \\ A_{21} & A_{22} \end{vmatrix} \begin{vmatrix} I & 0 \\ -A_{22}^{-1}A_{21} & I \end{vmatrix} = \begin{vmatrix} A_{11} - A_{12}A_{22}^{-1}A_{21} & A_{12} \\ 0 & A_{22} \end{vmatrix}$$

$$= |A_{22}||A_{11} - A_{12}A_{22}^{-1}A_{21}|.$$

Q.E.D.

Theorem 1.6.2. *Let the symmetric matrix A of dimension $p \times p$ be partitioned as*

$$A = \begin{pmatrix} A_{11} & A_{12} \\ A_{21} & A_{22} \end{pmatrix}$$

where A_{11}, A_{22} are square submatrices of dimensions $q \times q, (p-q) \times (p-q)$, respectively, and let A_{22} be nonsingular. Then $A_{11} - A_{12}A_{22}^{-1}A_{21}$ is a symmetric matrix of rank $r - (p-q)$ where r is the rank of A.

Proof. Since A is symmetric, $A_{11} - A_{12}A_{22}^{-1}A_{21}$ is obviously symmetric. Now

$$\text{rank } A = \text{rank}\left[\begin{pmatrix} I & -A_{12}A_{22}^{-1} \\ & I \end{pmatrix}\begin{pmatrix} A_{11} & A_{12} \\ A_{21} & A_{22} \end{pmatrix}\begin{pmatrix} I & 0 \\ -A_{22}^{-1}A_{21} & I \end{pmatrix}\right].$$

$$= \text{rank}\begin{pmatrix} A_{11} - A_{12}A_{22}^{-1}A_{21} & 0 \\ 0 & A_{22} \end{pmatrix}.$$

But A_{22} is nonsingular of rank $p - q$. Hence the rank of $A_{11} - A_{12}A_{22}^{-1}A_{21}$ is $r - (p - q)$
Q.E.D.

Theorem 1.6.3. *A symmetric matrix*

$$A = \begin{pmatrix} A_{11} & A_{12} \\ A_{21} & A_{22} \end{pmatrix}$$

of dimension $p \times p$ (A_{11} is of dimension $q \times q$) is positive definite if and only if $A_{11}, A_{22} - A_{21}A_{11}^{-1}A_{12}$ are positive definite.

Proof. Let $x = (x'_{(1)}, x'_{(2)})$ where $x'_{(1)} = (x_1, \ldots, x_q)$, $x'_{(2)} = (x_{q+1}, \ldots, x_p)$. Then

$$\begin{aligned} x'Ax &= (x_{(1)} + A_{11}^{-1}A_{12}x_{(2)})'A_{11}(x_{(1)} + A_{11}^{-1}A_{12}x_{(2)}) \\ &\quad + x'_{(2)}(A_{22} - A_{21}A_{11}^{-1}A_{12})x_{(2)}. \end{aligned} \quad (1.11)$$

Furthermore, if A is positive definite, then obviously A_{11} and A_{22} are both positive definite. Now from (1.11) if $A_{11}, A_{22} - A_{21}A_{11}^{-1}A_{12}$ are positive definite, then A is positive definite. Conversely, if A and consequently A_{11} are positive definite, then by taking $x(\neq 0)$ such that $x_{(1)} + A_{11}^{-1}A_{12}x_{(2)} = 0$ we conclude that $A_{22} - A_{21}A_{11}^{-1}A_{12}$ is positive definite. Q.E.D.

Theorem 1.6.4. *Let a positive definite matrix A be partitioned into submatrices $A_{ij}, i, j = 1, 2$, where A_{11} is a square submatrix, and let the inverse matrix $A^{-1} = B$ be similarly partitioned into submatrices $B_{ij}, i, j = 1, 2$. Then*

$$A_{11}^{-1} = B_{11} - B_{12}B_{22}^{-1}B_{21}, \quad A_{22}^{-1} = B_{22} - B_{21}B_{11}^{-1}B_{12}.$$

Proof. Since $AB = I$, we get

$$\begin{aligned} A_{11}B_{11} + A_{12}B_{21} &= I, & A_{11}B_{12} + A_{12}B_{22} &= 0, \\ A_{21}B_{11} + A_{22}B_{21} &= 0, & A_{21}B_{12} + A_{22}B_{22} &= I. \end{aligned}$$

Solving these matrix equations we obtain

$$A_{11}B_{11} - A_{11}B_{12}B_{22}^{-1}B_{21} = I, \quad A_{22}B_{22} - A_{22}B_{21}B_{11}^{-1}B_{12} = I,$$

or, equivalently,

$$A_{11}^{-1} = B_{11} - B_{12}B_{22}^{-1}B_{21}, \quad A_{22}^{-1} = B_{22} - B_{21}B_{11}^{-1}B_{12}.$$

Q.E.D.

From this it follows that $A_{11}^{-1}A_{12} = -B_{12}B_{22}^{-1}$, $B_{12} = -A_{11}^{-1}(A_{12})B_{22}$.

Theorem 1.6.5. *A symmetric positive definite quadratic form $x'Ax$, where $A = (a_{ij})$, can be transformed to $(Tx)'(Tx)$ where T is the unique upper triangular matrix with positive diagonal elements such that $A = T'T$.*

Vector and Matrix Algebra

Proof. Let $Q_p(x_1, \ldots, x_p) = x'Ax$. Then

$$Q_p(x_1, \ldots, x_p) = \left((a_{11})^{\frac{1}{2}} x_1 + \sum_{j=2}^{p} \frac{a_{1j}}{(a_{11})^{\frac{1}{2}}} x_j\right)^2$$

$$+ \sum_{j,k=2}^{p} \left(\frac{a_{11} a_{jk} - a_{1j} a_{1k}}{a_{11}}\right) x_j x_k \qquad (1.12)$$

$$= \left(a_{11}^{\frac{1}{2}} x_1 + \sum_{j=2}^{p} \frac{a_{1j}}{(a_{11})^{\frac{1}{2}}} x_j\right)^2$$

$$+ Q_{p-1}(x_2, \ldots, x_p).$$

Let

$$(a_{11})^{\frac{1}{2}} x_1 + \sum_{j=2}^{p} \frac{a_{1j}}{(a_{11})} x_j = \sum_{j=1}^{p} T_{1j} x_j.$$

Since Q_p is positive definite Q_{p-1} is also positive definite so that by continuing the procedure of completing the square, we can write

$$Q_p(x_1, \ldots, x_p) = \left(\sum_{j=1}^{p} T_{1j} x_j\right)^2 + \left(\sum_{j=2}^{p} T_{2j} x_j\right)^2 + \cdots + (T_{pp} x_p)^2 = (Tx)'(Tx)$$

where T is the unique upper triangular matrix

$$T = \begin{pmatrix} T_{11} & T_{12} & \cdots & T_{1p} \\ 0 & T_{22} & \cdots & T_{2p} \\ \vdots & \vdots & & \vdots \\ 0 & 0 & \cdots & T_{pp} \end{pmatrix}$$

with $T_{ii} > 0$, $i = 1, \ldots, p$. \hfill Q.E.D.

Thus a symmetric positive definite matrix A can be uniquely written as $A = T'T$ where T is the unique nonsingular upper triangular matrix with positive diagonal elements. From (1.12) it follows that

$$Q_p(x_1, \ldots, x_p) = \left((a_{pp})^{\frac{1}{2}} x_p + \sum_{j=p-1}^{1} \frac{a_{pj}}{(a_{pp})^{\frac{1}{2}}} x_j\right)^2 + Q_{p-1}(x_1, \ldots, x_{p-1})$$

so that we can write

$$Q_p(x_1,\ldots,x_p) + \left(\sum_{j=1}^{p} T_{pj}x_j\right)^2 + \left(\sum_{j=1}^{p-1} T_{p-1,j}x_j\right)^2 + \cdots + (T_{11}x_1)^2.$$

Hence, given any symmetric positive definite matrix A there exists a unique nonsingular lower triangular matrix T with positive diagonal elements, such that $A = T'T$. Let θ be an orthogonal matrix in the diagonal form. For any upper (lower) triangular matrix T, θT is also an upper (lower) triangular matrix and $T'T = (\theta T)'(\theta T)$. Thus given any symmetric positive definite matrix A, there exists a nonsingular lower triangular matrix T, not necessarily with positive diagonal elements, such that $A = T'T$. Obviously such decomposition is not unique.

Theorem 1.6.6. *Let $X = (X_1,\ldots,X_p)'$. There exists an orthogonal matrix θ of dimension $p \times p$ such that $\theta X = ((X'X)^{\frac{1}{2}}, 0, \ldots, 0)'$.*

Proof. Let

$$\theta = \begin{pmatrix} \dfrac{x_1}{(X'X)^{\frac{1}{2}}} & ,\ldots, & \dfrac{X_p}{(X'X)^{\frac{1}{2}}} \\ \theta_{21} & ,\ldots, & \theta_{2p} \\ \vdots & & \vdots \\ \theta_{p1} & ,\ldots, & \theta_{pp} \end{pmatrix}$$

be an orthogonal matrix of dimension $p \times p$ where the θ_{ij} are arbitrary. Let $Y = (Y_1,\ldots,Y_p)' = \theta X$. Then

$$Y_1 = (X'X)^{\frac{1}{2}},$$

$$Y_i = \sum_{j=1}^{p} \theta_{ij} X_j = 0, i > 1.$$

Q.E.D.

Example 1.6.1. Let

$$\Sigma = \begin{pmatrix} \sigma_{11} & \sigma_{12} & \cdots & \sigma_{1p} \\ \sigma_{12} & \sigma_{22} & \cdots & \sigma_{2p} \\ \vdots & \vdots & & \vdots \\ \sigma_{1p} & \sigma_{2p} & \cdots & \sigma_{pp} \end{pmatrix}, \quad T = \begin{pmatrix} t_{11} & 0 & \cdots & \cdots & 0 \\ t_{21} & t_{22} & 0 & \cdots & 0 \\ \vdots & \vdots & \vdots & & \vdots \\ t_{p1} & t_{p2} & \cdots & \cdots & t_{pp} \end{pmatrix}$$

Vector and Matrix Algebra

and $\Sigma = TT'$. Then

$$t_{11}^2 = \sigma_{11}, t_{11}t_{i1} = \sigma_{i1}, i = 1, \ldots, p;$$

$$t_{21}^2 + t_{22}^2 = \sigma_{22}, t_{21}t_{i1} + t_{22}t_{i2} = \sigma_{i2}, i = 2, \ldots, p.$$

Continuing in the same way for other rows we obtain

$$t_{i1} = \frac{\sigma_{i1}}{\sqrt{\sigma_{11}}}, i = 1, \ldots, p; \quad t_{jj} = \left(\sigma_{jj} - \sum_{k=1}^{j-1} t_{jk}^2\right)^{\frac{1}{2}},$$

$$t_{ij} = 0, j > i; \quad t_{ij} = \frac{\sigma_{ij} - \sum_{k=1}^{j-1} t_{jk}t_{ik}}{t_{jj}}$$

for $j \le i, j = 2, \ldots, p.$

1.7. SOME SPECIAL THEOREMS ON MATRIX DERIVATIVES

Let $x = (x_1, \ldots, x_p)'$ and let the partial derivative operator $\frac{\partial}{\partial x}$ be defined by

$$\frac{\partial}{\partial x} = \left(\frac{\partial}{\partial x_1}, \ldots, \frac{\partial}{\partial x_p}\right)'.$$

For any scalar function $f(x)$ of the vector x, the vector derivative of f is defined by

$$\frac{\partial f}{\partial x} = \left(\frac{\partial}{\partial x_1}, \ldots, \frac{\partial}{\partial x_p}\right)'.$$

Let

$$f(x) = x'Ax$$

where $A = (a_{ij})$ is a $p \times p$ matrix. Since

$$x'Ax = \begin{cases} a_{ii}x_i^2 + 2x_i \sum_{j \ne i} a_{ij}x_j + \sum_{\substack{k \ne i \\ l \ne i}} a_{kl}x_k x_l, \text{ if } A \text{ is symmetric,} \\ a_{ii}x_i^2 + x_i \sum_{j \ne i} a_{ij}x_j + x_j \sum_{j \ne i} a_{ji}x_i \\ + \sum_{\substack{k \ne i \\ l \ne i}} a_{kl}x_k x_l, \text{ if } A \text{ is not symmetric.} \end{cases}$$

We obtain

$$\frac{\partial f(x)}{\partial x_i} = \begin{cases} 2\sum_{j=1}^{p} a_{ij}x_j, \text{ if } A \text{ is symmetric,} \\ \sum_{j=1}^{p} a_{ij}x_j + \sum_{j=1}^{p} a_{ji}x_j, \text{ if } A \text{ is not symmetric.} \end{cases}$$

Hence

$$\frac{\partial f(x)}{\partial x} = \begin{cases} 2Ax \text{ if } A \text{ is symmetric,} \\ (A + A')x \text{ if } A \text{ is not symmetric.} \end{cases}$$

Let $A = (a_{ij})$ be a matrix of dimension $p \times p$. Denoting by A_{ij} the cofactor of a_{ij} we obtain $|A| = \sum_{i=1}^{p} a_{ij} A_{ij}$. Thus

$$\frac{\partial |A|}{\partial a_{ii}} = A_{ii}, \frac{\partial |A|}{\partial a_{ij}} = A_{ij}.$$

Let $f(x)$ be a scalar function of a $p \times q$ matrix variable $x = (x_{ij})$. The matrix derivative of f is defined by the matrix of partial derivatives

$$\frac{\partial f}{\partial x} = \left(\frac{\partial f}{\partial x_{ij}} \right).$$

From above it follows that

$$\frac{\partial |A|}{\partial A} = \begin{cases} |A|(A^{-1})', \text{ if } A \text{ is not symmetric} \\ |A|[2(A^{-1})' - \text{diag}(A^{-1})], \text{ if } A \text{ is symmetric.} \end{cases}$$

Hence

$$\frac{\partial \log |A|}{\partial A} = \frac{1}{|A|} \frac{\partial |A|}{\partial A} = \begin{cases} (A^{-1})', \text{ if } A \text{ is not symmetric} \\ 2(A^{-1})' - \text{diag}(A^{-1}) \text{ if } A \text{ is symmetric.} \end{cases}$$

The following results can be easily deduced.

Let $A = (a_{ij})$ be a $m \times p$ matrix and x be a $p \times m$ matrix. Then

$$\frac{\partial \text{tr}(Ax)}{\partial x} = A'$$

and, for $m = p$,

$$\frac{\partial \text{tr}(xx')}{\partial x} = \begin{cases} 2x, \text{ if } x \text{ is not symmetric,} \\ 2(x + x') - 2\text{diag}(x), \text{ if } x \text{ is symmetric.} \end{cases}$$

Theorem 1.7.1. *Let A be a symmetric and at least positive semidefinite matrix of dimension $p \times p$. The largest and the smallest values of $x'Ax/x'x$ for all $x \neq 0$ are the largest and the smallest characteristic roots of A, respectively.*

Proof. Let $x'Ax/x'x = \lambda$. Differentiating λ with respect to the components of x the stationary values of λ are given by the characteristic equation $(A - \lambda I)x = 0$. Eliminating x we get $|A - \lambda I| = 0$. Thus the values of λ are the characteristic

Vector and Matrix Algebra

roots of the matrix A and consequently the largest value of λ corresponds to the largest characteristic root of A, and the smallest value of λ corresponds to the smallest characteristic root of A. Q.E.D.

From this theorem it follows that if $g_1 \leq x'Ax/x'x \leq g_2$ for all $x \neq 0$, then $g_1 \leq \lambda_1 \leq \lambda_p \leq g_2$ where λ_1, λ_p are the smallest and the largest characteristic roots of A, respectively.

Theorem 1.7.2. *Let A be a symmetric and at least positive semidefinite matrix of dimension $p \times p$ and let B be a symmetric and positive definite matrix of the same dimension. The largest and the smallest values of $x'Ax/x'Bx$ for all $x \neq 0$ are the largest and the smallest roots respectively of the characteristic equation $|A - \lambda B| = 0$.*

Proof. Let $x'Ax/x'Bx = \lambda$. Differentiating λ with respect to the components of x, the stationary values of λ are given by the characteristic equation $(A - \lambda B)x = 0$; hence by eliminating x we conclude that the smallest and the largest values of λ are given by the smallest and the largest roots of the characteristic equation $|A - \lambda B| = 0$. Q.E.D.

If $g_1 \leq x'Ax/x'Bx \leq g_2$ for all $x \neq 0$, then $g_1 \leq \lambda_1 \leq \lambda_p \leq g_2$ where λ_1, λ_p are the smallest and the largest roots of the characteristic equation $|A - \lambda B| = 0$.

Example 1.7.1. Let A be a positive definite matrix of dimension $p \times p$ with characteristic roots $\lambda_1 \geq \lambda_2 \geq \cdots \geq \lambda_p > 0$ and corresponding normalized characteristic vectors $\theta_1, \theta_2, \ldots, \theta_p$. Let θ be the orthogonal matrix of dimension $p \times p$ with columns $\theta_1, \theta_2, \ldots, \theta_p$, Δ be the diagonal matrix of dimension $p \times p$ with diagonal elements $\lambda_1, \lambda_2, \ldots, \lambda_p$. Let $A^{\frac{1}{2}} = \theta \Delta^{\frac{1}{2}} \theta'$, $y = \theta'x$, $y = (y_1, \ldots, y_p)'$ such that $\Delta^{\frac{1}{2}} \Delta^{\frac{1}{2}} = \Delta$ with $\Delta^{\frac{1}{2}}$ diagonal. Now

$$\frac{x'Ax}{x'x} = \frac{x'A^{\frac{1}{2}}A^{\frac{1}{2}}x}{x'\theta\theta'x} = \frac{y'\Delta y}{y'y} = \frac{\sum_1^p \lambda_i y_i^2}{\sum_1^p y_i^2}.$$

Hence for all x orthogonal to $\theta_1, \ldots, \theta_k$ we get

$$0 = \theta_j'x = \theta_j'(\sum_1^p \theta_i y_i) = y_1 \theta_j'\theta_1 + \cdots + y_p \theta_j'\theta_p = y_j$$

for $j \leq k$ and

$$\frac{x'Ax}{x'x} = \frac{\sum_{k+1}^p \lambda_i y_i^2}{\sum_{k+1}^p y_i^2} \leq \frac{\lambda_{k+1} \sum_{k+1}^p y_i^2}{\sum_{k+1}^p y_i^2} = \lambda_{k+1}.$$

Taking $y_{k+1} = \lambda_{k+1}, y_{k+2} = \cdots = y_p = 0$ we get for all x orthogonal to $\theta_1, \ldots, \theta_k$

$$\max_x \frac{x'Ax}{x'x} = \lambda_{k+1}.$$

Let b and d be any two vectors of the same dimension $p \times 1$. Then

$$b'd = b'A^{\frac{1}{2}}A^{\frac{-1}{2}}d = (A^{\frac{1}{2}}b)'(A^{\frac{-1}{2}}d) \leq (b'Ab)(d'A^{-1}d),$$

the inequality is obtained by applying the Cauchy-Schwartz to the vectors $A^{\frac{1}{2}}b$ and $A^{\frac{-1}{2}}d$.

1.8. COMPLEX MATRICES

In this section we shall briefly discuss complex matrices, matrices with complex elements, and state some theorems without proof concerning these matrices which are useful for the study of complex Gaussian distributions. For a proof the reader is referred to MacDuffee (1946). The adjoint operator (conjugate transpose) will be denoted by an asterisk (*). The adjoint A^* of a complex matrix $A = (a_{ij})$ of dimension $p \times q$ is the $q \times p$ matrix $A^* = (\bar{a}_{ij})'$, where the over-bar ($\bar{}$) denotes the conjugate and the prime ($'$) denotes the transpose. Clearly for any two complex matrices A, B, $(A^*)^* = A$, $(AB)^* = B^*A^*$, provided AB is defined. A square complex matrix A is called unitary if $AA^* = I$ (real identity matrix) and it is called Hermitian if $A = A^*$. A square complex matrix is called normal if $AA^* = A^*A$. A Hermitian matrix A of dimension $p \times p$ is called positive definite (semidefinite) if for all complex non-null p-vectors ξ, $\xi^*A\xi > 0 (\geq 0)$. Since

$$(\xi^*A\xi)^* = \xi^*A\xi$$

for any Hermitian matrix A, the Hermitian quadratic form $\xi^*A\xi$ assumes only real values.

Theorem 1.8.1. *If A is an Hermitian matrix of dimension $p \times p$, there exists a unitary matrix U of dimension $p \times p$ such that U^*AU is a diagonal matrix whose diagonal elements $\lambda_1, \ldots, \lambda_p$ are the characteristic roots of A.*

Since $(U^*AU)^* = U^*AU$, it follows that all characteristic roots of a Hermitian matrix are real.

Theorem 1.8.2. *A Hermitian matrix A is positive definite if all its characteristic roots are positive.*

Vector and Matrix Algebra

Theorem 1.8.3. *Every Hermitian positive definite (semidefinite) matrix A is uniquely expressible as $A = BB^*$ where B is Hermitian positive definite (semidefinite).*

Theorem 1.8.4. *For every Hermitian positive definite matrix A there exists a complex nonsingular matrix B such that $BAB^* = I$.*

EXERCISES

1. Prove Theorem 1.1.2.
2. Show that for any basis $\alpha_1, \ldots, \alpha_k$ of V^k of rank k there exists an orthonormal basis $\gamma_1, \ldots, \gamma_k$ of V^k.
3. If $\alpha_1, \ldots, \alpha_k$ is a basis of a vector space V, show that no set of $k+1$ vectors in V is linearly independent.
4. Find the orthogonal projection of the vector $(1,2,3,4)$ on the vector $(1,0,1,1)$.
5. Find the number of linearly independent vectors in the set $(a, c, \ldots, c)'$, $(c, a, c, \ldots, c)', \ldots, (c, \ldots, c, a)'$ such that the sum of components of each vector is zero.
6. Let V be a set of vectors of dimension p and let V^+ be the set of all vectors orthogonal to V. Show that $(V^+)^+ = V$ if V is a linear subspace of V.
7. Let V_1 and V_2 be two linear subspaces containing the null vector of 0 and let V_i^+ denote the set of all vectors orthogonal to $V_i, i = 1, 2$. Show that $(V_1 \cup V_2)^+ = V_1^+ \cap V_2^+$.
8. Let $(\gamma_1, \ldots, \gamma_k)$ be an orthogonal basis of the subspace V^k of a vector space V^p. Show that it can be extended to an orthogonal basis $(\gamma_1, \ldots, \gamma_k, \gamma_{k+1}, \ldots, \gamma_p)$ of V^p.
9. Show that for any three vectors x, y, z in V^p, the function d, defined by
$$d(x, y) = \max_{1 \leq i \leq p} |x_i - y_i|,$$
satisfies
    (a) $d(x, y) = d(y, x) \geq 0$ (symmetry),
    (b) $d(x, z) \leq d(x, y) + d(y, z)$ (triangular inequality).
10. Let W be a vector subspace of the vector space V. Show that the rank of $W \leq$ rank of V.
11. (Cauchy-Schwarz inequality) Show that for any two vectors x, y in V^p,
$$(x'y) \leq \|x\| \|y\|.$$
12. (Triangle inequality) Show that for any two vectors x, y in V^p
$$\|x + y\| \leq \|x\| + \|y\|.$$

13 Let A, B be two positive definite matrices of the same dimension. Show that for $0 \leq \alpha \leq 1$,

$$|\alpha A + (1-\alpha)B| \geq |A|^\alpha |B|^{1-\alpha}.$$

14 (Skew matrix) A matrix A is skew if $A = -A'$. Show that
   (a) for any matrix A, AA is symmetric if A is skew,
   (b) the determinant of a skew matrix with an odd number of rows is zero,
   (c) the determinant of a skew symmetric matrix is nonnegative.

15 Show that for any square matrix A there exists an orthogonal matrix θ such that $A\theta$ is an upper triangular matrix.

16 (Idempotent matrix) A square matrix A is idempotent if $AA = A$. Show the following:
   (a) if A is idempotent and nonsingular, then $A = I$;
   (b) the characteristic roots of an idempotent matrix are either unity or zero;
   (c) if A is idempotent of rank r, then $\text{tr}A = r$;
   (d) let $A_1, \ldots, A_k$ be symmetric matrices of the same dimension; if $A_i A_j = 0 (i \neq j)$ and if $\sum_{i=1}^{k} A_i$ is idempotent, then show that A_i for each i is an idempotent matrix and rank $(\sum_{i=1}^{k} A_i) = \sum_{i=1}^{k} \text{rank}(A_i)$.

17 Show that for any lower triangular matrix A the diagonal elements are its characteristic roots.

18 Show that any orthogonal transformation may be regarded as the change of axes about a fixed origin.

19 Show that for any nonsingular matrix A of dimension $p \times p$ and non-null p-vector x,

$$x'(A + xx')^{-1}x = \frac{x'A^{-1}x}{1 + x'A^{-1}x}.$$

20 Let A be a nonsingular matrix of dimension $p \times p$ and let x, y be two non-null p-vectors. Show that

$$(A + xy')^{-1} = A^{-1} - \frac{(A^{-1}x)(y'A^{-1})}{1 + y'A^{-1}x}$$

21 Let X be a $p \times q$ matrix and let S be a $p \times p$ nonsingular matrix. Then show that $|XX' + S| = |S||I + X'S^{-1}X|$.

22 Let X be a $p \times p$ matrix. Show that the nonzero characteristic roots of $X'X$ are the same as those of XX'.

23 Let A, X be two matrices of dimension $q \times p$. Show that
   (a) $(\partial/\partial X)(\text{tr}A'X) = A$
   (b) $(\partial/\partial X)(\text{tr}AX') = A$ where $\partial/\partial X = (\partial/\partial x_{ij})$, $X = (x_{ij})$.

Vector and Matrix Algebra

24 For any square symmetric matrix A show that

$$\frac{\partial}{\partial A}(\text{tr}AA) = 2A.$$

25 Prove
   (a) $A - A(A + \Sigma)^{-1}A = (A^{-1} + \Sigma^{-1})^{-1}$ where A, Σ are both positive definite matrices.
   (b) $|I_p - \eta(I + \eta'\eta)^{-1}\eta'| = |Iq + \eta'\eta|^{-1}$ where η is a $p \times q$ matrix.

26 Let A be a $q \times p$ matrix of rank $q < p$. Show that $A = C(I_q, 0)\theta$ where C is a nonsingular matrix of dimension $q \times q$ and θ is an orthogonal matrix of dimension $p \times p$.

27 Let L be a class of non-negative definite symmetric $p \times p$ matrix and let J be a fixed nonsingular member of L. Show that if $tr\ J^{-1}B$ is maximized over all B in L by $B = J$, then $|B|$ is maximized by J.

REFERENCES

Basilevsky, A. (1983). *Applied Matrix Algebra*. New York: North Holland.

Giri, N. (1993). *Introduction to Probability and Statistics*. Revised and Expanded ed. New York: Marcel Dekker.

Giri, N. (1974). *Introduction to Probability and Statistics*. Part 1, Probability. New York: Marcel Dekker.

Graybill, F. (1969). *Introduction to Linear Statistical Models*. Vol. I, New York: McGraw-Hill.

MacDuffee, C. (1946). *The Theory of Matrices*. New York: Chelsea.

MacLane, S., Birkoff, G. (1967). *Algebra*. New York: Macmillan.

Markus, M., Mine, H. (1967). *Introduction to Linear Algebra*. New York: Macmillan.

Perlis, S. (1952). *Theory of Matrices*. Reading, Massachusetts: Addison Wesley.

Rao, C.R. (1973). *Linear Statistical Inference and its Applications*, 2nd ed. New York: Wiley.

2
Groups, Jacobian of Some Transformations, Functions and Spaces

2.0. INTRODUCTION

In multivariate analysis the most frequently used test procedures are often invariant with respect to a group of transformations, leaving the testing problems invariant. In such situations an application of group theory results leads us in a straightforward way to the desired test procedures (see Stein (1959)). In this chapter we shall describe the basic concepts and some basic results of group theory. Results on the Jacobian of some specific transformations which are very useful in deriving the distributions of multivariate test statistics are also discussed. Some basic materials on functions and spaces are given for better understanding of the materials presented here.

2.1. GROUPS

Definition 2.1.1. Group. A group is a nonempty set G of elements with an operation τ satisfying the following axioms:

1. O_1 For any $a, b \in G, a\tau b \in G$.
2. O_2 There exists a unit element $e \in G$ such that for all $a \in G, a\tau e \in G$.

3. O_3 For any $a, b, c \in G$, $(a\tau b)\tau c = a\tau(b\tau c)$.
4. O_4 For each $a \in G$, there exists $a^{-1} \in G$ such that $a\tau a^{-1} = e$.

The following properties follow directly from axioms $O_1 - O_4 (a, b \in G)$:

1. $a\tau a^{-1} = a^{-1}\tau a$,
2. $a\tau e = e\tau a$,
3. $a\tau x = b$ has the unique solution $x = a^{-1}\tau b$.

Note: For convenience we shall write $a\tau b$ as ab. The reader is cautioned not to confuse this with multiplication.

Definition 2.1.2. Abelian group. A group G is called Abelian if $ab = ba$ for $a, b \in G$.

Definition 2.1.3. Subgroup. If the restriction of the operation τ to a nonempty subset H of G satisfies the group axioms $O_1 - O_4$, then H is called a subgroup of G.

The following lemma facilitates verifying whether a subset of a group is a subgroup.

Lemma 2.1.1. *Let G be a group and $H \subset G$. Then H is a subgroup of G (i) if and only if $H \neq \phi$ (nonempty), (ii) if $a, b \in H$, then $ab^{-1} \in H$.*

Proof. If H satisfies (i) and (ii), then H is a group. For if $a \in H$, then by (ii) $aa^{-1} = e \in H$. Also if $b \in H$ then $b^{-1} = eb^{-1} \in H$. Hence $a, b \in H$ implies $a(b^{-1})^{-1} = ab \in H$. Axiom O_3 is true in H as it is true in G. Hence H is a group. Conversely, if H is a group, then clearly H satisfies (i) and (ii). Q.E.D.

2.2. SOME EXAMPLES OF GROUPS

Example 2.2.1. The additive group of real numbers is the set of all reals with the group operation $ab = a + b$.

Example 2.2.2. The multiplicative group of nonzero real numbers is the set of all nonzero reals with the group operation $ab = a$ multiplied by b.

Groups and Transformations

Example 2.2.3. Permutation group. Let X be a nonempty set and let G be the set of all one-to-one functions of X onto X. Define the group operation τ as follows: for $g_1, g_2 \in G, x \in X, (g_1 \tau g_2)(x) = g_1(g_2(x))$. Then G is a group and is called the permutation group.

Example 2.2.4. Let X be a linear space. Then under the operation of addition X is an Abelian group.

Example 2.2.5. Translation group. Let X be a linear space of dimension n and let $x_0 \in X$. Define $g_{x_0}(x) = x + x_0, x \in X$. The collection of all g_{x_0} forms an additive Abelian group.

Example 2.2.6. Full linear group. Let X be a linear space of dimension n. Let $G_l(n)$ denote the set of all nonsingular linear transformations of X onto X. $G_l(n)$ is obviously a group with matrix multiplication as the group operation and it is called the full linear group.

Example 2.2.7. Affine group. Let X be a linear space of dimension n and let $G_l(n)$ be the linear group. The affine group $(G_l(n), X)$ is the set of pairs $(g, x), g \in G_l(n), x \in X$, with the following operation: $(g_1, x_1)(g_2, x_2) = (g_1 g_2, g_1 x_2 + x_1)$. For the affine group the unit element is $(I, 0)$, where I is the identity matrix and $(g, x)^{-1} = (g^{-1}, -g^{-1} x)$.

Example 2.2.8. Unimodular group. The unimodular group is the subgroup of $G_l(n)$ such that g is in this group if and only if the determinant of g is ± 1.

Example 2.2.9. The set of all nonsingular lower (upper) triangular matrices of dimension n forms a group with the usual matrix multiplication as the group operation. Obviously the product of two nonsingular lower (upper) triangular matrices is a lower (upper) triangular matrix and the inverse of a nonsingular lower (upper) triangular matrix is a nonsingular lower (upper) triangular matrix. The unit element for this group is the identity matrix.

Example 2.2.10. The set of all orthogonal matrices of dimension n forms a group.

2.3. QUOTIENT GROUP, HOMOMORPHISM, ISOMORPHISM

Definition 2.3.1. Normal subgroup. A subgroup H of G is a normal subgroup if for all $h \in H$ and $g \in G, ghg^{-1} \in H$ or, equivalently, $gHg^{-1} = H$.

Definition 2.3.2. Quotient group. Let G be a group and let H be a normal subgroup of G. The set G/H is defined to be set of elements of the form $g_1H = \{g_1h | h \in H\}$, $g_1 \in G$. For $g_1, g_2 \in G$ we define $(g_1H)(g_2H)$ as the set of all elements obtained by multiplying all elements of g_1H by all elements of g_2H. With this operation defined on the elements of G/H, it is a group. We verify this as follows:

1. $g_1H = g_2H \Leftrightarrow g_2^{-1}g_1H = H \Leftrightarrow g_2^{-1}g_1 \in H \Leftrightarrow g_1 \in g_2H$.
2. Since H is a normal subgroup, we have for $g_1, g_2 \in G$, $g_2H = Hg_2$ and $(g_1H)(g_2H) = g_1(Hg_2H) = g_1(g_2H)H = g_1g_2H \in G/H$.
3. H is the identity element in $G/H(gHH = gH)$. The group G/H is the quotient group of G (mod H).

Example 2.3.1. The affine group (I, X), where X is a linear space of dimension n and I is the $n \times n$ identity matrix, is the normal subgroup of $(G_l(n), X)$. For $g \in G_l(n), x \in X$,

$$(g, x)(I, x)(g, x)^{-1}$$
$$= (g, x)(I, x)(g^{-1}, -g^{-1}x) = (g, x)(g^{-1}, -g^{-1}x + x) = (I, gx) \in (I, X). \tag{2.1}$$

Definition 2.3.3. Homomorphism. Let G and H be two groups. Then a mapping f of G into H is called a homomorphism if it preserves the group operation; i.e., for $g_1, g_2 \in G, f(g_1g_2) = f(g_1)f(g_2)$. This implies that if e is the identity element of G, then $f(e)$ is the identity element of H and $f(g_1^{-1}) = [f(g_1)]^{-1}$. For

(i) $f(g_1) = f(g_1e) = f(g_1)f(e)$,
(ii) $f(e) = f(g_1g_1^{-1}) = f(g_1)f(g_1^{-1})$.

If, in addition, f is a one to one mapping, it is called an isomorphism.

Definition 2.3.4. Direct products. Let G and H be groups and let $G \times H$ be the Cartesian product of G and H. With the operation $(g_1, h_1)(g_2, h_2) = (g_1g_2, h_1h_2)$, where $g_1, g_2 \in G, h_1, h_2 \in H$, and g_1g_2, h_1h_2 are the products in the groups G and H, respectively, $G \times H$ is a group and is known as the direct product of G and H.

Definition 2.3.5. The group G operates on the space X from the left if there exists a function on $G \times X$ to X whose value at $x \in X$ is denoted by gx such that

1. $ex = x$ for all $x \in X$ and e is the unit element of G;
2. for $g_1, g_2 \in G$ and $x \in X, g_1(g_2x) = g_1g_2(x)$.

Groups and Transformations

Note: (1) and (2) imply that $g \in G$ is one-to-one on X to X. To see this, suppose $gx_1 = gx_2 = y$. Then $g^{-1}(gx_1) = g^{-1}(gx_2) = g^{-1}y$. Using (1) and (2) we then have $x_1 = x_2$.

Definition 2.3.6. Let the group G operate from the left on the space X. G operates transitively on X if for every $x_1, x_2 \in X$, there exists a $g \in G$ such that $gx_1 = x_2$.

Example 2.3.2. Let X be the space of all $n \times n$ nonsingular matrices and let $G = G_l(n)$. Given any two points $x_1, x_2 \in X$, there exists a nonsingular matrix $g \in G$ such that $x_1 = gx_2$. In other words, G acts transitively on X.

Example 2.3.3. Let X be a linear space. $G_l(n)$ acts transitively on $X - \{0\}$.

2.4. JACOBIAN OF SOME TRANSFORMATIONS

Let $X_1, \ldots, X_n$ be a sequence of n continuous random variables with a joint probability density function $f_{X_1,\ldots,X_n}(x_1, \ldots, x_n)$. Let $Y_i = g_i(X_1, \ldots, X_n)$ be a set of continuous one to one transformations of the random variables $X_1, \ldots, X_n$. Let us assume that the functions $g_1, \ldots, g_n$ have continuous partial derivatives with respect to $x_1, \ldots, x_n$. Let the inverse function be denoted by $X_i = h_i(Y_1, \ldots, Y_n), i = 1, \ldots, n$. Denote by J the determinant of the $n \times n$ square matrix

$$\begin{pmatrix} \partial x_1/\partial y_1 & \cdots & \partial x_1/\partial y_n \\ \vdots & & \vdots \\ \partial x_n/\partial y_1 & \cdots & \partial x_n/\partial y_n \end{pmatrix}.$$

Then J is called the Jacobian of the transformation of $X_1, \ldots, X_n$ to $Y_1, \ldots, Y_n$. We shall assume that there exists a region R of points $(x_1, \ldots, x_n)$ on which J is different from zero. Let S be the image of R under the transformations. Then

$$\underbrace{\int \cdots \int_R}_{n \text{ integrals}} f_{X_1,\ldots,X_n}(x_1, \ldots, x_n) dx_1, \ldots, dx_n \quad (2.2)$$
$$= \int \cdots \int_S f_{X_1,\ldots,X_n}(h_1(y_1, \ldots, y_n), \ldots, h_n(y_1, \ldots, y_n))|J| dy_1, \ldots, dy_n.$$

From this it follows that the joint probability density function of the random variables $Y_1, \ldots, Y_n$ is given by

$$f_{Y_1,\ldots,Y_n}(y_1, \ldots, y_n) = \begin{cases} f_{X_1,\ldots,X_n}(h_1(y_1, \ldots, y_n), \ldots, h_n(y_1, \ldots, y_n))|J| \\ 0 \qquad \qquad \qquad \qquad \qquad \qquad \qquad \qquad \text{otherwise.} \end{cases}$$

We shall now state some theorems on the Jacobian (J), but will not give all the proofs. For further results on the Jacobian the reader is referred to Olkin (1962), Rao (1965), Roy (1957), and Nachbin (1965).

Theorem 2.4.1. *Let V be a vector space of dimension p. For $x, y \in V$, the Jacobian of the linear transformation $x \to y = Ax$, where A is a nonsingular matrix of dimension $p \times p$, is given by $|A|^{-1}$.*

Theorem 2.4.2. *Let the $p \times n$ matrix X be transformed to the $p \times n$ matrix $Y = AX$ where A is a nonsingular matrix of dimension $p \times p$. The Jacobian of this transformation is given by $|A|^{-n}$.*

Theorem 2.4.3. *Let a $p \times q$ matrix X be transformed to the $p \times q$ matrix $Y = AXB$ where A and B are nonsingular matrices of dimensions $p \times p$ and $q \times q$, respectively. Then the Jacobian of this transformation is given by $|A|^{-q}|B|^{-p}$.*

Theorem 2.4.4. *Let G_T be the multiplicative group of nonsingular lower triangular matrices of dimension $p \times p$. For $g = (g_{ij}), h = (h_{ij}) \in G_T$, the Jacobian of the transformation $g \to hg$ is $\prod_{i=1}^{p} (h_{ii})^{-i}$.*

Proof. Let $hg = c = (c_{ij})$. Obviously, $c \in G_T$ and $c_{ij} = \sum_{k=1}^{p} h_{ik} g_{kj}$ with $h_{ij} = 0$, $g_{ij} = 0$ if $i < j$. Then J^{-1} is given by the determinant of the $\frac{1}{2}p(p+1) \times \frac{1}{2}p(p+1)$ matrix

$$\begin{pmatrix} \partial c_{11}/\partial g_{11} & \partial c_{11}/\partial g_{21} & \cdots & \partial c_{11}/\partial g_{pp} \\ \partial c_{21}/\partial g_{11} & \partial c_{21}/\partial g_{21} & \cdots & \partial c_{21}/\partial g_{pp} \\ \vdots & \vdots & & \vdots \\ \partial c_{pp}/\partial g_{11} & \partial c_{pp}/\partial g_{21} & \cdots & \partial c_{pp}/\partial g_{pp} \end{pmatrix}$$

It is easy to see that this matrix is a lower triangular matrix with diagonal element

$$\frac{\partial c_{ij}}{\partial g_{ij}} = \begin{cases} h_{ii} & \text{if } i \geq j, \\ 0 & \text{otherwise.} \end{cases}$$

Thus among the diagonal elements h_{ii} is repeated i times. Hence the Jacobian is given by $\prod_{i=1}^{p}(h_{ii})^{-i}$. Q.E.D.

Corollary 2.4.1. *The Jacobian of the transformation $g \to gh$ is $\prod_{i=1}^{p}(h_{ii})^{i-p-1}$.*

Groups and Transformations

Proof. Let $gh = c = (c_{ij})$. Obviously c is a lower triangular matrix. Since

$$\frac{\partial c_{ij}}{\partial g_{ij}} = \begin{cases} h_{ij} & i \geq j \\ 0 & \text{otherwise,} \end{cases}$$

following the same argument as in Theorem 2.4.4 we conclude that the Jacobian is the determinant of a triangular matrix where h_{ii} is repeated $p + 1 - i$ times among its diagonal elements.. Hence the result. Q.E.D.

Theorem 2.4.5. *Let G_{UT} be the group of $p \times p$ nonsingular upper triangular matrices. For $g = (g_{ij}), h = (h_{ij}) \in G_{UT}$, the Jacobian of the transformation $g \to hg$ is $\prod_{i=1}^{p}(h_{ii})^{i-p-1}$.*

Proof. Let $hg = c = (c_{ij})$. Obviously c is an upper triangular matrix and $c_{ij} = \sum_{k=1}^{p} h_{ik} g_{kj}$ with $h_{ij} = 0, g_{ij} = 0$ if $i > j$. Then J^{-1} is given by the determinant of the matrix of dimension $\frac{1}{2}p(p+1) \times \frac{1}{2}p(p+1)$:

$$\begin{pmatrix} \partial c_{11}/\partial g_{11} & \partial c_{11}/\partial g_{12} & \cdots & \partial c_{11}/\partial g_{pp} \\ \partial c_{12}/\partial g_{11} & \partial c_{12}/\partial g_{12} & \cdots & \partial c_{12}/\partial g_{pp} \\ \vdots & \vdots & & \vdots \\ \partial c_{pp}/\partial g_{11} & \partial c_{pp}/\partial g_{12} & \cdots & \partial c_{pp}/\partial g_{pp} \end{pmatrix}.$$

Since

$$\frac{\partial c_{ij}}{\partial g_{ij}} = \begin{cases} h_{ii} & i \leq j \\ 0 & \text{otherwise,} \end{cases}$$

the preceding matrix is an upper triangular matrix such that among its diagonal elements h_{ii} is repeated $p + 1 - i$ times. Hence the Jacobian of this transformation is $\prod_{i=1}^{p}(h_{ii}^{-1})^{p+1-i}$. Q.E.D.

Corollary 2.4.2. *The Jacobian of the transformation $g \to gh$ is $\prod_{1}^{p}(h_{ii}^{-1})^{i}$.*

The proof follows from an argument similar to that of the theorem.

Theorem 2.4.6. *Let S be a symmetric positive definite matrix of dimension $p \times p$. The Jacobian of the transformation $S \to B$, where B is the unique lower triangular matrix with positive diagonal elements such that $S = BB'$, is $\prod_{i=1}^{p}(b_{ii})^{p+1-i}$.*

Theorem 2.4.7. *Let G_{BT} be the group of $p \times p$ lower triangular nonsingular matrices in block form, i.e., $g \in G_{BT}$,*

$$g = \begin{pmatrix} g_{(11)} & 0 & 0 & \cdots & 0 \\ g_{(21)} & g_{(22)} & 0 & \cdots & 0 \\ \vdots & \vdots & & & \vdots \\ g_{(k1)} & g_{(k2)} & g_{(k3)} & \cdots & g_{(kk)} \end{pmatrix},$$

where $g_{(ii)}$ are submatrices of g of dimension $d_i \times d_i$ such that $\sum_1^k d_i = p$. The Jacobian of the transformation $g \to hg$, $g, h \in G_{BT}$, is $\prod_{i=1}^k |h_{(ii)}^{-1}|^{\sigma_i}$ where $\sigma_i = \sum_{j=1}^i d_j$, $\sigma_0 = 0$. The Jacobian of the transformation $g \to gh$ is $\prod_{i=1}^k |h_{(ii)}^{-1}|^{p-\sigma_{i-1}}$.

Theorem 2.4.8. *Let G_{BUT} be the group of nonsingular upper triangular $p \times p$ matrices in block form, i.e., $g \in G_{BUT}$,*

$$g = \begin{pmatrix} g_{(11)} & g_{(12)} & \cdots & g_{(1k)} \\ 0 & g_{(22)} & \cdots & g_{(2k)} \\ \vdots & \vdots & & \vdots \\ 0 & 0 & 0 & g_{(kk)} \end{pmatrix},$$

where $g_{(ii)}$ are submatrices of dimension $d_i \times d_i$ and $\sum_{j=1}^k d_j = p$. For $g, h \in G_{BUT}$ the Jacobian of the transformation $g \to gh$ is $\prod_{j=1}^k |h_{(ii)}|^{\sigma_i}$ and that of $g \to hg$ is $\prod_{i=1}^k |h_{(ii)}|^{p-\sigma_{i-1}}$.

Theorem 2.4.9. *Let $S = (s_{ij})$ be a symmetric matrix of dimension $p \times p$. The Jacobian of the transformation $S \to CSC'$, where C is any nonsingular matrix of dimension $p \times p$, is $|C^{-1}|^{p+1}$.*

Proof. To prove this theorem it is sufficient to show that it holds for the elementary $p \times p$ matrices $E(ij), M_i(c)$, and $A(ij)$ where $E(ij)$ is the matrix obtained from the $p \times p$ identity matrix by interchanging the ith and the jth row; $M_i(c)$ is the matrix obtained from the $p \times p$ identity matrix by multiplying its ith row by the nonzero constant c; and $A(ij)$ is the matrix obtained from the $p \times p$ identity matrix by adding the jth row to the ith row. The fact that the theorem is valid for these matrices can be easily verified by the reader. For example, $M_i(c)SM_i(c)$ is obtained from S by multiplying s_{ii} by c^2 and s_{ij} by $c(i \neq j)$ so that the Jacobian is $c^{2+(p-1)} = c^{p+1}$. Q.E.D.

Groups and Transformations

Theorem 2.4.10. *Let $S = (s_{ij})$ be a symmetric positive definite matrix of dimension $p \times p$. The Jacobian of the transformation $S \to gSg'$, $g \in G_T$ is $|g^{-1}|^{p+1}$.*

Proof. Let $g = (g_{ij})$ with $g_{ij} = 0$ for $i < j$ and let $A = (a_{ij}) = gSg'$. Then $a_{ij} = \sum_{l,k} g_{il} s_{lk} g_{jk}$. Since

$$\frac{\partial a_{ij}}{\partial s_{kk}} = g_{ik} g_{jk}, \quad \frac{\partial a_{ij}}{\partial s_{lk}} = g_{il} g_{jk} + g_{ik} g_{jl},$$

J^{-1}, which is the determinant of the $\frac{1}{2}p(p+1) \times \frac{1}{2}p(p+1)$ lower triangular matrix

$$\begin{pmatrix} \partial a_{11}/\partial s_{11} & \partial a_{11}/\partial s_{12} & \cdots & \partial a_{11}/\partial s_{pp} \\ \vdots & \vdots & & \vdots \\ \partial a_{pp}/\partial s_{11} & \partial a_{pp}/\partial s_{12} & \cdots & \partial a_{pp}/\partial s_{pp} \end{pmatrix},$$

is equal to $\prod_{i=1}^{p} (g_{ii})^{p+1} = |g|^{p+1}$. Q.E.D.

Theorem 2.4.11. *Let $A = (a_{ij})$ be a $p \times p$ symmetric nonsingular matrix. The Jacobian of the transformation $A \to A^{-1}$ is $|A|^{-2p}$.*

Proof. Let $A^{-1} = B = (b_{ij})$. Since $BA = I$ we get

$$\left(\frac{\partial B}{\partial \theta}\right) A + B \left(\frac{\partial A}{\partial \theta}\right) = 0,$$

where

$$\left(\frac{\partial B}{\partial \theta}\right) = \begin{pmatrix} \frac{\partial b_{11}}{\partial \theta}, & \cdots, & \frac{\partial b_{1p}}{\partial \theta} \\ \cdot, & \cdots, & \cdot \\ \frac{\partial b_{p1}}{\partial \theta}, & \cdots, & \frac{\partial b_{pp}}{\partial \theta} \end{pmatrix}.$$

Hence

$$\left(\frac{\partial B}{\partial \theta}\right) = -B \left(\frac{\partial A}{\partial \theta}\right) B = -A^{-1} \left(\frac{\partial A}{\partial \theta}\right) A^{-1}.$$

Let $E_{\alpha\beta}$ be the $p \times p$ matrix whose all elements are zero except that the element of the αth row and the βth column is unity.

Taking $\theta = a_{\alpha\beta}$ we get

$$\left(\frac{\partial B}{\partial a_{\alpha\beta}}\right) = -BE_{\alpha\beta}B - b_{\cdot\beta}b_{\alpha\cdot}$$

where $b_{\alpha\cdot}$ and $b_{\cdot\beta}$ are the αth row and the βth column of B. Hence

$$\frac{\partial b_{ij}}{\partial a_{\alpha\beta}} = -b_{i\alpha}b_{\beta j}.$$

Thus the Jacobian is the determinant of the $p^2 \times p^2$ matrix

$$\left|\left(\frac{\partial b_{ij}}{\partial a_{\alpha\beta}}\right)\right| = |(b_{i\alpha}b_{\beta j})| = |(B \otimes B')|$$

$$= |B|^p|B'|^p = |B|^{2p}$$

$$= |A|^{-2p}.$$

Q.E.D.

2.5. FUNCTIONS AND SPACES

Let $\mathcal{X}, \mathcal{Y}$ be two arbitrary spaces and f be a function on $\mathcal{X}$ into $\mathcal{Y}$, written as $f: \mathcal{X} \to \mathcal{Y}$. The smallest closed subset of $\mathcal{X}$ on which f is different from zero is called the support of f. The function f is one-to-one (injective) if $f(x_1) = f(x_2)$ implies $x_1 = x_2$ for all $x_1, x_2 \in \mathcal{X}$. An one-to-one onto function is bijective. The inverse function f^{-1} of f is a set function defined by $f^{-1}(B) = \{x \in \mathcal{X} : f(x) \in B, B \subseteq \mathcal{X}\}$.

Definition 2.5.1. Continuous function. The function $f: \mathcal{X} \to \mathcal{Y}$ is continuous if $f^{-1}(B)$ for any open (closed) subset B of $\mathcal{Y}$ is an open (closed) subset of $\mathcal{X}$.

This definition corresponds to the $\epsilon - \delta$ definition of continuous functions of calculus: A function $f: R \to R$ (real line) is continuous at a point $b \in \mathcal{X}$ if for every $\epsilon > 0$ there exists a $\delta > 0$ such that $|x - b| < \delta$ implies $b - \delta < x < b + \delta$ and $|f(x) - f(b)| < \epsilon$ implies $f(b) - \epsilon < f(x) < f(b) + \epsilon$. The $\epsilon - \delta$ definition of continuity of f at b is equivalent to $x \in (b - \delta, b + \delta)$ implies $f(x) \in (f(b) - \epsilon, f(b) + \epsilon)$.

A function $f: \mathcal{X} \to \mathcal{Y}$ is continuous if it is continuous at every point of $\mathcal{X}$.

Definition 2.5.2. Composition. The composition of any two functions $f, g, f: \mathcal{X} \to \mathcal{Y}, g: \mathcal{Y} \to \mathcal{Z}$, is a function $g \circ f: \mathcal{X} \to \mathcal{Z}$ defined by $g \circ f(x) = g(f(x))$.

If b, g are both continuous so is $g \circ f$.

Definition 2.5.3. Vector space (Linear space). A vector space (Linear space) is a space $\mathcal{X}$ on which the sum $x + y$ and the scalar product cx, for c scalar and $x, y \in \mathcal{X}$, are defined.

A vector space with a defined metric is a metric space.

A function $f: \mathcal{X} \to \mathcal{Y}$ is called linear if

$$f(ax + by) = af(x) + bf(y)$$

for a, b scalars and $x, y \in \mathcal{X}$. A set A is called convex if

$$\alpha x + (1 - \alpha)y \in A$$

for $x, y \in A$ and $0 \le \alpha \le 1$.

Definition 2.5.4. Convex and Concave function. A real valued function f defined on a convex set A is convex

$$f(\alpha x + (1 - \alpha)y) \le \alpha f(x) + (1 - \alpha)f(y)$$

for $x, y \in A$ and $0 < \alpha < 1$ and is strictly convex if the inequality is strict for $x \ne y$. If

$$f(\alpha x + (1 - \alpha)y) \ge \alpha f(x) + (1 - \alpha)f(y),$$

f is called concave and is strictly concave if the inequality is strict for $x \ne y$.

Concave functions are bowl shaped and convex functions are upside-down bowl shaped.

Example 2.5.1. The function $f(x) = x^2$ or e^x for $x \in R$ is convex. The function $f(x) = -x^2, x \in R$ or $\log x, x \in (0, \infty)$ is convex. In R^2 with $x = (x_1, x_2)' \in R^2$, $x_1^2 + x_2^2 + x_1 x_2$ is strictly convex.

REFERENCES

Nachbin, L. (1965). *The Haar Integral*. Princeton, NJ: Van Nostrand-Reinhold.

Olkin, I. (1962). Note on the Jacobian of certain matrix transformations useful in multivariate analysis. *Biometrika* 40:43–46.

Rao, C. R. (1965). *Linear Statistical Inference and its Applications*. New York: Wiley.

Roy, S. N. (1957). *Some Aspects of Multivariate Analysis*. New York: Wiley.

Stein, C. (1959). Lecture Notes on Multivariate Analysis. Dept. of Statistics, Stanford Univ., California.

3
Multivariate Distributions and Invariance

3.0. INTRODUCTION

In this chapter we shall discuss the distribution of vector random variables and its properties. Most of the commonly used test criteria in multivariate analysis are invariant test procedures with respect to a certain group of transformations leaving the problem in question invariant. Thus to study the basic properties of such test criteria we will outline here the "principle of invariance" in some details. For further details the reader is referred to Eaton (1989); Ferguson (1969); Giri (1975, 1997); Lehmann (1959), and Wijsman (1990).

3.1. MULTIVARIATE DISTRIBUTIONS

By a multivariate distribution we mean the distribution of a random vector $X = (X_1, \ldots, X_p)'$, where $p(\geq 2)$ is arbitrary, whose elements X_i are univariate random variables with distribution function $F_{X_i}(x_i)$. Let $x = (x_1, \ldots, x_p)'$. The distribution function of X is defined by

$$F_X(x) = \text{prob}(X_1 \leq x_1, \ldots, X_p \leq x_p),$$

which is also written as

$$F_{X_1, X_2, \ldots, X_p}(x_1, x_2, \ldots, x_p)$$

to indicate the fact that it is the joint distribution of $X_1, \ldots, X_p$. If each X_i is a discrete random variable, then X is called a discrete random vector and its probability mass function is given by

$$p_X(x) = p_{X_1,\ldots,X_p}(x_1, \ldots, x_p) = \text{prob}(X_1 = x_1, \ldots, X_p = x_p).$$

It is also called the joint probability mass function of $X_1, \ldots, X_p$. If $F_X(x)$ is continuous in $x_1, \ldots, x_p$, $-\infty < x_i < \infty$ for all i, and if there exists a nonnegative function $f_{X_1,\ldots,X_p}(x_1, \ldots, x_p)$ such that

$$F_X(x) = \int_{-\infty}^{x_1} \cdots \int_{-\infty}^{x_p} f_X(y) dy_1, \ldots, dy_p \qquad (3.1)$$

where $y = (y_1, \ldots, y_p)'$, then $f_X(x)$ is called the probability density function of the continuous random vector X. [For clarity of exposition we have used $f_X(y)$ instead of $f_X(x)$ in (3.1).] If the components $X_1, \ldots, X_p$ are independent (statistically), then

$$F_X(x) = \prod_{i=1}^{p} F_{X_i}(x_i),$$

or, equivalently,

$$f_X(x) = \prod_{i=1}^{p} f_{X_i}(x_i), \qquad p_X(x) = \prod_{i=1}^{p} p_{X_i}(x_i).$$

Given $f_X(x)$, the marginal probability density function of any subset of X is obtained by integrating $f_X(x)$ over the domain of the variables not in the subset. For $q < p$

$$f_{X_1,\ldots,X_q}(x_1, \ldots, x_q) = \int \cdots \int f_X(x) dx_{q+1}, \ldots, dx_p. \qquad (3.2)$$

In the case of discrete $p_X(x)$, the marginal probability mass function of $X_1, \ldots, X_q$ is obtained from $p_X(x)$ by summing it over the domain of $X_{q+1}, \ldots, X_p$.

It is well-known that

(i) $\lim_{x_p \to \infty} F_X(x) = F_{X_1,\ldots,X_{p-1}}(x_1, \ldots, x_{p-1})$;
(ii) for each i, $1 \leq i \leq p$, $\lim_{x_i \to -\infty} F_X(x) = 0$;
(iii) $F_X(x)$ is continuous from above in each argument.

The notion of conditional probability of events can be used to obtain the conditional probability density function of a subset of components of X given that the variates of another subset of components of X have assumed constant specified values or have been constrained to lie in some subregion of the space described by their variate values. For a general discussion of this the reader is

Multivariate Distributions and Invariance

referred to Kolmogorov (1950). Given that X has a probability density function $f_X(x)$, the conditional probability density function of $X_1, \ldots, X_q$ where $X_{q+1} = x_{q+1}, \ldots, X_p = x_p$ is given by

$$f_{X_1,\ldots,X_q|X_{q+1},\ldots,X_p}(x_1, \ldots, x_q | x_{q+1}, \ldots, x_p) = \frac{f_X(x)}{f_{X_{q+1},\ldots,X_p}(x_{q=1}, \ldots, x_p)}, \qquad (3.3)$$

provided the marginal probability density function $f_{X_{q+1},\ldots,X_p}(x_{q+1}, \ldots, x_p)$ of $X_{q+1}, \ldots, X_p$ is not zero. For discrete random variables, the conditional probability mass function $p_{X_1,\ldots,X_q|X_{q+1},\ldots,X_p}(x_1, \ldots, x_q | x_{q+1}, \ldots, x_q)$ of $X_1, \ldots, X_q$ given that $X_{q+1} = x_{q+1}, \ldots, X_p = x_p$ can be obtained from (3.3) by replacing the probability density functions by the corresponding mass functions.

The mathematical expectation of a random matrix X

$$X = \begin{pmatrix} X_{11} & \cdots & X_{q1} \\ \vdots & & \vdots \\ X_{1p} & \cdots & X_{qp} \end{pmatrix}$$

of dimension $p \times q$ (the components X_{ij} are random variables) is defined by

$$E(X) = \begin{pmatrix} E(X_{11}) & \cdots & E(X_{q1}) \\ \vdots & & \vdots \\ E(X_{1p}) & \cdots & E(X_{qp}) \end{pmatrix} \qquad (3.4)$$

Since a random vector $X = (X_1, \ldots, X_p)'$ is a random matrix of dimension $p \times 1$, its mathematical expectation is given by

$$E(X) = (E(X_1), \ldots, E(X_p))'. \qquad (3.5)$$

Thus it follows that for any matrices A, B, C of real constants and for any random matrix X

$$E(AXB + C) = AE(X)B + C. \qquad (3.6)$$

Definition 3.1.1. For any random vector X, $\mu = E(X)$ and $\Sigma = E(X - \mu) \times (X - \mu)'$ are called, respectively, the mean and the covariance matrix of X.

Definition 3.1.2. For every real $t = (t_1, \ldots, t_p)'$, the characteristic function of any random vector X is defined by $\phi_X(t) = E(e^{it'X})$ where $i = (-1)^{1/2}$.
Since $E|e^{it'X}| = 1$, $\phi_X(t)$ always exists.

3.2. INVARIANCE IN STATISTICAL TESTING OF HYPOTHESES

Invariance is a mathematical term for symmetry and in practice many statistical testing problems exhibit symmetries. The notion of invariance in statistical tests is of old origin. The unpublished work of Hunt and Stein (see Lehmann (1959)) toward the end of World War II has given this principle strong support as to its applicability and meaningfulness in the framework of the general class of all statistical tests. It is now established as a very powerful tool for proving the admissibility and minimax property of many statistical tests. It is a generally accepted principle that if a problem with a unique solution is invariant under a certain transformation, then the solution should be invariant under that transformation. The main reason for the strong intuitive appeal of an invariant decision procedure is the feeling that there should be or exists a unique best way of analyzing a collection of statistical information. Nevertheless in cases in which the use of an invariant procedure conflicts violently with the desire to make a correct decision with high probability or to have a small expected loss, the procedure must be abandoned.

Let $\mathcal{X}$ be the sample space, let $\mathcal{A}$ be the σ-algebra of subsets of $\mathcal{X}$ (a class of subsets of $\mathcal{X}$ which contains $\mathcal{X}$ and is closed under complementation and countable unions), and let $\Omega = \{\theta\}$ be the parametric space. Denote by P the family of probability distributions P_θ on $\mathcal{A}$. We are concerned here with the problem of testing the null hypothesis $H_0 : \theta \in \Omega_{H_0}$ against the alternatives $H_1 : \theta \in \Omega_{H_1}$. The principle of invariance for testing problems involves transformations mainly on two spaces: the sample space $\mathcal{X}$ and the parametric space Ω. Between the two, the most basic is the transformation g on $\mathcal{X}$. The transformation on Ω is the transformation $\bar{g}$, induced by g on Ω. All transformations g, considered in the context of invariance, will be assumed to be

(i) one-to-one from $\mathcal{X}$ onto $\mathcal{X}$; i.e., for every $x_1 \in \mathcal{X}$ there exists $x_2 \in \mathcal{X}$ such that $x_2 = g(x_1)$ and $g(x_1) = g(x_2)$ implies $x_1 = x_2$.

(ii) bimeasurable, to ensure that whenever X is a random variable with values in $\mathcal{X}$, $g(X)$ (usually written as gX) is also a random variable with values in $\mathcal{X}$, $g(X)$ (usually written as gX) is also a random variable with values in $\mathcal{X}$ and for any set $A \in \mathcal{A}$, gA and $g^{-1}A$ (the image and the transformed set) both belongs to $\mathcal{A}$.

The induced transformation $\bar{g}$ corresponding to g on $\mathcal{X}$ is defined as follows:

If the random variable X with values in $\mathcal{X}$ has probability distribution P_θ, gX is also a random variable with values in $\mathcal{X}$, and has probability distribution $P_{\theta'}$, where $\theta' = \bar{g}\theta \in \Omega$. An equivalent way of stating this fact is (g^{-1} being the

Multivariate Distributions and Invariance

inverse transformation corresponding to g)

$$P_\theta(g^{-1}A) = P_{\bar{g}\theta}(A) \tag{3.7}$$

or

$$P_\theta(A) = P_{\bar{g}\theta}(gA) \tag{3.8}$$

for all $A \in \mathcal{A}$. In terms of mathematical expectation this is also equivalent to saying that for any integrable real-valued function ϕ

$$E_\theta(\phi(g^{-1}X)) = E_{\bar{g}\theta}(\phi(X)), \tag{3.9}$$

where E_θ refers to expectation when X has distribution P_θ.

If, in addition, all P_θ, $\theta \in \Omega$, are distinct, i.e., if $\theta_1 \neq \theta_2$, $\theta_1, \theta_2 \in \Omega$, implies $P_{\theta_1} \neq P_{\theta_2}$, then g determines $\bar{g}$ uniquely and the correspondence between g and $\bar{g}$ is a homomorphism.

The condition (3.7) or its equivalent is known as the condition of invariance of probability distributions with respect to the transformation g on $\mathcal{X}$.

Definition 3.2.1. *Invariance of the parametric space* Ω. The parametric space Ω remains invariant under a one-to-one transformation $g : \mathcal{X}$ onto $\mathcal{X}$ if the induced transformation $\bar{g}$ on Ω satisfies (i) $\bar{g}\theta \in \Omega$ for $\theta \in \Omega$, and (ii) for any $\theta' \in \Omega$ there exists a $\theta \in \Omega$ such that $\theta' = \bar{g}\theta$.

An equivalent way of writing (i) and (ii) is

$$\bar{g}\Omega = \Omega. \tag{3.10}$$

If the P_θ for different values of θ are distinct, then $\bar{g}$ is also one-to-one.

Given a set of transformations, each leaving Ω invariant, the following theorem will assert that we can always extend this set to a group G of transformations whose members also leave Ω invariant.

Theorem 3.2.1. *Let g_1 and g_2 be two transformations which leave Ω invariant. The transformations g_2g_1 and g_1^{-1} defined by $g_2g_1(x) = g_2(g_1(x))$, $g_1^{-1}g_1(x) = x$ for all $x \in \mathcal{X}$ leave Ω invariant and $\overline{g_2g_1} = \bar{g}_2\bar{g}_1$, $\overline{g_1^{-1}} = \bar{g}^{-1}$.*

Proof. If the random variable X with values in $\mathcal{X}$ has probability distribution P_θ, then for any transformation g, gX has probability distribution $P_{\bar{g}\theta}$ with $\bar{g}\theta \in \Omega$. Since $\bar{g}_1\theta \in \Omega$ and g_2 leaves Ω invariant, the probability distribution of $g_2g_1(X) = g_2(g_1(X))$ is $P_{\bar{g}_2\bar{g}_1\theta}$, $\bar{g}_2\bar{g}_1\theta \in \Omega$. Thus g_2g_1 leaves Ω invariant and obviously $\overline{g_2g_1} = \bar{g}_2\bar{g}_1$. The reader may find it instructive to verify the other assertion. Q.E.D.

Very often in statistical problems there exists a measure λ on $\mathcal{X}$ such that P_θ is absolutely continuous with respect to λ, so that we can write for every $A \in \mathcal{A}$,

$$P_\theta(A) = \int_A P_\theta(x) d\lambda(x). \tag{3.11}$$

It is possible to choose the measure λ such that it is left invariant under G i.e.,

$$\lambda(A) = \lambda(gA) \tag{3.12}$$

for all $A \in \mathcal{A}$ and all $g \in G$. Then the condition of invariance of distribution reduces to

$$p_{\bar{g}\theta}(x) = p_\theta(g^{-1}x)$$

for all $x \in \mathcal{X}$ and $g \in G$.

Example 3.2.1. Let $\mathcal{X}$ be the Euclidean space and G be the group of translations defined by

$$g_{x_1}(x) = x + x_1, \qquad x_1 \in \mathcal{X}, g \in G. \tag{3.13}$$

Here G acts transitively on $\mathcal{X}$. The n-dimensional Lebesgue measure λ is invariant under G and it is unique up to a positive multiplicative constant.

Let us now consider the problem of testing $H_0 : \theta \in \Omega_{H_0}$ against the alternatives $H_1 : \theta \in \Omega_{H_1}$, where Ω_{H_0} and Ω_{H_1} are disjoint subsets of Ω. Let G be a group of transformations which operates from the left on $\mathcal{X}$, satisfying conditions (3.7) and (3.10).

Definition 3.2.2. Invariance of statistical problems. The problem of testing $H_0 : \theta \in \Omega_{H_0}$ against $H_1 : \theta \in \Omega_{H_1}$ remains invariant with respect to G if

(i) for $g \in G, A \in \mathcal{A}, P_{\bar{g}\theta}(gA) = P_\theta(A)$, and
(ii) $\Omega_{H_0} = \bar{g}\Omega_{H_0}, \Omega_{H_1} = \bar{g}\Omega_{H_1}$.

Example 3.2.2. Let $X_1, \ldots, X_n$ be a random sample of size n from a normal distribution with mean μ and variance σ^2 and let $x_1, \ldots, x_n$ be sample

Multivariate Distributions and Invariance

observations. Denote by $\mathcal{X}$ the space of all values $(x_1, \ldots, x_n)$. Let

$$\bar{x} = \frac{1}{n} \sum_{i=1}^{n} x_i,$$

$$s^2 = \frac{1}{n} \sum_{i=1}^{n} (x_i - \bar{x})^2,$$

$$\bar{X} = \frac{1}{n} \sum_{i=1}^{n} X_i,$$

$$S^2 = \frac{1}{n} \sum_{i=1}^{n} (X_i - \bar{X})^2, \qquad \theta = (\mu, \sigma^2).$$

The parametric space Ω is given by

$$\Omega = \{\theta = (\mu, \sigma^2), -\infty < \mu < \infty, \sigma^2 > 0\}$$

and let

$$\Omega_{H_0} = \{(0, \sigma^2) : \sigma^2 > 0\}, \qquad \Omega_{H_1} = \{(\mu, \sigma^2) : \mu \neq 0, \sigma^2 > 0\}.$$

The group of transformations G which leaves the problem invariant is the group of scale changes

$$X_i \to aX_i, \qquad i = 1, \ldots, n$$

with $a \neq 0$ and $\bar{g}\theta = (a\mu, a^2\sigma^2)$. Obviously $\bar{g}\Omega = \Omega$ for all $g \in G$. Since $x = (x_1, \ldots, x_n) \in g^{-1}A$ implies $gx \in A$ we have, with $y_i = ax_i, i = 1, \ldots, n$

$$P_\theta(g^{-1}A) = \int_{g^{-1}A} \frac{1}{(2\pi)^{n/2}(\sigma^2)^{n/2}} \exp\left[-\frac{1}{2} \sum_{i=1}^{n} \frac{(x_i - \mu)^2}{\sigma^2}\right] dx_1, \ldots, dx_n$$

$$= \int_A \frac{1}{(2\pi)^{n/2}(a^2\sigma^2)^{n/2}} \exp\left[-\frac{1}{2a^2\sigma^2} \sum_{i=1}^{n} (y_i - a\mu)^2\right] dy_1, \ldots, dy_n$$

$$= P_{\bar{g}\theta}(A).$$

Furthermore $\bar{g}\Omega_{H_0} = \Omega_{H_0}, \bar{g}\Omega_{H_1} = \Omega_{H_1}$.

If a statistical problem remains invariant under a group of transformations G operating on the sample space $\mathcal{X}$, it is then natural to restrict attention to statistical test ϕ which are also invariant under G, i.e.,

$$\phi(x) = \phi(gx), \qquad x \in \mathcal{X}, g \in G$$

Definition 3.2.3. Invariant function A function $T(x)$ defined on $\mathcal{X}$ is invariant under the group of transformations G if $T(x) = T(gx)$ for all $x \in \mathcal{X}, g \in G$.

Definition 3.2.4. Maximal invariant A function $T(x)$ defined on $\mathcal{X}$ is a maximal invariant under G if (i) $T(x) = T(gx), x \in \mathcal{X}, g \in G$, and (ii) $T(x) = T(y)$ for $x, y \in \mathcal{X}$ implies that there exists a $g \in G$ such that $y = gx$.

The reader is referred to Lehmann (1959) for the interpretation of invariant function and maximal invariant in terms of partition of the sample space.

Let $\mathcal{Y}$ be a space and let B be the σ-algebra of subsets of $\mathcal{Y}$. Suppose $T(x)$ is a measurable mapping from $\mathcal{X}$ into $\mathcal{Y}$. Let h be a one-to-one function on $\mathcal{Y}$ to $\mathcal{Z}$. If $T(x)$ with values in $\mathcal{Y}$ is a maximal invariant on $\mathcal{X}$, then $T \circ h$ is a maximal invariant on $\mathcal{X}$ with values in $\mathcal{Z}$. This fact is often used to write a maximal invariant in a convenient form.

Let $\phi(x)$ be a statistical test (probability of rejecting H_0 when x is observed). For a nonrandomized test $\phi(x)$ takes values 0 or 1. Suppose $\phi(x), x \in \mathcal{X}$ is invariant under a group of transformations G, operating from the left on $\mathcal{X}$. A useful characterization of $\phi(x)$ in terms of the maximal invariant $T(x)$ (under G) on $\mathcal{X}$ is given by the following theorem.

Theorem 3.2.2. *A test $\phi(x)$ is invariant under G if and only if there exists a function h such that $\phi(x) = h(T(x))$.*

Proof. Let $\phi(x) = h(T(x))$. Obviously

$$\phi(x) = h(T(x)) = h(T(gx)) = \phi(gx)$$

for $x \in \mathcal{X}, g \in G$. Conversely, if $\phi(x)$ is invariant under G and $T(x) = T(y), x, y \in \mathcal{X}$, then there exists a $g \in \mathcal{G}$ such that $y = gx$ and therefore $\phi(x) = \phi(y)$. Q.E.D.

In general h may not be a Borel measurable function. However, if the range of T is Euclidean and T is Borel measurable, then h is Borel measurable. See, for example, Blackwell (1956).

Let $\bar{G}$ be the group of induced (induced by G) transformations on Ω. We define on Ω, as on $\mathcal{X}$, a maximal invariant on Ω with respect to $\bar{G}$.

Theorem 3.2.3. *The distribution of $T(X)$ with values in the space of $\mathcal{Y}$, where X is a random variable with values in $\mathcal{X}$, depends on Ω only through $\nu(\theta)$.*

Proof. Suppose $\nu(\theta_1) = \nu(\theta_2), \theta_1, \theta_2 \in \Omega$. Since $\nu(\theta)$ is a maximal invariant on Ω under $\bar{G}$, there exists a $\bar{g} \in \bar{G}$ such that $\theta_2 = \bar{g}\theta_1$. Now for any measurable set C in $\mathcal{B}$ [by (3.7)]
$P_{\theta_1}(T(X) \in C) = P_{\theta_1}(T(gX) \in C) = P_{\bar{g}\theta_1}(T(X) \in C) = P_{\theta_2}(T(X) \in C)$. Q.E.D.

Example 3.2.3. Consider Example 3.2.2. Here

$$T(x) = \sqrt{n}\bar{x}\left[\sum_{i=1}^{n}\frac{(x_i - \bar{x})^2}{n-1}\right]^{-1/2}, \quad \nu(\theta) = (\sqrt{n}\mu)/\sigma = \lambda$$

where λ is an arbitrary designation. The probability density function of T is given by (see Giri (1993))

$$f_T(t) = \frac{(n-1)^{(n-1)/2}}{\Gamma((n-1)/2)}\frac{\exp(-\lambda^2/2)}{(n-1+t^2)^{n/2}}\sum_{j=0}^{\infty}\Gamma\left(\frac{n+j}{2}\right)\frac{\lambda^j}{j!}\left(\frac{2t^2}{n-1+t^2}\right)^{j/2}.$$

For point estimation problems the term "equivariant" is used instead of invariant.

Definition 3.2.5. Equivariant estimator. A point estimator defined on χ is equivariant under the group of transformations G on χ if $T(gx) = gT(x)$ for all $x \in \chi$ and $g \in G$.

Example 3.2.4. Let X be $N(\theta, 1)$. $T(x)$ is an equivariant point estimator of θ if and only if

$$T(x + g) = T(x) + g$$

for all $x, g \in R^1$. Taking $g = -x$ we conclude that T is equivariant if and only if $T(x) = x + a$ where a is some fixed real number.

The unique maximum likelihood estimator is an equivariant estimator.

3.3. ALMOST INVARIANCE AND INVARIANCE

To study the relative performances of different test criteria we need to compare their power functions. Thus it is of interest to study the implication of the invariance of power functions of the tests rather than the tests themselves. Since the power function of invariant tests depends only on the maximal invariant on Ω, any invariant test has invariant power functions. The converse that if the power function of a test ϕ is invariant under the induced group $\bar{G}$, i.e.,

$$E_\theta \phi(g^{-1}X) = E_{\bar{g}\theta}\phi(X), \qquad (3.14)$$

then the test ϕ is invariant under G, does not always hold well. To investigate this further we need to define the notions of almost invariance and equivalence to an invariant test.

Definition 3.3.1. Equivalence to an invariant test. Let G be a group of transformations satisfying (3.8) and (3.10). A test $\psi(x)$, $x \in \mathcal{X}$, is equivalent to an invariant test $\phi(x)$, $x \in \mathcal{X}$, with respect to the group of transformations G if

$$\phi(x) = \psi(x) \quad \text{for all} \quad x \in \mathcal{X} - N$$

where $P_\theta(N) = 0$ for $\theta \in \Omega$.

Definition 3.3.2. Almost invariance. Let G be a group of transformations on $\mathcal{X}$ satisfying (3.8) and (3.10). A test $\phi(x)$ is said to be almost invariant with respect to G if for $g \in G$, $\phi(x) = \phi(gx)$ for all $\mathcal{X} - Ng$ where $P_\theta(Ng) = 0$, $\theta \in \Omega$.

It is tempting to conjecture that an almost invariant test is equivalent to an invariant test. If any test ψ is equivalent to an invariant test ϕ, then it is almost invariant. For example, take

$$Ng = N \bigcup (g^{-1}N).$$

Obviously $x \in \mathcal{X} - Ng$ implies $x \in \mathcal{X} - N$ and $gx \in \mathcal{X} - N$. Hence for $x \in \mathcal{X} - Ng$, $\psi(x) = \phi(x) = \phi(gx) = \psi(gx)$. Since $P_\theta(g^{-1}N) = P_{\bar{g}\theta}(N) = 0$, $P_\theta(Ng) = 0$. Conversely, if the group G is countable, for any almost invariant test ψ, take

$$N = \bigcup_{g \in G} Ng,$$

where $\psi(x) = \psi(gx)$, $x \in \mathcal{X} - Ng$, $g \in G$, so that $P_\theta(N) = 0$. Then $\psi(x) = \psi(gx)$, $x \in \mathcal{X} - N$. Now define $\phi(x)$ such that

$$\phi(x) = \begin{cases} 1 & \text{if } x \in N \\ \psi(x) & \text{if } x \in \mathcal{X} - N. \end{cases}$$

Obviously $\phi(x)$ is an invariant function and $\psi(x)$ is equivalent to an invariant test. (Note that $gN = N$, $g \in G$.) If the group G is uncountable, such a result does not, in general, hold well.

Let $\mathcal{X}$ be the sample space and let $\mathcal{A}$ be the σ-field of subsets of $\mathcal{X}$. Suppose that G is a group of transformations operating on $\mathcal{X}$ and that $\mathcal{B}$ is a σ-field of subsets of G. Let for any $A \in \mathcal{A}$, the set of pairs (x, g), such that $gx \in A$, belong to $\mathcal{A} \times \mathcal{B}$. Suppose further that there exists a σ-finite measure μ (i.e., for $B_1, B_2, \ldots$ in $\mathcal{B}$ such that $\cup B_i = G$ and $\mu(B_i) < \infty$ for all i) on G such that

$$\mu(B) = 0 \quad \text{implies} \quad \mu(Bg) = 0$$

for all $g \in G$. Then any almost invariant function on $\mathcal{X}$ with respect to G is equivalent to an invariant function with respect to G. For a proof of this result the reader is referred to (Lehmann (1959), p. 225). This requirement is satisfied in

Multivariate Distributions and Invariance

particular when

$$\mu(Bg) = \mu(B), \qquad B \in \mathcal{B}, g \in G.$$

In other words, μ is a σ-finite right invariant measure. Such a right invariant measure μ exists for a large number of groups.

Example 3.3.1. Let $G = E^p$ (Euclidean p-space) where the group operation is addition. The Lebesgue measure in the space of E^p is the right invariant measure. Since G is Abelian, the right invariant measure is also left invariant, i.e., $\mu(gB) = \mu(B)$ for $g \in G$.

Example 3.3.2. Let G be the positive half of the real line with multiplication as the group operation. The right invariant measure μ is given by ($B \in \mathcal{B}$)

$$\mu(B) = \int_B \frac{dg}{g}.$$

Example 3.3.3. Let G be the multiplicative group of $p \times p$ nonsingular real matrices $g = (g_{ij})$. Write

$$dg = \prod_{i,j} dg_{ij}.$$

The right invariant measure μ on G is given by

$$\mu(B) = \int_B \frac{dg}{|\det g|^p}.$$

This follows from the fact that the Jacobian of the transformation

$$g \to gh, \qquad g, h \in G,$$

is $(\det(h))^{-p}$. Furthermore it is also left invariant.

Example 3.3.4. Let G be the group of affine transformations of the real line R onto itself, i.e., $g \in G$ has the form (a, b) such that for $x \in R$

$$(a, b)x = ax + b.$$

Here the group operation is defined by

$$g_1 g_2 = (a_1 a_2, a_1 b_2 + b_1)$$

where $g_1 = (a_1, b_1)$, $g_2 = (a_2, b_2)$. The Jacobian of the transformation $g_1 \to g_1 g_2$ is $(a_2)^{-1}$. So the right invariant measure μ on G is given by

$$\mu(B) = \int_B \frac{da\,db}{a}.$$

Note that the left invariant measure in this case is given by

$$\mu(B) = \int_B \frac{da\,db}{a^2}$$

Example 3.3.5. Let G be the group of affine transformations of V^p (a real vector space of dimension p) onto itself; i.e., for $x \in V^p$, $g = (c, b) \in G$ where $c \in G_l(p)$, the group of $p \times p$ nonsingular real matrices, and b is a p-vector,

$$gx = cx + b.$$

The group operation in G is defined by

$$g_1 g_2 = (c_1 c_2, c_1 b_2 + b_1)$$

where $g_1 = (c_1, b_1)$, $g_2 = (c_2, b_2)$, $b_1, b_2 \in V^p$, and $c_1, c_2 \in G_l(p)$. The right invariant measure μ on G is defined by

$$\mu(B) = \int_B \frac{dc\,db}{|\det c|^p}.$$

The left invariant measure in this case is given by

$$\mu(B) = \int_B \frac{dc\,db}{|\det c|^{p+1}}.$$

Multivariate Distributions and Invariance

Example 3.3.6. Let G_T be the multiplicative group of $p \times p$ nonsingular lower triangular matrices g, given by

$$g = \begin{pmatrix} g_{11} & 0 & 0 & \cdots & 0 \\ g_{21} & g_{22} & 0 & \cdots & 0 \\ \vdots & \vdots & & & \vdots \\ g_{k1} & g_{k2} & g_{k3} & \cdots & g_{kk} \end{pmatrix}$$

where g_{ii} is a submatrix of g of dimension $d_i \times d_i$ such that $\sum_1^k d_i = p$. The right invariant measure μ on G is given by

$$\mu(B) = \int_B \frac{dg}{\prod_{i=1}^k |\det g_{ii}|^{p-\sigma_{i-1}}}$$

where $\sigma_i = \sum_{j=1}^i d_j$ with $\sigma_0 = 0$. The left invariant measure on G is given by

$$\mu(B) = \int_B \frac{dg}{\prod_{i=1}^k |\det g_{ii}|^{\sigma_i}}$$

For further results on invariant measure the reader is referred to Nachbin (1965).

It is now evident that any almost invariant test function with respect to a group of transformations G on $\mathcal{X}$ has an invariant power function with respect to the induced group $\bar{G}$ on Ω. The converse of this is not true in general. However, in cases in which prior to the application of invariance the problem can be reduced to one based on a sufficient statistic on the sample space whose distributions constitute a boundedly complete family, the converse is true.

Let T be sufficient for $\{P_\theta, \theta \in \Omega\}$ and let the distribution $\{P_\theta^T, \theta \in \Omega\}$ of T be boundedly complete; i.e., for any bounded function $g(T)$ of T, if

$$E_\theta^{g(T)} \equiv 0 \tag{3.15}$$

for all $\theta \in \Omega$, then $g(T) = 0$ almost everywhere with respect to the probability measure P_θ^T. For any almost invariant test function $\psi(T)$ with respect to the group of transformations G on the space of the sufficient statistic T we have, for $g \in G$,

$$E_\theta \psi(T) = E_\theta \psi(gT) = E_{\bar{g}\theta} \psi(T).$$

Conversely, if

$$E_\theta \psi(T) = E_{\bar{g}\theta} \psi(T),$$

then for $g \in G$ [note $gT(x) = T(gx)$],

$$E_\theta \psi(T) = E_\theta \psi(gT)$$

or, equivalently,

$$E_\theta(\psi(gT) - \psi(T)) \equiv 0$$

for all $\theta \in \Omega$. Since the distribution of T is boundedly complete, we obtain

$$\psi(t) = \psi(gt)$$

almost everywhere with respect to the probability measure P_θ^T.

Since for any test $\phi(x)$ on the original sample space $\mathcal{X}$,

$$\psi(t) = E(\phi(X)|T = t)$$

is also a test based on the sufficient statistic T with the same power function as that of $\phi(x)$, we can conclude that if there exists a uniformly most powerful almost invariant test among all tests based on the sufficient statistic T, then that test is uniformly most powerful among all tests based on the original observations x and its power function depends only on the maximal invariant on the parametric space Ω.

Example 3.3.7. Consider Example 3.2.3. Let us first show that the sufficient statistic $(\bar{X}, S^2)$ for (μ, σ^2) is boundedly complete. The joint probability density function of $(\bar{X}, S^2)$ is given by (see Giri (1993))

$$f_{\bar{X},S^2}(\bar{x}, s^2) = \frac{K}{(\sigma^2)^{n/2}} \exp\{-\tfrac{1}{2}\sigma^2(ns^2 + n(\bar{x}-\mu)^2)\}(ns^2)^{(n-3)/2}$$

where

$$K = \sqrt{n}/[(2\pi)^{1/2} 2^{(n-1)/2} \Gamma((n-1)/2)].$$

For any bounded function $g(\bar{x}, s^2)$

$$E(g(\bar{X}, S^2)) = K \int \frac{g(\bar{x}, s^2)}{(\sigma^2)^{n/2}} \exp\left\{-\frac{1}{2\sigma^2}(ns^2 + n(\bar{x}-\mu)^2)\right\}(ns^2)^{(n-3)/2} ds^2 d\bar{x}.$$

Let $1/\sigma^2 = 1 - 2\theta$ and $(1 - 2\theta)^{-1} t = \mu$. Then

$$E(g(\bar{X}, S^2)) = K \int g(\bar{x}, s^2)(1 - 2\theta)^{n/2}(ns^2)^{(n-3)/2}$$

$$\times \exp\{-\tfrac{1}{2}[(1-2\theta)(ns^2 + n\bar{x}^2) - 2nt\bar{x} + nt^2/(1-2\theta)]\} ds^2 d\bar{x} \quad (3.16)$$

If

$$E(g(\bar{X}, S^2)) \equiv 0$$

Multivariate Distributions and Invariance

for all (μ, σ^2), then from (3.16) we obtain

$$K \int g(\bar{x}, ns^2 + n\bar{x}^2 - n\bar{x}^2)(ns^2)^{(n-3)/2}$$

$$\times \exp\{-\tfrac{1}{2}(ns^2 + n\bar{x}^2) + \theta(ns^2 + n\bar{x}^2) + nt\bar{x}\} ds^2 d\bar{x} \equiv 0.$$

This is the Laplace transformation of

$$g(\bar{x}, ns^2 + n\bar{x}^2 - n\bar{x}^2) K (ns^2)^{(n-3)/2} \exp\{-\tfrac{1}{2}(n\bar{x}^2 + ns^2)\}$$

with respect to the variables $n\bar{x}$, $ns^2 + n\bar{x}^2$. Since this is zero for all (μ, σ^2), we obtain

$$g(\bar{x}, s^2) = 0$$

except for a set of $(\bar{x}, s^2)$ with probability measure 0. So the distribution of $(\bar{X}, S^2)$ is boundedly complete. Second, from Example 3.2.3 we can conclude that for testing $H_0 : \mu = 0$, the test which rejects H_0 whenever $|t| \geq t_{1-\alpha/2}$, where $t_{1-\alpha/2}$ is the upper $1 - \alpha/2$ percent point of the central t-distribution with $n - 1$ degrees of freedom, is uniformly most powerful among all test whose power function depends only on $\sqrt{n}\mu/\sigma$.

3.4. SUFFICIENCY AND INVARIANCE

It is well known that some simplification is introduced in a testing problem by characterizing the statistical tests as a function of the sufficient statistic and thus reducing the dimension of the sample space to the dimension of the space of the sufficient statistic. On the other hand, invariance by reducing the dimension of the sample space to that of the space of the maximal invariant also shrinks the parametric space. Thus a question naturally arises: Is it possible to use both principles simultaneously and if so in what order, i.e., first sufficiency and then invariance, or first invariance and then sufficiency. Under certain conditions this reduction can be done by using both principles, and the order in which the reduction is made is immaterial in such cases. The reader is referred to Hall et al. (1965) for these conditions and some related results.

One can also avoid the task of verifying these conditions by replacing the sample space by the space of the sufficient statistic before looking for the group of transformations which leave the problem invariant and then look for the group of transformations on the space of the sufficient statistic that leave the problem invariant.

3.5. UNBIASEDNESS AND INVARIANCE

The discussions presented in this and the following section are sketchy. For further study and details relevant references are given.

In testing statistical hypotheses, the principle of unbiasedness plays an important role in deriving a suitable test statistic in complex situations involving composite hypotheses. A size α test ϕ is said to be unbiased for testing $H_0 : \theta \in \Omega_{H_0}$ against $H_1 : \theta \in \Omega_{H_1}$ if $E_\theta \phi(X) \geq \alpha$ for $\theta \in \Omega_{H_1}$. In many such problems the principle of unbiasedness and the principle of invariance seem to complement each other in the sense that each is successful in the cases in which the other is not. For example, it is well known that a uniformly most powerful unbiased test exists for testing the hypothesis $H_0 : \sigma^2 = \sigma_0^2$ (specified) against the alternatives $H_1 : \sigma^2 \neq \sigma_0^2$ in a normal distribution with mean μ whereas the principle of invariance does not reduce the problem sufficiently far to ensure the existence of a uniformly most powerful invariant test. On the other hand, for problems involving general linear hypotheses there exists a uniformly most powerful invariant test (F-test) but no uniformly most powerful unbiased test exists if the null hypothesis has more than one degree of freedom. However, if both principles can be applied successfully, then they lead to the same (almost everywhere) optimum test. Consider the problem of testing $H_0 : \theta \in \Omega_{H_0}$ against the alternatives $H_1 : \theta \in \Omega_{H_1}$. Let us assume that it is invariant under the group of transformations G. Let C_α be the class of unbiased tests of size $\alpha (0 < \alpha < 1)$. For any test $\phi(x)$ define the test function ϕg by

$$\phi g(x) = \phi(gx), \quad x \in \mathcal{X}, g \in G.$$

Obviously $\phi \in C_\alpha$ if and only if $\phi g \in C_\alpha$. Thus if the test ϕ^* is a unique (up to measure 0) uniformly most powerful unbiased test for this problem, then

$$E_\theta(\phi^* g(X)) = E_{\bar{g}\theta}(\phi^*(X)) = \sup_{\phi \in C_\alpha} E_{\bar{g}\theta}(\phi(X)) = \sup_{\phi g \in C_\alpha} E_\theta(\phi(g(X)))$$

$$= \sup_{\phi \in C_\alpha} E_\theta(\phi g(X)) = E_\theta \phi^*(X).$$

Thus ϕ^* and $\phi^* g$ have the same power function. Hence under the assumption of completeness of the sufficient statistic, ϕ^* is almost invariant. Therefore if there exists a uniformly most powerful almost invariant test ϕ^{**}, we have

$$E_\theta \phi^{**}(X) \geq E_\theta \phi^*(X) \tag{3.17}$$

for $\theta \in \Omega_{H_1}$. Comparing this with the trivial level α invariant test $\phi(x) = \alpha$, we conclude that ϕ^{**} is also unbiased, and hence

$$E_\theta \phi^{**}(X) \leq E_\theta \phi^*(X) \tag{3.18}$$

Multivariate Distributions and Invariance

for $\theta \in \Omega_{H_1}$. Thus from (3.17) and (3.18) it follows that ϕ^* and ϕ^{**} have the same power function. Since ϕ^* is unique $\phi^* = \phi^{**}$ almost everywhere.

Thus for a testing problem which is invariant under G, if there exists a unique uniformly most powerful unbiased test ϕ^* and if there exists a unique uniformly most powerful almost invariant test ϕ^{**}, then $\phi^* = \phi^{**}$ almost everywhere.

3.6. INVARIANCE AND OPTIMUM TESTS

Apart from the important fact that the performance of an invariant test is independent of the nuisance parameters, a powerful support of the principle comes from the famous unpublished Hunt-Stein theorem which asserts that under certain conditions on the group G there exists an invariant test which is minimax among all size α tests. It is well known that given any test function on the sample space we can always replace it by a test which depends only on the sufficient statistic such that both have the same power function. Such a result is too strong to expect from the maximal invariant statistic on the sample space. The appropriate weakening of this property and the conditions under which it holds constitute the Hunt-Stein theorem which asserts that for testing $H_0 : \theta \in \Omega_{H_0}$ against $H_1 : \theta \in \Omega_{H_1}$ (which is invariant under G), under certain conditions on the group G, given any test function ϕ on the sample space $\mathcal{X}$, there exists an invariant test ψ such that

$$\sup_{\theta \in \Omega_{H_0}} E_\theta \phi \geq \sup_{\theta \in \Omega_{H_0}} E_\theta \psi, \qquad \inf_{\theta \in \Omega_{H_1}} E_\theta \phi \leq \inf_{\theta \in \Omega_{H_1}} E_\theta \psi. \qquad (3.19)$$

In other words, ψ performs at least as well as ϕ in the worst possible cases. For the exact statement of this theorem the reader is referred to (Lehmann (1959) p. 335).

This method has been successfully used by Giri et al. (1963), Giri and Kiefer (1964a), Linnik et al. (1966), and Salaevskii (1968) to solve the long time open problem of the minimax character of Hotelling's T^2 test, and by Giri and Kiefer (1964b) to prove the minimax character of the R^2 test in some special cases.

It may be remarked here that the conditions of Hunt-Stein's theorem, whether algebraic or topological, are almost entirely on the group and are nonstatistical in nature. For verifying the admissibility of statistical tests through invariance the situation is more complicated. Aside from the trivial case of compact groups only the one-dimensional translation parameter case has been studied by Lehmann and Stein (1953). If G is a finite or a compact group, the most powerful invariant test is admissible. For other groups statistical structure plays an important role.

For further relevant results in this context the reader is referred to Kiefer (1957, 1966), Ghosh (1967), and Pitman (1939).

3.7. MOST STRINGENT TESTS AND INVARIANCE

Consider the problem of testing $H_0 : \theta \in \Omega_{H_0}$ against the alternatives $H_1 : \theta \in \Omega_{H_1}$ where $\Omega_{H_0} \cap \Omega_{H_1}$ is a null set. Let Q_α denote the class of all level α tests of H_0 and let

$$\beta_\alpha^*(\theta) = \sup_{\phi \in Q_\alpha} E_\theta \phi, \qquad \theta \in \Omega_{H_1}.$$

$\beta_\alpha^*(\theta)$ is called the envelope power function and it is the maximum power that can be obtained at level α against the alternative θ.

Definition 3.7.1. Most stringent test. A test ϕ that minimizes $\sup_{\theta \in \Omega_{H_1}} (\beta_\alpha^*(\theta) - E_\theta(\phi))$ is said to be most stringent. In other words, it minimizes the maximum shortcomings.

If the testing problem is invariant under a group of transformations G and if there exists a uniformly most powerful almost invariant test ϕ^* with respect to G such that the group satisfies the conditions of the Hunt-Stein theorem (see Lehmann (1959), p. 336), then ϕ^* is most stringent. For details and further reading in this context the reader is referred to Lehmann (1959a), Kiefer (1958), and Giri and Kiefer (1964a).

3.8. LOCALLY BEST AND UNIFORMLY MOST POWERFUL INVARIANT TESTS

Let X be a random variable (vector or matrix valued) with probability density function $f_X(x|\theta)$, $\theta \in \Omega$. Consider the problem of testing the null hypothesis $H_0 : \theta \in \Omega_0$ against the alternatives $H_1 : \theta \in \Omega_1$ where Ω_0 and Ω_1 are disjoint subsets of Ω. Assume that the problem of testing H_0 against H_1 is invariant under the group G of transformations g, transforming $x \to gx$. Let $T(x)$ be a maximal invariant under G in the sample space $\mathcal{X}$ of X whose distribution depends on the corresponding maximal invariant $\nu(\theta)$ in the parametric space Ω. Any invariant test depends on X only through $T(X)$ and its power depends only on $\nu(\theta)$.

Definition 3.8.1. *Uniformly most powerful invariant test.* An invariant test ϕ^* of size α is uniformly most powerful for testing H_0 against H_1 if its power

$$E_\theta \phi^* \geq E_\theta \phi, \theta \in \Omega_1$$

for any other invariant test ϕ of the same size.

Multivariate Distributions and Invariance

Definition 3.8.2. *Locally best invariant test.* An invariant test ϕ^* of size α is locally best invariant for testing H_0 against H_1 if there exists an open neighborhood $\tilde{\Omega}_1$ of Ω_0 such that its power

$$E_\theta \phi^* \geq E_\theta \phi, \; \theta \in \tilde{\Omega}_1 - \Omega_0$$

for any other invariant test ϕ of the same size.

3.9. RATIO OF DISTRIBUTIONS OF MAXIMAL INVARIANT, STEIN'S THEOREM

Invariant tests depend on observations only through a maximal invariant in the sample space. To find the optimum invariant test we need to find the form of the maximal invariant explicitly and its distribution. In many multivariate testing problems the explicit forms of maximal invariants are not easy to obtain. Stein (1956) gave a representation of the ratio of densities of a maximal invariant with respect to a group G of transformations g, leaving the testing problem invariant. To state this representation we require the following concepts.

Definition 3.9.1. Relatively left invariant measure. Let G be a locally compact group and let $\mathbb{B}$ be the σ-algebra of compact subsets of G. A measure ν on $(G, \mathbb{B})$ is relatively left invariant with left multiplier $\chi(g)$ if

$$\nu(gB) = \chi(g)\nu(B), \; B \in \mathbb{B}, g \in G.$$

Examples of such locally compact topological groups includes E^p, the linear group $G_l(p)$, the affine group, the group $G_T(p)$ of $p \times p$ nonsingular lower triangular matrices and the group $G_{UT}(p)$ of $p \times p$ upper triangular nonsingular matrices. The multiplier $\chi(g)$ is a continuous homomorphism from $G \to R^+$. In other words

$$\chi(g_1 g_2) = \chi(g_1)\chi(g_2) \quad for \quad g_1, g_2 \in G.$$

From this it follows that

$$\chi(e) = 1, \qquad \chi(g^{-1}) = 1/\chi(g)$$

where e is the identity element of G and $g \in G$. If ν is a relatively left invariant with left multiplier $\chi(g)$ then $\chi(g)\nu(dg)$ is a left invariant measure.

For example the Lebesgue measure $dg, g \in G_l(p)$ is relatively left invariant with $\chi(g)$ = the absolute value of $\det(g)$, which is the Jacobian of the inverse transformation $Y = gX \to X, X \in \mathcal{X} = E^p$.

Definition 3.9.2. A group G acts topologically on $\mathcal{X}$ if the function $f : G \times \mathcal{X} \to \mathcal{X}$; given by $f(g, x) = gx, g \in G, x \in \mathcal{X}$; is continuous.

For example if we define gx to be matrix product of g with the vector x, then $G_l(p)$ acts topologically on E^p.

Definition 3.9.3. (Cartan G-space). Let G act topologically on $\mathcal{X}$. Then $\mathcal{X}$ is called a Cartan G-space if for every $x \in \mathcal{X}$, there exists a neighborhood V of x such that $(V, V) = \{g \in G | (gV) \cap V \neq \phi\}$ has a compact closure.

Definition 3.9.4. (Proper action). Let G be a group of transformations acting topologically from the left on the space χ and let h be a mapping on

$$G \times \mathcal{X} \to \mathcal{X} \times \mathcal{X}$$

given by

$$h(g, x) = (gx, x), x \in \mathcal{X}, g \in G.$$

The group G acts properly on $\mathcal{X}$ if for every compact

$$C \subset \mathcal{X} \times \mathcal{X}$$

$h^{-1}(C)$ is compact. If G acts properly on $\mathcal{X}$ then $\mathcal{X}$ is a called Cartan G-space. The action is proper if for every pair (A, B) of compact subsets of $\mathcal{X}$

$$((A, B)) = \{g \in G | (gA) \cap B \neq \phi\}$$

is closed. If G acts properly on $\mathcal{X}$ then $\mathcal{X}$ is a Cartan G-space. It is not known if the converse is true. Wijsman (1967) has studied the properness of several groups of transformations used in multivariate testing problems. We refer to this paper and the references contained therein for the verification of these two concepts.

Theorem 3.9.1. *(Stein (1956)). Let G be a group of transformations g operating on a topological space $(\mathcal{X}, \mathbb{A})$ and λ a measure on $\mathcal{X}$ which is left-invariant under G. Suppose that there are two given probability densities p_1, p_2 with respect to λ such that*

$$P_1(A) = \int_A p_1(x) d\lambda(x)$$

$$P_2(A) = \int_A p_2(x) d\lambda(x)$$

Multivariate Distributions and Invariance

for $A \in \mathbb{A}$ and P_1, P_2 are absolutely continuous. Let $T(X): \mathcal{X} \to \mathcal{X}$ be a maximal invariant under G. Denote by P_i^*, the distribution of $T(X)$ when X has distribution $P_i, i = 1, 2$. Then under certain conditions

$$\frac{dP_2^*(T)}{dP_1^*(T)} = \frac{\int_G p_2(gx) d\mu(g)}{\int_G p_1(gx) d\mu(g)} \qquad (3.20)$$

where μ is a left invariant Haar measure on G.

An alternative form of (3.20) is given by

$$\frac{dP_2^*(T)}{dP_1^*(T)} = \frac{\int_G f_2(gx) \mathcal{X}(g) d\nu(g)}{\int_G f_1(gx) \mathcal{X}(g) d\nu(g)} \qquad (3.21)$$

where $f_i(gx), i = 1, 2$ denote the probability density function with respect to relatively invariant measure ν with left multiplier $\mathcal{X}(g)$. Stein gave the statement of Theorem 3.9.1 without giving explicitly the conditions under which it holds. However this theorem was successfully used by Giri (1961, 1964, 1965) and Schwartz (1967). Schwartz (1967) gave also a set of conditions (rather complicated) which must be satisfied for this theorem to hold. Wijsman (1967a) gave a sufficient condition for this theorem using the concept of Cartan G-space. Koehn (1970) gave a generalization of the results of Wijsman (1967). Bonder (1976) gave a condition for (3.21) through topological arguments. Anderson (1982) obtained certain conditions for the validity of (3.20) in terms of "proper action" on groups. Wijsman (1985) studied the properness of several groups of transformations commonly used for invariance in multivariate testing problems.

The presentation of materials in this section is very sketchy. We refer to references cited above for further reading and for the proof of Theorem 3.9.1.

3.10. DERIVATION OF LOCALLY BEST INVARIANT TESTS (LBI)

Let $\mathcal{X}$ be the sample space of X and let G be a group of transformations g acting on the left of $\mathcal{X}$. Assume that the problem of testing $H_0: \theta \in \Omega_0$ against the alternative $H_1: \theta \in \Omega_1$ is invariant under G, transforming $X \to gX$ and let $T(X)$ be a maximal invariant on $\mathcal{X}$ under G. The ratio R of the distributions of $T(X)$, for $\theta_1 \in \Omega_1, \theta_0 \in \Omega_0$ (by (3.21)) is given by

$$R = \frac{dP_{\theta_1}(T)}{dP_{\theta_0}(T)} = D^{-1} \int_G f_{\theta_1}(gx) \mathcal{X}(g) d\nu(g) \qquad (3.22)$$

where

$$D = \int_G f_{\theta_0}(gx)\mathcal{X}(g)d\nu(g).$$

Let

$$f_\theta(x) = \beta(\theta)q(\psi(x|\theta)), \quad \theta \in \Omega \tag{3.23}$$

where $\beta(\theta)$, ψ and q are known functions and q is $[0, \infty)$ to $[0, \infty)$. For multivariate normal distributions $q(z) = \exp(-z)$ and $\psi(x|0)$ is a quadratic function of x. Assuming q and β are continuously twice differentiable we expand $f_{\theta_1}(x)$

$$\begin{aligned}f_{\theta_1}(x) = \beta(\theta_1)\{&q(\psi(x|\theta_0)) + q^{(1)}(\psi(x|\theta_0))[\psi(x|\theta_1) - \psi(x|\theta_0)] \\ &+ \tfrac{1}{2}q^{(2)}(z)[\psi(x|\theta_1) - \psi(x|\theta_0)]^2 + o(\|\theta_1 - \theta_0\|)\}\end{aligned} \tag{3.24}$$

where $\beta(\theta_1) = \beta(\theta_0) + o(\|\theta_1 - \theta_0\|)$; $z = \alpha\psi(x|\theta_0) + (1 - \alpha)\psi(x|\theta_1), 0 \leq \alpha \leq 1$, $\|\theta_1 - \theta_0\|$ is the norm of $\theta_1 - \theta_0$ and $q^{(i)}(x) = d^i(q)/dx^i$. From (3.22) and (3.24)

$$R = 1 + D^{-1}\int_G q^{(1)}(\psi(gx|\theta_0))[\psi(gx|\theta_1) \\ - \psi(gx|\theta_0)]\chi(g)\nu(dg) + M(x, \theta_1, \theta_0) \tag{3.25}$$

where M is the remainder term.
Assumptions

1. The second term in the right-hand side of (3.25) is a function $\lambda(\theta_1, \theta_0)S(x)$, where $S(x)$ is a function of $T(x)$.
2. Any invariant test $\phi(X)$ of size α satisfies $E_{\theta_0}(\phi(X)M(X, \theta_1, \theta_0)) = o(\|\theta_1 - \theta_0\|)$ uniformly in ϕ.

Under above assumptions the power function $E_{\theta_1}(\phi(X))$ satisfies

$$E_{\theta_1}(\phi) = \alpha + E_{\theta_0}(\phi(X)\lambda(\theta_1, \theta_0)S(X)) + o(\|\theta_1 - \theta_0\|) \tag{3.26}$$

By the Neyman-Pearson lemma the test based $S(x)$ is LBI.

The following simple characterization of LBI test has been given by Giri (1968). Let $R_1, \ldots, R_p$ be maximal invariant in the sample space and let $\theta_1, \ldots, \theta_p$ be the corresponding maximal invariant in the parametric space. For notational convenience we shall write $(R_1, \ldots, R_p)$ as a vector R and $(\theta_1, \ldots, \theta_p)$ as a vector θ though R and θ may very well be diagonal matrices with diagonal elements $R_1, \ldots, R_p$ and $\theta_1, \ldots, \theta_p$ respectively.

For fixed θ suppose that $p(r; \theta)$ is the pdf of R with respect to the Lebesgue measure. For testing $H_0 : \theta = \theta_0 = (\theta_1^0, \ldots, \theta_p^0)$ against alternatives

Multivariate Distributions and Invariance

$H_1 : \theta = (\theta_1, \ldots, \theta_p) \neq \theta_0$, suppose that

$$\frac{p(r;\theta)}{p(r;\theta_0)} = 1 + \sum_{i=1}^{p}(\theta_i - \theta_0)[g(\theta, \theta_0) + K(\theta, \theta_0)U(r)] + B(r; \theta, \theta_0) \quad (3.27)$$

where $g(\theta, \theta_0)$ and $K(\theta, \theta_0)$ are bounded for θ in the neighborhood of θ_0; $K(\theta, \theta_0) > 0$ $B(r; \theta, \theta_0) = o(\sum_{i=1}^{p}(\theta_i - \theta_i^0))$; $U(R)$ is bounded and has continuous distribution function for each θ in Ω. If (3.27) is satisfied we say that a test is LBI for testing H_0 against H_1 if its rejection region is given by

$$U(r) \geq C$$

where the constant C depends on the level α of the test.

EXERCISES

1 Let $\{P_\theta, \theta \in \Omega\}$, the family of distributions on $(\mathcal{X}, \mathcal{A})$, be such that each P_θ is absolutely continuous with respect to a σ-finite measure μ; i.e., if $\mu(A) = 0$ for $A \in \mathcal{A}$, then $P_\theta(A) = 0$. Let $p_\theta = \partial P_\theta / \partial \mu$ and define the measure μg^{-1} for $g \in G$, the group of transformations on $\mathcal{X}$, by

$$\mu g^{-1}(A) = \mu(g^{-1}A).$$

Suppose that
(a) μ is absolutely continuous with respect to μg^{-1} for all $g \in G$;
(b) $p_\theta(x)$ is absolutely continuous in θ for all x;
(c) Ω is separable;
(d) the subspaces Ω_{H_0} and Ω_{H_1} are invariant with respect to G. Then show that

$$\sup_{\Omega_{H_1}} p_\theta(x) / \sup_{\Omega_{H_0}} p_\theta(x)$$

is almost invariant with respect to G.

2 Let $X_1, \ldots, X_n$ be a random sample of size n from a normal population with unknown mean μ and variance σ^2. Find the uniformly most powerful invariant test of $H_0 : \sigma^2 < \sigma_0^2$ (specified) against the alternatives $\sigma^2 > \sigma_0^2$ with respect to the group of transformations which transform $X_i \to X_i + c$, $-\infty < c < \infty$, $i = 1, \ldots, n$.

3 Let $X_1, \ldots, X_n$ be a random sample of size n_1 from a normal population with mean μ and variance σ_1^2; and let $Y_1, \ldots, Y_{n_2}$ be a random sample of size n_2 from another normal population with mean ν and variance σ_2^2. Let

$(X_1, \ldots, X_{n_1})$ be independent of $(Y_1, \ldots, Y_{n_2})$. Write

$$\bar{X} = \frac{1}{n_1} \sum_{i=1}^{n_1} X_i,$$

$$S_1^2 = \sum_{i=1}^{n_1} (X_i - \bar{X})^2,$$

$$\bar{Y} = \frac{1}{n_2} \sum_{i=1}^{n_2} Y_i,$$

$$S_2^2 = \sum_{i=1}^{n_2} (Y_i - \bar{Y})^2.$$

The problem of testing $H_0 : \sigma_1^2/\sigma_2^2 \leq \lambda_0$ (specified) against the alternatives $H_1 : \sigma_1^2/\sigma_2^2 > \lambda_0$ remains invariant under the group of transformations

$$\bar{X} \to \bar{X} + c_1, \quad \bar{Y} \to \bar{Y} + c_2, \quad S_1^2 \to S_1^2, \quad S_2^2 \to S_2^2,$$

where $-\infty < c_1, c_2 < \infty$ and also under the group of common scale changes

$$\bar{X} \to a\bar{X}, \quad \bar{Y} \to a\bar{Y}, \quad S_1^2 \to a^2 S_1^2, \quad S_2^2 \to a^2 S_2^2,$$

where $a > 0$. A maximal invariant under these two groups of transformations is

$$F = \frac{S_1^2}{n_1 - 1} \Big/ \frac{S_2^2}{n_2 - 1}.$$

Show that for testing H_0 against H_1 the test which rejects H_0 whenever $F \geq C_\alpha$, where C_α is a constant such that $P(F \geq C_\alpha) = \alpha$ when H_0 is true, is the uniformly most powerful invariant. Is it uniformly most powerful unbiased for testing H_0 against H_1?

4 In exercise 3 assume that $\sigma_1^2 = \sigma_2^2$. Let $S^2 = S_1^2 + S_2^2$.
   (a) The problem of testing $H_0 : \nu - \mu \leq 0$ against the alternatives $H_1 : \nu - \mu > 0$ is invariant under the group of transformations

$$\bar{X} \to \bar{X} + c, \quad \bar{Y} \to \bar{Y} + c, \quad S^2 \to S^2,$$

where $-\infty < c < \infty$, and also under the group of transformations

$$\bar{X} \to a\bar{X}, \quad \bar{Y} \to a\bar{Y}, \quad S^2 \to a^2 S^2,$$

$0 < a < \infty$. Find the uniformly most powerful invariant test with respect to these transformations.

(b) The problem of testing $H_0 : \nu - \mu = 0$ against the alternatives $H_1 : \nu - \mu \neq 0$ is invariant under the group of affine transformations

$$X_i \to aX_i + b, \qquad Y_j = aY_j + b,$$

$a \neq 0, -\infty < b < \infty, i = 1, \ldots, n_1, j = 1, \ldots, n_2$. Find the uniformly most powerful test of H_0 against H_1 with respect to this group of transformations.

5. (*Linear hypotheses*) Let $Y_1, \ldots, Y_n$ be independently distributed normal random variables with a common variance σ^2 and with means

$$E(Y_i) = \begin{cases} \mu_i, & i = 1, \ldots, s(s < n) \\ 0, & i = s+1, \ldots, n \end{cases}$$

and let $\delta^2 = \sum_{i=1}^r \mu_i^2/\sigma^2$. Show that the test which rejects $H_0 : \mu_1 = \cdots = \mu_r = 0, r < s$, whenever

$$W = \frac{\sum_{i=1}^r Y_i^2}{r} \bigg/ \frac{\sum_{i=s+1}^n Y_i^2}{n-s} \geq k,$$

where the constant k is determined so that the probability of rejection is α whenever H_0 is true, is uniformly most powerful among all tests whose power function depends only on δ^2.

6. (*General linear hypotheses*) Let $X_1, \ldots, X_n$ be n independently distributed normal random variables with mean $\xi_i, i = 1, \ldots, n$ and common variance σ^2. Assume that $\xi = (\xi_1, \ldots, \xi_n)$ lies in a linear subspace of Π_Ω of dimension $s < n$. Show that the problem of testing $H_0 : \xi \in \Pi_\omega \subset \Pi_\Omega$ can be reduced to Exercise 5 by means of an orthogonal transformation. Find the test statistic W (of Exercise 5) in terms of $X_1, \ldots, X_n$.

7. (*Analysis of variance, one-way classification*) Let $Y_{ij}, j = 1, \ldots, n_i$, $i = 1, \ldots, k$, be independently distributed normal random variables with means $E(Y_{ij}) = \mu_i$ and common variance σ^2. Let $H_0 : \mu_1 = \cdots = \mu_k$. Identify this as a problem of general linear hypotheses. Find the uniformly most powerful invariant test with respect to a suitable group of transformations.

8. In Example 3.2.3 show that for testing $H_0 : \mu = 0$ against $H_1 : \delta^2 > 0$, student's test is minimax. Is it stringent for H_0 against H_1?

REFERENCES

Anderson, S. A. (1982). Distribution of maximal invariants using quotient measures. *Ann. Statist.* 10:955–961.

Bonder, J. V. (1976). Borel cross-section and maximal invariants. *Ann. Statist.* 4:866–877.

Blackwell, D. (1956). On a class of probability spaces. In: Proc. Berkeley Symp. Math. Statist. Probability. 3rd. Univ. of California Press, Berkeley, California.

Eaton, M. L. (1989). *Group Invariance Applications in Statistics*. Institute of Mathematical Statistics and American Statistical Association: USA.

Ferguson, T. S. (1969). *Mathematical Statistics*. New York: Academic Press.

Ghosh, J. K. (1967). Invariance in Testing and Estimation. Indian Statist. Inst., Calcutta, Publ. No. SM67/2.

Giri, N. (1961). On the Likelihood Ratio Tests of Some Multivariate Problems. Ph.D. thesis, Stanford Univ.

Giri, N. (1964). On the likelihood ratio test of a normal multivariate testing problem. *Ann. Math. Statist.* 35:181–189.

Giri, N. (1965). On the likelihood ratio test of a normal multivariate testing problem II. *Ann. Math. Statist.* 36:1061–1065.

Giri, N. (1968). Locally and asymptotically minimax tests of a multivariate problem. *Ann. Math. Statist.* 39:171–178.

Giri, N. (1993). *Introduction to Probability and Statistics*, 2nd ed., Revised and Expanded. New York: Marcel Dekker.

Giri, N. (1975). *Invariance and Minimax Statistical Tests*. India: The Univ. Press of Canada and Hindusthan Publ. Corp.

Giri, N. (1997). *Group Invariance in Statistical Inference*. Singapore: World Scientific.

Giri, N., Kiefer, J. (1964a). Local and asymptotic minimax properties of multivariate tests. *Ann. Math. Statist.* 35:21–35.

Giri, N., Kiefer, J. (1964b). Minimax character of R^2-test in the simplest case. *Ann. Math. Statist.* 35:1475–1490.

Giri, N., Kiefer, J., Stein, C. (1963). Minimax character of Hotelling's T^2-test in the simplest case. *Ann. Math. Statist.* 34:1524–1535.

Hall, W. J., Wijsman, R. A., Ghosh, J. K. (1965). The relationship between sufficiency and invariance with application in sequential analysis. *Ann. Math. Statist.* 36:575–614.

Halmos, P. R. (1958). *Measure Theory*. Princeton, NJ: Van Nostrand-Reinhold.

Kiefer, J. (1957). Invariance sequential estimation and continuous time processes. *Ann. Math. Statist.* 28:675–699.

Kiefer, J. (1958). On the nonrandomized optimality and randomized nonoptimality of symmetrical designs. *Ann. Math. Statist.* 29:675–699.

Kiefer, J. (1966). Multivariate optimality results. In: Krishnaiah, P. R., ed. *Multivariate Analysis*. Academic Press: New York.

Koehn, U. (1970). Global cross-sections and densities of maximal invariants. *Ann. Math. Statist.* 41:2046–2056.

Kolmogorov, A. N. (1950). *Foundations of the Theory of Probability*. Chelsea: New York.

Lehmann, E. L. (1959). *Testing Statistical Hypotheses*. New York: Wiley.

Lehmann, E. L. (1959a). Optimum invariant tests. *Ann. Math. Statist.* 30:881–884.

Lehmann, E. L., Stein, C. (1953). The admissibility of certain invariant statistical tests involving a translation parameter. *Ann. Math. Statist.* 24:473–479.

Linnik, Ju, V., Pliss, V. A., Salaevskii, O. V. (1966). *Sov. Math. Dokl.* 7:719.

Nachbin, L. (1965). *The Haar Integral*. Princeton, NJ: Van Nostrand-Reinhold.

Pitman, E. J. G. (1939). Tests of hypotheses concerning location and scale parameters. *Biometrika* 31:200–215.

Salaevskii, O. V. (1968). Minimax character of Hotelling's T^2-test. *Sov. Math. Dokl.* 9:733–735.

Schwartz, R. (1967). Local minimax tests. *Ann. Math. Statist.* 38:340–360.

Stein, C. (1956). Some Problems in Multivariate Analysis, Part I. Stanford University Technical Report, No. 6, Stanford, Calif.

Wijsman, R. A. (1967). Cross-sections of orbits and their application to densities of maximal invariants. In: Proc. Fifth Berk Symp. Math. Statist. Prob. Vol. 1, University of California Press, pp. 389–400.

Wijsman, R. A. (1967a). General proof of termination with probability one of invariant sequential probability ratio tests based on multivariate observations. *Ann. Math. Statist.* 38:8–24.

Wijsman, R. A. (1985). Proper action in steps, with application to density ratios of maximal invariants. *Ann. Statist.* 13:395–402.

Wijsman, R. A. (1990). *Invariance Measures on Groups and Their Use in Statistics*. USA: Institute of Mathematical Statistics.

4
Properties of Multivariate Distributions

4.0. INTRODUCTION

We will first define the multivariate normal distribution in the classical way by means of its probability density function and study some of its basic properties. This definition does not include the cases in which the covariance matrix is singular and also the cases in which the dimension of the random vector is countable or uncountable. We will then define multivariate normal distribution in the general way to include such cases. A number of characterizations of the multivariate normal distribution will also be given in order to enable the reader to study this distribution in Hilbert and Banach spaces.

The complex multivariate normal distribution plays an important role in describing the statistical variability of estimators and of functions of the elements of a multiple stationary Gaussian time series. This distribution is also useful in analyzing linear models with complex covariance structures which arise when they are invariant under cyclic groups. We treat it here along with some basic properties.

Multivariate normal distribution has many advantages from the theoretical viewpoints. Most elegant statistical theories are centered around this distribution. However, in practice, it is hard to ascertain if a sample of observation is drawn from a multivariate normal population or not. Sometimes, it is advantageous to consider a family of distributions having certain similar properties. The family of elliptically symmetric distributions include, among others, the multivariate normal, the compound multivariate normal, the multivariate t-distribution and the multivariate Cauchy distribution. For all probability density functions in this family the shapes of contours of equal densities are elliptical. We shall treat it here along with some of its basic properties.

Another deviation from the multivariate normal family is the family of multivariate exponential power distributions where the multivariate normal distribution is enlarged through the introduction of an additional parameter θ and the deviation from the multivariate normal family is described in terms of θ. This family (problem 25) includes the multivariate normal family ($\theta = 1$), multivariate double exponential family ($\theta = \frac{1}{2}$) and the asymptotically uniform distributions ($\theta \to \infty$). The univariate case ($p = 1$) is often treated in Bayesian inference (Box and Tiao (1973)).

4.1. MULTIVARIATE NORMAL DISTRIBUTION (CLASSICAL APPROACH)

Definition 4.1.1. *Multivariate normal distribution.* A random vector $X = (X_1, \ldots, X_p)'$ taking values $x = (x_1, \ldots, x_p)'$ in E^p (Euclidean space of dimension p) is said to have a p-variate normal distribution if its probability density function can be written as

$$f_X(x) = \frac{1}{(2\pi)^{p/2}|\Sigma|^{\frac{1}{2}}} \exp\left\{-\frac{1}{2}(x-\mu)'\Sigma^{-1}(x-\mu)\right\}, \tag{4.1}$$

where $\mu = (\mu_1, \ldots, \mu_p)' \in E^p$ and Σ is a $p \times p$ symmetric positive definite matrix.

In what follows a random vector will always imply a real vector unless it is specifically stated otherwise.

We show now that $f_X(x)$ is an honest probability density function of X. Since Σ is positive definite, $(x-\mu)'\Sigma^{-1}(x-\mu) \geq 0$ for all $x \in E^p$ and $\det(\Sigma) > 0$. Hence $f_X(x) \geq 0$ for all $x \in E^p$. Furthermore, since Σ is a $p \times p$ positive definite matrix there exists a $p \times p$ nonsingular matrix C such that $\Sigma = CC'$. Let $y = C^{-1}x$. The Jacobian of the transformation $x \to y = C^{-1}x$ is $\det C$. Writing $v = (v_1, \ldots, v_p)' = C^{-1}\mu$, we obtain

$$\int_{E^p} \frac{1}{(2\pi)^{p/2}(\det \Sigma)^{\frac{1}{2}}} \exp\left\{-\frac{1}{2}(x-\mu)'\Sigma^{-1}(x-\mu)\right\} dx$$

$$= \int_{E^p} \frac{1}{(2\pi)^{p/2}} \exp\left\{-\frac{1}{2}(y-v)'(y-v)\right\} dy$$

$$= \prod_{i=1}^{p} \int_{-\infty}^{\infty} \frac{1}{(2\pi)^{1/2}} \exp\left\{-\frac{1}{2}(y_i - v_i)^2\right\} dy_i$$

$$= 1.$$

Properties of Multivariate Distributions

Theorem 4.1.1. *If the random vector X has a multivariate normal distribution with probability density function $f_X(x)$, then the parameters μ and Σ are given by $E(X) = \mu$, $E(X - \mu)(X - \mu)' = \Sigma$.*

Proof. The random vector $Y = C^{-1}X$, with $\Sigma = CC'$, has probability density function

$$f_Y(y) = \prod_{i=1}^{p} \frac{1}{(2\pi)^{1/2}} \exp\left\{-\frac{1}{2}(y_i - v_i)^2\right\}$$

Thus $E(Y) = (E(Y_1), \ldots, E(Y_p))' = v = C^{-1}\mu$,

$$\text{cov}(Y) = E(Y - v)(Y - v)' = I. \tag{4.2}$$

From this

$$E(C^{-1}X) = C^{-1}E(X) = C^{-1}\mu$$
$$E(C^{-1}X - C^{-1}\mu)(C^{-1}X - C^{-1}\mu)' = C^{-1}E(X - \mu)(X - \mu)'C^{-1'} = I.$$

Hence

$$E(X) = \mu, \quad E(X - \mu)(X - \mu)' = \Sigma.$$

Q.E.D.

We will frequently write Σ as

$$\Sigma = \begin{pmatrix} \sigma_1^2 & \sigma_{12} & \cdots & \sigma_{1p} \\ \sigma_{21} & \sigma_2^2 & \cdots & \sigma_{2p} \\ \vdots & \vdots & & \\ \sigma_{p1} & \sigma_{p2} & \cdots & \sigma_p^2 \end{pmatrix} \quad \text{with } \sigma_{ij} = \sigma_{ji}$$

The fact that Σ is symmetric follows from the identity $E[(X - \mu)(X - \mu)']' = E(X - \mu)(X - \mu)'$.

The term covariance matrix is used here instead of the matrix of variances and covariances of the components.

We will now prove some basic characteristic properties of multivariate normal distributions in the following theorems.

Theorem 4.1.2. *If the covariance matrix of a normal random vector $X = (X_1, \ldots, X_p)'$ is a diagonal matrix, then the components of X are independently distributed normal variables.*

Proof. Let

$$\Sigma = \begin{pmatrix} \sigma_1^2 & 0 & \cdots & 0 \\ 0 & \sigma_2^2 & \cdots & 0 \\ \vdots & \vdots & & \vdots \\ 0 & 0 & \cdots & \sigma_p^2 \end{pmatrix}$$

Then

$$(x - \mu)'\Sigma^{-1}(x - \mu) = \sum_{i=1}^{p}\left(\frac{x_i - \mu_i}{\sigma_i}\right)^2, \quad \det \Sigma = \prod_{i=1}^{p}\sigma_i^2$$

Hence

$$f_X(x) = \prod_{i=1}^{p}\frac{1}{(2\pi)^{1/2}\sigma_i}\exp\left\{-\frac{1}{2}\left(\frac{x_i - \mu_i}{\sigma_i}\right)^2\right\},$$

which implies that the components are independently distributed normal random variables with means μ_i and variance σ_i^2. Q.E.D.

It may be remarked that the converse of this theorem holds for any random vector X. The theorem does not hold if X is not a normal vector. The following theorem is a generalization of the above theorem to two subvectors.

Theorem 4.1.3. *Let* $X = (X'_{(1)}, X'_{(2)})'$, $X_{(1)} = (X_1, \ldots, X_q)'$, $X_{(2)} = (X_{q+1}, \ldots, X_p)'$, *let* μ *be similarly partitioned as* $\mu = (\mu'_{(1)}, \mu'_{(2)})'$, *and let* Σ *be partitioned as*

$$\Sigma = \begin{pmatrix} \Sigma_{11} & \Sigma_{12} \\ \Sigma_{21} & \Sigma_{22} \end{pmatrix}$$

where Σ_{11} *is the upper left-hand corner submatrix of* Σ *of dimension* $q \times q$. *If* X *has normal distribution with means* μ *and positive definite covariance matrix* Σ *and* $\Sigma_{12} = 0$, *then* $X_{(1)}, X_{(2)}$ *are independently normally distributed with means* $\mu_{(1)}, \mu_{(2)}$ *and covariance matrices* Σ_{11}, Σ_{22} *respectively.*

Proof. Under the assumption that $\Sigma_{12} = 0$, we obtain

$$(x - \mu)'\Sigma^{-1}(x - \mu) = (x_{(1)} - \mu_{(1)})'\Sigma_{11}^{-1}(x_{(1)} - \mu_{(1)}) + (x_{(2)} - \mu_{(2)})'$$
$$\times \Sigma_{22}^{-1}(x_{(2)} - \mu_{(2)}), \quad \det \Sigma = (\det \Sigma_{11})(\det \Sigma_{22}).$$

Properties of Multivariate Distributions

Hence

$$f_X(x) = \frac{1}{(2\pi)^{q/2}(\det \Sigma_{11})^{1/2}} \exp\left\{-\frac{1}{2}(x_{(1)} - \mu_{(1)})' \Sigma_{11}^{-1}(x_{(1)} - \mu_{(1)})\right\}$$

$$\times \frac{1}{(2\pi)^{(p-q)/2}(\det \Sigma_{22})^{1/2}} \exp\left\{-\frac{1}{2}(x_{(2)} - \mu_{(2)})' \Sigma_{22}^{-1}(x_{(2)} - \mu_{(2)})\right\}$$

and the result follows. Q.E.D.

This theorem can be easily extended to the case where X is partitioned into more than two subvectors, to get the result that any two of these subvectors are independent if and only if the covariance between them is zero. An important reproductive property of the multivariate normal distribution is given in the following theorem.

Theorem 4.1.4. *Let $X = (X_1, \ldots, X_p)'$ with values x in E^p be normally distributed with mean μ and positive definite covariance matrix Σ. Then the random vector $Y = CX$ with values $y = Cx$ in E^p where C is a nonsingular matrix of dimension $p \times p$ has p-variate normal distribution with mean $C\mu$ and covariance matrix $C\Sigma C'$.*

Proof. The Jacobian of the transformation $x \to y = Cx$ is $(\det C)^{-1}$. Hence the probability density function of Y is given by

$$f_Y(y) = f_x(C^{-1}y)(\det C)^{-1} = \frac{1}{(2\pi)^{p/2}(\det C\Sigma C')^{1/2}}$$

$$\times \exp\left\{-\frac{1}{2}(y - C\mu)'(C\Sigma C')^{-1}(y - C\mu)\right\}$$

Thus Y has p-variate normal distribution with mean $C\mu$ and positive definite covariance matrix $C\Sigma C'$. Q.E.D.

Theorem 4.1.5. *Let $X = (X'_{(1)}, X'_{(2)})'$ be distributed as $N_p(\mu, \Sigma)$ where $X_{(1)}, X_{(2)}$ are as defined in Theorem 4.1.3. Then*

(a) *$X_{(1)}, X_{(2)} - \Sigma_{21}\Sigma_{11}^{-1}X_{(1)}$ are independently normally distributed with means $\mu_{(1)}, \mu_{(2)} - \Sigma_{21}\Sigma_{11}^{-1}\mu_{(1)}$ and positive definite covariance matrices $\Sigma_{11}, \Sigma_{22.1} = \Sigma_{22} - \Sigma_{21}\Sigma_{11}^{-1}\Sigma_{12}$ respectively.*
(b) *The marginal distribution of $X_{(1)}$ is q-variate normal with mean $\mu_{(1)}$ and covariance matrix Σ_{11}.*
(c) *The condition distribution of $X_{(2)}$ given $X_{(1)} = x_{(1)}$ is normal with mean $\mu_{(2)} + \Sigma_{21}\Sigma_{11}^{-1}(x_{(1)} - \mu_{(1)})$ and covariance matrix $\Sigma_{22.1}$.*

Proof.

(a) Let

$$Y = \begin{pmatrix} Y_{(1)} \\ Y_{(2)} \end{pmatrix} = \begin{pmatrix} X_{(1)} \\ X_{(2)} - \Sigma_{21}\Sigma_{11}^{-1}X_{(1)} \end{pmatrix}.$$

Then

$$Y = \begin{pmatrix} I_1 & 0 \\ -\Sigma_{21}\Sigma_{11}^{-1} & I_2 \end{pmatrix} \begin{pmatrix} X_{(1)} \\ X_{(2)} \end{pmatrix} = CX$$

where I_1 and I_2 are identity matrices of dimensions $q \times q$ and $(p-q) \times (p-q)$ respectively and

$$C = \begin{pmatrix} I_1 & 0 \\ -\Sigma_{21}\Sigma_{11}^{-1} & I_2 \end{pmatrix}.$$

Obviously C is a nonsingular matrix. By Theorem 4.1.4 Y has p-variate normal distribution with mean

$$C\mu = \begin{pmatrix} \mu_{(1)} \\ \mu_{(2)} - \Sigma_{21}\Sigma_{11}^{-1}\mu_{(1)} \end{pmatrix}$$

and covariance matrix

$$C\Sigma C' = \begin{pmatrix} \Sigma_{11} & 0 \\ 0 & \Sigma_{22.1} \end{pmatrix}.$$

Hence by Theorem 4.1.3 we get the result.
(b) It follows trivially from part (a).
(c) The Jacobian of the inverse transformation $Y = CX$ is unity. From (a) the probability density function of X can be written as

$$f_X(x) = \frac{\exp\{-\frac{1}{2}(x_{(1)} - \mu_{(1)})'\Sigma_{11}^{-1}(x_{(1)} - \mu_{(1)})\}}{(2\pi)^{q/2}(\det \Sigma_{11})^{1/2}}$$

$$\times \frac{1}{(2\pi)^{(p-q)/2}(\det \Sigma_{22.1})^{1/2}} \qquad (4.3)$$

$$\times \exp\left\{-\frac{1}{2}(x_{(2)} - \mu_{(2)} - \Sigma_{21}\Sigma_{11}^{-1}(x_{(1)} - \mu_{(1)}))'\right.$$

$$\left.\Sigma_{22.1}^{-1}(x_{(2)} - \mu_{(2)} - \Sigma_{21}\Sigma_{11}^{-1}(x_{(1)} - \mu_{(1)}))\right\}$$

Properties of Multivariate Distributions

Hence the results. Q.E.D.

Thus it is interesting to note that if X has p-variate normal distribution, the marginal distribution of any subvector of X is also a multivariate normal and the conditional distribution of any subvector given the values of the remaining subvector is also a multivariate normal.

Example 4.1.1. Bivariate normal. Let

$$\Sigma = \begin{pmatrix} \sigma_1^2 & \rho\sigma_1\sigma_2 \\ \rho\sigma_1\sigma_2 & \sigma_2^2 \end{pmatrix}$$

with $\sigma_1^2 > 0$, $\sigma_2^2 > 0$, $-1 < \rho < 1$. Since $\det \Sigma = \sigma_1^2 \sigma_2^2 (1 - \rho^2) > 0$, Σ^{-1} exists and is given by

$$\Sigma^{-1} = \begin{pmatrix} \dfrac{1}{\sigma_1^2} & \dfrac{-\rho}{\sigma_1\sigma_2} \\ \dfrac{-\rho}{\sigma_1\sigma_2} & \dfrac{1}{\sigma_2^2} \end{pmatrix} \dfrac{1}{1-\rho^2}.$$

Furthermore, for $x = (x_1, x_2)' \neq 0$

$$x'\Sigma x = (\sigma_1 x_1 + \rho\sigma_2 x_2)^2 + (1-\rho^2)\sigma_2^2 x_2^2 > 0.$$

Hence Σ is positive definite. With $\mu = (\mu_1, \mu_2)'$,

$$(x-\mu)'\Sigma^{-1}(x-\mu) = \frac{1}{1-\rho^2}\left[\left(\frac{x_1-\mu_1}{\sigma_1}\right)^2 + \left(\frac{x_2-\mu_2}{\sigma_2}\right)^2 - 2\rho\left(\frac{x_1-\mu_1}{\sigma_1}\right)\left(\frac{x_2-\mu_2}{\sigma_2}\right)\right].$$

The probability density function of a bivariate normal random variable with values in E^2 is

$$\frac{1}{2\pi\sigma_1\sigma_2(1-\rho^2)^{1/2}} \times \exp\left\{-\frac{1}{2(1-\rho^2)}\left[\left(\frac{x_1-\mu_1}{\sigma_1}\right)^2 + \left(\frac{x_2-\mu_2}{\sigma_2}\right)^2 - 2\rho\left(\frac{x_1-\mu_1}{\sigma_1}\right)\left(\frac{x_2-\mu_2}{\sigma_2}\right)\right]\right\}.$$

The coefficient of correlation between X_1 and X_2 is

$$\frac{\mathrm{cov}(X_1, X_2)}{(\mathrm{var}(X_1)\mathrm{var}(X_2))^{1/2}} = \rho.$$

If $\rho = 0$, X_1, X_2 are independently normally distributed with means μ_1, μ_2 and variances σ_1^2, σ_2^2, respectively. If $\rho > 0$, then X_1, X_2 are positively related; and if $\rho < 0$, then X_1, X_2 are negatively related.

The marginal distributions of X_1 and of X_2 are both normal with means μ_1 and μ_2, and with variances σ_1^2 and σ_2^2, respectively. The conditional probability density of X_2 given $X_1 = x_1$ is a normal with

$$E(X_2|X_1 = x_1) = \mu_2 + \rho\left(\frac{\sigma_2}{\sigma_1}\right)(x_1 - \mu_1), \text{ var}(X_2|X_1 = x_1) = \sigma_2^2(1 - \rho^2).$$

Figures 4.1 and 4.2 give the graphical presentation of the bivariate normal distribution and its contours.

We now give an example to show that the normality of marginal distributions does not necessarily imply the multinormality of the joint distribution though the converse is always true.

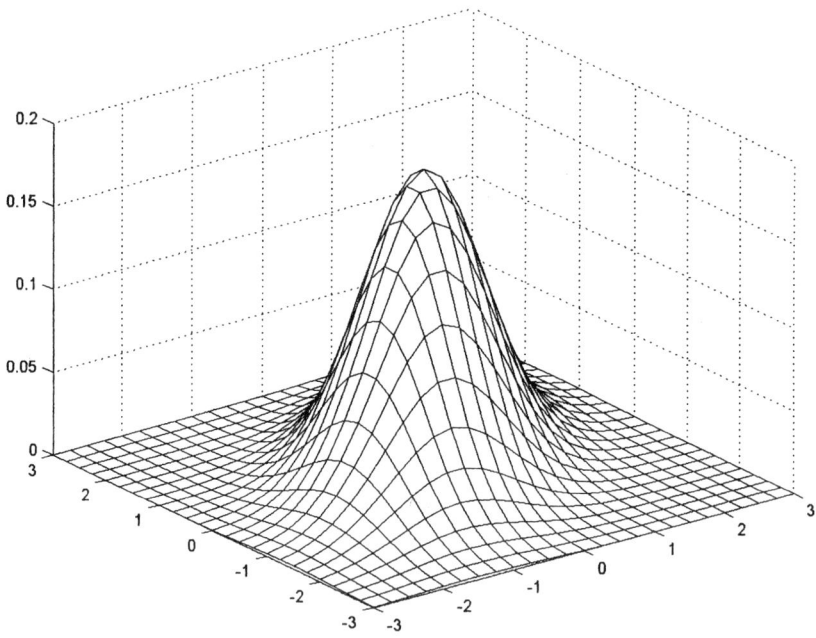

Figure 4.1. Bivariate normal with mean 0 and $\Sigma = \begin{pmatrix} 1 & \frac{1}{2} \\ \frac{1}{2} & 1 \end{pmatrix}$.

Properties of Multivariate Distributions

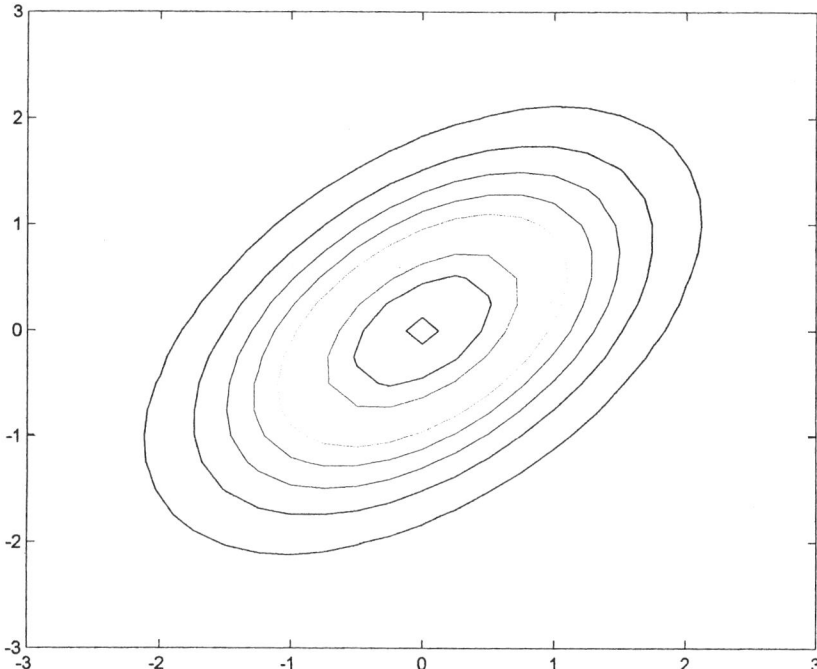

Figure 4.2. Contours of bivariate normal in Figure 4.1.

Example 4.1.2. Let

$$f(x_1, x_2 | \rho_1) = \frac{1}{2\pi(1-\rho_1^2)^{1/2}} \exp\left\{-\frac{1}{2(1-\rho_1^2)}(x_1^2 + x_2^2 - 2\rho_1 x_1 x_2)\right\}$$

$$f(x_1, x_2 | \rho_2) = \frac{1}{2\pi(1-\rho_2^2)^{1/2}} \exp\left\{-\frac{1}{2(1-\rho_2^2)}(x_1^2 + x_2^2 - 2\rho_2 x_1 x_2)\right\}$$

be two bivariate normal probability functions with 0 means, unit variances and different correlation coefficients. Let

$$f(x_1, x_2) = \frac{1}{2} f(x_1, x_2 | \rho_1) + \frac{1}{2} f(x_1, x_2 | \rho_2).$$

Obviously $f(x_1, x_2)$ is not a bivariate normal density function though the marginals of X_1 and of X_2 are both normals.

Theorem 4.1.6. *Let $X = (X_1, \ldots, X_p)'$ be normally distributed with mean μ and positive definite covariance matrix Σ. The characteristic function of the* *random vector X is given by*

$$\phi_X(t) = E(e^{it'X}) = \exp\left\{it'\mu - \frac{1}{2}t'\Sigma t\right\} \tag{4.4}$$

where

$$t = (t_1, \ldots, t_p)' \in E^p, i = (-1)^{1/2}.$$

Proof. Since Σ is positive definite, there exists a nonsingular matrix C such that $\Sigma = CC'$. Write $y = C^{-1}x$, $\alpha = (\alpha_1, \ldots, \alpha_p)' = C't$, $\vartheta = C^{-1}\mu = (\vartheta_1, \ldots, \vartheta_p)'$. Then

$$E(e^{it'X}) = \int_{E^p} (2\pi)^{-p/2} \exp\left\{i\alpha'y - \frac{1}{2}(y - \vartheta)'(y - \vartheta)\right\} dy$$

$$= \prod_{j=1}^{p} \int (2\pi)^{1/2} \exp\left\{i\alpha_j y_j - \frac{1}{2}(y_j - \vartheta_j)^2\right\} dy_j$$

$$= \prod_{j=1}^{p} \exp\left\{i\alpha_j \vartheta_j - \frac{1}{2}\alpha_j^2\right\} = \exp\left\{i\alpha'\vartheta - \frac{1}{2}\alpha'\alpha\right\}$$

$$= \exp\left\{it'\mu - \frac{1}{2}t'\Sigma t\right\}$$

as the characteristic function of a univariate normal random variable is $\exp\{it\mu - \frac{1}{2}t^2\sigma^2\}$. Q.E.D.

Since the characteristic function determines uniquely the distribution function it follows from (4.4) that the p-variate normal distribution is completely specified by its mean vector μ and covariance matrix Σ. We shall therefore use the notation $N_p(\mu, \Sigma)$ for the density function of a p-variate normal random vector involving parameter μ, Σ whenever Σ is positive definite.

In Theorem 4.1.4 we have shown that if C is a nonsingular matrix then CX is a p-variate normal whenever X is a p-variate normal. The following theorem will assert that this restriction on C is not essential.

Theorem 4.1.7. *Let $X = (X_1, \ldots, X_p)'$ be distributed as $N_p(\mu, \Sigma)$ and let $Y = AX$ where A is a matrix of dimension $q \times p$ of rank $q(q < p)$. Then Y is distributed as $N_q(A\mu, A\Sigma A')$.*

Properties of Multivariate Distributions

Proof. Let C be a nonsingular matrix of dimension $p \times p$ such that

$$C = \begin{pmatrix} A \\ B \end{pmatrix},$$

where B is a matrix of dimension $(p-q) \times p$ of rank $p-q$, and let $Z = BX$. Then by Theorem 4.1.4. $\begin{pmatrix} Y \\ Z \end{pmatrix}$ has p-variate normal distribution with mean

$$C\mu = \begin{pmatrix} A\mu \\ B\mu \end{pmatrix}$$

and covariance matrix

$$C \Sigma C' = \begin{pmatrix} A\Sigma A' & A\Sigma B' \\ B\Sigma A' & B\Sigma B' \end{pmatrix}.$$

By Theorem 4.1.5(b) we get the result. Q.E.D.

This theorem tells us that if X is distributed as $N_p(\mu, \Sigma)$, then every linear combination of X has a univariate normal distribution. We will now show that if, for a random vector X with mean μ and covariance matrix Σ, every linear combination of the components of X having a univariate normal distribution, then X has a multivariate normal distribution.

Theorem 4.1.8. *Let $X = (X_1, \ldots, X_p)'$. If every linear combination of the components of X is distributed as a univariate normal, then X is distributed as a p-variate normal.*

Proof. For any nonnull fixed real p-vector L, let $L'X$ have a univariate normal with mean $L'\mu$ and variance $L'\Sigma L$. Then for any real t the characteristic function of $L'X$ is

$$\phi(t, L) = E(e^{itL'X}) = \exp\left\{ itL'\mu - \frac{1}{2}t^2 L'\Sigma L \right\}.$$

Hence

$$\phi(1, L) = E(e^{iL'X}) = \exp\left\{ iL'\mu - \frac{1}{2}L'\Sigma L \right\},$$

which as a function of L is the characteristic function of X. By the inversion theorem of the characteristic function (see Giri (1993), or Giri (1974)) the probability density function of X is $N_p(\mu, \Sigma)$. Q.E.D.

Motivated by Theorem 4.1.7 and Theorem 4.1.8 we now give a more general definition of the multivariate normal distribution.

Definition 4.1.2. *Multivariate normal distribution.* A p-variate random vector X with values in E^p is said to have a multivariate normal distribution if and only if every linear combination of the components of X has a univariate normal distribution.

When Σ is nonsingular, this definition is equivalent to that of the multivariate normal distribution given in 4.1.1. If X has a multivariate normal distribution according to Definition 4.1.2, then each component X_i of X is distributed as univariate normal so that $-\infty < E(X_i) < \infty$, $\text{var}(X_i) < \infty$, and hence $\text{cov}(X_i, X_i) = \sigma_i^2$, $\text{cov}(X_i, X_j) = \sigma_{ij}$. Then $E(X), \text{cov}(X)$ exist and we denote them by μ, Σ respectively.

In Definition 4.1.2 it is not necessary that Σ be positive definite; it can be semipositive definite also.

Definition 4.1.2 can be extended to the definition of a normal probability measure on Hilbert and Banach spaces by demanding that the induced distribution of every linear functional be univariate normal. The reader is referred to Fréchet (1951) for further details. One other big advantage of Definition 4.1.2 over Definition 4.1.1 is that certain results of univariate normal distribution can be immediately generalized to the multivariate case. Readers may find it instructive to prove Theorems 4.1.1–4.1.8 by using Definition 4.1.2. As an illustration let us prove Theorem 4.1.3 and then Theorem 4.1.7.

Proof of Theorem 4.1.3. For any nonzero real p-vector $L = (l_1, \ldots, l_p)'$ the characteristic function of $L'X$ is

$$\phi(t, L) = \exp\left\{itL'\mu - \frac{1}{2}t^2 L'\Sigma L\right\}. \tag{4.5}$$

Write $L = (L'_{(1)}, L'_{(2)})'$ where $L_{(1)} = (l_1, \ldots, l_q)'$.

Then

$$L'\mu = L'_{(1)}\mu_{(1)} + L'_{(2)}\mu_{(2)}, \quad L'\Sigma L = L'_{(1)}\Sigma_{11}L_{(1)} + L'_{(2)}\Sigma_{22}L_{(2)}$$

Hence

$$\phi(t, L) = \exp\left\{itL'_{(1)}\mu_{(1)} - \frac{1}{2}t^2 L'_{(1)}\Sigma_{11}L_{(1)}\right\}$$

$$\times \exp\left\{itL'_{(2)}\mu_{(2)} - \frac{1}{2}t^2 L'_{(2)}\Sigma_{22}L_{(2)}\right\}$$

Properties of Multivariate Distributions

In other words the characteristic function of X is the product of the characteristic functions of $X_{(1)}$ and $X_{(2)}$ and each one is the characteristic function of a multivariate normal distribution. Hence Theorem 4.1.3 is proved.

Proof of Theorem 4.1.7. Let $Y = AX$. For any fixed nonnull vector L,

$$L'Y = (L'A)X.$$

By Definition 4.1.2 $L'AX$ has univariate normal distribution with mean $L'A\mu$ and variance $L'A\Sigma A'L$. Since L is arbitrary, this implies that Y has q-variate normal distribution with mean $A\mu$ and covariance matrix $A\Sigma A'$. Q.E.D.

Using Definition 4.1.2 we need to establish the existence of the probability density function of the multivariate normal distribution. Let us now examine the following question: Does Definition 4.1.2 always guarantee the existence of the probability density function? If not, under what conditions can we determine explicitly the probability density function?

Evidently Definition 4.1.2 does not restrict the covariance matrix to be positive definite. If Σ is a nonnegative definite of rank q, then for any real nonnull vector L, $L'\Sigma L$ can be written as

$$L'\Sigma L = (\alpha'_1 L)^2 + \cdots + (\alpha'_q L)^2 \tag{4.6}$$

where $\alpha_i = (\alpha_{i1}, \ldots, \alpha_{ip})'$ $i = 1, \ldots, q$ are linearly independent vectors. Hence the characteristic function of X can be written as

$$\exp\left\{iL'\mu - \frac{1}{2}\sum_{j=1}^{q}(\alpha'_j L)^2\right\}. \tag{4.7}$$

Now $\exp\{iL'\mu\}$ is the characteristic function of a p-dimensional random variable Z_0 which assumes value μ with probability one and $\exp\{-\frac{1}{2}\sum_{j=1}^{q}(\alpha'_j L)^2\}$ is the characteristic function of a p-dimensional random variable

$$Z_i = (\alpha_{i1}U_i, \ldots, \alpha_{ip}U_i)'$$

where $U_1, \ldots, U_q$ are independently, identically distributed (real) random variables with mean zero and variance unity.

Theorem 4.1.9. *The random vector* $X = (X_1, \ldots, X_p)'$ *has p-variate normal distribution with mean* μ *and with covariance matrix* Σ *of rank* $q(q \leq p)$ *if and only if*

$$X = \mu + \alpha U, \ \alpha\alpha' = \Sigma,$$

where α *is a* $p \times q$ *matrix of rank* q *and* $U = (U_1, \ldots, U_q)'$ *has q-variate normal distribution with mean 0 and covariance matrix I (identity matrix).*

Proof. Let $X = \mu + \alpha U$, $\alpha \alpha' = \Sigma$, and U be normally distributed with mean 0 and covariance matrix I. For any nonnull fixed real p-vector L,

$$L'X = L'\mu + (L'\alpha)U.$$

But $(L'\alpha)U$ has univariate normal distribution with mean zero and variance $L'\alpha\alpha'L$. Hence $L'X$ has univariate normal distribution with mean $L'\mu$ and variance $L'\alpha\alpha'L$. Since L is arbitrary, by Definition 4.1.2 X has p-variate normal distribution with mean μ and covariance matrix $\Sigma = \alpha\alpha'$ of rank q.

Conversely, if the rank of Σ is q and X has a p-variate normal distribution with mean μ and covariance matrix Σ, then from (4.7) we can write

$$X = Z_0 + Z_1 + \cdots + Z_q = \mu + \alpha U,$$

satisfying the conditions of the theorem. Q.E.D.

4.1.1. Some Characterizations of the Multivariate Normal Distribution

We give here only two characterizations of the multivariate normal distribution which are useful for our purpose. For other characterizations we refer to the book by Kagan et al. (1972).

Before we begin to discuss characterization results we need to state the following results due to Cramer (1937) regarding univariate random variables.

If the sum of two independent random variables X, Y is normally distributed, then each one is normally distributed. For a proof of this the reader is referred to Cramer (1937). The following characterizations of the multivariate normal distribution are due to Basu (1955).

Theorem 4.1.10. *If X, Y are two independent p-vectors and if $X + Y$ has a p-variate normal distribution, then both X, Y have p-variate normal distribution.*

Proof. Since $X + Y$ has a p-variate normal distribution, for any nonnull p-vector L, $L'(X + Y) = L'X + L'Y$ has univariate normal distribution. Since $L'X$ and $L'Y$ are independent, by Cramer's result, $L'X$, $L'Y$ are both univariate normal random variables. This, by Definition 4.1.2, implies that both X, Y have p-variate normal distribution. Q.E.D.

Properties of Multivariate Distributions

Theorem 4.1.11. *Let $X_1, \ldots, X_n$ be a set of mutually independent p-vectors and let*

$$X = \sum_{i=1}^{n} a_i X_i, \; Y = \sum_{i=1}^{n} b_i X_i$$

where $a_1, \ldots, a_n; b_1, \ldots, b_n$ are two sets of real constants.

(a) If $X_1, \ldots, X_n$ are identically normally distributed p-vectors and if $\sum_{i=1}^{n} a_i b_i = 0$, then X and Y are independent.
(b) If X and Y are independently distributed, then each X_i for which $a_i b_i \neq 0$ has p-variate normal distribution.

Note: Part (b) of this theorem is a generalization of the Darmois-Skitovic theorem which states that if $X_1, \ldots, X_n$ are independently distributed random variables, then the independence of $\sum_{i=1}^{n} a_i X_i$, $\sum_{i=1}^{n} b_i X_i$, implies that each X_i is normally distributed provided $a_i b_i \neq 0$ (See Darmois (1953), Skitovic (1954), or Basu (1951)).

Proof.

(a) For any nonnull p-vector L

$$L'X = a_1(L'X_1) + \cdots + a_n(L'X_n).$$

If $X_1, \ldots X_n$ are independent and identically distributed normal random vectors, then $L'X_1, \ldots, L'X_n$ are independently and identically distributed normal random variables and hence $L'X$ has univariate normal distribution for all L. This implies that X has a p-variate normal distribution. Similarly Y has a p-variate normal distribution. Furthermore, the joint distribution of X, Y is a 2p-variate normal. Now

$$\text{cov}(X, Y) = \Sigma_i a_i b_i \, \text{cov}(X_i) = \Sigma \cdot 0 = 0.$$

Thus X, Y are independent.

(b) For any nonnull real p-vector L

$$L'X = \sum_{i=1}^{n} a_i(L'X_i), \; L'Y = \sum_{i=1}^{n} b_i(L'X_i).$$

Since $L'X_i$ are independent random variables, independence of $L'X, L'Y$ and $a_i b_i \neq 0$ implies $L'X_i$ has a univariate normal distribution. Since L is arbitrary, X_i has a p-variate normal distribution.

Q.E.D.

4.2. COMPLEX MULTIVARIATE NORMAL DISTRIBUTION

A complex random variable Z with values in $\mathbb{C}$ (field of complex numbers) is written as $Z = X + iY$ where $i = \sqrt{-1}$, X, Y are real random variables. The expected value of Z is defined by

$$E(Z) = E(X) + iE(Y), \qquad (4.8)$$

assuming both $E(X)$ and $E(Y)$ exist. The variance of Z is defined by

$$\begin{aligned}\operatorname{var}(Z) &= E(Z - E(Z))(Z - E(Z))^*, \\ &= \operatorname{var}(X) + \operatorname{var}(Y)\end{aligned} \qquad (4.9)$$

where $(Z - E(Z))^*$ denote the adjoint of $(Z - E(Z))$, i.e. the conjugate and transpose of $(Z - E(Z))$.

Note that for 1-dimensional variables the transpose is superfluous. It follows that for $a, b \in \mathbb{C}$

$$\begin{aligned}\operatorname{var}(aZ + b) &= E((a(Z - E(Z))(a(Z - E(Z))^*) \\ &= aa^* \operatorname{var}(Z).\end{aligned}$$

The covariance of any two complex random variables Z_1, Z_2 is defined by

$$\operatorname{cov}(Z_1, Z_2) = E(Z_1 - E(Z_1))(Z_2 - E(Z_2))^*. \qquad (4.10)$$

Theorem 4.2.1. *Let $Z_1, \ldots, Z_n$ be a sequence of n complex random variables. Then*

(a) $\operatorname{cov}(Z_1, Z_2) = (\operatorname{cov}(Z_2, Z_1))^*$.
(b) *For $a_1, \ldots, a_n \in \mathbb{C}, b_1, \ldots, b_n \in \mathbb{C}$,*

$$\operatorname{cov}\left(\sum_{j=1}^n a_j Z_j, \sum_{j=1}^n b_j Z_j\right)$$

$$= \sum_{j=1}^n a_j \bar{b}_j \operatorname{var}(Z_j) + 2 \sum_{j<k} a_j \bar{b}_k \operatorname{cov}(Z_j, Z_k).$$

Properties of Multivariate Distributions

Proof.

(a) Let $Z_j = X_j + iY_j, j = 1, \ldots, n,$ where $X_1, \ldots, X_n; Y_1, \ldots, Y_n$ are real random variables.

$$\text{cov}(Z_1, Z_2) = \text{cov}(X_1, X_2) + \text{cov}(Y_1, Y_2)$$
$$+ i(\text{cov}(Y_1, X_2) - \text{cov}(X_1, Y_2)),$$
$$\text{cov}(Z_2, Z_1) = \text{cov}(X_1, X_2) + \text{cov}(Y_1, Y_2)$$
$$- i(\text{cov}(Y_1, X_2) - \text{cov}(X_1, Y_2)).$$

Hence the result.

(b)
$$\text{cov}\left(\sum_j a_j Z_j, \sum_j b_j Z_j\right) = E\left(\left(\sum_j a_j(Z_j - E(Z_j))\right)\right.$$
$$\left.\left(\sum_j b_j(Z_j - E(Z_j))\right)^*\right)$$
$$= \sum_j a_j b_j^* \text{var}(Z_j) + 2\sum_{j<k} a_j b_k^* \text{cov}(Z_j, Z_k).$$

Q.E.D.

A p-variate complex random vector with values in $\mathbb{C}^p$

$$Z = (Z_1, \ldots, Z_p)',$$

with $Z_j = X_j + iY_j$, is a p-tuple of complex random variables $Z_1, \ldots, Z_p$. The expected value of Z is

$$E(Z) = (E(Z_1), \ldots, E(Z_p))'. \quad (4.11)$$

The complex covariance of Z is defined by

$$\Sigma = E(Z - E(Z))(Z - E(Z))^*. \quad (4.12)$$

Since $\Sigma^* = \Sigma$, Σ is a Hermitian matrix.

Definition 4.2.1. Let $Z = X + iY \in \mathbb{C}^p$ be a complex vector of p-dimension

$$[Z] = \begin{pmatrix} X \\ Y \end{pmatrix} \in E^{2p}. \quad (4.13)$$

This representation defines an isomorphism between $\mathbb{C}^p$ and E^{2p}.

Theorem 4.2.2. *Let $\mathbb{C}_p$ be the space of $p \times p$ complex matrices and let $C = A + iB \in \mathbb{C}_p$ where A and B are $p \times p$ real matrices. For $Z \in \mathbb{C}^p$*

$$[CZ] = \langle C \rangle [Z], \tag{4.14}$$

where

$$\langle C \rangle = \begin{pmatrix} A & -B \\ B & A \end{pmatrix}. \tag{4.15}$$

Proof.

$$[CZ] = [(A + iB)(X + iY)] = [AX - BY + i(BX + AY)]$$

$$= \begin{pmatrix} AX - BY \\ BX + AY \end{pmatrix} = \langle C \rangle [Z].$$

Q.E.D.

Definition 4.2.2. A univariate complex normal random variable is a complex random variable $Z = X + iY$ such that the distribution of $[Z] = \binom{X}{Y}$ is a bivariate normal.

The probability density function of Z can be written as

$$f_Z(z) = \frac{1}{\pi \text{var}(Z)} \exp\left\{ -\frac{(z-\alpha)(z-\alpha)^*}{\text{var}(Z)} \right\} = \frac{1}{\pi(\text{var}(X) + \text{var}(Y))}$$

$$\times \exp\left\{ -\frac{(x-\mu)^2 - (y-\nu)^2}{\text{var}(X) + \text{var}(Y)} \right\}$$

where $\alpha = \mu + i\nu = E(Z)$.

Definition 4.2.3. A p-variate complex random vector

$$Z = (Z_1, \ldots, Z_p)',$$

with $Z_j = X_j + iY_j$ is a p-tuple of complex normal random variables $Z_1, \ldots, Z_p$ such that the real $2p$-vector $(X_1, \ldots, X_p, Y_1, \ldots, Y_p)'$ has a $2p$-variate normal distribution.

Let

$$\alpha = E(Z) = E(X) + iE(Y) = \mu + i\nu, \quad \Sigma = E(Z - \alpha)(Z - \alpha)^*,$$

where $\Sigma \in \mathbb{C}_p$ is a positive definite Hermitian matrix; $\alpha \in \mathbb{C}^p$; $\mu, \nu \in E^p$; $X = (X_1, \ldots, X_p) \in E^p$; $Y = (Y_1, \ldots, Y_p)' \in E^p$. The joint probability density

Properties of Multivariate Distributions

function of X, Y can be written as

$$f_Z(z) = \frac{1}{\pi^p \det(\Sigma)} \exp\{-(z-\alpha)^*\Sigma^{-1}(z-\alpha)\}$$

$$= \frac{1}{\pi^p \left(\det \begin{pmatrix} 2\Gamma & -2\Delta \\ 2\Delta & 2\Gamma \end{pmatrix}\right)^{\frac{1}{2}}} \quad (4.16)$$

$$\times \exp\left\{-\begin{pmatrix} x-\mu \\ y-\nu \end{pmatrix}' \begin{pmatrix} 2\Gamma & -2\Delta \\ 2\Delta & 2\Gamma \end{pmatrix}^{-1} \begin{pmatrix} x-\mu \\ y-\nu \end{pmatrix}\right\}$$

where $\Sigma = 2\Gamma + i2\Delta$, Γ is a positive definite matrix and $\Delta = -\Delta'$ (skew symmetric). Hence $E(X) = \mu$, $E(Y) = \nu$ and the covariance matrix of $\binom{X}{Y}$ is given by

$$\begin{pmatrix} \Gamma & -\Delta \\ \Delta & \Gamma \end{pmatrix}$$

Thus if Z has the probability density function $f_Z(z)$ given by (4.16), then $E(Z) = \mu + i\vartheta$, $\text{cov}(Z) = \Sigma$.

Example 4.2.1. Bivariate complex normal. Here

$$Z = (Z_1, Z_2)', Z_1 = X_1 + iY_1, Z_2 = X_2 + iY_2, E(Z)$$
$$= \alpha = (\alpha_1, \alpha_2)' = (\mu_1, \mu_2)' + i(\nu_1, \nu_2)'.$$

Let

$$\text{cov}(Z_j, Z_k) = \begin{cases} \sigma_k^2 & \text{if } j = k, \\ (\alpha_{jk} + i\beta_{jk})\sigma_j\sigma_k & \text{if } j \neq k. \end{cases}$$

Hence

$$\Sigma = \begin{pmatrix} \sigma_1^2 & (\alpha_{12} + i\beta_{12})\sigma_1\sigma_2 \\ (\alpha_{12} - i\beta_{12})\sigma_1\sigma_2 & \sigma_2^2 \end{pmatrix},$$

$$\det \Sigma = \sigma_1^2\sigma_2^2 - (\alpha_{12}^2 + \beta_{12}^2)\sigma_1^2\sigma_2^2,$$

$$\Sigma^{-1} = \frac{1}{(1 - \alpha_{12}^2 - \beta_{12}^2)\sigma_1^2\sigma_2^2}$$

$$\times \begin{pmatrix} \sigma_2^2 & -(\alpha_{12} + i\beta_{12})\sigma_1\sigma_2 \\ -(\alpha_{12} - i\beta_{12})\sigma_1\sigma_2 & \sigma_2^2 \end{pmatrix}.$$

Thus

$$f_Z(z) = \frac{1}{\pi^2(1-\alpha_{12}^2-\beta_{12}^2)} \exp[((1-\alpha_{12}^2-\beta_{12}^2)\sigma_1^2\sigma_2^2)^{-1}$$
$$\times \{-\sigma_2^2(z_1-\alpha_1)^*(z_1-\alpha_1) + \sigma_1^2(z_2-\alpha_2)^*(z-\alpha_2)^*(z-\alpha_2)$$
$$-2\sigma_1\sigma_2(\alpha_{12}+i\beta_{12})(z_1-\alpha_1)^*(z_2-\alpha_2)\}]$$

The numerator inside the braces can be expressed as

$$\sigma_2^2[(x_1-\mu_1)^2+(y_1-\nu_1)^2] + \sigma_1^2[(x_2-\mu_2)^2+(y_2-\nu_2)^2] - 4\sigma_1\sigma_2$$
$$\times [\alpha_{12}((x_1-\mu_1)(x_2-\mu_2)+(y_1-\nu_1)(y_2-\nu_2))$$
$$+ \beta_{12}((x_1-\mu_1)(y_2-\mu_2)-(x_2-\mu_2)(y_1-\mu_1))].$$

The special case of the probability density function of the complex random vector Z given in (4.16) with the added restriction

$$E(Z-\alpha)(Z-\alpha)' = 0 \qquad (4.17)$$

is of considerable interest in the literature. This condition implies that the real and imaginary parts of different components are pairwise independent and the real and the imaginary parts of the same components are independent with the same variance.

With the density function $f_Z(z)$ in (4.16) one can obtain results analogous to Theorems 4.1.1–4.1.5 for the complex case. We shall prove below three theorems which are analogous to Theorems 4.1.3–4.1.5.

Theorem 4.2.3. *Let $Z = (Z_1, \ldots, Z_p)'$ with values in $\mathbb{C}^p$ be distributed as the complex p-variate normal with mean α and Hermitian positive definite covariance matrix Σ. Then CZ, where C is a complex nonsingular matrix of dimension $p \times p$, has a complex p-variate normal distribution with mean $C\alpha$ and Hermitian positive definite covariance matrix $C\Sigma C^*$.*

Proof.

$$[CZ] = \begin{pmatrix} AX-BY \\ BX+AY \end{pmatrix} = \begin{pmatrix} A & -B \\ B & A \end{pmatrix}\begin{pmatrix} X \\ Y \end{pmatrix}$$
$$= \langle C \rangle [Z]$$

Properties of Multivariate Distributions

Since $[Z]$ is distributed as

$$N_{2p}\left[\alpha, \begin{pmatrix} \Gamma & -\Delta \\ \Delta & \Gamma \end{pmatrix}\right], \quad \begin{pmatrix} A & -B \\ B & A \end{pmatrix}[Z]$$

is distributed as $2p$-variate normal with mean

$$\begin{pmatrix} A\mu - B\nu \\ B\mu + A\nu \end{pmatrix}$$

and $2p \times 2p$ covariance matrix

$$\begin{pmatrix} A\Gamma A' - B\Delta A' + A\Delta B' - B\Gamma B' & -B\Gamma A' - B\Delta B' + A\Gamma B' - A\Delta A' \\ B\Gamma A' + B\Delta B' - A\Gamma B' + A\Delta A' & A\Gamma A' - B\Delta A' + A\Delta B' + B\Gamma B' \end{pmatrix}$$

Hence CZ is distributed as p-variate complex normal with mean $C\alpha$ and $p \times p$ complex covariance matrix

$$2(A\Gamma A' - B\Delta A' + B\Gamma B' + A\Delta B')$$
$$+ i2(B\Gamma A' + A\Delta A' - A\Gamma B' + B\Delta B') = C\Sigma C^*.$$

Q.E.D.

Theorem 4.2.4. *Let $Z = (Z'_{(1)}, Z'_{(2)})'$, where $Z_{(1)} = (Z_1, \ldots, Z_q)'$, $Z_{(2)} = (Z_{q+1}, \ldots, Z_p)'$ be distributed as p-variate complex normal with mean $\alpha = (\alpha'_{(1)}, \alpha'_{(2)})'$ and positive definite Hermitian covariance matrix Σ and let Σ be partitioned as*

$$\Sigma = \begin{pmatrix} \Sigma_{11} & \Sigma_{12} \\ \Sigma_{21} & \Sigma_{22} \end{pmatrix}$$

where Σ_{11} is the upper left-hand corner submatrix of dimension $q \times q$. If $\Sigma_{12} = 0$, then $Z_{(1)}, Z_{(2)}$ are independently distributed complex normal vectors with means $\alpha_{(1)}, \alpha_{(2)}$ and Hermitian covariance matrices Σ_{11}, Σ_{22} respectively.

Proof. Under the assumption that $\Sigma_{12} = 0$, we obtain

$$(z - \alpha)^* \Sigma^{-1}(z - \alpha) = (z_{(1)} - \alpha_{(1)})^* \Sigma_{11}^{-1}(z_{(1)} - \alpha_{(1)})$$
$$+ (z_{(2)} - \alpha_{(2)})^* \Sigma_{22}^{-1}(z_{(2)} - \alpha_{(2)}),$$
$$\det \Sigma = (\det \Sigma_{11})(\det \Sigma_{22}).$$

Hence

$$f_Z(z) = \frac{1}{\pi^q \det \Sigma_{11}} \exp\{-(z_{(1)} - \alpha_{(1)})^* \Sigma_{11}^{-1}(z_{(1)} - \alpha_{(1)})\}$$

$$\times \frac{1}{\pi^{(p-q)} \det \Sigma_{22}} \exp\{-(z_{(2)} - \alpha_{(2)})^* \Sigma_{22}^{-1}(z_{(2)} - \alpha_{(2)})\}$$

and the result follows. Q.E.D.

Theorem 4.2.5. *Let* $Z = (Z'_{(1)}, Z'_{(2)})'$, *where* $Z_{(1)}, Z_{(2)}$ *are as defined in Theorem 4.2.4.*

(a) $Z_{(1)}, Z_{(2)} - \Sigma_{21}\Sigma_{11}^{-1}Z_{(1)}$ *are independently distributed complex normal random vectors with means* $\alpha_{(1)}, \alpha_{(2)} - \Sigma_{21}\Sigma_{11}^{-1}\alpha_{(1)}$ *and positive definite Hermitian covariance matrixes* $\Sigma_{11}, \Sigma_{22.1} = \Sigma_{22} - \Sigma_{21}\Sigma_{11}^{-1}\Sigma_{12}$ *respectively.*
(b) *The marginal distribution of* $Z_{(1)}$ *is a q-variate complex normal with means* $\alpha_{(1)}$ *and positive definite Hermitian covariance matrix* Σ_{11}.
(c) *The conditional distribution of* $Z_{(2)}$ *given* $Z_{(1)} = z_{(1)}$ *is complex normal with mean* $\alpha_{(2)} + \Sigma_{21}\Sigma_{11}^{-1}(z_{(1)} - \alpha_{(1)})$ *and positive definite Hermitian covariance matrix* $\Sigma_{22.1}$.

Proof. (a) Let

$$U = \begin{pmatrix} U_1 \\ U_2 \end{pmatrix} = \begin{pmatrix} Z_{(1)} \\ Z_{(2)} - \Sigma_{21}\Sigma_{11}^{-1}Z_{(1)} \end{pmatrix}.$$

Then

$$U = \begin{pmatrix} I_1 & 0 \\ -\Sigma_{21}\Sigma_{11}^{-1} & I_2 \end{pmatrix} \begin{pmatrix} Z_{(1)} \\ Z_{(2)} \end{pmatrix} = CZ$$

where I_1 and I_2 are identity matrices of dimensions $q \times q$ and $(p-q) \times (p-q)$ respectively and C is a complex nonsingular matrix. By theorem 4.2.3 U has a p-variate complex normal distribution with mean

$$C\alpha = \begin{pmatrix} \alpha_{(1)} \\ \alpha_{(2)} - \Sigma_{21}\Sigma_{11}^{-1}\alpha_{(1)} \end{pmatrix}$$

and (Hermitian) complex covariance matrix

$$C\Sigma C^* = \begin{pmatrix} \Sigma_{(11)} & 0 \\ 0 & \Sigma_{22.1} \end{pmatrix}.$$

By Theorem 4.2.4 we get the result.

Properties of Multivariate Distributions

(*b*) and (*c*). They follow from part (*a*) above. Q.E.D.

The characteristic function of Z is given by

$$E \exp\{iR(t^*Z)\} = \exp\{iR(t^*\alpha) - t^*\Sigma t\} \quad (4.18)$$

for $t \in \mathbb{C}^p$ and R denotes the real part of a complex number. As in the real case we denote a p-variate complex normal with mean α and positive definite Hermitian matrix Σ by $CN_p(\alpha, \Sigma)$.

From Theorem 4.2.3 we can define a p-variate complex normal distribution in the general case as follows.

Definition 4.2.4. A complex random p-vector Z with values in $\mathbb{C}^p$ has a complex normal distribution if, for each $a \in \mathbb{C}^p$, a^*Z has a univariate complex normal distribution.

4.3. SYMMETRIC DISTRIBUTION: ITS PROPERTIES AND CHARACTERIZATIONS

In multivariate statistical analysis multivariate normal distribution plays a very dominant role. Many results relating to univariate normal statistical inference have been successfully extended to the multivariate normal distribution. In practice, the verification of the assumption that a given set of data arises from a multivariate normal population is cumbersome. A natural question thus arises how sensitive these results are to the assumption of multinormality. In recent years one such investigation involves in considering a family of density functions having many similar properties as the multinormal. The family of elliptically symmetric distributions contains probability density functions whose contours of equal probability have elliptical shapes. In recent years this family is becoming increasingly popular because of its frequent use in "filtering and stochastic control" (Chu (1973)), "random signal input" (McGraw and Wagner (1968)), "stock market data analysis" (Zellner (1976)) and because some optimum results of statistical inference in the multivariate normal preserves their properties for all members of the family. The family of "spherically symmetric distributions" is a special case of this family.

They contain the multivariate student-t, compound (or scale mixed) multinormal, contaminated normal, multivariate normal with zero mean vector and covariance matrix I among others.

It is to be pointed out that these families do not possess all basic requirements for an ideal statistical inference. For example the sample observations are not independent, in general, for all members of these families.

4.3.1. Elliptically and Spherically Symmetric Distribution (Univariate)

Definition 4.3.1.1. Elliptically symmetric distribution (univariate). A random vector $X = (X_1, \ldots, X_p)'$ with values x in R^p is said to have a distribution belonging to the family of elliptically symmetric distributions (univariate) with location parameter $\mu = (\mu_1, \ldots, \mu_p)' \in R^p$ and scale matrix Σ (symmetric positive definite) if its probability density function (pdf), if it exists, can be expressed as a function of the quadratic form $(x - \mu)'\Sigma^{-1}(x - \mu)$ and is given by

$$f_X(x) = (\det \Sigma)^{-\frac{1}{2}} q((x - \mu)'\Sigma^{-1}(x - \mu)), \qquad (4.19)$$

where q is a function on $[0, \infty)$ satisfying $\int_{R^p} q(y'y) dy = 1$ for $y \in R^p$.

Definition 4.3.1.2. Spherically symmetric distribution (univariate). A random vector $X = (X_1, \ldots, X_p)'$ is said to have a distribution belonging to the family of spherically symmetric distributions if X and OX have the same distributions for all $p \times p$ orthogonal matrices O.

Let $X = (X_1, \ldots, X_p)'$ be a random vector having elliptically symmetric pdf (4.19) and let $Y = C^{-1}(X - \mu)$ where C is a $p \times p$ nonsingular matrix satisfying $\Sigma = CC'$ (by Theorem 1.5.5). The pdf of Y is given by

$$\begin{aligned} f_Y(y) &= (\det \Sigma)^{-1/2} q((Cy)'\Sigma^{-1}(Cy)) \det C \\ &= (\det(C^{-1}\Sigma C^{-1'}))^{-1/2} q(y'(C^{-1}\Sigma C^{-1'})^{-1} y) \qquad (4.20) \\ &= q(y'y) \end{aligned}$$

as the Jacobian of the transformation $X \to C^{-1}(X - \mu) = Y$ is $\det(C)$. Furthermore

$$(Oy)'Oy = y'y$$

for all $p \times p$ orthogonal matrix O and the Jacobian of the transformation $Y \to OY$ is unity. Hence $f_Y(y)$ is the pdf of a spherically symmetric distribution.

We denote the elliptically symmetric pdf (4.19) of a random vector X by $E_p(\mu, \Sigma, q)$ and the spherically symmetric pdf (4.20) of a random vector Y by $E_p(0, I, q)$. When the mention of q is unnecessary we will omit q in the notations.

4.3.2. Examples of $E_p(\mu, \Sigma, q)$

Example 4.3.2.1. Multivariate normal $N_p(\mu, \Sigma)$. The pdf of $X = (X_1, \ldots, X_p)'$ is

$$f_X(x) = (2\pi)^{\frac{-p}{2}} (\det \Sigma)^{-\frac{1}{2}} etr\left(-\frac{1}{2}(x - \mu)'\Sigma^{-1}(x - \mu)\right)$$

for $x \in R^p$. Here $q(z) = (2\pi)^{-p/2} \exp(-\frac{1}{2}z), z \geq 0$.

Properties of Multivariate Distributions

Example 4.3.2.2. Multivariate student-t with m degrees of freedom. The *pdf* of $X = (X_1, \ldots, X_p)'$ is

$$f_X(x) = \frac{\Gamma(\frac{1}{2}(m+p))(\det \Sigma)^{-\frac{1}{2}}}{(\pi m)^{\frac{1}{2}p}\Gamma(\frac{1}{2}m)}\left(1 + \frac{1}{m}(x-\mu)'\Sigma^{-1}(x-\mu)\right)^{-\frac{1}{2}(m+p)}. \qquad (4.21)$$

Here

$$q(z) = \frac{\Gamma(\frac{1}{2}(m+p))}{(\pi m)^{\frac{1}{2}p}\Gamma(\frac{1}{2}m)}\left(1 + \frac{z}{m}\right)^{-\frac{1}{2}(m+p)}.$$

We will denote the *pdf* of the multivariate student-t with m degrees of freedom and with parameter (μ, Σ) by $t_p(\mu, \Sigma, m)$ in order to distinguish it from the multivariate student-t based on spherically symmetric distribution with *pdf* given by

$$f_Y(y) = \frac{\Gamma(\frac{1}{2}(m+p))}{(\pi m)^{\frac{1}{2}p}\Gamma(\frac{1}{2}m)}\left(1 + \frac{1}{m}y'y\right)^{-\frac{1}{2}(m+p)}. \qquad (4.22)$$

where: $Y = \Sigma^{-\frac{1}{2}}(X - \mu)$. Since the Jacobian of the transformation $X \to Y$ is $(\det \Sigma)^{\frac{1}{2}}$ the *pdf* of Y is given by (4.22).

To prove $f_Y(y)$ or $f_X(x)$ is a *pdf* we use the identity

$$\int_{-\infty}^{\infty}(1+x^2)^{-n}dx = \int_0^{\infty}(1+y)^{-n}y^{\frac{1}{2}-1}dy$$

$$= \int_0^1 (1-u)^{n+\frac{1}{2}-1}u^{\frac{1}{2}-1}du$$

$$= \frac{\Gamma(n+\frac{1}{2})\Gamma(\frac{1}{2})}{\Gamma(n+1)}.$$

Let A be a $k \times p$ matrix of rank $k(k \le p)$ and let C be a $p \times p$ nonsingular matrix such that

$$C = \begin{pmatrix} A \\ B \end{pmatrix}$$

where B is a $(p-k) \times p$ matrix of rank $p-k$. Then

$$Z = CX = \begin{pmatrix} AX \\ BX \end{pmatrix}$$

is distributed as $t_p(C\mu, C\Sigma C', m)$ and

$$C\mu = \begin{pmatrix} A\mu \\ B\mu \end{pmatrix}, \quad C\Sigma C' = \begin{pmatrix} A\Sigma A' & A\Sigma B' \\ B\Sigma A' & B\Sigma B' \end{pmatrix}.$$

Using problem 24 we get $Y = AX$ is distributed as $t_k(A\mu, A\Sigma A', m)$.

Figures 4.3 and 4.4 give the graphical representation of the bivariate student-t with 2 degrees of freedom and its contour.

Example 4.3.2.3. Scale mixed (compound) multivariate normal. Let $X = (X_1, \ldots, X_p)'$ be a random vector with *pdf*

$$f_X(x) = \int_0^\infty (2\pi z)^{\frac{-1}{2}p} (\det \Sigma)^{-\frac{1}{2}}$$

$$\times \exp\left\{-\frac{1}{2}(x-\mu)'\Sigma^{-1}(x-\mu)z^{-1}\right\} dG(z)$$

where Z is a positive random variable with distribution function G. The

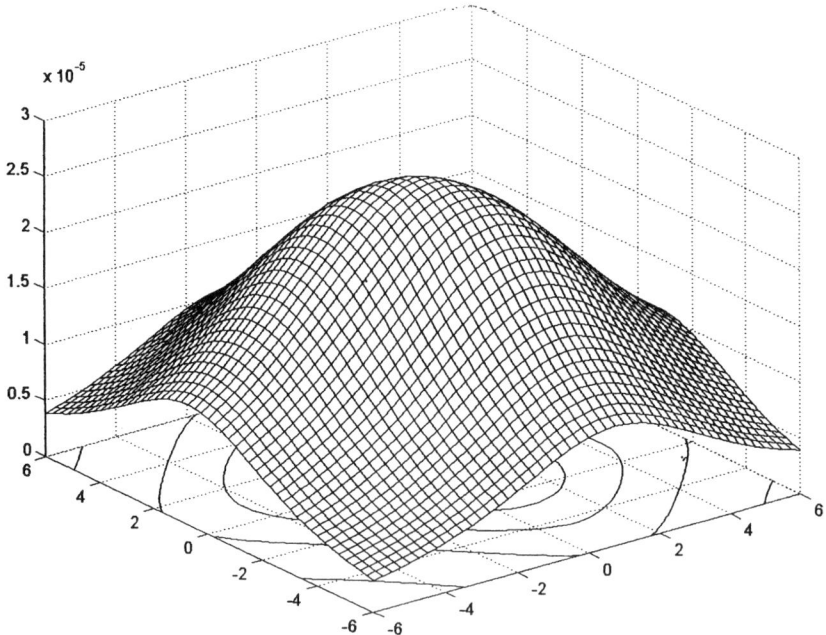

Figure 4.3. Bivariate student-t with 2 degrees of freedom and $\mu = 0$, $\Sigma = \begin{pmatrix} 1 & \frac{1}{2} \\ \frac{1}{2} & 1 \end{pmatrix}$.

Properties of Multivariate Distributions

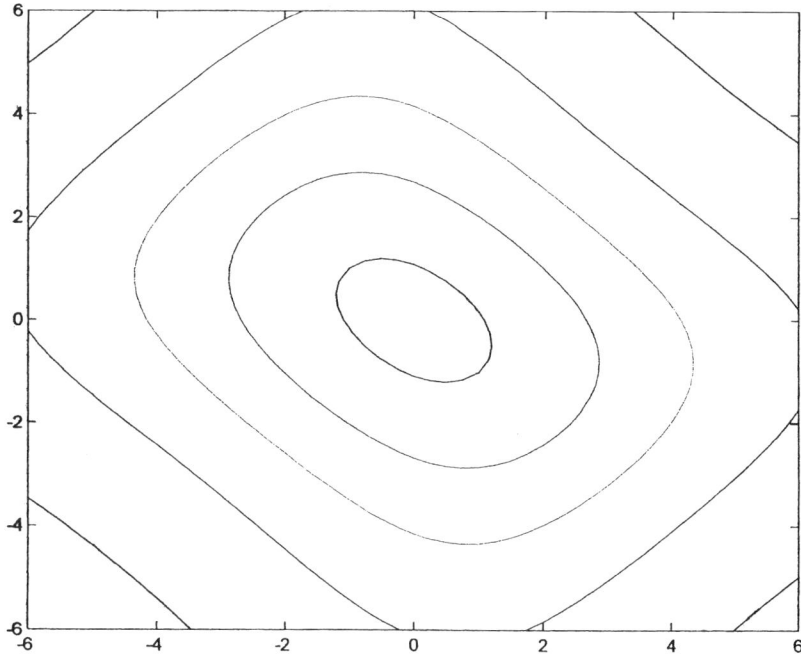

Figure 4.4. Contours of bivariate student-t in Figure 4.3.

multivariate t distribution given in (4.21) can be obtained from this by taking G to be the inverse gamma, given by,

$$\frac{dG(z)}{dz} = \frac{(\frac{1}{2})^{\frac{1}{2}m}}{\Gamma(\frac{1}{2}m)} z^{\frac{1}{2}m-1} \exp\{-m/2z\}.$$

4.3.3. Examples of $E_p(0, I)$

Example 4.3.3.1. Contaminated normal. Let $X = (X_1, \ldots, X_p)'$. The *pdf* of a contaminated normal is given by

$$f_X(x) = \frac{\alpha}{(2\pi)^{\frac{1}{2}p}(z_1^2)^{\frac{1}{2}p}} \exp\left\{-\frac{x'x}{2z_1^2}\right\}$$
$$+ \frac{(1-\alpha)}{(2\pi)^{\frac{1}{2}p}} (z_2^2)^{\frac{1}{2}p} \exp\left\{-\frac{x'x}{2z_2^2}\right\}$$

with $0 \leq \alpha \leq 1$, $z_i^2 > 0$, $i = 1, 2$.

Example 4.3.3.2. Multivariate student-t with m degrees of freedom. Its *pdf* is given by (4.22).

4.3.4. Basic Properties of $E_p(\mu, \Sigma, q)$ and $E_p(0, I, q)$

Theorem 4.3.4.1. *Let $X = (X_1, \ldots, X_p)'$ be distributed as $E_p(\mu, \Sigma, q)$. Then*

(a) $E(X) = \mu$ *for all q,*
(b) $Cov(X) = E(X - \mu)(X - \mu)' = K_q \Sigma$ *where K_q is a positive constant depending on q,*
(c) *the correlation matrix $R = (R_{ij})$ with*

$$R_{ij} = \frac{Cov(X_i, X_j)}{[var(X_i)var(X_j)]^{\frac{1}{2}}} \qquad (4.23)$$

for all members of $E_p(\mu, \Sigma, q)$ are identical.

Proof.

(a) Let $Z = C^{-1}(X - \mu)$ where C is a $p \times p$ nonsingular matrix such that $\Sigma = CC'$ and let $Y = X - \mu$. Since the Jacobian of the transformation $Y \to Z$ is $\det C$ and $q(Z'Z)$ is an even function of Z we get

$$E(X - \mu) = \int_{R^p} (x - \mu)(\det \Sigma)^{-\frac{1}{2}}$$

$$\times q((x - \mu)' \Sigma^{-1}(x - \mu)) \, dx$$

$$= \int_{R^p} y(\det \Sigma)^{-\frac{1}{2}} q(y' \Sigma^{-1} y) \, dy$$

$$= C \int_{R^p} z q(z'z) \, dz$$

$$= 0.$$

Hence $E(X) = \mu$ for all q.

(b)

$$E(X - \mu)(X - \mu)' = \int_{R^p} yy'(\det \Sigma)^{\frac{1}{2}} q(y' \Sigma^{-1} y) \, dy$$

$$= C \left[\int_{R^p} (zz') q(z'z) \, dz \right] C'.$$

Properties of Multivariate Distributions

Using the fact that $q(z'z)$ is an even function of $z = (z_1, \ldots, z_p)'$ we conclude that

$$E(z_i z_j) = \begin{cases} p^{-1} K_q, & \text{for all } i = j, \\ 0, & \text{for all } i \neq j, \end{cases}$$

where K_q is a positive constant depending on q. Hence

$$\int_{R^p} (zz') q(zz') \, dz = p^{-1} K_q I \tag{4.24}$$

where I is the $p \times p$ identity matrix. This implies that

$$K_q = \int_{R^p} tr(zz') q(zz') \, dz. \tag{4.25}$$

Let

$$L = trZZ', \qquad e_i = \frac{Z_i^2}{L}, \qquad i = 1, \ldots, p.$$

We will prove in Theorem 6.12.1 that L is independent of $(e_1, \ldots, e_p)$ and the joint pdf of $(e_1, \ldots, e_p)$ is Dirichlet $D(\frac{1}{2}, \ldots, \frac{1}{2})$ with pdf

$$f(e_1, \ldots, e_p) = \frac{\Gamma(\frac{1}{2} p)}{(\Gamma(\frac{1}{2}))^p} \left[\prod_{i=1}^{p-1} e_i^{\frac{1}{2}-1} \right] \left(1 - \sum_{1}^{p-1} e_i \right)^{\frac{1}{2}-1} \tag{4.26}$$

with $0 \leq e_i \leq 1$, $\sum_{i=1}^{p} e_i = 1$ and the pdf of L is

$$f_L(l) = \frac{\pi^{\frac{1}{2} p}}{\Gamma\left(\frac{p}{2}\right)} l^{\frac{1}{2}p - 1} q(l). \tag{4.27}$$

From (4.24) and (4.25) we get

$$K_q = E(L). \tag{4.28}$$

Hence $Cov(X) = p^{-1} K_q \Sigma$.
(c) Since the covariance of X depends on q through K_q and the K_q factor cancels in R_{ij} we get part (c). Q.E.D.

Example 4.3.4.1. Consider Example 4.3.1.1. Here

$$q(l) = (2\pi)^{-\frac{1}{2}p} \exp\left\{ -\frac{1}{2} l \right\}.$$

Hence

$$E(L) = \frac{\pi^{\frac{1}{2}p}}{(2\pi)^{\frac{p}{2}}\Gamma(\frac{1}{2}p)} \int_0^\infty l^{\frac{1}{2}(p+2)-1} \exp\left(-\frac{1}{2}l\right) dl$$

$$= p.$$

Thus $E(X) = \mu$, $Cov(X) = \Sigma$.

Example 4.3.4.2. Consider Example 4.3.2.2. Here

$$q(l) = \frac{\Gamma(\frac{1}{2}(m+p))}{(\pi m)^{\frac{1}{2}p}\Gamma(\frac{1}{2}m)} \left(1 + \frac{l}{m}\right)^{-\frac{1}{2}(m+p)}.$$

From (4.27)

$$f_L(l) = \frac{\pi^{\frac{1}{2}p}}{\Gamma(\frac{1}{2}p)} l^{\frac{1}{2}p-1} q(l). \tag{4.29}$$

Hence

$$E(L) = \frac{\Gamma(\frac{1}{2}(m+p))}{\Gamma(\frac{1}{2}m)\Gamma(\frac{1}{2}p)} \int_0^\infty \left(\frac{l}{m}\right)^{\frac{1}{2}(p+2)-1} \left(1 + \frac{l}{m}\right)^{-\frac{1}{2}(m+p)} dl$$

$$= \frac{m\Gamma(\frac{1}{2}(m+p))}{\Gamma(\frac{1}{2}m)\Gamma(\frac{1}{2}p)} \int_0^1 u^{\frac{1}{2}(p+2)-1} (1+u)^{-\frac{1}{2}(m+p)} du$$

$$= \frac{m\Gamma(\frac{1}{2}(m+p))}{\Gamma(\frac{1}{2}m)\Gamma(\frac{1}{2}p)} \int_0^1 z^{\frac{1}{2}(p+2)-1} (1-z)^{-\frac{1}{2}(m-2)} dz$$

$$= \frac{mp}{(m-2)}.$$

Hence $E(X) = \mu$, $Cov(X) = \frac{m}{m-2}\Sigma$ with $m > 2$.

Theorem 4.3.4.2. Let $X = (X_1, \ldots, X_p)'$ be distributed as $E_p(\mu, \Sigma, q)$. Then $Y = CX + b$, where C is a $p \times p$ nonsingular matrix and $b \in R^p$, is distributed as $E_p(C\mu + b, C\Sigma C', q)$.

Proof. The *pdf* of X is

$$f_X(x) = (\det \Sigma)^{\frac{-1}{2}} q((x-\mu)'\Sigma^{-1}(x-\mu)).$$

Properties of Multivariate Distributions

Since the Jacobian of the transformation $X \to Y = CX + b$ is $(\det C)^{-1}$ we get

$$f_Y(y) = (\det \Sigma)^{-\frac{1}{2}} q((C^{-1}(y-b) - \mu)' \Sigma^{-1}(C^{-1}(y-b) - \mu))(\det C)^{-1}$$

$$= (\det(C\Sigma C'))^{-\frac{1}{2}} q((y - C\mu - b)(C\Sigma C')^{-1}(y - C\mu - b)).$$

Q.E.D.

Example 4.3.4.3. Let X be distributed as $E_p(\mu, \Sigma, q)$ and let $Y = C(X - \mu)$ where C is a $p \times p$ nonsingular matrix such that $C\Sigma C' = I$. Then Y is distributed as $E_p(0, I, q)$.

Theorem 4.3.4.3. Let $X = (X_1, \ldots, X_p)' = (X'_{(1)}, X'_{(2)})$ with $X_{(1)} = (X_1, \ldots, X_{p_1})'$, $X_{(2)} = (X_{p_1+1}, \ldots, X_p)'$ be distributed as $E_p(\mu, \Sigma, q)$. Let $\mu = (\mu_1, \ldots, \mu_p)'$ and Σ be similarly partitioned as

$$\mu = (\mu'_{(1)}, \mu'_{(2)})', \quad \Sigma = \begin{pmatrix} \Sigma_{11} & \Sigma_{12} \\ \Sigma_{21} & \Sigma_{22} \end{pmatrix}$$

where Σ_{11} is the $p_1 \times p_1$ upper left-hand corner submatrix of Σ. Then

(a) the marginal distribution of $X_{(1)}$ is elliptically symmetric $E_{p_1}(\mu_{(1)}, \Sigma_{11}, \bar{q})$ where $\bar{q}$ is a function on $[0, \infty)$ satisfying

$$\int_{R^{p_2}} \bar{q}(w'_{(1)} w_{(1)}) dw_{(1)} = 1;$$

(b) the conditional distribution of $X_{(2)}$ given $X_{(1)} = x_{(1)}$ is elliptically symmetric $E_{p-p_1}(\mu_{(2)} + \Sigma_{21} \Sigma_{11}^{-1}(x_{(1)} - \mu_{(1)}), \Sigma_{22.1}, \tilde{q})$ where $\Sigma_{22.1} = \Sigma_{22} - \Sigma_{21} \Sigma_{11}^{-1} \Sigma_{12}$ and $\tilde{q}$ is a function on $[0, \infty)$ satisfying

$$\int_{R^{p-p_1}} \tilde{q}(w'_{(2)} w_{(2)}) dw_{(2)} = 1.$$

Proof. Since Σ is positive definite, by Theorem 1.6.5 there exists a $p \times p$ nonsingular lower triangular matrix T in the block form

$$T = \begin{pmatrix} T_{11} & 0 \\ T_{21} & T_{22} \end{pmatrix}$$

where T_{11} is a $p_1 \times p_1$ matrix such that $T\Sigma T' = I_p$. This implies that $T_{11} \Sigma_{11} T'_{11} = I_{p_1}$. Let $Y = T(X - \mu)$ be similarly partitioned as X. From Theorem 4.3.3.2 the *pdf* of Y is

$$f_Y(y) = q(y'y) = q(y'_{(1)} y_{(1)} + y'_{(2)} y_{(2)}).$$

Thus

$$f_{Y_{(1)}}(y_{(1)}) = \int q(y'_{(1)}y_{(1)} + y'_{(2)}y_{(2)})dy_{(2)} \qquad (4.30)$$

$$= \bar{q}(y'_{(1)}y_{(1)}).$$

Obviously $\bar{q}$ is from $[0, \infty)$ to $[0, \infty)$ satisfying

$$\int \bar{q}(y'_{(1)}y_{(1)})dy_{(1)} = 1$$

and $\bar{q}$ is determined only by the functional form of q and by the number of components in the vector $X_{(1)}$ and does not depend on μ, Σ. Hence all marginal *pdf* of any dimension do not differ in their functional form. Since $Y_{(1)} = T_{11}(X_{(1)} - \mu_{(1)})$ we obtain from (4.30)

$$\begin{aligned} f_{X_{(1)}}(x_{(1)}) &= f_{Y_{(1)}}(T_{11}(x_{(1)} - \mu_{(1)}))(\det \Sigma_{11})^{-\frac{1}{2}} \\ &= (\det \Sigma_{11})^{-\frac{1}{2}}\bar{q}((x_{(1)} - \mu_{(1)})'\Sigma_{11}^{-1}(x_{(1)} - \mu_{(1)})). \end{aligned} \qquad (4.31)$$

(b) Let

$$u = x_{(2)} - \mu_{(2)} - \Sigma_{21}\Sigma_{11}^{-1}(x_{(1)} - \mu_{(1)}).$$

Then

$$(x - \mu)'\Sigma^{-1}(x - \mu) = (x_{(1)} - \mu_{(1)})'\Sigma_{11}^{-1}(x_{(1)} - \mu_{(1)}) + u'\Sigma_{22.1}^{-1}u.$$

Let

$$W_{(1)} = \Sigma_{11}^{-\frac{1}{2}}(X_{(1)} - \mu_{(1)})$$

$$W_{(2)} = \Sigma_{22.1}^{-\frac{1}{2}}U.$$

The joint *pdf* of $(W_{(1)}, W_{(2)})$ is given by

$$f_{W_{(1)}, W_{(2)}}(w_{(1)}, w_{(2)}) = q(w'_{(1)}w_{(1)} + w'_{(2)}w_{(2)}).$$

The marginal *pdf* of $W_{(1)}$ is

$$f_{W_{(1)}}(w_{(1)}) = \bar{q}(w'_{(1)}w_{(1)})$$

$$= \int q(w'_{(1)}w_{(1)} + w'_{(2)}w_{(2)})dw_{(2)}.$$

Properties of Multivariate Distributions

Hence the conditional *pdf* of $W_{(2)}$ given $W_{(1)} = w_{(1)}$ is

$$f_{W_{(2)}|W_{(1)}}(w_2|w_1) = \frac{q(w'_{(1)}w_{(1)} + w'_{(2)}w_{(2)})}{\bar{q}(w'_{(1)}w_{(1)})}$$

$$= \tilde{q}(w'_{(2)}w_{(2)}).$$

Obviously $\tilde{q}$ is a function from $[0, \infty)$ to $[0, \infty)$ satisfying

$$\int \tilde{q}(w'_{(2)}w_{(2)})dw_{(2)} = 1.$$

Thus the conditional *pdf* of $X_{(2)}$ given that $X_{(1)} = x_{(1)}$ is

$$f_{X_{(2)}|X_{(1)}}(x_{(2)}|x_{(1)}) = (\det(\Sigma_{22.1}))^{-\frac{1}{2}}$$

$$\times \tilde{q}[(x_{(2)} - \mu_{(2)} - \Sigma_{21}\Sigma_{11}^{-1}(x_{(1)} - \mu_{(1)}))'$$

$$\times \Sigma_{22.1}^{-1}(x_{(2)} - \mu_{(2)} - \Sigma_{21}\Sigma_{11}^{-1}(x_{(1)} - \mu_{(1)}))]$$

which is $E_{p-p_1}(\mu_{(2)} + \Sigma_{21}\Sigma_{11}^{-1}(x_{(1)} - \mu_{(1)}), \Sigma_{22.1}, \tilde{q})$. Q.E.D.

Using Theorem 4.3.4.1 we obtain from Theorem 4.3.4.3

$$E(X_{(2)}|X_{(1)} = x_{(1)}) = \mu_{(2)} + \Sigma_{21}\Sigma_{11}^{-1}(x_{(1)} - \mu_{(1)}),$$
$$Cov(X_{(2)}|X_{(1)} = x_{(1)}) = K_{\bar{q}}(x_{(1)})\Sigma_{22.1} \qquad (4.32)$$

where K is a real valued function of $x_{(1)}$.

Theorem 4.3.4.4. Let $X = (X_1, \ldots, X_p)'$ be distributed as $E_p(\mu, \Sigma, q)$ and let $Y = DX$, where D is a $m \times p$ matrix of rank $m \leq p$. Then Y is distributed as $E_m(D\mu, D\Sigma D', \bar{q})$, where $\bar{q}$ is a function from $[0, \infty)$ to $[0, \infty)$ satisfying

$$\int_{R^m} \bar{q}(u'u)du = 1.$$

Proof. Let A be a $(p - m) \times p$ matrix such that $C = \binom{D}{A}$ is a $p \times p$ nonsingular matrix. Then from Theorem 4.3.4.2 CX is distributed as $E_p(C\mu, C\Sigma C', q)$. But

$$CX = \begin{pmatrix} DX \\ AX \end{pmatrix}, \quad C\mu = \begin{pmatrix} D\mu \\ A\mu \end{pmatrix}, \quad C\Sigma C = \begin{pmatrix} D\Sigma D' & D\Sigma A' \\ A\Sigma D' & A\Sigma A' \end{pmatrix}.$$

From Theorem 4.3.4.3 we get the theorem. Q.E.D.

Theorem 4.3.4.5. *Let $X = (X_1, \ldots, X_p)'$ be distributed as $E_p(\mu, \Sigma, q)$. The characteristic function of X is $E(e^{it'X}) = \exp\{it'\mu\}\psi(t'\Sigma t)$ for some function ψ on $[0, \infty)$, $t = (t_1, \ldots, t_p)' \in R^p$ and $i = \sqrt{-1}$.*

Proof. Let $Y = \Sigma^{-\frac{1}{2}}(X - \mu)$, $\alpha = \Sigma^{\frac{1}{2}}t$. Then

$$E(e^{it'X}) = E(e^{it'\mu}e^{it'(X-\mu)}).$$

$$= \exp\{it'\mu\} \int e^{i\alpha'y} q(y'y) dy.$$

Using Theorem 1.6.6 we can find a $p \times p$ orthogonal matrix 0 such that

$$0\alpha = ((\alpha'\alpha)^{\frac{1}{2}}, 0, \ldots, 0)'.$$

Let $Z = (Z_1, \ldots, Z_p)' = 0Y$. Hence

$$E(e^{it'X}) = \exp\{it'\mu\} \int e^{i(\alpha'\alpha)^{\frac{1}{2}}Z_1} q(z'z) dz$$

$$= \exp\{it'\mu\}\psi(\alpha'\alpha)$$
$$= \exp\{it'\mu\}\psi(t'\Sigma t)$$

for some function ψ on $[0, \infty)$. Q.E.D.

Example 4.3.4.4. Consider Example 4.3.2.1. Here

$$f_X(x) = (\det \Sigma)^{-\frac{1}{2}} q((x-\mu)'\Sigma^{-1}(x-\mu))$$

with

$$q(z) = (2\pi)^{-\frac{1}{2}p} \exp\left\{-\frac{1}{2}z\right\}.$$

Let $Y = T(X - \mu) = (Y_1, \ldots, Y_p)' = (Y'_{(1)}, Y'_{(2)})'$, $Y_{(1)} = (Y_1, \ldots, Y_{p_1})'$ where T is given in Theorem 4.3.4.3. We get

$$q(y'y) = q(y'_{(1)}y_{(1)} + y'_{(2)}y_{(2)})$$

$$= (2\pi)^{-\frac{1}{2}p} \exp\left\{-\frac{1}{2}(y'_{(1)}y_{(1)} + y'_{(2)}y_{(2)})\right\}.$$

Properties of Multivariate Distributions

Hence with $p_2 = p - p_1$

$$\bar{q}(y'_{(1)}y_{(1)}) = (2\pi)^{-\frac{1}{2}p_1} \exp\left\{-\frac{1}{2}y'_{(1)}y_{(1)}\right\}$$

$$\times \int (2\pi)^{-\frac{1}{2}p_2} \exp\left\{-\frac{1}{2}y'_{(2)}y_{(2)}\right\} dy_{(2)}.$$

Thus

$$f_{X_{(1)}}(x_{(1)}) = (2\pi)^{-\frac{1}{2}p_2}(\det \Sigma_{11})^{-\frac{1}{2}}$$

$$\times \exp\left\{-\frac{1}{2}(x_{(1)} - \mu_{(1)})'\Sigma^{-1}(x_{(1)} - \mu_{(1)})\right\}.$$

From Theorem 4.3.4.3

$$\tilde{q}(w'_{(2)}w_{(2)}) = \frac{(2\pi)^{-\frac{1}{2}p} \exp\{-\frac{1}{2}(w'_{(1)}w_{(1)} + w'_{(2)}w_{(2)})\}}{(2\pi)^{-\frac{1}{2}p_1} \exp\{-\frac{1}{2}w'_{(1)}w_{(1)}\}}$$

$$= (2\pi)^{-\frac{1}{2}p_2} \exp\left\{-\frac{1}{2}w'_{(2)}w_{(2)}\right\}.$$

Hence

$$f_{X_{(2)}|X_{(1)}}(x_{(2)}|x_{(1)}) = (2\pi)^{-\frac{1}{2}p_2}(\det(\Sigma_{22.1}))^{-\frac{1}{2}}$$

$$\times \exp\left\{-\frac{1}{2}(x_{(2)} - \mu_{(2)} - \Sigma_{21}\Sigma_{11}^{-1}(x_{(1)} - \mu_{(1)}))'\Sigma_{22.1}^{-1}\right. \quad (4.33)$$

$$\times \left.\left(x_{(2)} - \mu_{(2)} - \Sigma_{21}\Sigma_{11}^{-1}(x_{(1)} - \mu_{(1)})\right)\right\}.$$

Example 4.3.4.5. Consider Example 4.3.2.2. Let $X = (X_1, \ldots, X_p)' = (X'_{(1)}, X'_{(2)})'$, $X_{(1)} = (X_1, \ldots, X_{p_1})'$, $p = p_1 + p_2$. The marginal pdf of $X_{(1)}$ is

$$f_{X_{(1)}}(x_{(1)}) = \frac{\Gamma(\frac{1}{2}(m+p_1))}{\Gamma(\frac{1}{2}m)(m\pi)^{\frac{1}{2}p_1}} (\det \Sigma_{11})^{-\frac{1}{2}}$$

$$\times \left(1 + \frac{1}{m}(x_{(1)} - \mu_{(1)})' \Sigma_{11}^{-1}(x_{(1)} - \mu_{(1)})\right)^{-\frac{1}{2}(m+p_1)}, \quad (4.34)$$

$$f_{X_{(2)}|X_{(1)}}(x_{(2)}|x_{(1)}) = \frac{\Gamma(\frac{1}{2}(m+p_2))}{\Gamma(\frac{1}{2}m)(m\pi)^{\frac{1}{2}p_2}} (\det \Sigma_{22.1})^{-\frac{1}{2}}$$

$$\times \left[1 + \frac{1}{m}(x_{(2)} - \mu_{(2)} - \Sigma_{21}\Sigma_{11}^{-1}(x_{(1)} - \mu_{(1)}))' \Sigma_{22.1}^{-1}\right.$$

$$\times \left. (x_{(2)} - \mu_{(2)} - \Sigma_{21}\Sigma_{11}^{-1}(x_{(1)} - \mu_{(1)}))\right]^{-\frac{1}{2}(m+p_2)}. \quad (4.35)$$

4.3.5. Multivariate Normal Characterization of $E_p(\mu, \Sigma, q)$

We now give several normal characterization of the elliptically symmetric probability density functions.

Theorem 4.3.5.1. *Let $X = (X_1, \ldots, X_p)'$ have the distribution $E_p(\mu, \Sigma, q)$. If the marginal probability density function of any subvector of X is multinormal then X is distributed as $N_p(\mu, \Sigma)$.*

Proof. Let $X = (X'_{(1)}, X'_{(2)})'$, $X_{(1)} = (X_1, \ldots, X_{p_1})'$ and let $t = (t_1, \ldots, t_p)'$ be similarly partitioned as X. From Theorem 4.3.4.5

$$E(\exp\{it'X\}) = \exp\{it'\mu\}\psi(t'\Sigma t).$$

From this it follows that

$$E(\exp\{it'_{(1)}X_{(1)}\}) = \exp\{it'_{(1)}\mu_{(1)}\}\psi(t'_{(1)}\Sigma_{11}t_{(1)}).$$

Thus the characteristic function of X has the same functional form as that of the characteristic function of $X_{(1)}$. If $X_{(1)}$ is distributed as $N_{p_1}(\mu_{(1)}, \Sigma_{11})$ then

$$\psi(t'_{(1)}\Sigma_{11}t_{(1)}) = \exp\left\{-\frac{1}{2}t'_{(1)}\Sigma_{11}t_{(1)}\right\}.$$

Properties of Multivariate Distributions

Thus

$$\psi(t'\Sigma t) = \exp\left\{-\frac{1}{2}t'\Sigma t\right\},$$

which implies that X is distributed as $N_p(\mu, \Sigma)$. Q.E.D.

Theorem 4.3.5.2. *Let $X = (X_1, \ldots, X_p)'$ have the distribution $E_p(\mu, \Sigma, q)$ and let Σ be a diagonal matrix with diagonal elements $\sigma_1^2, \ldots, \sigma_p^2$. If $X_1, \ldots, X_p$ are independent then X is distributed as $N_p(\mu, \Sigma)$.*

Proof. Let $Y = (Y_1, \ldots, Y_p)' = X - \mu$. Using Theorem 4.3.4.5 the characteristic function of Y is given by

$$E(\exp\{it'Y\}) = \psi(t'\Sigma t)$$

$$= \psi\left(\sum_{j=1}^{p} t_j^2 \sigma_j^2\right).$$

If $Y_1, \ldots, Y_p$ are independent

$$\psi\left(\sum_{j=1}^{p} t_j^2 \sigma_j^2\right) = \psi\left(\sum_{j=1}^{p} \alpha_j^2\right) \quad (4.36)$$

$$= \prod_{j=1}^{p} \psi(\alpha_j^2)$$

where $\alpha_j = t_j \sigma_j$. The equation (4.36) is known as the Hamel equation and has the solution

$$\psi(x) = \exp\{kx\} \quad (4.37)$$

for some constant k. Since Σ is positive definite and the right side of (4.37) is the characteristic function we must have $k < 0$. This implies that Y is distributed as $N_p(0, \Sigma)$ or equivalently X is distributed as $N_p(\mu, \Sigma)$. Q.E.D.

Theorem 4.3.5.3. *Let $X = (X_1, \ldots, X_p)' = (X'_{(1)}, X'_{(2)})'$ with $X_{(1)} = (X_1, \ldots, X_{p_1})'$ distributed as $E_p(\mu, \Sigma, q)$. If the conditional distribution of $X_{(2)}$ given $X_{(1)} = x_{(1)}$ is multinormal for any p_1 then X is distributed as $N_p(\mu, \Sigma)$.*

Proof. From Theorem 4.3.4.3 part (b) the conditional *pdf* of $W_{(2)}$ given $W_{(1)} = w_{(1)}$ is given by

$$f_{W_{(2)}|W_{(1)}}(w_{(2)}|w_{(1)}) = \tilde{q}(w'_{(2)}w_{(2)})$$

$$= \frac{q(w'_{(1)}w_{(1)} + w'_{(2)}w_{(2)})}{\bar{q}(w'_{(1)}w_{(1)})}.$$

Let us assume that

$$\tilde{q}(w'_{(2)}w_{(2)}) = (2\pi)^{-\frac{1}{2}(p-p_1)} \exp\left\{-\frac{1}{2}w'_{(2)}w_{(2)}\right\}.$$

Hence we get

$$q(w'_{(1)}w_{(1)} + w'_{(2)}w_{(2)}) = \bar{q}(w'_{(1)}w_{(1)})$$
$$\times (2\pi)^{-\frac{1}{2}(p-p_1)} \exp\left\{-\frac{1}{2}w'_{(2)}w_{(2)}\right\}.$$

Since for the conditional distribution of $W_{(2)}$, $W_{(1)}$ is fixed and the joint *pdf* of $W_{(1)}$ and $W_{(2)}$ is

$$f_{W_{(1)},W_{(2)}}(w_{(1)}, w_{(2)}) = q(w'_{(1)}w_{(1)} + w'_{(2)}w_{(2)})$$

$$= \bar{q}(w'_{(1)}w_{(1)})(2\pi)^{-\frac{1}{2}(p-p_1)} \exp\left\{-\frac{1}{2}w'_{(2)}w_{(2)}\right\},$$

we conclude that

$$f_{W_{(1)},W_{(2)}}(w_{(1)}, w_{(2)}) = (2\pi)^{-\frac{1}{2}p} \exp\left\{-\frac{1}{2}(w'_{(1)}w_{(1)} + w'_{(2)}w_{(2)})\right\}.$$

From this it follows that X is distributed as $N_p(\mu, \Sigma)$. Q.E.D.

4.3.6. Elliptically Symmetric Distribution Multivariate

Let $X = (X_{ij}) = (X_1, \ldots, X_N)'$, where $X_i = (X_{i1}, \ldots, X_{ip})'$, be a $N \times p$ random matrix with

$$E(X_i) = \mu_i = (\mu_{i1}, \ldots, \mu_{ip})'.$$

Definition 4.3.6.1. Elliptically symmetric distribution (multivariate). A $N \times p$ matrix X with values $x \in E^{Np}$ is said to have a distribution belonging to the family of elliptically symmetric distribution (multivariate) with location parameter $\mu = (\mu_1, \ldots, \mu_N)'$ and scale matrix $\Delta = dig(\Sigma_1, \ldots, \Sigma_N)$ if its

Properties of Multivariate Distributions

probability density function (assuming it exists) can be written as

$$f_X(x) = (\det \Delta)^{-1/2} q\left(\sum_{i=1}^{N}(x_i - \mu_i)'\Sigma_i^{-1}(x_i - \mu_i)\right), \quad (4.38)$$

where q is a function on $[0, \infty)$ of the sum of N quadratic forms $(x_i - \mu_i)'\Sigma_i^{-1}(x_i - \mu_i), i = 1, \ldots, N$.

Let us define for any $N \times p$ random matrix $X = (X_1, \ldots, X_N)', X_i = (X_{i1}, \ldots, X_{ip})', i = 1, \ldots, N$

$$v = vec(x) = (x_1', \ldots, x_N')'. \quad (4.39)$$

It is a $Np \times 1$ vector. In terms of v, μ and Δ we can rewrite (4.38) as a elliptically symmetric distribution (univariate) as follows:

$$f_X(x) = (\det \Delta)^{-1/2} q((v - \delta)' \Delta^{-1}(v - \delta)) \quad (4.40)$$

where $\delta = vec(\mu)$.

We now define another convenient way of rewriting (4.40) in terms of tensor product of matrices.

Definition 4.3.6.2. *Tensor product.* Let $\alpha = (\alpha_{ij})$, $\beta = (\beta_{ij})$ be two matrices of dimensions $N \times m$ and $l \times k$ respectively. The tensor product $\alpha \otimes \beta$ of α, β is the matrix $\alpha \otimes \beta$ given by

$$\alpha \otimes \beta = \begin{pmatrix} \alpha_{11}\beta & \alpha_{12}\beta & ,\ldots, & \alpha_{1m}\beta \\ \alpha_{21}\beta & \alpha_{22}\beta & ,\ldots, & \alpha_{2m}\beta \\ \cdot & \cdot & \ldots & \cdot \\ \alpha_{N1}\beta & \alpha_{N2}\beta & ,\ldots, & \alpha_{Nm}\beta \end{pmatrix}$$

The following theorem gives some basic properties of the vec operation and the tensor product.

Theorem 4.3.6.1. *Let α, β, γ be arbitrary matrices. Then,*

(a) $(\alpha \otimes \beta) \otimes \gamma = \alpha \otimes (\beta \otimes \gamma)$;
(b) $(\alpha \otimes \beta)' = \alpha' \otimes \beta'$;
(c) *if α, β are orthogonal matrices, then $\alpha \otimes \beta$ is also an orthogonal matrix;*
(d) *if α, β are square matrices of the same dimension* $\operatorname{tr}(\alpha \otimes \beta) = (\operatorname{tr} \alpha)(\operatorname{tr} \beta)$;
(e) *if α, β are nonsingular square matrices, then* $(\alpha \otimes \beta)^{-1} = \alpha^{-1} \otimes \beta^{-1}$;
(f) *if α, β are nonsingular square matrices, then* $(\alpha \otimes \beta)' = \alpha' \otimes \beta'$;
(g) *if α, β are positive definite matrices, then $\alpha \otimes \beta$ is also a positive definite matrix;*
(h) *if α, β are square matrices of dimensions $p \times p$ and $q \times q$ respectively, then* $\det(\alpha \otimes \beta) = (\det \alpha)^q (\det \beta)^p$;

(i) let $\alpha, \mathbf{X}, \beta$ be matrices of dimensions $q \times p, p \times N, N \times r$ respectively, then $vec(\alpha \mathbf{X} \beta) = (\beta' \otimes \alpha) vec(X)$;

(j) let α, β be matrices of dimensions $p \times q, q \times r$ respectively, then

$$vec(\alpha\beta) = (I_r \otimes \alpha)vec(\beta) = (\beta' \otimes I_p)vec(\alpha)$$
$$= (\beta' \otimes \alpha)vec(I_q)$$

where I_N is the $N \times N$ identity matrix;

(k) $tr(\alpha \mathbf{X} \beta) = vec\alpha'(I \otimes X)vec(\beta)$.

The proofs are straightforward and are left for the readers.

If X has the probability density function given in (4.38) then its characteristic function has the form

$$\exp\left\{i \sum_{j=1}^N t'_j \mu_j\right\} \psi\left(\sum_{j=1}^N t'_j \Sigma_j t_j\right) \tag{4.41}$$

for some function ψ on $[0, \infty)$ and $t_j \in E^p, j = 1, \ldots, N$.

Example 4.3.6.1. Let $X^\alpha = (X_{\alpha 1}, \ldots, X_{\alpha p})', \alpha = 1, \ldots, N$, be independently distributed normal random vectors with the same mean μ and the same covariance matrix Σ and let

$$X = (X^1, \ldots, X^N)', x = (x^1, \ldots, x^N)'$$

Then the probability density function X is

$$f_X(x) = \prod_{\alpha=1}^N f_{X^\alpha}(x^\alpha)$$

$$= (2\pi)^{-\frac{Np}{2}} (\det \Sigma)^{-N/2} \exp\left\{-\frac{1}{2} tr \Sigma^{-1} (\sum_{\alpha=1}^N (x^\alpha - \mu)(x^\alpha - \mu)')\right\}$$

$$= (2\pi)^{-\frac{Np}{2}} (\det(\Sigma \otimes I))^{-\frac{1}{2}}$$

$$\times \exp\left\{-\frac{1}{2} tr(\Sigma \otimes I)^{-1}(x - e \otimes \mu)(x - e \otimes \mu)'\right\}$$

where $e = (1, \ldots, 1)'$ is an N vector with all components equal to unity and I is the $N \times N$ identity matrix.

Properties of Multivariate Distributions

Example 4.3.6.2.

$$f_X(x) = (\det \Sigma)^{-\frac{1}{2}N} q(\Sigma_{j=1}^{N}(x_j - \mu)'\Sigma^{-1}(x_j - \mu)) \qquad (4.42)$$

and $q(z) = (2\pi)^{-\frac{1}{2}Np} \exp\{-\frac{1}{2}z\}, z \geq 0$.

Then $X_1, \ldots, X_N$ are independently and identically distributed $N_p(\mu, \Sigma)$. We denote the *pdf* (4.42) by $E_{Np}(\mu, \Sigma, q)$.

Example 4.3.6.3. Let X be distributed as $E_{N_p}(\mu, \Sigma, q)$ and let $Y = (Y_1, \ldots, Y_N)' = ((X_1 - \mu), \ldots, (X_N - \mu))'\Sigma^{-\frac{1}{2}}$ and $Y_i = (Y_{i1}, \ldots, Y_{ip})'$. Then

$$f_Y(y) = q\left(\sum_{i=1}^{N} y_i' y_i\right) = q(y'y).$$

Let $e = (1, \ldots, 1)'$ be a $N \times 1$ vector, $\mu = (\mu_1, \ldots, \mu_p)' \in R^p$. The *pdf* of X having distribution $E_{Np}(\mu, \Sigma, q)$ can also be written as

$$f_X(x) = (\det \Sigma)^{-\frac{1}{2}N}(q(\text{tr}\Sigma^{-1}(x - e\mu')'(x - e\mu')). \qquad (4.43)$$

Obviously $E_{Np}(0, \Sigma, q)$ satisfies the condition that X and OX, where O is a $N \times N$ orthogonal matrix, have the same distribution. This follows from the fact that the Jacobian of the transformation $X \to OX$ is unity.

Definition 4.3.6.3. $E_{Np}(0, \Sigma, q)$ is called the *pdf* of a spherically symmetric (multivariate) distribution.

4.3.7. Singular Symmetrical Distributions

In this section we deal with the case where Σ is not of full rank. Suppose that the rank of Σ is $k (k < p)$. We consider the family of elliptically symmetric distributions $E_p(\mu, \Sigma, q)$ with rank of $\Sigma = k < p$ and prove Theorem 4.3.7.1 (below). For this we need the following stochastic representation due to Schoenberg (1938) (without proof).

Definition 4.3.7.1. The generalized inverse (g-inverse) of a $m \times n$ matrix A is an $n \times m$ matrix A^- such that $X = A^- Y$ is a solution of the equation $AX = Y$.

Obviously A^- is not unique and $A^- = A^{-1}$ if A is nonsingular. A necessary and sufficient condition for A^- to be the g-inverse of A is $AA^-A = A$. We refer to Rao and Mitra (1971) for results of g-inverse.

Lemma 4.3.7.1. *Schoenberg (1938). If $\phi_k, k \geq 1$, is the class of all functions ϕ on $[0, \infty)$ to R^1 such that $\phi(\|t\|^2), t \in R^k$, is a characteristic function, then*

$\phi \in \phi_k$ if and only if

$$\phi(u) = \int_0^\infty \Omega_k(r^2 u) dF(r), u \geq 0 \qquad (4.44)$$

for some distribution F on $[0, \infty)$, where $\Omega_k(\|t\|^2), t \in R^k$, is the characteristic function of a k-dimensional random vector U which is uniformly distributed on the unit sphere in R^k. Also $\phi_k \downarrow \phi_\infty$ and $\phi \in \phi_\infty$ if and only if ϕ is given by (4.44) with $\Omega_k(r^2 x) = \exp\{-\frac{1}{2} r^2 x\}$.

Theorem 4.3.7.1. $X = (X_1, \ldots, X_p)'$ is distributed as $E_p(\mu, \Sigma, q)$ with rank $(\Sigma) = k < p$ if and only if X is distributed as $\mu + RAU$ where $R \geq 0$ is distributed independently of $U = (U_1, \ldots, U_k)'$, $\Sigma = AA'$ with A a $p \times k$ matrix of rank k, and the distribution F of R is related to ϕ as in (4.44).

Proof. The if part follows from (4.44). Since $A^-(X - \mu) = Y$ has the characteristic function $\phi(\|t\|^2), t \in R^k$ we conclude that $\phi \in \phi_k$ and using (4.44) we get Y is distributed as RU which implies that X is distributed as $\mu + RAU$. This proves the only if part of the Theorem. Q.E.D.

4.4. CONCENTRATION ELLIPSOID AND AXES (MULTIVARIATE NORMAL)

It may be observed that the probability density function [given in Eq. (4.1)] of a p-variate normal distribution is constant on the ellipsoid

$$(x - \mu)' \Sigma^{-1} (x - \mu) = C$$

in E^p for every positive constant C. The family of ellipsoids obtained by varying $C (C > 0)$ has the same center μ, their shapes and orientation are determined by Σ and their sizes for a given Σ are determined by C. In particular,

$$(x - \mu)' \Sigma^{-1} (x - \mu) = p + 2 \qquad (4.45)$$

is called the concentration ellipsoid of X (see Cramer (1946)).

It may be verified that the probability density function defined by the uniform distribution

$$f_X(x) = \begin{cases} \left\{\dfrac{\Gamma(\frac{1}{2} p + 1)}{(\det \Sigma)^{1/2} ((p+2)\pi)^{p/2}}\right\} & \text{if } (x - \mu)' \Sigma^{-1}(x - \mu) \leq p + 2, \\ 0 & \text{otherwise,} \end{cases} \qquad (4.46)$$

has the same mean $E(X) = \mu$ and the same covariance matrix

$$E(X - \mu)(X - \mu)' = \Sigma$$

Properties of Multivariate Distributions

of the p-variate normal distribution. Representing any line through the center μ to the surface of the ellipsoid

$$(x - \mu)'\Sigma^{-1}(x - \mu) = C$$

by its coordinates on the surface, the principal axis of the ellipsoid

$$(x - \mu)'\Sigma^{-1}(x - \mu) = C$$

will have its coordinates x which maximize its squared half length

$$(x - \mu)'(x - \mu)$$

subject to the restriction that

$$(x - \mu)'\Sigma^{-1}(x - \mu) = C.$$

Using the Lagrange multiplier λ we can conclude that the coordinates of the first (longest) principal axis must satisfy

$$(I - \lambda\Sigma^{-1})(x - \mu) = 0$$

or, equivalently

$$(\Sigma - \lambda I)(x - \mu) = 0 \tag{4.47}$$

From (4.47) the squared length of the first principal axis of the ellipsoid

$$(x - \mu)'\Sigma^{-1}(x - \mu) = C$$

for fixed C, is equal to

$$4(x - \mu)'(x - \mu) = 4\lambda_1(x - \mu)'\Sigma^{-1}(x - \mu) = 4\lambda_1 C$$

where λ_1 is the largest characteristic root of Σ. The coordinates of x, specifying the first principal axis, are proportional to the characteristic vector corresponding to λ_1. Thus the position of the first principal axis of the ellipsoid

$$(x - \mu)'\Sigma^{-1}(x - \mu) = C$$

is specified by the direction cosines which are the elements of the normalized characteristic vector corresponding to the largest characteristic root of Σ.

The second (longest) axis has the orientation given by the characteristic vector corresponding to the second largest characteristic root of Σ. In Chapter 1 we have observed that if the characteristic roots of Σ are all different, then the corresponding characteristic vectors are all orthogonal and hence in this case the positions of the axes are uniquely specified by p mutually perpendicular axes. But if any two successive roots of Σ (in descending order of magnitude) are equal, the ellipsoid is circular through the plane generated by the corresponding characteristic vectors. However, two perpendicular axes can be constructed for the common root, though their position through the circle is hardly unique. If γ_i is a characteristic root of Σ of multiplicity r_i, then the ellipsoid has a hyperspherical shape in the r_i-dimensional subspace.

Thus for any p-variate normal random vector X we can define a new p-variate normal random vector $Y = (Y_1, \ldots, Y_p)'$ whose elements Y_i have values on the ellipsoid by means of the transformation (called the principal axis transformations)

$$Y = A'(x - \mu) \qquad (4.48)$$

where the columns of A are the normalized characteristic vectors α_i of Σ. If the characteristic roots of Σ are all different or if the characteristic vectors corresponding to the multiple characteristic roots of Σ have been constructed to be orthogonal, then the covariance of the principal axis variates Y is a diagonal matrix whose diagonal elements are the characteristic roots of Σ. Thus the principal axis transformation of the p-variate normal vector X results in uncorrelated variates whose variances are proportional to axis length of any specified ellipsoid.

Example 4.4.1. Consider Example 4.1.1 with $\mu = 0$ and $\sigma_1^2 = \sigma_2^2 = 1$. The characteristic roots of Σ are $\gamma_1 = 1 + \rho$, $\gamma_2 = 1 - \rho$, and the corresponding characteristic vectors are

$$\left(\frac{1}{\sqrt{2}}, \frac{1}{\sqrt{2}}\right)\left(\frac{1}{\sqrt{2}}, -\frac{1}{\sqrt{2}}\right).$$

If $\rho > 0$, the first principal axis (major axis) is $y_2 = y_1$ and the second axis (minor axis) is $y_2 = -y_1$. For $\rho < 0$ the first principal axis is $y_2 = -y_1$ and the second axis is $y_2 = y_1$.

4.5. REGRESSION, MULTIPLE AND PARTIAL CORRELATION

We define these concepts in details dealing with the multivariate normal distribution though they apply to other multivariate distributions defined above.

We observed in Theorem 4.1.5 that the conditional probability density function of $X_{(2)}$ given that $X_{(1)} = x_{(1)}$, is a $(p-q)$-variate normal with mean $\mu_{(2)} + \Sigma_{21}\Sigma_{11}^{-1}(x_{(1)} - \mu_{(1)})$ and covariance matrix $\Sigma_{22.1} = \Sigma_{22} - \Sigma_{21}\Sigma_{11}^{-1}\Sigma_{12}$. The matrix $\Sigma_{21}\Sigma_{11}^{-1}$ is called the matrix of regression coefficients of $X_{(2)}$ on $X_{(1)} = x_{(1)}$.

The quantity

$$E(X_{(2)}|X_{(1)} = x_{(1)}) = \mu_{(2)} + \Sigma_{21}\Sigma_{11}^{-1}(x_{(1)} - \mu_{(1)}) \qquad (4.49)$$

Properties of Multivariate Distributions

is called the regression surface of $X_{(2)}$ on $X_{(1)}$. This is a linear regression since it depends linearly on $x_{(1)}$. This is used to predict $X_{(2)}$ from the observed value $x_{(1)}$ of $X_{(1)}$. It will be shown in Theorem 4.5.1 that among all linear combinations $\alpha X_{(1)}$, the one that minimizes

$$\text{var}(X_{q+i} - \alpha X_{(1)})$$

is the linear combination $\beta_{(i)} \Sigma_{11}^{-1} X_{(1)}$ where α is a row vector and $\beta_{(i)}$ denotes the ith row of the matrix Σ_{21}.

Since (4.49) holds also for the multivariate complex (Theorem 4.2.4) and for $E_p(\mu, \Sigma)$ (Theorem 4.3.4.3) the same definition applies for these two families of distributions.

The regression terminology is due to Galton (1889) who first introduced it in his studies of the correlation (association) between diameter of seeds of parents and daughters of sweet peas and between heights of fathers and sons. He observed that the heights of sons of either unusually short or tall fathers tend more closely to the average height than their deviant father's values did to the mean for their generation; the daughters of dwarf peas are less dwarfish and the daughters of giant peas are less giant than their respective parents. Galton called his phenomenon "regression (or reversion) to mediocrity" and the parameters of the linear relationship as regression parameters.

From (4.49) it follows that

$$E(X_{q+i} | X_{(1)} = x_{(1)}) = \mu_{q+i} + \beta_{(i)}(x_{(1)} - \mu_{(1)})$$

where $\beta_{(i)}$ denotes the ith row of $(p-q) \times q$ matrix Σ_{21}.

Furthermore the covariance between X_{q+1} and $\beta_{(i)} \Sigma_{11}^{-1} X_{(1)}$ is given by

$$E((X_{q+i} - \mu_{q+i})(\beta_{(i)} \Sigma_{11}^{-1}[X_{(1)} - \mu_{(1)}])') = E(X_{q+i} - \mu_{q+i})$$
$$\times (X_{(1)} - \mu_{(1)})' \Sigma_{11}^{-1} \beta'_{(i)} = \beta_{(i)} \Sigma_{11}^{-1} \beta'_{(i)}$$

and

$$\text{var}((X_{q+i}) = \sigma^2_{q+i}, \text{var}(\beta_{(i)} \Sigma_{11}^{-1} X_{(1)})$$
$$= E(\beta_{(i)} \Sigma_{11}^{-1}(X_{(1)} - \mu_{(1)})$$
$$(X_{(1)} - \mu_{(1)})' \Sigma_{11}^{-1} \beta'_{(i)} = \beta'_{(i)} \Sigma_{11}^{-1} \beta_{(i)}$$

The coefficient of correlation between X_{q+i} and $\beta_{(i)} \Sigma_{11}^{-1} X_{(1)}$ is defined by the positive square root of $\beta'_{(i)} \Sigma_{11}^{-1} \beta_{(i)}$ and is written as

$$\rho = \sqrt{\frac{\beta'_{(i)} \Sigma_{11}^{-1} \beta_{(i)}}{\sigma^2_{q+i}}} \qquad (4.50)$$

Its square ρ^2 is known as the coefficient of determination. Note that $0 \le \rho \le 1$ unlike an ordinary correlation coefficient.

Definition 4.5.1. *Multiple correlation.* The term ρ, as defined above, is called the multiple correlation between the component X_{q+i} of $X_{(2)}$ and the linear function $(\beta_{(i)} X_{(1)})$. *Note*: For the p-variate complex normal distribution $\rho^2 = \beta_{(i)}^* \Sigma_{11}^{-1} \beta_{(i)}$.

The following theorem will show that the multiple correlation coefficient ρ is the correlation between X_{q+i} and its best linear predictor.

Theorem 4.5.1. *Of all linear combinations $\alpha X_{(1)}$ of $X_{(1)}$, the one that minimizes the variance of $(X_{q+i} - \alpha X_{(1)})$ and maximizes the correlation between X_{q+i} and $\alpha X_{(1)}$ is the linear function $\beta_{(i)} \Sigma_{11}^{-1} X_{(1)}$.*

Proof. Let $\beta = \beta_{(i)} \Sigma_{11}^{-1}$. Then

$$\text{var}(X_{q+i} - \alpha X_{(1)}) = E(X_{q+i} - \mu_{q+i} - \alpha(X_{(1)} - \mu_{(1)}))^2$$

$$= E((X_{q+i} - \mu_{q+i})$$

$$- \beta(X_{(1)} - \mu_{(1)}))^2 + E((\beta - \alpha)(X_{(1)} - \mu_{(1)}))^2$$

$$+ 2E(X_{q+i} - \mu_{q+i} - \beta(X_{(1)} - \mu_{(1)}))((\beta - \alpha)(X_{(1)} - \mu_{(1)}))'.$$

But $E(\beta(X_{(1)} - \mu_{(1)})(X_{q+i} - \mu_{q+i}) = \beta \beta'_{(i)} = \beta_{(i)} \Sigma_{11}^{-1} \beta'_{(i)}$; $E((X_{q+i} - \mu_{q+i}) - \beta(X_{(1)} - \mu_{(1)}))(X_{(1)} - \mu_{(1)})' = \beta'_{(i)} - \beta'_{(i)} = 0;$

$$E((\beta - \alpha)(X_{(1)} - \mu_{(1)}))^2 = (\beta - \alpha) E(X_{(1)} - \mu_{(1)})(X_{(1)} - \mu_{(1)})'(\beta - \alpha)'$$

$$= (\beta - \alpha) \Sigma_{11} (\beta - \alpha)';$$

$$E(X_{q+i} - \mu_{q+i}) - \beta(X_{(1)} - \mu_{(1)}))^2 = \sigma^2_{q+i} - 2\beta_{(i)} \Sigma_{11}^{-1} \beta'_{(i)} + \beta_{(i)} \Sigma_{11}^{-1} \beta'_{(i)}$$

$$= \sigma^2_{q+i} - \beta_{(i)} \Sigma_{11}^{-1} \beta'_{(i)}.$$

Hence

$$\text{var}(X_{q+i} - \alpha X_{(1)}) = \sigma^2_{q+i} - \beta_{(i)} \Sigma_{11}^{-1} \beta'_{(i)} + (\beta - \alpha) \Sigma_{11} (\beta - \alpha)'.$$

Properties of Multivariate Distributions

Since Σ_{11} is positive definite, $(\beta - \alpha)\Sigma_{11}(\beta - \alpha)' \geq 0$ and is equal to zero if $\beta = \alpha$. Thus $\beta X_{(1)}$ is the linear function such that $X_{q+i} - \beta X_{(1)}$ has the minimum variance.

We now consider the correlation between X_{q+i} and $\alpha X_{(1)}$ and show that this correlation is maximum when $\alpha = \beta$. For any nonzero scalar C, $C \alpha X_{(1)}$ is a linear function of $X_{(1)}$. Hence

$$E(X_{q+i} - \mu_{q+i} - \beta(X_{(1)} - \mu_{(1)}))^2 \geq E(X_{q+i} - \mu_{q+i} - C\alpha(X_{(1)} - \mu_{(1)}))^2 \quad (4.51)$$

Dividing both sides of (4.51) by $\sigma_{q+i}[E(\beta(X_{(1)} - \mu_{(1)}))^2]^{1/2}$ and choosing

$$C = \left[\frac{E(\beta(X_{(1)} - \mu_{(1)}))^2}{E(\alpha(X_{(1)} - \mu_{(1)}))^2} \right]^{1/2}$$

we get from (4.51)

$$\frac{E(X_{q+i} - \mu_{q+i})(\beta(X_{(1)} - \mu_{(1)}))}{\sigma_{q+i}[E(\beta(X_{(1)} - \mu_{(1)}))^2]^{1/2}} \geq \frac{E(X_{q+i} - \mu_{q+i})(\alpha(X_{(1)} - \mu_{(1)}))}{\sigma_{q+i}[E(\alpha(X_{(1)} - \mu_{(1)}))^2]^{1/2}}$$

Q.E.D.

Definition 4.5.2. Partial correlation coefficient. Let $\sigma_{ij \cdot 1, \ldots, q}$ be the (i, j)th element of the matrix $\Sigma_{22 \cdot 1} = \Sigma_{22} - \Sigma_{21} \Sigma_{11}^{-1} \Sigma_{12}$ of dimension $(p - q) \times (p - q)$. Then

$$\rho_{ij \cdot 1, \ldots, q} = \frac{\sigma_{ij \cdot 1, \ldots, q}}{(\sigma_{ii \cdot 1, \ldots, q} \sigma_{jj \cdot 1, \ldots, q})^{\frac{1}{2}}} \quad (4.52)$$

is called the partial correlation coefficient (of order q) between the components X_{q+i} and X_{q+j} when $X_1, \ldots, X_q$ are held fixed.

Thus the partial correlation is the correlation between two variables when the combined effects of some other variables of them are eliminated.

We would now like to find a recursive relation to compute the partial correlation of order k (say) from the partial correlations of order $(k - 1)$.

Let $X = (X_1, \ldots, X_p)'$ be normally distributed with mean μ and positive definite covariance matrix Σ. Write

$$X = (X'_{(1)}, X'_{(2)}, X'_{(3)})'$$

where

$$X_{(1)} = (X_1, \ldots, X_{p_1})', X_{(2)} = (X_{p_1+1}, \ldots, X_{p_1+p_2})',$$
$$X_{(3)} = (X_{p_1+p_2+1}, \ldots, X_p)',$$

and

$$\Sigma = \begin{pmatrix} \Sigma_{11} & \Sigma_{12} & \Sigma_{13} \\ \Sigma_{21} & \Sigma_{22} & \Sigma_{23} \\ \Sigma_{31} & \Sigma_{32} & \Sigma_{33} \end{pmatrix},$$

where Σ_{ij} are submatrices of Σ of dimensions $p_i \times p_j$, $i, j = 1, 2, 3$ satisfying $p_1 + p_2 + p_3 = p$. From Theorem 4.1.5(c)

$$\text{cov}\left(\begin{pmatrix} X_{(2)} \\ X_{(3)} \end{pmatrix} \Big| X_{(1)} = x_{(1)} \right) = \begin{pmatrix} \Sigma_{22} & \Sigma_{23} \\ \Sigma_{32} & \Sigma_{33} \end{pmatrix} - \begin{pmatrix} \Sigma_{21} \\ \Sigma_{31} \end{pmatrix} \Sigma_{11}^{-1} (\Sigma_{12} \Sigma_{13}).$$

Following the same argument we can deduce that

$\text{cov}(X_{(3)} | X_{(2)} = x_{(2)}, X_{(1)} = x_{(1)})$

$$= \Sigma_{33} - (\Sigma_{31} \Sigma_{32}) \begin{pmatrix} \Sigma_{11} & \Sigma_{12} \\ \Sigma_{21} & \Sigma_{22} \end{pmatrix}^{-1} \begin{pmatrix} \Sigma_{13} \\ \Sigma_{23} \end{pmatrix}$$

$$= (\Sigma_{33} - \Sigma_{31} \Sigma_{11}^{-1} \Sigma_{13})$$
$$\quad - (\Sigma_{32} - \Sigma_{31} \Sigma_{11}^{-1} \Sigma_{12})(\Sigma_{22} - \Sigma_{21} \Sigma_{11}^{-1} \Sigma_{12})^{-1} (\Sigma_{23} - \Sigma_{21} \Sigma_{11}^{-1} \Sigma_{12}).$$

Now taking $p_1 = q - 1, p_2 = 1, p_3 = p - q$ we get for the (i, j)th element $i, j = q + 1, \ldots, p$,

$$\sigma_{ij \cdot 1, \ldots, q} = \sigma_{ij \cdot 1, \ldots, q-1} - \frac{\sigma_{iq \cdot 1, \ldots, q-1} \sigma_{jq \cdot 1, \ldots, q-1}}{\sigma_{qq \cdot 1, \ldots, q-1}}. \quad (4.53)$$

If $j = i$, we obtain

$$\sigma_{ii \cdot 1, \ldots, q} = \sigma_{ii \cdot 1, \ldots, q-1}(1 - \rho_{iq \cdot 1, \ldots, q-1}^2).$$

Hence from (4.53) we obtain

$$\rho_{ij \cdot 1, \ldots, q} = \frac{\rho_{ij \cdot 1, \ldots, q-1} - \rho_{iq \cdot 1, \ldots, q-1} \rho_{jq \cdot 1, \ldots, q-1}}{[(1 - \rho_{iq \cdot 1, \ldots, q-1}^2)(1 - \rho_{jq \cdot 1, \ldots, q-1}^2)]^{1/2}} \quad (4.54)$$

In particular,

$$\rho_{34.12} = \frac{\rho_{34.1} - \rho_{32.1} \rho_{42.1}}{[(1 - \rho_{32.1}^2)(1 - \rho_{42.1}^2)]^{1/2}}$$

and

$$\rho_{23.1} = \frac{\rho_{23} - \rho_{21} \rho_{31}}{[(1 - \rho_{21}^2)(1 - \rho_{31}^2)]^{1/2}}$$

Properties of Multivariate Distributions

where $\rho_{ij} = \text{cov}(X_i, X_j)$. Thus if all partial correlations of certain order are zero, then all higher order partial correlations must be zero.

In closing this section we observe that in the case of the p-variate and normal and the p-variate complex normal distributions $\rho^2 = 0$ implies the independence of X_{q+i} and $X_{(1)}$. But this does not hold, in general, for the family of elliptically symmetric distributions.

4.5.1. Regressions and Correlations in Symmetric Distributions

We discuss in brief analogous results for the family of symmetric distributions. Let $X = (X_1, \ldots, X_p)' = (X_1, X'_{(2)})'$ with

$$X_{(2)} = (X_2, \ldots, X_p)', \mu = (\mu_1, \ldots, \mu_p)' = (\mu_1, \mu'_{(2)})', \Sigma = \begin{pmatrix} \Sigma_{11} & \Sigma_{12} \\ \Sigma_{21} & \Sigma_{22} \end{pmatrix},$$

where Σ_{22} is the lower right-hand corner $(p-1) \times (p-1)$ submatrix of Σ. From Theorem 4.3.4.3, the conditional distribution of X_1 given $X_{(2)} = x_{(2)}$ is $E_1(\mu_1 + \Sigma_{12}\Sigma_{22}^{-1}(x_{(2)} - \mu_{(2)}), \Sigma_{11} - \Sigma_{12}\Sigma_{22}^{-1}\Sigma_{21}, \tilde{q})$ if X is distributed as $E_p(\mu, \Sigma, q)$. Thus

$$\text{var}(X_1|X_{(2)} = x_{(2)}) = K_{\tilde{q}}(x_{(2)})(\Sigma_{11} - \Sigma_{12}\Sigma_{22}^{-1}\Sigma_{21})$$

where $K_{\tilde{q}}(x_{(2)})$ is a function of $x_{(2)}$ and it depends on q. The regression of X_1 on $X_{(2)}$ is given by

$$E(X_1|X_{(2)} = x_{(2)}) = \mu_1 + \Sigma_{12}\Sigma_{22}^{-1}(x_{(2)} - \mu_{(2)}).$$

It does not depend on q. Let $\beta = \Sigma_{12}\Sigma_{22}^{-1}$. The multiple correlation ρ between X_1 and $X_{(2)}$ is given by

$$\begin{aligned}
\rho &= \rho(X_1, \beta X_{(2)}) \\
&= \frac{E(X_1 - \mu_1)(\beta(X_{(2)} - \mu_{(2)}))'}{[\text{var}(X_1)\text{var}(\beta X_{(2)})]^{\frac{1}{2}}} \\
&= \frac{E(X_1 - \mu_1)(X_{(2)} - \mu_{(2)})'\beta'}{[\text{var}(X_1)(\beta E(X_{(2)} - \mu_{(2)})(X_{(2)} - \mu_{(2)})'\beta')]^{\frac{1}{2}}} \quad (4.55) \\
&= \frac{K_{\tilde{q}}(\Sigma_{12}\Sigma_{22}^{-1}\Sigma_{21})}{K_{\tilde{q}}[\Sigma_{11}(\Sigma_{12}\Sigma_{22}^{-1}\Sigma_{21})]^{\frac{1}{2}}} \\
&= \left[\frac{\Sigma_{12}\Sigma_{22}^{-1}\Sigma_{21}}{\Sigma_{11}}\right]^{\frac{1}{2}}
\end{aligned}$$

Obviously p does not depend on q in $E_p(\mu, \Sigma, q)$. The proof of Theorem 4.5.1 for $E_p(\mu, \Sigma, q)$ is straightforward.

Let $X = (X_1, \ldots, X_p)' = (X'_{(1)}, X'_{(2)})'$ with $X_{(1)} = (X_1, \ldots, X_{p_1})'$ be distributed as $E_p(\mu, \Sigma, q)$. From Theorem 4.3.4.3 it follows (changing the role of $X_{(1)}$ by $X_{(2)}$ and vice versa) that the conditional distribution of $X_{(1)}$ given $X_{(2)} = x_{(2)}$ is elliptically symmetric with mean $\mu_{(1)} + \Sigma_{12}\Sigma_{22}^{-1}(x_{(2)} - \mu_{(2)})$ and covariance matrix $K_{\tilde{q}}(x_{(2)})(\Sigma_{11} - \Sigma_{12}\Sigma_{22}^{-1}\Sigma_{21})$ where $K_{\tilde{q}}(x_{(2)})$ is a positive constant related to the conditional distribution $\tilde{q}$. Partial correlation coefficients as defined in (4.52) do not depend on q of $E_p(\mu, \Sigma, q)$.

4.6. CUMULANTS AND KURTOSIS

Let Y be a random variable with characteristic function $\phi_Y(t)$, and let $\mu_j = E(Y - E(Y))^j, j = 1, 2, \ldots$ be the jth central moment of Y. The coefficients

$$\beta_1 = \frac{\mu_3^2}{\mu_2^3}, \beta_2 = \frac{\mu_4}{\mu_2^2} \quad (4.56)$$

are called measures of skewness and kurtosis, respectively, of the distribution of Y. For the univariate normal distribution with mean μ and variance σ^2, $\beta_1 = 0, \beta_2 = 3$.

Assuming that all moments of Y exist, the cumulants κ_j of order j of the distribution of Y are the coefficients κ_j in

$$\log \phi_Y(t) = \sum_{j=0}^{\infty} \kappa_j \frac{(it)^j}{j!}$$

In terms of the raw moments $m_j = E(Y^j), j = 1, 2, \ldots$ the first four cumulants are given by

$$\begin{aligned} \kappa_1 &= m_1 \\ \kappa_2 &= m_2 - m_1^2 \\ \kappa_3 &= m_3 - 3m_2 m_1 + 2m_1^3 \\ \kappa_4 &= m_4 - 4m_1 m_3 - 3m_2^2 + 12m_2 m_1^2 - 6m_1^4. \end{aligned} \quad (4.57)$$

We now define cumulants for the multivariate distributions. Let $\phi_X(t), t = (t_1, \ldots, t_p)' \in R^p$, be the characteristic function of the random vector $X = (X_1, \ldots, X_p)'$.

Properties of Multivariate Distributions

Definition 4.6.1. Cumulants. Assuming that all moments of the distribution of X exist, the cumulants of the distribution of X are coefficients $\kappa_{r_1\ldots r_p}^{1\ldots p}$ in

$$\log \phi_X(t) = \sum_{r_1\ldots r_p=0}^{\infty} \kappa_{r_1\ldots r_p}^{1\ldots p} \frac{(it_1)^{r_1}\ldots(it_p)^{r_p}}{r_1!\ldots r_p!}. \qquad (4.58)$$

The superscript on κ refers to coordinate variables $X_1, \ldots, X_p$ and the subscript on κ refers to the order of the cumulant.

Example 4.6.1. Let $X = (X_1, \ldots, X_p)'$ be distributed as $N_p(\mu, \Sigma)$ with $\mu = (\mu_1, \ldots, \mu_p)'$ and $\Sigma = (\sigma_{ij})$. Then

$$E(X_i - \mu_i)(X_j - \mu_j) = \sigma_{ij};$$
$$E(X_i - \mu_i)(X_j - \mu_j)(X_l - \mu_l) = 0, i \neq j \neq l;$$
$$E(X_i - \mu_i)(X_j\mu_j)(X_e - \mu_e)(X_m - \mu_m)$$
$$= \sigma_{ij}\sigma_{lm} + \sigma_{il}\sigma_{jm} + \sigma_{im}\sigma_{jl}.$$

Hence

$$\kappa_{10\ldots 0}^{12\ldots p} = \mu_1, \kappa_{010\ldots 0}^{12\ldots p} = \mu_2, \ldots, \kappa_{0\ldots 01}^{12\ldots p} = \mu_p,$$

$$\kappa_{20\ldots 0}^{12\ldots p} = \sigma_{11}, \kappa_{020\ldots 0}^{12\ldots p} = \sigma_{22}, \ldots, \kappa_{0\ldots 02}^{12\ldots p} = \sigma_{pp},$$

$$\kappa_{110\ldots 0}^{12\ldots p} = \sigma_{12}, \qquad \text{and etc.}$$

and all cumulants for which $\sum_{i=1}^{p} r_i > 2$ are zero. Let $X = (X_1, \ldots, X_p)'$, be distributed as $E_p(\mu, \Sigma, q)$. ▷From Theorem 4.3.4.5, the characteristic function of X is given by $\phi_X(t) = \exp\{it'\mu\}\psi(t'\Sigma t)$ for some function ψ on $[0, \infty)$ and $t \in R^p$. The covariance matrix of X is given by

$$\Delta = E(X-\mu)(X-\mu)' = -2\psi'(0)\Sigma = (\sigma_{ij}) \quad \text{(say)} \qquad (4.59)$$

where $\psi'(0) = (\partial/\partial t)\psi(t'\Sigma t)|_{t=0}$. Assuming the existence of the moments of fourth order and differentiating $\log \phi_X(t)$, it is easy to verify that the marginal distribution of each component of X has zero skewness and the same kurtosis

$$\beta_2 = \frac{3[\psi''(0) - \psi'(0)]^2}{(\psi'(0))^2} = 3\kappa \quad \text{(say)} \qquad (4.60)$$

All fourth order cumulants are

$$\kappa_{1111}^{ijlm} = \kappa(\sigma_{ij}\sigma_{lm} + \sigma_{il}\sigma_{jm} + \sigma_{im}\sigma_{jl}). \qquad (4.61)$$

For more relevant basic results in the context of elliptically symmetric distributions we refer to Cambanis et al. (1981); Das Cupta et al. (1972); Dawid (1977, 1978); Giri (1993, 1996); Kariya and Sinha (1989); Kelkar (1970).

4.7. THE REDUNDANCY INDEX

Let $X_{(1)} : p \times 1, X_{(2)} : q \times 1$ be two random vectors with $E(X_{(i)}) = \mu_{(i)}, i = 1, 2$ and covariance $\Sigma_{ij} = E[(X_{(i)} - \mu_{(i)})(X_{(j)} - \mu_{(j)})', i, j = 1, 2$. The population redundancy index ρI, introduced by Stewart and Love (1968) and generalized by Gleason (1976) is given by

$$\rho I = \frac{tr \Sigma_{12} \Sigma_{22}^{-1} \Sigma_{21}}{tr \Sigma_{11}} \tag{4.62}$$

It is related to the prediction of $X_{(1)}$ by $X_{(2)}$ by multiple linear regression. Lazraq and Cléroux (2002) gives an up-to-date reference on this index. It is evident that $0 \le \rho I \le 1$. ρI equals to the squared simple correlation coefficient if $p = q = 1$ and it reduces to the square multiple correlation if $p = 1$ and $q > 1$.

EXERCISES

1. Find the mean and the covariance matrix of the random vector $X = (X_1, X_2)'$ with probability density function $f_X(x) = (1/2\pi) \exp\{-\frac{1}{2}(2x_1^2 + x_2^2 + 2x_1x_2 - 22x_1 - 14x_2 + 65)\}$ and $x \in E^2$.
2. Show that if the sum of two independent random variables is normally distributed, then each one is normally distributed.
3. Let $X = (X_1, X_2)'$ be a random vector with the moment generating function

   $$E(\exp(t_1X_1 + t_2X_2)) = a(\exp(t_1 + t_2) + 1)) + b(\exp(t_1) + \exp(t_2)),$$

   where a, b are positive constants satisfying $a + b = \frac{1}{2}$. Find the covariance of matrix of X.
4. (Intraclass covariance). Let $X = (X_1, \ldots, X_p)'$ be distributed as $N_p(\mu, \Sigma)$

   $$\Sigma = \sigma^2 \begin{pmatrix} 1 & \rho & \cdots & \rho \\ \rho & 1 & \cdots & \rho \\ \cdot & \cdots & \cdots & \cdot \\ \rho & \cdots & \cdot & 1 \end{pmatrix},$$

   where $\sigma^2 > 0, -1/(p-1) < \rho < 1$.

Properties of Multivariate Distributions

(a) Show that $\det \Sigma = (\sigma^2)^p (1 + (p-2)\rho)(1-\rho)^{p-1}$.

(b) Show that

$$\Sigma^{-1} = (\sigma^{ij}), \quad \sigma^{ii} = \frac{(1+(p-2)\rho)}{\sigma^2(1-\rho)(1+(p-1)\rho)},$$

$$\sigma^{ij} = -\rho/(\sigma^2(1+(p-1)\rho)(1-\rho)), \quad i, j = 1, \ldots, p.$$

(c) Write down the probability density function of X in terms of μ, ρ and σ^2.

(d) Find the joint probability density function of $(X_1 + X_2, X_1 - X_2)'$.

5. Let $X = (X_1, X_2)'$ be distributed as $N_2(0, \Sigma)$ with

$$\Sigma = \begin{pmatrix} 1 & \rho \\ \rho & 1 \end{pmatrix}, \quad -1 < \rho < 1.$$

Show that

(a) $P(X_1 X_2 > 0) = \frac{1}{2} + (1/\pi) \sin^{-1} \rho$.

(b) $P(X_1 X_2 < 0) = (1/\pi) \cos^{-1} \rho$.

In Theorem 4.1.5 show that

(a) The marginal distribution of $X_{(2)}$ is $N_{p-q}(\mu_{(2)}, \Sigma_{22})$.

(b) The conditional distribution of $X_{(1)}$ given $X_{(2)} = x_{(2)}$ is q-variate normal with mean $\mu_{(1)} + \Sigma_{12}\Sigma_{22}^{-1}(x_{(2)} - \mu_{(2)})$ and covariance matrix $\Sigma_{11} - \Sigma_{12}\Sigma_{22}^{-1}\Sigma_{21}$.

(c) Show that

$$x'\Sigma^{-1}x = (x_{(1)} - \Sigma_{12}\Sigma_{22}^{-1}x_{(2)})'(\Sigma_{11} - \Sigma_{12}\Sigma_{22}^{-1}\Sigma_{21.})^{-1}$$

$$\times (x_{(1)} - \Sigma_{12}\Sigma_{22}^{-1}x_{(2)}) + x'_{(2)}\Sigma_{22}^{-1}x_{(2)}.$$

6. Let X_i, $i = 1, \ldots, n$ be independently distributed $N_p(\mu_i, \Sigma_i)$. Show that the distribution of $\Sigma_i a_i X_i$ where $a_1, \ldots, a_n$ are real, is distributed as $N_p(\Sigma_i a_i \mu_i, \Sigma_i a_i^2 \Sigma_i)$.

7. Let $X = (X_1, \ldots, X_p)'$ be distributed as $N_p(\mu, \Sigma)$ where $\mu = (\mu_1, \ldots, \mu_p)'$, $\Sigma = (\sigma_{ij})$ and let $Y = (Y_1, \ldots, Y_p)'$ where $Y_i = X_{p+1-i}$, $i = 1, \ldots, p$. Find the probability density function of Y.

8. (*The best linear predictor*). Let $X = (X_1, \ldots, X_p)'$ be a random vector with $E(X) = 0$ and covariance Σ. Show that among all functions g of $X_2, \ldots, X_p$

$$E(X_1 - g(X_2, \ldots, X_p))^2$$

is minimum when

$$g(x_2, \ldots, x_p) = E(X_1 | X_2 = x_2, \ldots, X_p = x_p).$$

9. Let $Z_1, \ldots, Z_n$ be independently distributed complex normal random variables with $E(Z_j) = \alpha_j$, $\text{var}(Z_j) = \sigma_j^2$, $j = 1, \ldots, n$. Show that for $a_j, b_j \in \mathbb{C}$, $j = 1, \ldots, n$, $\sum_{j=1}^{n}(a_j Z_j + b_j)$ is distributed as a complex normal with mean $\sum_{j=1}^{n}(a_j \alpha_j + b_j)$ and variance $\sum_{j=1}^{n}(a_j \bar{a}_j \sigma_j^2)$.

10. In Theorem 4.2.5 find
    (a) The marginal distribution of $Z_{(2)}$;
    (b) The conditional distribution of $Z_{(1)}$ given $Z_{(2)} = z_{(2)}$.

11. Let Z be distributed as a p-variate complex normal. Show that its characteristic function is given by (4.18).

12. Let $X = (X_1, \ldots, X_p)'$ be distributed as $E_p(\mu, \Sigma, q)$ with characteristic function

$$e^{it'\mu}\Psi(t'\Sigma t), \, t \in E^p,$$

where Ψ is a function on $[0, \infty)$. Show that $E(X) = \mu$,

$$\text{cov}(X) = -\Psi'(0)\Sigma$$

where

$$\Psi'(0) = \frac{\partial \psi(t'\Sigma t)}{\partial t}\bigg|_{t=0}.$$

13. In Theorem 4.3.4.3 show that
    (a) the marginal distribution of $X_{(2)}$ is $E_{p-q}(\mu_{(2)}, \Sigma_{22}, \bar{q})$ with $\int_{R^{p_2}} \bar{q}(w'_{(2)} w_{(2)}) dw_{(2)} = 1$,
    (b) find the conditional distribution of $X_{(1)}$ given $X_{(2)} = x_{(2)}$.

14. Let $X = (X_1, X_2, X_3)'$ be a random vector whose first and second moments are assumed known. Show that among all linear functions $a + bX_2 + cX_3$, the linear function that minimizes

$$E(X_1 - a - bX_2 - cX_3)^2$$

is given by

$$E(X_1) + \beta(X_2 - E(X_2)) + \gamma(X_3 - E(X_3))$$

where

$$\beta = \mathrm{cov}(X_1, X_2)\sigma_{11} + \mathrm{cov}(X_1, X_3)\sigma_{12},$$
$$\gamma = \mathrm{cov}(X_1, X_2)\sigma_{21} + \mathrm{cov}(X_1, X_3)\sigma_{22},$$
$$\sigma_{11} = \mathrm{var}(X_3)/\Delta, \ \sigma_{22} = \mathrm{var}(X_2)/\Delta,$$
$$\sigma_{12} = \sigma_{21} = -\mathrm{cov}(X_2, X_3)/\Delta,$$
$$\Delta = \mathrm{var}(X_2)\mathrm{var}(X_3)(1 - \rho^2(X_2, X_3)),$$

and $\rho(X_2, X_3)$ is the coefficient of correlation between X_2, and X_3.

15 (*Residual variates*). Let $X = (X'_{(1)}, X'_{(2)})'$, be a p-dimensional normal random vector with mean $\mu = (\mu'_{(1)}, \mu'_{(2)})'$ and covariance matrix

$$\Sigma = \begin{pmatrix} \Sigma_{11} & \Sigma_{12} \\ \Sigma_{21} & \Sigma_{22} \end{pmatrix}$$

where $\Sigma_{11} = \mathrm{cov}(X_{(1)})$. The random vector

$$X_{1.2} = X_{(1)} - \mu_{(1)} - \Sigma_{12}\Sigma_{22}^{-1}(X_{(2)} - \mu_{(2)})$$

is called the residual variates since it represents the discrepancies of the elements of $X_{(1)}$ from their values as predicted from the mean vector of the conditional distribution of $X_{(1)}$ given $X_{(2)} = x_{(2)}$. Show that
(a) $E(X_{(1)} - \mu_{(1)})X'_{1.2} = \Sigma_{11} - \Sigma_{12}\Sigma_{22}^{-1}\Sigma_{21}$,
(b) $E(X_{(2)} - \mu_{(2)})X'_{1.2} = 0$.

16 Show that the multiple correlation coefficient $\rho_{1(2,\ldots,p)}$ of X_1 on $X_2, \ldots, X_p$ of the normal vector $X = (X_1, \ldots, X_p)'$ satisfies

$$1 - \rho^2_{1(2,\ldots,p)} = (1 - \rho^2_{12})(1 - \rho^2_{13.2})\cdots(1 - \rho^2_{1p.2,3,\ldots,p-1}).$$

17 Show that the multiple correlation $\rho_{1(2,\ldots,j)}$ between X_1 and $(X_2, \ldots, X_j)$, $j = 2, \ldots, p$, satisfy

$$\rho^2_{1(2)} \leq \rho^2_{1(23)} \leq \cdots \leq \rho^2_{1(2,\ldots,p)}.$$

In other words, the multiple correlation cannot be reduced by adding to the set of variables on which the dependence of X_1 has to be measured.

18 Let the covariance matrix of a four dimensional normal random vector $X = (X_1, \ldots, X_4)'$ be given by

$$\Sigma = \sigma^2 \begin{pmatrix} 1 & \rho & \rho^2 & \rho^3 \\ \rho & 1 & \rho & \rho^2 \\ \rho^2 & \rho & 1 & \rho \\ \rho^3 & \rho^2 & \rho & 1 \end{pmatrix}.$$

Find the partial correlation coefficient between the $(i+1)$th and $(i-1)$th component of X when the ith component is held fixed.

19 Let $X = (X_1, X_2, X_3)'$ be normally distributed with mean 0 and covariance matrix

$$\Sigma = \begin{pmatrix} 3 & 1 & 1 \\ 1 & 3 & 1 \\ 1 & 1 & 3 \end{pmatrix}.$$

Show that the first principal axis of its concentration ellipsoid passes through the point $(1, 1, 1)$.

20 (*Multinomial distribution*). Let $X = (X_1, \ldots, X_p)'$ be a discrete p-dimensional random vector with probability mass function

$$p_{X_1,\ldots,X_p}(x_1, \ldots, x_p) = \begin{cases} \dfrac{n!}{x_1! \ldots x_p!} \prod_{i=1}^{p} p_i^{x_i} & \text{if } 0 \leq x_i \leq n \text{ for all } n, \sum_{1}^{p} x_i = n; \\ 0 & \text{otherwise,} \end{cases}$$

where $p_i \geq 0$, $\sum_{1}^{p} p_i = 1$.

(a) Show that

$$E(X_i) = np_i, \ \text{var}(X_i) = np_i(1-p_i),$$
$$\text{cov}(X_i, X_j) = -np_i p_j \, (i \neq j).$$

(b) Find the characteristic function of X.
(c) Show that the marginal probability mass function of

$$X_{(1)} = (X_1, \ldots, X_q)', q \leq p,$$

is given by

$$p_{X_1,\ldots,X_q}(x_1,\ldots,x_q) = \frac{n!}{x_1!\ldots x_q!(n-n_o)!}$$

$$\times \prod_{i=1}^{q} p_i^{x_i}(1 - p_1 - \cdots - p_q)^{n-n_o} \text{ if } \sum_{1}^{q} x_i = n_o.$$

(d) Find the conditional distribution of X_1 given $X_3 = x_3, \ldots, X_q = x_q$.
(e) Show that the partial correlation coefficient is

$$\rho_{12.3,\ldots,q} = \left[\frac{p_1 p_2}{(1 - p_2 - p_3 - \cdots - p_q)(1 - p_1 - p_3 - \cdots - p_q)}\right]^{1/2}$$

(f) Show that the squared multiple correlation between X_1 and $(X_2, \ldots, X_p)'$ is

$$\rho = \frac{p_1(p_2 + \cdots + p_q)}{(1 - p_1)(1 - p_2 - \cdots - p_q)}.$$

(g) Let $Y_i = (X_i - np_i)/\sqrt{n}$. Show that as $n \to \infty$ the distribution of $(Y_1, \ldots, Y_{p-1})'$ tends to a multivariate normal distribution. Find its mean and its covariance matrix.

21 (*The multivariate log-normal distribution*). Let $X = (X_1, \ldots, X_p)'$ be normally distributed with mean μ and positive definite (symmetric) covariance matrix $\Sigma = (\sigma_{ij})$. For any random vector $Y = (Y_1, \ldots, Y_p)'$ let us define

$$\log Y = (\log Y_1, \ldots, \log Y_p)'$$

and let $\log Y_i = X_i, i = 1, \ldots, p$. Then Y is said to have a p-variate log-normal distribution with probability density function

$$f_Y(y) = (2\pi)^{-p/2}(\det \Sigma)^{-1/2} \left(\prod_{i=1}^{p} y_i^{-1}\right)$$

$$\times \exp\left\{-\frac{1}{2}(\log y - \mu)'\Sigma^{-1}(\log y - \mu)\right\}$$

when $y_i > 0, i = 1, \ldots, p$ and is zero otherwise.

(a) Show that for any positive integer r

$$E(Y_i^r) = \exp\left\{r\mu_i + \frac{1}{2}r^2\sigma_{ii}\right\},$$

$$\text{var}(Y_i) = \exp\{2\mu_i + 2\sigma_{ii}\} - \exp\{2\mu_i + \sigma_{ii}\},$$

$$\text{cov}(Y_i Y_j) = \exp\left\{\mu_i + \mu_j + \frac{1}{2}(\sigma_{ii} + \sigma_{jj})\right\}$$

$$= \exp\left\{\mu_i + \mu_j + \frac{1}{2}(\sigma_{ii} + \sigma_{jj})\right\}.$$

(b) Find the marginal probability density function of $(Y_1, \ldots, Y_q) q < p$.

22 (*The multivariate beta (Dirichlet) distribution*). Let $\mathbf{X} = (X_1, \ldots, X_p)'$ be a p-variate random vector with values in the simplex

$$S = \left\{ \mathbf{x} = (x_1, \ldots, x_p)' : x_i \geq 0 \quad \text{for all} \quad \sum_1^p x_i \leq 1 \right\}.$$

X has a multivariate beta distribution with parameters $v_1, \ldots, v_{p+1}$, $v_i > 0$, if its probability density function is given by

$$f_X(x) = \begin{cases} \dfrac{\Gamma(v_1 + \cdots + v_{p+1})}{\Gamma(v_1)\cdots\Gamma(v_{p+1})} \left(\prod_{i=1}^p x^{v_i-1}\right) \left(1 - \sum_1^p x_i\right)^{v_{p+1}-1} & \text{if } x \in S, \\ 0 & \text{otherwise.} \end{cases}$$

(a) Show that

$$E(X_i) = \frac{v_i}{v_1 + \cdots + v_{p+1}}, \quad i = 1, \ldots, p,$$

$$\text{var}(X_i) = \frac{v_i(v_1 + \cdots + v_{p+1} - v_i)}{(v_1 + \cdots + v_{p+1})^2(1 + v_1 + \cdots + v_{p+1})},$$

$$\text{cov}(X_i X_j) = \frac{-v_i v_j}{(v_1 + \cdots + v_{p+1})^2(1 + v_1 + \cdots + v_{p+1})} \quad (i \neq j).$$

(b) Show that the marginal probability density function of $X_1, \ldots, X_q$ is a multivariate beta with parameters $v_1, \ldots, v_q, v_{q+1} + \cdots + v_{p+1}$.

23 Let $Z = (Z_1, \ldots, Z_p)'$ be distributed as $E_p(0, I)$ and let $L = Z'Z$, $e_i^2 = Z_i^2/L$, $i = 1, \ldots, p$.

Properties of Multivariate Distributions

   (a) Show that L is independent of $(e_1, \ldots, e_p)$.
   (b) The joint distribution of $e_1, \ldots, e_p$ is Dirichlet as given in (4.26).
24. (*Multivariate Student t-distribution*). The random vector $X = (X_1, \ldots, X_p)'$ has a p-variate Student t-distribution with N degrees of freedom if the probability density function of X can be written as

$$f_X(x) = C(\det \Sigma)^{-1/2}(N + (x - \mu)'\Sigma^{-1}(x - \mu))^{-(N+p)/2}$$

where $x \in E^p$, $\mu = (\mu_1, \ldots, \mu_p)'$, Σ is a symmetric positive definite matrix of dimension $p \times p$ and

$$C = \frac{N^{N/2}\Gamma\left(\dfrac{N+p}{2}\right)}{\pi^{p/2}\Gamma\left(\dfrac{N}{2}\right)}.$$

   (a) Show that

$$E(X) = \mu \text{ if } N > 1, \; \text{cov}(X) = [N/(N-2)]\Sigma, \; N > 2.$$

   (b) Show that the marginal probability density function of $(X_1, \ldots, X_q)'$, $q < p$, is distributed as a q-variate Student t-distribution.

25. (*Multivariate exponential power distribution*). Let $X = (X_1, \ldots, X_p)'$ has a p-variate exponential power distribution if its probability density function is given by

$$f_X(x) = \frac{C(\theta)}{(\det \Sigma)^{\frac{1}{2}}} \exp\left[-\frac{1}{2}\{(x - \mu)'\Sigma^{-1}(x - \mu)\}^\theta\right],$$

$\theta \in R^+ = \{r \in R, r > 0\}$, $\mu \in R^p$, $\Sigma > 0$ and $C(\theta)$ is a positive constant. Show that

$$C(\theta) = \theta\Gamma\left(\frac{p}{2}\right)[2^{1/\theta}\pi]^{\frac{-p}{2}}[\Gamma(p/2\theta)]^{-1}.$$

REFERENCES

Basu, D. (1951). On the independence of linear functions of independent chance variables. *Bull. Int. Statis. Inst.* 33:83–86.

Basu, D. (1955). A note on the multivariate extension of some theorems related to the univariate normal distribution. *Sanklya* 17:221–224.

Box, G. E. P., Tiao, G. C. (1973). *Bayesian Inference in Statistical Analysis.* Reading: Addison-Wesley.

Cambanis, S., Huang, S., Simons, C. (1981). On the theory of elliptically contoured distributions. *J. Multivariate Anal.* 11:368–375.

Chu, K'ai-Ching (1973). Estimation and decision for linear systems with elliptical random processes. *IEEE Trans. Automatic Control*, 499–505.

Cramer, H. (1937). *Random Variables and Probability Distribution* (Cambridge Tracts, No. 36). London and New York: Cambridge Univ. Press.

Cramer, H. (1946). *Mathematical Methods of Statistics.* Princeton, New Jersey: Princetown Univ. Press.

Darmois, G. (1953). Analyse générale des liaisons stochastiques, étude particulière de l'analyse factorielle linéaire. *Rev. Inst. Int. Statist.* 21:2–8.

Das Cupta, S., Eaton, M. L., Olkin, I., Pearlman, M., Savage, L. J., Sobel, M. (1972). Inequalities on the probability content of convex regions for elliptically contoured distributions. In: Proc. Sixth Berkeley Symp. Math. Statist. Prob. 2, Univ. of California Press, Berkeley, pp. 241–264.

Dawid, A. P. (1978). Extendibility of spherical matrix distributions. *J. Multivariate. Anal.* 8:559–566.

Dawid, A. P. (1977). Spherical matrix distributions and a multivariate model. *J. Roy. Statist. Soc. B*, 39:254–261.

Fréchet, M. (1951). Généralisation de la loi de probabilité de Laplace. *Ann. Inst. Henri Poincaré* 12:Fasc 1.

Galton, F. (1889). *Natural Inheritance.* New York: MacMillan.

Giri, N. (1974). *An Introduction to Probability and Statistics.* Part 1 Probability. New York: Dekker.

Giri, N. (1993). *Introduction to Probability and Statistics*, 2nd ed, revised and expanded. New York: Dekker.

Giri, N. (1996). *Multivariate Statistical Analysis.* New York: Dekker.

Gleason, T. (1976). On redundancy in canonical analysis. *Psycho. Bull.* 83:1004–1006.

Kagan, A., Linnik, Yu. V., Rao, C. R. (1972). *Characterization Problems of Mathematical Statistics* (in Russian). U.S.S.R.: Academy of Sciences.

Kariya, T., Sinha, B. (1989). *Robustness of Statistical Tests.* New York: Academic Press.

Kelkar, D. (1970). Distribution theory of spherical distribution and a location scale parameter generalization. *Sankhya A*, 43:419–430.

Lazraq, A., Cléroux, R. (2002). Statistical inference concerning several redundancy indices. *J. Multivariate Anal.* (to appear).

McGraw, D. K., Wagner, J. F. (1968). Elliptically symmetric distributions. *IEEE Trans Infor.* 110–120.

Nimmo-Smith, I. (1979). Linear regression and sphericity. *Biometrika*, 66:390–392.

Rao, C. R., Mitra, S. K. (1971). *Generalized Inverse of Matrices and Its Applications*. New York: John Wiley.

Shoenberg, I. J. (1938). Metric space and completely monotone functions. *Ann. Math.* 39:811–841.

Skitovic, V. P. (1954). Linear combinations of independent random variables and the normal distribution law, *Izϑ. Akad. Nauk. SSSR Ser. Mat.* 18:185–200.

Stewart, D., Love, W. (1968). A general canonical index. *Psycho. Bull.* 70:160–163.

Zellner, A. (1976). Bayesian and non-Bayesian analysis of the regression model with multivariate T-error.term. *J.A.S.A.*, 440–445.

5
Estimators of Parameters and Their Functions

5.0. INTRODUCTION

We observed in Chapter 4 that the probability density function (when it exists) of the multivariate normal distribution, the multivariate complex normal distribution and the elliptically symmetric distribution depends on the parameter μ and Σ. In this chapter we will estimate these parameters and some of their functions, namely the multiple correlation coefficient, partial correlation coefficients of different orders, and regression coefficients on the basis of information contained in a random sample of size N from the multivariate normal and the multivariate complex normal distribution, and, elliptically symmetric distribution (multivariate) where $\mu_1 = \cdots = \mu_N = \mu$ and $\Sigma_1 = \cdots = \Sigma_N = \Sigma$. Equivariant estimation under curved models will be treated in this chapter.

The method of maximum likelihood (Fisher, 1925) has been very successful in finding suitable estimators of parameters in many problems. Under certain regularity conditions on the probability density function, the maximum likelihood estimator is strongly consistent in large samples (Wald (1943); Wolfowitz (1949); LeCam (1953); Bahadur (1960)). Under such conditions, if the dimension p of the random vector is not large, it seems likely that the sample size N occuring in practice would usually be large enough for this optimum result to hold. However, if p is large it may be that the sample size N needs to be extremely large for this result to apply; for example, there are cases where N/p^3 must be large. The fact that the maximum likelihood is not universally good has been demonstrated by Basu (1955), Neyman and Scott (1948), and Kiefer and Wolfowitz (1956), among others.

In recent years methods of multivariate Bayesian analysis have proliferated through virtually all aspects of multivariate Bayesian analysis. Berger (1993) gave a brief summary of current subjects in multivariate Bayesian analysis along with recent references on this topic.

5.1. MAXIMUM LIKELIHOOD ESTIMATORS OF μ, Σ IN $N_p(\mu, \Sigma)$

Let $x^\alpha = (x_{\alpha 1}, \ldots, x_{\alpha p})'$, $\alpha = 1, \ldots, N$ be a sample of size N from a normal distribution $N_p(\mu, \Sigma)$ with mean μ and positive definite covariance matrix Σ, and let

$$\bar{x} = \sum_{\alpha=1}^{N} x^\alpha / N, \quad s = \sum_{\alpha=1}^{N} (x^\alpha - \bar{x})(x^\alpha - \bar{x})'$$

We are interested here in finding the maximum likelihood estimates of (μ, Σ). The likelihood of the sample observations x^α, $\alpha = 1, \ldots, N$ is given by

$$L(x^1, \ldots, x^N) = (2\pi)^{-Np/2} (\det \Sigma)^{-N/2} \exp\left\{-\frac{1}{2} \sum_{\alpha=1}^{N} (x^\alpha - \mu)' \Sigma^{-1} (x^\alpha - \mu)\right\}$$

$$= (2\pi)^{-Np/2} (\det \Sigma)^{-N/2} \exp\left\{-\frac{1}{2} \operatorname{tr} \Sigma^{-1} \sum_{\alpha=1}^{N} (x^\alpha - \mu)(x^\alpha - \mu)'\right\}$$

$$= (2\pi)^{-Np/2} (\det \Sigma)^{-N/2} \exp\left\{-\frac{1}{2} \operatorname{tr} \Sigma^{-1} (s + N(\bar{x} - \mu)(\bar{x} - \mu)')\right\}$$

as

$$\sum_{\alpha=1}^{N} (x^\alpha - \mu)(x^\alpha - \mu)'$$

$$= \sum_{\alpha=1}^{N} (x^\alpha - \bar{x})(x^\alpha - \bar{x})' + N(\bar{x} - \mu)(\bar{x} - \mu)' + 2 \sum_{\alpha=1}^{N} (x^\alpha - \bar{x})(\bar{x} - \mu)'$$

$$= \sum_{\alpha=1}^{N} (x^\alpha - \bar{x})(x^\alpha - \bar{x})' + N(\bar{x} - \mu)(\bar{x} - \mu)'$$

Estimators of Parameters and Their Functions

Since Σ is positive definite

$$N(\bar{x} - \mu)'\Sigma^{-1}(\bar{x} - \mu) \geq 0$$

for all $\bar{x} - \mu$ and is zero if and only if $\bar{x} = \mu$. Hence

$$L(x^1, \ldots, x^N \mu, \Sigma) \leq (2\pi)^{-Np/2}(\det \Sigma)^{-N/2} \exp\left\{-\frac{1}{2}\operatorname{tr} \Sigma^{-1} s\right\}$$

and the equality holds if $\mu = \bar{x}$. Thus $\bar{x}$ is the maximum likelihood estimator of μ for all Σ. We will assume throughout that $N > p$; the reason for such an assumption will be evident from Lemma 5.1.2. Given x^α, $\alpha = 1, \ldots, N$, L is a function of μ and Σ only and we will denote it simply by $L(\mu, \Sigma)$. Hence

$$L(\hat{\mu}, \Sigma) = (2\pi)^{-Np/2}(\det \Sigma)^{-N/2} \exp\left\{-\frac{1}{2}\operatorname{tr} \Sigma^{-1} s\right\} \quad (5.1)$$

We now prove three Lemmas which are useful in the sequel and for subsequent presentations.

Lemma 5.1.1. *Let A be any symmetric positive definite matrix and let*

$$f(A) = c(\det A)^{N/2} \exp\left\{-\frac{1}{2}\operatorname{tr} A\right\}$$

where c is a positive constant. Then $f(A)$ has a maximum in the space of all positive definite matrices when $A = NI$, where I is the identity matrix of dimension $p \times p$.

Proof. Clearly

$$f(A) = c \prod_{i=1}^{p}(\theta_i^{N/2} \exp\{-\theta_i/2\})$$

where $\theta_1, \ldots, \theta_p$ are the characteristic roots of the matrix A. But this is maximum when $\theta_1 = \cdots = \theta_p = N$, which holds if and only if $A = NI$. Hence $f(A)$ is maximum if $A = NI$.   Q.E.D.

Lemma 5.1.2. *Let $X^\alpha = (X_{\alpha 1}, \ldots, X_{\alpha p})'$, $\alpha = 1, \ldots, N$, be independently distributed normal random vectors with the same mean vector μ and the same positive definite covariance matrix Σ, and let*

$$S = \sum_{\alpha=1}^{N}(X^\alpha - \bar{X})(X^\alpha - \bar{X})' \quad \text{where} \quad \bar{X} = \frac{1}{N}X^\alpha.$$

Then

(a) $\bar{X}, S$ are independent, $\sqrt{N}\bar{X}$ has a p-dimensional normal distribution with mean $\sqrt{N}\mu$ and covariance matrix Σ, and S is distributed as

$$\sum_{\alpha=1}^{N-1} Z^\alpha Z^{\alpha\prime}$$

where Z^α, $\alpha = 1, \ldots, N-1$ are independently distributed normal p-vectors with the same mean $\mathbf{0}$ and the same covariance matrix Σ;
(b) S is positive definite with probability one if and only if $N > p$.

Proof.

(a) Let $\mathbf{0}$ be an orthogonal matrix of dimension $N \times N$ of the form

$$\mathbf{0} = \begin{pmatrix} 0_{11} & \cdots & 0_{1N} \\ \vdots & & \vdots \\ 0_{N-11} & \cdots & 0_{N-1N} \\ 1/\sqrt{N} & \cdots & 1/\sqrt{N} \end{pmatrix}$$

The last row of $\mathbf{0}$ is the equiangular vector of unit length. Since X^α, $\alpha = 1, \ldots, N$ are independent,

$$E(X^\alpha - \mu)(X^\beta - \mu)' = \begin{cases} 0 & \text{if } \alpha \neq \beta \\ \Sigma & \text{if } \alpha = \beta \end{cases}.$$

Let

$$Z^\alpha = \sum_{\beta=1}^{N} 0_{\alpha\beta} X^\beta, \quad \alpha = 1, \ldots, N.$$

The set of vectors Z^α, $\alpha = 1, \ldots, N$, has a joint normal distribution because the entire set of components is a set of linear combinations of the components of the set of vectors X^α, which has a joint normal distribution.

Estimators of Parameters and Their Functions

Now

$$E(Z^N) = \sqrt{N}\mu,$$

$$E(Z^\alpha) = \sum_{\beta=1}^{N} \mu 0_{\alpha\beta} = \sqrt{N}\mu \sum_{\beta=1}^{N} 0_{\alpha\beta} \frac{1}{\sqrt{N}} = 0, \; \alpha \neq N,$$

$$\operatorname{cov}(Z^\alpha Z^\gamma) = \sum_{\beta=1}^{N} 0_{\alpha\beta} 0_{\gamma\beta} E(X^\beta - \mu)(X^\beta - \mu)' = \begin{cases} 0 & \text{if } \alpha \neq \gamma \\ \Sigma & \text{if } \alpha = \gamma \end{cases}.$$

Furthermore,

$$\sum_{\alpha=1}^{N} Z^\alpha Z^{\alpha'} = \sum_{\alpha=1}^{N} \sum_{\beta=1}^{N} 0_{\alpha\beta} X^\beta \sum_{\gamma=1}^{N} 0_{\alpha\gamma} X'_\gamma$$

$$= \sum_{\beta=1}^{N} \sum_{\gamma=1}^{N} \left(\sum_{\alpha=1}^{N} 0_{\alpha\beta} 0_{\alpha\gamma} \right) X^\beta X^\gamma = \sum_{\beta=1}^{N} X^\beta X^{\beta'}.$$

Thus it is evident that $Z^\alpha, \alpha = 1, \ldots, N$ are independent and Z^α, $\alpha = 1, \ldots, N-1$, are normally distributed with mean 0 and covariance matrix Σ. Since

$$Z^N = \sqrt{N}\bar{X}$$

$$S = \sum_{\alpha=1}^{N} X^\alpha X^{\alpha'} - Z^N Z^{N'} = \sum_{\alpha=1}^{N-1} Z^\alpha Z^{\alpha'},$$

we conclude that $\bar{X}, S$ are independent, Z^N has p-variate normal distribution with mean $\sqrt{N}\mu$ and covariance matrix Σ and S is distributed as $\sum_{\alpha=1}^{N-1} Z^\alpha Z^{\alpha'}$.

(b) Let $B = (Z^1, \ldots, Z^{N-1})$. Then $S = BB'$ where B is a matrix of dimension $p \times (N-1)$. This part will be proved if we can show that B has rank p with probability one if and only if $N > p$. Obviously by adding more columns to B we can not diminish its rank and if $N \leq p$, then the rank of B is less than p. Thus it will suffice to show that B has rank p with probability one when $N - 1 = p$.

For any set of $(p-1)$ p-vectors $(\alpha^1, \ldots, \alpha^{p-1})$ in E^p let $S(\alpha^1, \ldots, \alpha^{p-1})$ be the subspace spanned by $\alpha^1, \ldots, \alpha^{p-1}$. Since Σ is $p \times p$ nonsingular, for any given

$\alpha^1, \ldots, \alpha^{p-1}$,

$$P\{Z^i \in S(\alpha^1, \ldots, \alpha^{p-1})\} = 0.$$

Now, as $Z^1, \ldots, Z^p$ are independent and identically distributed random p-vectors,

$P\{Z^1, \ldots, Z^p$ are linearly dependent$\}$

$$\leq \sum_{i=1}^{p} P\{Z^i \in S(Z^1, \ldots, Z^{i-1}, Z^{i+1}, \ldots, Z^p)\}$$

$$= pP\{Z^1 \in S(Z^2, \ldots, Z^p)\}$$

$$= pE[P\{Z^1 \in S(Z^2, \ldots, Z^p)|Z^2 = z^2, \ldots, Z^p = z^p\}]$$

$$= pE(0) = 0,$$

Q.E.D.

This lemma is due to Dykstra (1970). A similar proof also appears in the lecture notes of Stein (1969).

This result depends heavily on the normal distribution of $Z^1, \ldots, Z^p$. Subsequently Eaton and Pearlman (1973) have given conditions in the case of a random matrix whose columns are independent but not necessarily normal or identically distributed.

Note: The distribution of S is called the Wishart distribution with parameter Σ and degrees of freedom $N - 1$. We will show in Chapter 6 that its probability density function is given by

$$\frac{(\det s)^{(N-p-2)/2} \exp\{-\frac{1}{2}\operatorname{tr} \Sigma^{-1} s\}}{2^{(N-1)p/2} \pi^{p(p-1)/4} (\det \Sigma)^{(N-1)/2} \prod_{i=1}^{p} \Gamma((N-i)/2)}. \quad (5.2)$$

for S positive definite and zero otherwise.

The following lemma gives an important property, usually called the invariance property of the method of maximum likelihood in statistical estimation. Briefly stated, if $\hat{\theta}$ is a maximum likelihood estimator of $\theta \in \Omega$, then $f(\hat{\theta})$ is a maximum likelihood estimator of $f(\theta)$, where $f(\theta)$ is some function of θ.

Lemma 5.1.3. *Let $\theta \in \Omega$ (an interval in a K-dimensional Euclidean space) and let $L(\theta)$ denote the likelihood function—a mapping from Ω to the real line R.*

Assume that the maximum likelihood estimator $\hat{\theta}$ of θ exists so that $\hat{\theta} \in \Omega$ and $L(\hat{\theta}) \geq L(\theta)$ for all $\theta \in \Omega$. Let f be any arbitrary transformation mapping Ω to

Estimators of Parameters and Their Functions

Ω^* *(an interval in an r-dimensional Euclidean space, $1 \leq r \leq k$). Then $f(\hat{\theta})$ is a maximum induced likelihood estimator of $f(\theta)$.*

Proof. Let $w = f(\theta)$ is a function, $f(\hat{\theta})$ is a unique number ($\hat{w}$) of Ω^*. For each $w \in \Omega^*$

$$F(w) = \{\theta, \theta \in \Omega \quad \text{such that} \quad f(\theta) = w\},$$
$$M(w) = \sup_{\theta \in F(w)} L(\theta).$$

The function $M(w)$ on Ω^* is the induced likelihood function of $f(\theta)$. Clearly $\{F(w) : \omega \in \Omega^*\}$ is a partition of Ω and $\hat{\theta}$ belongs to one and only one set of the partition. Let us denote this set by $F(\hat{w})$. Moreover

$$L(\hat{\theta}) = \sup_{\theta \in F(\hat{w})} L(\theta) = M(\hat{w}) \leq \sup_{w \in \Omega^*} M(w) = \sup_{\theta \in w} L(\theta) = L(\hat{\theta}),$$

and $\quad M(\hat{w}) = \sup_{\Omega^*} M(w).$

Hence $\hat{w}$ is a maximum likelihood estimator of $f(\theta)$. Since $\hat{\theta} \in F(\hat{w})$ we get $f(\hat{\theta}) = \hat{w}$. Q.E.D.

From this it follows that if $\hat{\theta} = (\hat{\theta}_1, \ldots, \hat{\theta}_k)$ is a maximum likelihood estimator of $\theta = (\theta_1, \ldots, \theta_k)$ and if the transformation

$$\theta \to (f_1(\theta), \ldots, f_k(\theta))$$

is one to one, then $f_1(\hat{\theta}), \ldots, f_k(\hat{\theta})$ are the maximum likelihood estimators of $f_1(\theta), \ldots, f_k(\theta)$ respectively. Furthermore, if $\hat{\theta}_1, \ldots, \hat{\theta}_k$ are unique, then $f_1(\hat{\theta}), \ldots, f_k(\hat{\theta})$ are also unique.

Since $N > p$ by assumption, from Lemma 5.1.2 we conclude that s is positive definite. Hence we can write $s = \alpha\alpha'$ where α is a nonsingular matrix of dimension $p \times p$. From (5.1) we can write

$$L(\hat{\mu}, \Sigma) = (2\pi)^{-Np/2}(\det \Sigma)^{-N/2} \exp\left\{-\frac{1}{2}\operatorname{tr} \Sigma^{-1} s\right\}$$

$$= (2\pi)^{-Np/2}(\det s)^{-N/2}(\det(\alpha'\Sigma^{-1}\alpha))^{N/2} \exp\left\{-\frac{1}{2}\operatorname{tr} \alpha'\Sigma^{-1}\alpha\right\}$$

Using Lemmas 5.1.1 and 5.1.3 we conclude that

$$\alpha'(\hat{\Sigma})^{-1}\alpha = NI$$

or equivalently, $\hat{\Sigma} = s/N$. Hence the maximum likelihood estimator of μ is $\bar{X}$ and that of Σ is S/N.

5.1.1. Maximum Likelihood Estimator of Regression, Multiple and Partial Correlation Coefficients, Redundancy Index

Let the covariance matrix of the random vector $X = (X_1, \ldots, X_p)'$ be denoted by

$$\Sigma = (\sigma_{ij})$$

with $\sigma_{ii} = \sigma_i^2$. Then

$$\rho_{ij} = \sigma_{ij}/\sigma_i \sigma_j$$

is called the Pearson correlation between the ith and jth components of the random vector X. (Karl Pearson, 1986, gave the first justification for the estimate of ρ_{ij}.)

Write $s = (s_{ij})$. The maximum likelihood estimate of σ_{ij}, on the basis of observations $x^\alpha = (x_{\alpha 1}, \ldots, x_{\alpha p})'$, $\alpha = 1, \ldots, N$, is $(1/N)s_{ij}$. Since $\mu_i = \mu_i$ and $\sigma_i^2 = \sigma_{ii}$, $\rho_{ij} = \sigma_{ij}/(\sigma_{ii}, \sigma_{jj})^{1/2}$ is a function of the σ_{ij}, the maximum likelihood estimates of μ_i, σ_i^2, and ρ_{ij} are

$$\hat{\mu}_i = \bar{x}_i, \quad \hat{\sigma}_i^2 = \frac{1}{N}\sum_{\alpha=1}^{N}(x_{\alpha i} - \bar{x}_i)^2 = \frac{s_i^2}{N},$$

$$\hat{\rho}_{ij} = \frac{s_{ij}}{(s_i^2 s_j^2)^{1/2}} = \frac{\sum_{\alpha=1}^{N}(x_{\alpha i} - \bar{x}_i)(x_{\alpha j} - \bar{x}_j)}{\left(\sum_{\alpha=1}^{N}(x_{\alpha i} - \bar{x}_i)^2\right)^{1/2}\left(\sum_{\alpha=1}^{N}(x_{\alpha j} - \bar{x}_j)^2\right)^{1/2}} \quad (5.3)$$

$$= \frac{\sum_{\alpha=1}^{N}(x_{\alpha i} - \bar{x}_i)x_{\alpha j}}{\left(\sum_{\alpha=1}^{N}(x_{\alpha i} - \bar{x}_i)^2\right)^{1/2}\left(\sum_{\alpha=1}^{N}(x_{\alpha j} - \bar{x}_j)^2\right)^{1/2}} = r_{ij} \quad \text{(say)}.$$

Let $X = (X_1, \ldots, X_p)'$ be normally distributed with mean vector $\mu = (\mu_1, \ldots, \mu_p)'$ and positive definite covariance matrix Σ. We observed in Chapter 4 that the regression surface of $X_{(2)} = (X_{q+1}, \ldots, X_p)'$ On $X_{(1)} = (X_1, \ldots, X_q)' = x_{(1)} = (x_1, \ldots, x_q)'$ is given by

$$E(X_{(2)}|X_{(1)} = x_{(1)}) = \mu_{(2)} + \beta(x_{(1)} - \mu_{(1)})$$

where

$$\beta = \Sigma_{21}\Sigma_{11}^{-1}$$

Estimators of Parameters and Their Functions

is the matrix of regression coefficients of $X_{(2)}$ on $X_{(1)} = x_{(1)}$ and Σ, μ are partitioned as

$$\Sigma = \begin{pmatrix} \Sigma_{11} & \Sigma_{12} \\ \Sigma_{21} & \Sigma_{22} \end{pmatrix}, \mu = (\mu'_{(1)}, \mu'_{(2)})', \mu_{(1)} = (\mu_1, \ldots, \mu_q)',$$

with Σ_{11} the upper left-hand corner submatrix of Σ of dimension $q \times q$. Let

$$s = \sum_{\alpha=1}^{N} (x^\alpha - \bar{x})(x^\alpha - \bar{x})'$$

be similarly partitioned as

$$s = \begin{pmatrix} s_{11} & s_{12} \\ s_{21} & s_{22} \end{pmatrix}.$$

From Lemma 5.1.3 we obtain the following theorem.

Theorem 5.1.1. *On the basis of observations* $x^\alpha = (x_{\alpha 1}, \ldots, x_{\alpha p})'$, $\alpha = 1, \ldots, N$, *from the p-dimensional normal distribution with mean* μ *and positive definite covariance matrix* Σ, *the maximum likelihood estimates of* $\beta, \Sigma_{22.1}$ *and* Σ_{11} *are given by* $\hat{\beta} = s_{21}s_{11}^{-1}, \hat{\Sigma}_{22.1} = (1/N)(s_{22} - s_{21}s_{11}^{-1}s_{12})$, $\hat{\Sigma}_{11} = s_{11}/N$.

Let $s_{ij \cdot 1, \ldots, q}$ be the (i,j)th element of the matrix $s_{22} - s_{22}s_{11}^{-1}s_{12}$ of dimension $(p-q) \times (p-q)$. From Theorem 5.1.1 the maximum likelihood estimate of the partial correlation coefficient between the components X_i and $X_j (i \neq j), i, j = q+1, \ldots, p$ of X, when $X_{(1)} = (X_1, \ldots, X_q)'$ is held fixed, is given by

$$\hat{\rho}_{ij \cdot 1, \ldots, q} = \frac{s_{ij \cdot 1, \ldots, q}}{(s_{ii \cdot 1, \ldots, q})^{1/2}(s_{jj \cdot 1, \ldots, q})^{1/2}} = r_{ij \cdot 1, \ldots, q} \tag{5.4}$$

where $r_{ij \cdot 1, \ldots, q}$ is an arbitrary designation.

In Chapter 4 we defined the multiple correlation coefficient between the ith component X_{q+i} of $X_{(2)} = (X_{q+1}, \ldots, X_p)'$ and $X_{(1)}$ as

$$\rho = \left(\frac{\beta'_{(i)} \Sigma_{11}^{-1} \beta_{(i)}}{\sigma^2_{q+i}} \right)^{1/2}$$

where $\beta_{(i)}$ is the ith row of the submatrix Σ_{21} of dimension $(p-q) \times q$ of Σ.

If $q = p - 1$, then the multiple correlation coefficient between X_p and $(X_1, \ldots, X_{p-1})'$ is

$$\rho = \left(\frac{\Sigma_{21} \Sigma_{11}^{-1} \Sigma_{12}}{\Sigma_{22}} \right)^{1/2}.$$

Since $(\rho, \Sigma_{21}, \Sigma_{11})$ is a one-to-one transformation of Σ, the maximum likelihood estimator of ρ is given by

$$\hat{\rho} = \left(\frac{s_{21} S_{11}^{-1} s_{12}}{s_{22}} \right)^{1/2} = r$$

where r is an arbitrary designation. Since S is positive definite $R^2 \geq 0$. Furthermore

$$1 - R^2 = \frac{S_{22} - S_{21} S_{11}^{-1} S_{12}}{S_{22}} = \frac{\det S}{S_{22}(\det S_{11})}.$$

In the general case the maximum likelihood estimate of ρ is obtained by replacing the parameters by their maximum likelihood estimates.

In Chapter 4 we have defined the redundancy index ρI by

$$\rho I = \frac{\operatorname{tr} \Sigma_{12} \Sigma_{22}^{-1} \Sigma_{21}}{\operatorname{tr} \Sigma_{11}}$$

between two random vectors $X_{(1)} : p \times 1$ and $X_{(2)} : q \times 1$ with $E(X_{(i)}) = \mu_{(i)}$, $i = 1, 2$, $E(X_{(i)} - \mu_{(i)})(X_{(j)} - \mu_{(j)})' = \Sigma_{ij}$, $i, j = 1, 2$. It is related to the prediction of $X_{(1)}$ by $X_{(2)}$ by means of multivariate regression, given by,

$$E(X_{(1)} - \mu_{(1)} | X_{(2)} = x_{(2)}) = B(x_{(2)} - \mu_{(2)})$$

where B, the $q \times p$ regression matrix. Let $x^\alpha = (x_{(1)}^{\alpha'}, x_{(2)}^{\alpha'})'$, $\alpha = 1, \ldots, N$ be a random sample of size N from $X = (X_{(1)}', X_{(2)}')'$. Write $N \bar{X}_{(i)} = \sum_{\alpha=1}^{N} X_{(i)}^\alpha$, $S_{ij} = \sum_{i=1}^{N} (X_{(i)}^\alpha - \bar{X}_{(i)})(X_{(j)}^\alpha - \bar{X}_{(j)})'$, $i, j = 1, 2$. The least square estimate of B is $\hat{B} = S_{12} S_{22}^{-1}$. The sample estimate of ρI is RI given by

$$RI = \frac{\operatorname{tr} S_{12} S_{22}^{-1} S_{21}}{\operatorname{tr} S_{11}}$$

It is called the sample redundancy index.

5.2. CLASSICAL PROPERTIES OF MAXIMUM LIKELIHOOD ESTIMATORS

5.2.1. Unbiasedness

Let $X^\alpha = (X_{\alpha 1}, \ldots, X_{\alpha p})'$, $\alpha = 1, \ldots, N$, be independently and identically distributed normal p-vectors with the same mean vector μ and the same positive definite covariance matrix Σ and let $N > p$. The maximum likelihood estimator of μ is

$$\bar{X} = \frac{1}{N} \sum_{\alpha=1}^{N} X^\alpha$$

and that of Σ is

$$\frac{S}{N} = \frac{1}{N} \sum_{\alpha=1}^{N} (X^\alpha - \bar{X})(X^\alpha - \bar{X})'$$

Furthermore we have observed that S is distributed independently of $\bar{X}$ as

$$S = \sum_{\alpha=1}^{N-1} Z^\alpha Z^{\alpha'}$$

where $Z^\alpha = (Z_{\alpha 1}, \ldots, Z_{\alpha p})'$, $\alpha = 1, \ldots, N-1$, are independently and identically distributed normal p-vectors with the same mean vector 0 and the same positive definite covariance matrix Σ. Since

$$E(\bar{X}) = \mu, \quad E\left(\frac{S}{N-1}\right) = \frac{1}{N-1} \sum_{\alpha=1}^{N-1} E(Z^\alpha Z^{\alpha'}) = \Sigma,$$

we conclude that $\bar{X}$ is an unbiased estimator of μ and $S/(N-1)$ is an unbiased estimator of Σ.

5.2.2. Sufficiency

A statistic $T(X^1, \ldots, X^N)$, which is a function of the random sample X^α, $\alpha = 1, \ldots, N$, only, is said to be sufficient for a parameter θ if the conditional distribution of $X^1, \ldots, X^N$ given $T = t$ does not depend on θ and it is said to be minimal sufficient for θ if the sample space of X^α, $\alpha = 1, \ldots, N$, cannot be reduced beyond that of $T(X^1, \ldots, X^N)$ without losing sufficiency. Explicit procedures for obtaining minimal sufficient statistics are given by Lehmann and Scheffie (1950) and Bahadur (1955). It has been established that the sufficient statistic obtained through the following Fisher-Neyman factorization theorem is minimal sufficient.

Fisher-Neyman Factorization Theorem

Let

$$X^\alpha = (X_{\alpha 1}, \ldots, X_{\alpha p})'$$

$\alpha = 1, \ldots, N$, be a random sample of size N from a distribution with probability density function $f_X(x|\theta)$, $\theta \in \Omega$. The statistic $T(X^1, \ldots, X^N)$ is sufficient for θ if and only if we can find two nonnegative functions $g_T(t|\theta)$ (not necessarily a probability density function) and $K(X^1, \ldots, X^N)$ such that

$$\prod_{\alpha=1}^{N} f_{X^\alpha}(x^\alpha) = g_T(t|\theta) K(x^1, \ldots, x^N)$$

where $g_T(t|\theta)$ depends on $x^1, \ldots, x^N$ only through $T(x^1, \ldots, x^N)$ and depends on θ, and K is independent of θ.

For a proof of this theorem the reader is referred to Giri (1993) or (1975) or to Halmos and Savage (1949) for a general proof involving some deeper theorems of measure theory.

If X^α, $\alpha = 1, \ldots, N$, is a random sample of size N from the p-dimensional normal distribution with mean μ and positive definite covariance matrix Σ, the joint probability density function of X^α, $\alpha = 1, \ldots, N$ is given by

$$(2\pi)^{-Np/2} (\det \Sigma)^{-N/2} \exp\left\{-\frac{1}{2} \text{tr}(\Sigma^{-1} s + N\Sigma^{-1}(\bar{x} - \mu)(\bar{x} - \mu)')\right\}.$$

Using the Fisher-Neyman factorization theorem we conclude that $(\bar{X}, S)$ is a minimal sufficient statistic for (μ, Σ). In the sequel we will use sufficiency to indicate minimal sufficiency.

5.2.3. Consistency

A real valued estimator T_N (function of a random sample of size N) is said to be weakly consistent for a parametric function $g(\theta)$, $\theta \in \Omega$, if T_N converges to $g(\theta)$, $\theta \in \Omega$ in probability i.e. for every $\epsilon > 0$

$$\lim_{N \to \infty} P\{|T_N - g(\theta)| < \varepsilon\} = 1$$

and is strongly consistent if

$$P\{\lim_{N \to \infty} T_N = g(\theta)\} = 1$$

In the case of a normal univariate random variable with mean μ and variance σ^2, the sample mean $\bar{X}$ of a random sample $X_1, \ldots, X_N$ of size N is both weakly and strongly consistent (see Giri, 1993). When the estimator T_N is a random matrix

Estimators of Parameters and Their Functions

there are various ways of defining the stochastic convergence $T_N \to g(\theta)$. Let

$$T_N = (T_{ij}(N)), \quad g(\theta) = (g_{ij}(\theta))$$

be matrices of dimension $p \times q$. For any matrix $A = (a_{ij})$ let us define two different norms

$$N_1(A) = \operatorname{tr} AA', \qquad N_2(A) = \max_{ij} |a_{ij}|.$$

Some alternative ways of defining the convergence of T_N to $g(\theta)$ are

1. $T_{ij}(N)$ converges stochastically to $g_{ij}(\theta)$ for all i, j.
2. $N_1(T_N - g(\theta))$ converges stochastically to zero. (5.5)
3. $N_2(T_N - g(\theta))$ converges stochastically to zero.

It may be verified that these three different ways of defining stochastic convergence are equivalent. We shall establish stochastic convergence by using the first criterion.

To show that $\bar{X}$ converges stochastically to $\mu = (\mu_1, \ldots, \mu_p)'$, $S/(N-1)$ converges stochastically to $\Sigma = (\sigma_{ij})$, $\sigma_{ii} = \sigma_i^2$, we need to show that $\bar{X}_i$ converges stochastically to μ_i for all i, $S_{ij}/(N-1)$ converges stochastically to σ_{ij} for all (i, j), where $\bar{X} = (\bar{X}_1, \ldots, \bar{X}_p)'$ and $S = (S_{ij})$. Since

$$\bar{X}_i = \frac{1}{N} \sum_{\alpha=1}^{N} X_{\alpha i}$$

where $X_{\alpha i}, \alpha = 1, \ldots, N$, are independently and identically distributed normal random variables with mean μ_i and variance σ_i^2, using the Chebychev inequality and the Kolmogorov theorem (see Giri, 1993), we conclude that $\bar{X}_i$ is both weakly and strongly consistent for μ_i, $i = 1, \ldots, p$. Thus $\bar{X}$ is a consistent estimator of μ.

From Lemma 5.1.2 S can be written as

$$S = \sum_{\alpha=1}^{N-1} Z^\alpha Z^{\alpha'}$$

where $Z^\alpha, \alpha = 1, \ldots, N-1$ are independently and identically distributed normal p-vectors with mean 0 and positive definite covariance matrix Σ. Hence

$$\frac{S_{ij}}{N-1} = \frac{1}{N-1} \sum_{\alpha=1}^{N-1} Z_{\alpha i} Z_{\alpha j} = \frac{1}{N-1} \sum_{\alpha=1}^{N-1} Z_\alpha(i, j)$$

where $Z_\alpha(i,j) = Z_{\alpha i} Z_{\alpha j}$. Obviously $Z_\alpha(i,j)$, $\alpha = 1, \ldots, N-1$, are independently and identically distributed random variables with

$$E(Z_\alpha(i,j)) = \sigma_{ij}$$

$$\text{var}(Z_\alpha(i,j)) = E(Z_{\alpha i}^2 Z_{\alpha j}^2) - E^2(Z_{\alpha i} Z_{\alpha j})$$

$$\leq (E(Z_{\alpha i}^4) E(Z_{\alpha j}^4))^{1/2} - \sigma_{ij}^2 \leq \sigma_i^2 \sigma_j^2 (3 - \rho_{ij}^2) < \infty$$

where ρ_{ij} is the coefficient of correlation between the ith and the jth component of Z_α. Now applying the Chebychev inequality and the Kolmogorov theorem we conclude that $S_{ij}/(N-1)$ is weakly and strongly consistent for σ_{ij} for all i,j.

5.2.4. Completeness

Let T be a continuous random variable (univariate or multivariate) with probability density function $f_T(t|\theta)$, $\theta \in \Omega$—the parametric space. The family of probability density functions $\{f_T(t|\theta), \theta \in \Omega\}$ is said to be complete if for any real valued function $g(T)$

$$E_\theta(g(T)) = 0 \tag{5.6}$$

for every $\theta \in \Omega$ implies that $g(T) = 0$ for all values of T for which $f_T(t|\theta)$ is greater than zero for some $\theta \in \Omega$. If the family of probability density functions of a sufficient statistic is complete, we call it a complete sufficient statistic.

We would like to show that $(\bar{X}, S)$ is a complete sufficient statistic for (μ, Σ). From (5.2) the joint probability density function of $\bar{X}, S$ is given by

$$c(\det \Sigma)^{-\frac{1}{2}N}(\det s)^{(N-p-2)/2} \exp\left\{-\frac{1}{2}\text{tr }\Sigma^{-1}s + N(\bar{x} - \mu)'\Sigma^{-1}(\bar{x} - \mu)\right\} \tag{5.7}$$

where

$$c^{-1} = 2^{Np/2} \pi^{p(p+1)/4} N^{-p/2} \prod_{i=1}^p \Gamma\left(\frac{N-i}{2}\right).$$

For any real valued function $g(\bar{X}, S)$ of $(\bar{X}, S)$

$$Eg(\bar{X}, S) = c \int g(\bar{x}, s)(\det \Sigma)^{-N/2}(\det s)^{(N-p-2)/2}$$
$$\times \exp\left\{-\frac{1}{2}\text{tr}(\Sigma^{-1}s + N(\bar{x} - \mu)'\Sigma^{-1}(\bar{x} - \mu)\right\} d\bar{x}\, ds \tag{5.8}$$

Estimators of Parameters and Their Functions

where $d\bar{x} = \Pi_i d\bar{x}_i$, $ds = \Pi_{ij} ds_{ij}$. Write $\Sigma^{-1} = I - 2\theta$ where I is the identity matrix of dimension $p \times p$ and θ is symmetric. Let

$$\mu = (I - 2\theta)^{-1}\alpha.$$

If $Eg(\bar{X}, S) = 0$ for all (μ, Σ), then from (5.8) we get

$$c \int g(\bar{x}, s)(\det(I - 2\theta))^{N/2}(\det s)^{(N-p-2)/2}$$

$$\times \exp\left\{-\frac{1}{2}[\text{tr}(I - 2\theta)(s + N\bar{x}\bar{x}') - 2N\alpha'\bar{x} + N\alpha'(I - 2\theta)^{-1}\alpha]\right\}$$

$$\times d\bar{x}\, ds = 0,$$

or

$$c \int g(\bar{x}, s + N\bar{x}\bar{x}' - N\bar{x}\bar{x}')(\det s)^{(N-p-2)/2}$$

$$\times \exp\left\{-\frac{1}{2}[\text{tr}(s + N\bar{x}\bar{x}') + \text{tr}\,\theta(s + N\bar{x}\bar{x}') + N\alpha'\bar{x}]\right\} \quad (5.9)$$

$$\times d\bar{x}\, ds = 0$$

identically in θ and α. We now identify (5.9) as the Laplace transform of

$$cg(\bar{x}, s + N\bar{x}\bar{x}' - N\bar{x}\bar{x}')(\det s)^{(N-p-2)/2} \exp\left\{-\frac{1}{2}\text{tr}(s + N\bar{x}\bar{x}')\right\}$$

with respect to variables $N\bar{x}, s + N\bar{x}\bar{x}'$. Since this is identically equal to zero for all α and θ we conclude that $g(\bar{x}, s) = 0$ except possibly for a set of values of $(\bar{X}, S)$ with probability measure 0. In other words, $(\bar{X}, S)$ is a complete sufficient statistic for (μ, Σ).

5.2.5. Efficiency

Let $X^\alpha = (X_{\alpha 1}, \ldots, X_{\alpha p})'$, $\alpha = 1, \ldots, N$ be a random sample of size N from a distribution with probability density function $f_X(x|\theta)$, $\theta \in \Omega$. Assume that $\theta = (\theta_1, \ldots, \theta_k)'$ and Ω is E^k (Euclidean k-space) or an interval in E^k.

Consider the problem of estimating parametric functions

$$g(\theta) = (g_1(\theta), \ldots, g_r(\theta))'.$$

We shall denote an estimator

$$T(X^1, \ldots, X^N) = (T_1(X^1, \ldots, X^N), \ldots, T_r(X^1, \ldots, X^N))'$$

simply by $T = (T_1, \ldots, T_r)'$.

An unbiased estimator T of $g(\theta)$ is said to be an efficient estimator of $g(\theta)$ if for any other unbiased estimator U of $g(\theta)$

$$\text{cov}(T) \leq \text{cov}(U) \quad \text{for all} \quad \theta \in \Omega \tag{5.10}$$

in the sense that $\text{cov}(U) - \text{cov}(T)$ is nonnegative definite for all $\theta \in \Omega$.

The efficient unbiased estimator of $g(\theta)$ can be obtained by the following two methods.

Generalized Rao-Cramer Inequality for a Vector Parameter

Let

$$L(\theta) = L(x^1, \ldots, x^N | \theta) = \prod_{\alpha=1}^{N} f_{X^\alpha}(x^\alpha | \theta)$$

$$P_{ij} = -\frac{\partial^2 \log L(\theta)}{\partial \theta_i \partial \theta_j}, \quad I_{ij} = E(P_{ij}).$$

The $k \times k$ matrix

$$I = (I_{ij}) \tag{5.11}$$

is called the Fisher information measure on θ or simply the information matrix (provided the P_{ij} exist).

For any unbiased estimator T^{**} of $g(\theta)$ let us assume that

$$\frac{\partial}{\partial \theta_j} \int T_i^{**} L(x^1, \ldots, x^N | \theta) dx^1, \ldots, dx^N = \int T_i^{**} \frac{\partial L(\theta)}{\partial \theta_j} dx^1, \ldots, dx^N$$

$$= \frac{\partial g_i(\theta)}{\partial \theta_j}, i = 1, \ldots, r, j = 1, \ldots, k,$$

and let

$$\Delta = \left(\frac{\partial g_i(\theta)}{\partial \theta_j}\right) \tag{5.12}$$

be a matrix of dimension $r \times k$. Then it can be verified that (see, e.g., Rao (1965))

$$\text{cov}(T^{**}) - \Delta I^{-1} \Delta' \tag{5.13}$$

is nonnegative definite. Since $\Delta I^{-1} \Delta'$ is defined independently of any estimation procedure it follows that for any unbiased estimator T^{**} of $g(\theta)$

$$\text{var}(T_i^{**}) \geq \sum_{m=1}^{k} \sum_{n=1}^{k} I^{mn} \frac{\partial g_i}{\partial \theta_m} \frac{\partial g_i}{\partial \theta_n}, i = 1, \ldots, r, \tag{5.14}$$

Estimators of Parameters and Their Functions   147

where $I^{-1} = (I^{mn})$. Hence the efficient unbiased estimator of $g(\theta)$ is an estimator T (if it exists) such that

$$\text{cov}(T) = \Delta I^{-1} \Delta'. \tag{5.15}$$

If $g(\theta) = \theta$, then Δ is the identity matrix and the covariance of the efficient unbiased estimator is I^{-1}. From (5.13) it follows that if for any unbiased estimator $T = (T_1, \ldots, T_r)'$ of $g(\theta)$

$$\text{var}(T_i) = \sum_{m=1}^{k} \sum_{n=1}^{k} I^{mn} \frac{\partial g_i}{\partial \theta_m} \frac{\partial g_i}{\partial \theta_n}, \quad i = 1, \ldots, r, \tag{5.16}$$

then

$$\text{cov}(T_i, T_j) = \sum_{m=1}^{k} \sum_{n=1}^{k} I^{mn} \frac{\partial g_i}{\partial \theta_m} \frac{\partial g_j}{\partial \theta_n} \quad \text{for all} \quad i \neq j. \tag{5.17}$$

Thus (5.16) implies that

$$\text{cov}(T) = \Delta I^{-1} \Delta'. \tag{5.18}$$

Thus any unbiased estimator of $g(\theta)$ is efficient if (5.16) holds. Now we would like to establish that (5.16) holds well if

$$T_i = g_i(\theta) + \sum_{j=1}^{k} \xi_{ij} \frac{1}{L(\theta)} \frac{\partial L(\theta)}{\partial \theta_j}, \quad i = 1, \ldots, r, \tag{5.19}$$

where $\xi_i = (\xi_{i1}, \ldots, \xi_{ik})' = \text{const} \times I^{-1} \beta_i$ with

$$\beta_i = \left(\frac{\partial g_i(\theta)}{\partial \theta_1}, \ldots, \frac{\partial g_i(\theta)}{\partial \theta_k} \right)'.$$

To do that let

$$U = T_i - g_i(\theta), \quad W = \sum_{j=1}^{k} \xi_{ij} \frac{1}{L(\theta)} \frac{\partial L(\theta)}{\partial \theta_j}$$

where $\xi_i = (\xi_{i1}, \ldots, \xi_{ik})'$ is a constant nonnull vector which is independent of x^α, $\alpha = 1, \ldots, N$, but possibly dependent on θ. Since

$$\int \frac{\partial}{\partial \theta_i} L(x^1, \ldots, x^N | \theta) dx^1, \ldots, dx^N = \frac{\partial}{\partial \theta_i} \int L(x^1, \ldots, x^N | \theta) dx^1, \ldots, dx^N$$

$$= 0 \quad \text{for all} \quad i,$$

we get $E(W) = 0$. Also $E(U) = 0$. The variances and covariance of U, W are given by

$$\text{var}(U) = \text{var}(T_i),$$

$$\text{var}(W) = \text{var}\left(\sum_{j=1}^{k} \xi_{ij} \frac{1}{L(\theta)} \frac{\partial L(\theta)}{\partial \theta_j}\right)$$

$$= \sum_{j=1}^{k} \sum_{j'=1}^{k} \xi_{ij} \xi_{ij'} \text{cov}\left(\frac{1}{L(\theta)} \frac{\partial L(\theta)}{\partial \theta_j}, \frac{1}{L(\theta)} \frac{\partial L(\theta)}{\partial \theta_{j'}}\right)$$

$$= \sum_{j=1}^{k} \sum_{j'=1}^{k} \xi_{ij} \xi_{ij'} I_{jj'} = \xi_i' I \xi_i,$$

$$\text{cov}(U, W) = \sum_{j=1}^{k} \xi_{ij} E\left((T_i - g_i(\theta)) \frac{1}{L(\theta)} \frac{\partial L(\theta)}{\partial \theta_j}\right)$$

$$= \sum_{j=1}^{k} \xi_{ij} E\left(T_i \frac{1}{L(\theta)} \frac{\partial L(\theta)}{\partial \theta_j}\right)$$

$$= \sum_{j=1}^{k} \xi_{ij} \frac{\partial g_i(\theta)}{\partial \theta_j} = \xi_i' \beta_i$$

where $\beta_i = (\partial g_i(\theta)/\partial \theta_1, \ldots, \partial g_i(\theta)/\partial \theta_k)'$. Now applying the Cauchy-Schwarz inequality, we obtain

$$(\xi_i' \beta_i)^2 \leq \text{var}(T_i)(\xi_i' I \xi_i),$$

which implies that

$$\text{var}(T_i) \geq \frac{(\xi_i' \beta_i)^2}{\xi_i' I \xi_i}.$$

Since ξ_i is arbitrary (nonnull), this implies

$$\text{var}(T_i) \geq \sup_{\xi_i \neq 0} \frac{(\xi_i' \beta_i)^2}{\xi_i' I \xi_i} = \beta_i' I^{-1} \beta_i \quad (5.20)$$

and the supremum is attained when

$$\xi_i = cI^{-1} \beta_i = \xi_i^0,$$

where c is constant and ξ_i^0 is an arbitrary designation.

Estimators of Parameters and Their Functions

The equality in (5.20) holds if and only if

$$U = \text{const} \times W = \text{const} \times \sum_{j=1}^{k} \xi_{ij}^0 \frac{1}{L(\theta)} \frac{\partial L(\theta)}{\partial \theta_j}$$

with probability 1, i.e.,

$$T_i = g_i(\theta) + \sum_{j=1}^{k} \xi_{ij} \frac{1}{L(\theta)} \frac{\partial L(\theta)}{\partial \theta_j}$$

with probability 1 where $\xi_i = \text{const} \times I^{-1} \beta_i$.

To prove that the sample mean $\bar{X}$ is efficient for μ we first observe that $\bar{X}$ is unbiased for μ. Let

$$\Sigma^{-1} = (\sigma^{ij}), \theta = (\mu_1, \ldots, \mu_p, \sigma^{11}, \ldots, \sigma^{pp})'$$

where θ is a vector of dimension $p(p+3)/2$. Let

$$g(\theta) = (g_1(\theta), \ldots, g_p(\theta))' = (\mu_1, \ldots, \mu_p)'.$$

Take $T_i = \bar{X}_i$, $g_i(\theta) = \mu_i$. The likelihood of $x^1, \ldots, x^N$ is

$$L(x^1, \ldots, x^N | \theta) = L(\theta)$$

$$= (2\pi)^{-Np/2} (\det \Sigma^{-1})^{N/2}$$

$$\times \exp\left\{-\frac{1}{2} \text{tr}(\Sigma^{-1} s + N\Sigma^{-1}(\bar{x} - \mu)(\bar{x} - \mu)')\right\}.$$

Hence

$$\frac{\partial \log L}{\partial \mu_i} = N\sigma^{ii}(\bar{x}_i - \mu_i) + N \sum_{j(\neq i)} \sigma^{ij}(\bar{x}_j - \mu_j),$$

$$\frac{\partial g_i(\theta)}{\partial \mu_j} = \begin{cases} 1 & \text{if } j = i, \\ 0 & \text{if } j \neq i, \end{cases}$$

$$\frac{\partial g_i(\theta)}{\partial \sigma^{i'j'}} = 0 \quad \text{for all} \quad i', j', i.$$

Hence

$$\beta_i = (0, \ldots, 0, 1, 0, \ldots, 0)', i = 1, \ldots, p,$$

which is a unit vector with unity as its ith coordinate. Since

$$\frac{\partial^2 \log L(\theta)}{\partial \mu_i^2} = -N\sigma^{ii}, \quad \frac{\partial^2 \log L(\theta)}{\partial \mu_i \partial \mu_j} = -N\sigma^{ij},$$

we get, for $i \neq j$, $\ell', \ell = 1, \ldots, p$,

$$E\left(-\frac{\partial^2 \log L(\theta)}{\partial \mu_i^2}\right) = -N\sigma^{ii}, \quad E\left(-\frac{\partial^2 \log L(\theta)}{\partial \mu_i \partial \mu_j}\right) = N\sigma^{ij},$$

$$E\left(-\frac{\partial^2 \log L(\theta)}{\partial \mu_i \partial \sigma^{\ell\ell'}}\right) = 0.$$

Thus, the information matrix I is given by

$$I = \begin{pmatrix} N\Sigma^{-1} & 0 \\ 0 & A \end{pmatrix}$$

where A is a nonsingular matrix of dimension $\frac{1}{2}p(p+3) \times \frac{1}{2}p(p+3)$. (It is not necessary, in this context, to evaluate A specifically.) So

$$I^{-1}\beta_i = (1/N)(\sigma_{1i}, \ldots, \sigma_{pi}, 0, \ldots, 0)'.$$

Choosing $\xi_{(i)} = I^{-1}\beta_i$, we obtain

$$\sum_{j=1}^{k} \xi_{ij} \frac{1}{L(\theta)} \frac{\partial L(\theta)}{\partial \theta_j} = (\bar{x}_i - \mu_i)(\sigma^{1i}\sigma_{1i} + \cdots + \sigma^{pi}\sigma_{pi})$$

$$+ \sum_{j(\neq i)} (\bar{x}_j - \mu_j)(\sigma^{1j}\sigma_{1i} + \cdots + \sigma^{pj}\sigma_{pi}) = \bar{x}_i - \mu_i$$

since $\Sigma^{-1}\Sigma$ is the identity matrix. Hence we conclude that $\bar{X}$ is efficient for μ.

Second Method

Let $T^* = (T_1*, \ldots, T_k*)'$ be a sufficient (minimal) estimator of θ and let the distribution of T^* be complete. Given any unbiased estimator $T^{**} = (T_1^{**}, \ldots, T_r^{**})'$ of $g(\theta)$, the estimator

$$T = E(T^{**}|T^*) = (E(T_1^{**}|T^*), \ldots, E(T_r^{**}|T^*))'$$

is at least as good as T^{**} for $g(\theta)$, in the sense that $\text{cov}(T^{**}) - \text{cov}(T)$ is nonnegative definite for all $\theta \in \Omega$.

This follows from the fact that for any nonnull vector L, $L'T^{**}$ is an unbiased estimator of the parametric function $L'g(\theta)$ and by the Rao-Blackwell theorem

Estimators of Parameters and Their Functions

(see Giri, 1993), the estimator

$$L'T = E(L'T^{**}|T^*)$$

is at least as good as $L'T^{**}$ for all θ. Since this holds well for all $L \neq 0$ it follows that $\text{cov}(T^{**}) - \text{cov}(T)$ is nonnegative definite for all θ. Thus given any unbiased estimator T^{**} of $g(\theta)$ which is not a function of T^*, the estimator T is better than T^{**}. Hence in our search for efficient unbiased estimators we can restrict attention to unbiased estimators which are functions of T^* alone. Furthermore, if $f(T^*)$ and $g(T^*)$ are two unbiased estimators of $g(\theta)$, then

$$E_\theta(f(T^*) - g(T^*)) \equiv 0 \tag{5.21}$$

for all $\theta \in \Omega$. Since the distribution of T^* is complete (5.21) will imply $f(T^*) - g(T^*) = 0$ almost everywhere. Thus we conclude that there exists a unique unbiased efficient estimator of $g(\theta)$ and this is obtained by exhibiting a function of T^* which is unbiased for $g(\theta)$.

We established earlier that $(\bar{X}, S)$ is a complete sufficient statistic of (μ, Σ) of the p-variate normal distribution. Since $E(\bar{X}) = \mu$ and $E(S/(N-1)) = \Sigma$, it follows that $\bar{X}$ and $S/(N-1)$ are unbiased efficient estimators of μ and Σ, respectively.

5.3. BAYES, MINIMAX, AND ADMISSIBLE CHARACTERS

Let $\mathcal{X}$ be the sample space and let A be the σ-algebra of subsets of $\mathcal{X}$, and let $P_\theta, \theta \in \Omega$, be the probability on $(\mathcal{X}, A)$, where Ω is an interval in E^p. Let D be the set of all possible estimators of θ. A function

$$L(\theta, d), \theta \in \Omega, d \in D,$$

defined on $\Omega \times D$, represents the loss of erroneously estimating θ by d. (It may be remarked that d is a vector quantity.) Let

$$R(\theta, d) = E_\theta(L(\theta, d)) = \int L(\theta, d(x)) f_X(x|\theta) dx \tag{5.22}$$

where $f_X(x|\theta)$ denotes the probability density function of X with values $x \in \mathcal{X}$, corresponding to P_θ with respect to the Lebesgue measure dx. $R(\theta, d)$ is called the risk function of the estimator $d \in D$ for the parameter $\theta \in \Omega$. Let $h(\theta), \theta \in \Omega$, denote the prior probability density on Ω. The posterior probability density function of θ given that $X = x$ is given by

$$h(\theta|x) = \frac{f_X(x|\theta)h(\theta)}{\int f_X(x|\theta)h(\theta)d\theta} \tag{5.23}$$

The prior risk [Bayes risk of d with respect to the prior $h(\theta)$] is given by

$$R(h, d) = \int R(\theta, d) h(\theta) d(\theta). \tag{5.24}$$

If $R(\theta, d)$ is bounded, we can interchange the order of integration in $R(h, d)$ and obtain

$$R(h, d) = \int \left\{ \int L(\theta, d(x)) f_X(x|\theta) dx \right\} h(\theta) d\theta$$
$$= \int \tilde{f}(x) \left\{ \int L(\theta, d(x)) h(\theta|x) d\theta \right\} dx \tag{5.25}$$

where

$$\tilde{f}(x) = \int f_X(x|\theta) h(\theta) d\theta. \tag{5.26}$$

The quantity

$$\int L(\theta, d(x)) h(\theta|x) d\theta \tag{5.27}$$

is called the posterior risk of d given $X = x$ (the posterior conditional expected loss).

Definition 5.3.1. *Bayes estimator.* A Bayes estimator of θ with respect to the prior density $h(\theta)$ is the estimator $d_0 \in D$ which takes the value $d_0(x)$ for $X = x$ and minimizes the posterior risk given $X = x$. In other words, for every $x \in \mathcal{X}$, $d_0(x)$ is defined as

$$\int L(\theta, d_o(x)) h(\theta|x) d\theta = \inf_{d \in D} \int L(\theta, d) h(\theta|x) d\theta \tag{5.28}$$

Note:

i. It is easy to check that the Bayes estimator d_0 also minimizes the prior risk.
ii. The Bayes estimator is not necessarily unique. However, if $L(\theta, d)$ is strictly convex in d for given θ, then d_0 is essentially unique.

For a thorough discussion of this the reader is referred to Berger (1980) or Ferguson (1967). Raiffa and Schlaifer (1961) have discussed in considerable detail the problem of choosing prior distributions for various models.

Let $X^\alpha = (X_{\alpha 1}, \ldots, X_{\alpha p})'$, $\alpha = 1, \ldots, N$, be a sample of size N from a p-dimensional normal distribution with mean μ and positive definite covariance

Estimators of Parameters and Their Functions

matrix Σ. Let
$$X = (X^1, \ldots, X^N), x = (x^1, \ldots, x^N).$$

Then
$$f_X(x) = (2\pi)^{-Np/2} (\det \Sigma)^{-N/2}$$
$$\times \exp\left\{-\frac{1}{2} \mathrm{tr}(\Sigma^{-1} s + N\Sigma^{-1}(\bar{x} - \mu)(\bar{x} - \mu)')\right\} \quad (5.29)$$

Let
$$L(\theta, d) = (\mu - d)'(\mu - d). \quad (5.30)$$

The posterior risk
$$E((\mu - d)'(\mu - d)|X = x) = E(\mu' \mu | X = x) = 2d' E(\mu | X = x) + d' d$$

is minimum when
$$d(x) = E(\mu | X = x).$$

In other words, the Bayes estimator is the mean of the marginal posterior density function of μ. Since
$$\frac{\partial^2 E((\mu - d)'(\mu - d)|X = x)}{\partial d' \partial d} = 2I,$$

$E(\mu | X = x)$ actually corresponds to the minimum value.

Let us take the prior as
$$h(\theta) = h(\mu, \Sigma) = K(\det \Sigma)^{-(\nu+1)/2}$$
$$\times \exp\left\{-\frac{1}{2}[(\mu - a)'\Sigma^{-1}(\mu - a)b + \mathrm{tr}\,\Sigma^{-1} H]\right\} \quad (5.31)$$

where $b > 0$, $\nu > 2p$, H is a positive definite matrix, and K is the normalizing constant. From (5.29) and (5.31) we get

$$h(\theta|X = x) = K'(\det \Sigma)^{-(N+\nu+1)/2} \exp\left\{-\frac{1}{2}\mathrm{tr}\,\Sigma^{-1}[s + H + (N+b)\right.$$
$$\times \left(\mu - \frac{N\bar{x} + ab}{N+b}\right)\left(\mu - \frac{N\bar{x} + ab}{N+b}\right)' \quad (5.32)$$
$$\left.+ \frac{Nb}{N+b}(\bar{x} - a)(\bar{x} - a)'\right]\right\}$$

where K' is a constant. Using (5.2), we get from (5.32)

$$h(\mu|X=x) = C\left[\det\left(s+H+\frac{Nb}{N+b}\right)\left(s+H+\frac{Nb}{N+b}(\bar{x}-a)(\bar{x}-a)'\right.\right.$$
$$\left.\left.+(N+b)\left(\mu-\frac{N\bar{x}+ab}{N+b}\right)\left(\mu-\frac{N\bar{x}+ab}{N+b}\right)'\right)\right]^{-(N+v-p)/2}$$

$$= C\frac{\left[\det\left(s+H+\frac{Nb}{N+b}(\bar{x}-a)(\bar{x}-a)'\right)\right]^{-(N+v-p)/2}}{\left[1+(N+b)\left(\mu-\frac{N\bar{x}+ab}{N+b}\right)' \times \right.}$$

$$\left(s+H+\frac{Nb}{N+b}(\bar{x}-a)(\bar{x}-a)'\right)^{-1}\left(\mu-\frac{N\bar{x}+ab}{N+b}\right)\right]^{(N-v-p)/2}$$

(5.33)

where C is a constant.

From Exercise 4.15 it is easy to calculate that

$$E(\mu|X=x) = (N\bar{x}+ab)/(N+b),$$

which is the Bayes estimate of μ for the prior (5.31).

For estimating Σ by an estimator d let us consider the loss function

$$L(\theta,d) = tr(\Sigma-d)(\Sigma-d). \tag{5.34}$$

The posterior risk with respect to this loss function is given by

$$E(tr\ \Sigma\Sigma|X=x) - 2E(tr\ d\Sigma|X=x) + tr\ dd \tag{5.35}$$

The posterior risk is minimized (see Exercise 1.14) when

$$d = E(\Sigma|X=x) \tag{5.36}$$

From (5.32), integrating out μ, the marginal posterior probability density function of Σ is given by

$$h(\Sigma|X=x) = K(\det \Sigma^{-1})^{(N+v)/2}$$
$$\times \exp\left\{-\frac{1}{2}tr\ \Sigma^{-1}\left(s+H+\frac{Nb}{N+b}(\bar{x}-a)(\bar{x}-a)'\right)\right\} \tag{5.37}$$

Estimators of Parameters and Their Functions

where K is the normalizing constant independent of Σ. Identifying the marginal distribution of Σ as an inverted Wishart distribution,

$$W^{-1}\left(s + H + \frac{Nb}{N+b}(\bar{x}-a)(\bar{x}-a)', p, N+\nu\right)$$

we get from Exercise 5.8

$$E(\Sigma|X=x)$$
$$= \frac{s + H + [Nb/(N+b)](\bar{x}-a)(\bar{x}-a)'}{N+\nu-2p-2}, \quad N+\nu-2p > 2 \tag{5.38}$$

as the Bayes estimate of Σ for the prior (5.31).

Note. If we work with the diffuse prior $h(\theta) \propto (\det \Sigma)^{(-p+1)/2}$ which is obtained from (5.31) by putting $b = 0$, $H = 0$, $\nu = p$, and which ceases to be a probability density on Ω, we get $\bar{x}$ and $s/(N-p-2)$, $(N > p+2)$, as the Bayes estimates of μ and Σ, respectively. Such estimates are called generalized Bayes estimates. Thus for the multivariate normal distribution, the maximum likelihood estimates of μ and Σ are not exactly Bayes estimates.

Definition 5.3.2. *Extended Bayes estimator.* An estimator $d_0 \in D$ is an extended Bayes estimator for $\theta \in \Omega$ if it is ϵ-Bayes for every $\epsilon > 0$; i.e., given any $\epsilon > 0$, there exists a prior $h_\epsilon(\theta)$ on Ω such that

$$E_{h_\epsilon(\theta)}(R(\theta, d_0)) \leq \inf_{d \in D} E_{h_\epsilon(\theta)}(R(\theta, d)) + \epsilon. \tag{5.39}$$

Definition 5.3.3. *Minimax estimator.* An estimator $d^* \in D$ is minimax for estimating $\theta \in \Omega$ if

$$\sup_{\theta \in \Omega} R(\theta, d^*) = \inf_{d \in D} \sup_{\theta \in \Omega} R(\theta, d). \tag{5.40}$$

In other words, the minimax estimator protects against the largest possible risk when θ varies over Ω.

To show that $\bar{X}$ is minimax for the mean μ of the normal distribution (with known covariance matrix Σ) with respect to the loss function

$$L(\mu, d) = (\mu - d)' \Sigma^{-1} (\mu - d), \tag{5.41}$$

we need the following theorem.

Theorem 5.3.1. *An extended Bayes estimator with constant risk is minimax.*

Proof. Let $d_0 \in D$ be such that $R(\theta, d_0) = C$, a constant for all $\theta \in \Omega$, and let d_0 also be an extended Bayes estimator; i.e., given any $\epsilon > 0$, there exists a prior density $h_\epsilon(\theta)$ on Ω such that

$$E_{h_\epsilon(\theta)} R(\theta, d_0) \leq \inf_{d \in D} E_{h_\epsilon(\theta)}\{R(\theta, d)\} + \epsilon. \tag{5.42}$$

Suppose d_0 is not minimax; then there exists an estimator $d^* \in D$ such that

$$\sup_{\theta \in \Omega} R(\theta, d^*) < \sup_{\theta \in \Omega} R(\theta, d_0) = C. \tag{5.43}$$

This implies that

$$\sup_{\theta \in \Omega} R(\theta, d^*) \leq C - \epsilon_0 \quad \text{for some } \epsilon_0 > 0, \tag{5.44}$$

or

$$R(\theta, d^*) \leq C - \epsilon_0 \quad \text{for all } \theta \in \Omega,$$

which implies

$$E\{R(\theta, d^*)\} \leq C - \epsilon_0,$$

where the expectation is taken with respect to any prior distribution over Ω. From (5.42) and (5.44) we get for every $\epsilon > 0$ and the corresponding prior density $h_\epsilon(\theta)$ over Ω

$$C - \epsilon \leq \inf_{d \in D} E_{h_\epsilon(\theta)}(R(\theta, d)) \leq E_{h_\epsilon(\theta)}(R(\theta, d^*)) \leq C - \epsilon_0,$$

which is a contradiction for $0 < \epsilon < \epsilon_0$. Hence d_0 is minimax. Q.E.D.

We first show that $\bar{X}$ is the rninimax estimator for μ when Σ is known. Let $X^\alpha = (X_{\alpha 1}, \ldots, X_{\alpha p})$, $\alpha = 1, \ldots, N$, be independently distributed normal vectors with mean μ and with a known positive definite covariance matrix Σ. Let

$$X = (X^1, \ldots, X^N), \bar{X} = \frac{1}{N}\sum_{\alpha=1}^N X^\alpha, S = \sum_{\alpha=1}^N (X^\alpha - \bar{X})(X^\alpha - \bar{X})'.$$

Assume that the prior density $h(\mu)$ of μ is a p-variate normal with mean 0 and covariance matrix $\sigma^2 \Sigma$ with $\sigma^2 > 0$. The joint probability density function of X

Estimators of Parameters and Their Functions

and μ is given by

$$h(\mu, x) = (2\pi)^{-(N+1)p/2}(\det \Sigma)^{-(N+1)/2}(\sigma^2)^{-p/2}$$

$$\times \exp\left\{-\frac{1}{2}\operatorname{tr}\Sigma^{-1}s - \frac{1}{2}N(\bar{x}-\mu)'\Sigma^{-1}(\bar{x}-\mu) - \frac{1}{2\sigma^2}\mu'\Sigma^{-1}\mu\right\}$$

$$= (2\pi)^{-(N+1)p/2}(\det \Sigma)^{-(N+1)/2}(\sigma^2)^{-p/2}$$

$$\times \exp\left\{-\frac{1}{2}\operatorname{tr}\Sigma^{-1}s\right\}\exp\left\{-\frac{1}{2}\left(\frac{N}{N\sigma^2+1}\right)\bar{x}'\Sigma^{-1}\bar{x}\right\}$$

$$\times \exp\left\{-\frac{1}{2}\left(N + \frac{1}{\sigma^2}\right)\left(\mu - \frac{N\bar{x}}{N+1/\sigma^2}\right)'\right.$$

$$\left.\times \Sigma^{-1}\left(\mu - \frac{N\bar{x}}{N+1/\sigma^2}\right)\right\}.$$

(5.45)

From above the marginal probability density function of X is

$$(2\pi)^{-Np/2}(\det \Sigma)^{-N/2}(1 + N\sigma^2)^{-p/2}$$

$$\exp\{-(N/2)\bar{x}'\Sigma^{-1}\bar{x}(N\sigma^2+1)^{-1}\}.$$

(5.46)

From (5.45) and (5.46) the posterior probability density function of μ, given $X = x$, is a p-variate normal with mean $N(N + 1/\sigma^2)^{-1}\bar{x}$ and covariance matrix $(N + 1/\sigma^2)^{-1}\Sigma$. The Bayes risk of $N(N + 1/\sigma^2)^{-1}\bar{X}$ with respect to the loss function given in (5.41) is

$$E\{(\mu - N(N + 1/\sigma^2)^{-1}\bar{x})'\Sigma^{-1}(\mu - N(N + 1/\sigma^2)^{-1}\bar{x})|X = x\}$$

$$= E\{\operatorname{tr} \Sigma^{-1}(\mu - N(N + 1/\sigma^2)^{-1}\bar{x})(\mu - N(N + 1/\sigma^2)^{-1}\bar{x})'|X = x\} \quad (5.47)$$

$$= p(N + 1/\sigma^2)^{-1}.$$

Thus, although $\bar{X}$ is not a Bayes estimator of μ with respect to the prior density $h(\mu)$, it is almost Bayes in the sense that the Bayes estimators $N(N + 1/\sigma^2)^{-1}\bar{X}$ which are Bayes with respect to the prior density $h(\mu)$ [with the loss function as given in (5.41)], tend to $\bar{X}$ as $\sigma^2 \to \infty$. Furthermore, since $N(N + 1/\sigma^2)\bar{X}$ is Bayes with respect to the prior density $h(\mu)$, we obtain

$$\inf_{d \in D} E_{h(\mu)}(R(\mu, d)) = E_{h(\mu)}(R(\mu, N(N + 1/\sigma^2)^{-1}\bar{X}) = p(N + 1/\sigma^2)^{-1}.$$

To show that $\bar{X}$ is extended Bayes we first compute

$$E_{h(\mu)}(R(\mu, \bar{X})) = p/N.$$

Hence

$$E_{h(\mu)}R(\mu, \bar{X}) = \inf_{d \in D} E_{h(\mu)}(R(\mu, d)) + \epsilon.$$

where $\epsilon = p/N(N\sigma^2 + 1) > 0$. Thus $\bar{X}$ is ϵ-Bayes for every $\epsilon > 0$.

Also, $\bar{X}$ has constant risk and hence, by Theorem 5.3.1, $\bar{X}$ is minimax for estimating μ when Σ is known.

We now show that $\bar{X}$ is minimax for estimating μ with loss function (5.41) when Σ is unknown. Let $\Sigma^{-1} = R$. Suppose that the joint prior density $\Pi(\theta)$ of (μ, R) is given by (5.31), which implies that the conditional prior of μ given R is $N_p(a, \Sigma^{-1}b)$ and the marginal prior of R is a Wishart $W_p(\alpha, H)$, with α degrees of freedom and parameter H with $v = 2\alpha - p > p$. From this it follows that the posterior joint density $\Pi(\theta | X = x)$ of (μ, R) is the product of the conditional posterior density $\Pi(\mu | R = r, X = x)$ given $R = r, X = x$ and the marginal posterior density $\Pi(r | X = x)$ of R given $X = x$ where

$$\Pi(\mu | R = r, X = x) \text{ is } N_p\left(\frac{N\bar{x} + ab}{N + b}, (b+n)^{-1}\Sigma\right), \quad (5.48)$$

$$\Pi(r | X = x) \text{ is } W_p(\alpha + n, \Sigma^*),$$

with $\Sigma^* = H + s + \dfrac{Nb}{N+b}(\bar{x} - a)(\bar{x} - a)'$. Thus the Bayes estimator of μ is given by

$$\frac{N\bar{x} + ab}{N + b}.$$

For the loss (5.41) its risk is $p(N+b)^{-1}$. Hence, taking the expectation with respect to $\Pi(\theta)$, we obtain

$$E(R(\theta, \bar{X})) - \inf_{d \in D} R(\mu, d) = \frac{p}{N} - \frac{p}{b+N} = \frac{bp}{N(b+N)} = \epsilon > 0.$$

Thus $\bar{X}$ is ϵ-Bayes for every $\epsilon > 0$. Since $\bar{X}$ has constant risk p/N we conclude that $\bar{X}$ is minimax when Σ is unknown.

5.3.1. Admissibility of Estimators

An estimator $d_1 \in D$ is said to be as good as $d_2 \in D$ for estimating θ if

$$R(\theta, d_1) \leq R(\theta, d_2)$$

for all $\theta \in \Omega$.

An estimator $d_1 \in D$ is said to be better than or strictly dominates $d_2 \in D$ if

$$R(\theta, d_1) \leq R(\theta, d_2)$$

for all $\theta \in \Omega$ with strict inequality for at least one $\theta \in \Omega$.

Definition 5.3.4. *Admissible estimator.* An estimator $d^* \in D$, which is not dominated by any other estimator in D, is called admissible.

For further studies on Bayes, minimax and admissible estimators the reader is referred to Brandwein and Strawderman (1990), Berger (1980), Stein (1981) and Ferguson (1967) among others.

Admissible Estimation of Mean

It is well known that if the dimension p of the normal random vector is unity, the sample mean $\bar{X}$ is minimax and admissible for the population mean with the squared error loss function (see for example Giri (1993)). As we have seen earlier for general p the sample mean $\bar{X}$ is minimax for the population mean with the quadratic error loss function. However, Stein (1956) has shown that the square error loss function and $\Sigma = I$ (identity matrix), $\bar{X}$ is admissible for $p = 2$ and it becomes inadmissible for $p \geq 3$. He showed that estimators of the form

$$\left(1 - \frac{a}{b + \|\bar{X}\|^2}\right)\bar{X} \tag{5.49}$$

dominates $\bar{X}$ for a sufficiently small a and b sufficiently large for $p \geq 3$. James and Stein (1961) improved this result and showed that even with one observation on the random vector X having p-variate normal distribution with mean μ and covariance matrix $\Sigma = I$ the class of estimators

$$\left(1 - \frac{a}{\|X\|^2}\right)X, \quad 0 < a < 2(p-2),$$

dominates X for $p \geq 3$. Their results led many researchers to work in this area which produced an enormous amount of rich literature of considerable importance in statistical theory and practice. Actually this estimator is a special case of the more general estimator

$$\left(1 - \frac{a}{X'\Sigma^{-1}X}\right)X \tag{5.50}$$

where X has normal distribution with mean μ and known positive definite covariance matrix Σ. We now add Stein's proof of inadmissibility of one observation for the mean vector μ from the p-dimensional normal distribution ($p \geq 3$) with mean μ and known covariance matrix I under the squared-error loss function. His method of proof depends on the Poisson approximation of the noncentral chi-square. In 1976 Stein (published 1981) gave a new method based on the "unbiased estimation of risk" which simplifies the computations considerably. This method depends on Lemma 5.3.2 to be proved later in this section.

Let

$$d_1 = X, \quad d_2 = \left(1 - \frac{p-2}{X'X}\right)X.$$

Now using (5.41)

$$R(\mu, d_1) - R(\mu, d_2)$$

$$= E_\mu\left\{\left(2 - \frac{p-2}{X'X}\right)(p-2) - 2\frac{(p-2)X'\mu}{X'X}\right\}. \tag{5.51}$$

To achieve our goal we need the following lemma.

Lemma 5.3.1. *Let $X = (X_1, \ldots, X_p)'$ be normally distributed with mean μ and covariance matrix I. Then*

$$\text{(a)} \quad E\left(\frac{\mu'X}{X'X}\right) = E_{\delta^2}\left(\frac{2\lambda}{p - 2 + 2\lambda}\right), \tag{5.52}$$

$$\text{(b)} \quad E\left(\frac{p-2}{X'X}\right) = E_{\delta^2}\left(\frac{p-2}{p - 2 + 2\lambda}\right) \tag{5.53}$$

where $\delta^2 = \mu'\mu$ and λ is a Poisson random variable with parameter $\frac{1}{2}\delta^2$.

Proof.

(a) Let $Y = OX$ where O is an orthogonal $p \times p$ matrix such that $Y_1 = \mu'X/\delta$. It follows that

$$E\left(\frac{\mu'X}{X'X}\right) = \delta E\left(\frac{Y_1}{Y'Y}\right) \tag{5.54}$$

Estimators of Parameters and Their Functions

where $Y = (Y_1, \ldots, Y_p)'$ and $Y_1, \ldots, Y_p$ are independently distributed normal random variables with unit variance and $E(Y_1) = \delta$, $E(Y_i) = 0, i > 1$. The conditional probability density function of Y_1, given $Y'Y = v$, is given by

$$f_{Y_1|Y'Y}(y_1|v) = \begin{cases} \dfrac{K \exp\{\delta y_1\}(v - y_1^2)^{(p-3)/2}}{\sum_{j=0}^{\infty} \dfrac{(\delta^2/2)^j v^{(p+2j-2)/2}}{2^{(p+2j)/2}\Gamma((p+2j)/2)}}, & \text{if } y_1^2 < v, \\ 0, & \text{otherwise} \end{cases} \tag{5.55}$$

where K is the normalizing constant independent of μ. From (5.55) we get

$$\int_{y_1^2 \leq v} K \exp\{\delta y_1\}(v - y_1^2)^{(p-3)/2} dy_1 = \sum_{j=0}^{\infty} \frac{(\delta^2/2)^j v^{(p+2j)/2-1}}{2^{(p+2j)/2}\Gamma((p+2j)/2)j!} \tag{5.56}$$

identically in $\mu \in \Omega$. Differentiating (5.56) with respect to δ

$$E(Y_1|Y'Y = v) = \delta \frac{\sum_{j=0}^{\infty} \dfrac{(\delta^2/2)^j v^{(p+2j)/2}}{j! 2^{(p+2+2j)/2}\Gamma((p+2+2j)/2)}}{\sum_{j=0}^{\infty} \dfrac{(\delta^2/2)^j v^{(p+2j)/2-1}}{j! 2^{(p+2j)/2}\Gamma((p+2j)/2)}} \tag{5.57}$$

The probability density function of $Y'Y$ is given by

$$f_{Y'Y}(v) = \exp\left(-\frac{1}{2}\delta^2\right) \sum_{j=0}^{\infty} \frac{(\delta^2/2)^j}{j!} \frac{e^{-v/2} v^{(p+2j)/2-1}}{2^{(p+2j)/2}\Gamma((p+2j)/2)} \tag{5.58}$$

which is gamma $G(\frac{1}{2}, p/2 + \lambda)$ where λ is a Poisson random variable with parameter $\frac{1}{2}\delta^2$. From (5.57–5.58) we obtain

$$\begin{aligned} E\left(\frac{Y_1}{Y'Y}\right) &= E\left(\{E(Y_1|Y'Y = v)\}\frac{1}{v}\right) \\ &= \delta \exp\left\{-\frac{1}{2}\delta^2\right\} \sum_{j=0}^{\infty} \frac{(\delta^2/2)^j}{j!} \\ &\quad \times \left(\int_0^{\infty} \frac{e^{-v/2} v^{(p+2j)/2-1} dv}{2^{(p+2+2j)/2}\Gamma((p+2+2j)/2)}\right) \\ &= \delta \exp\left\{-\frac{1}{2}\delta^2\right\} \sum_{j=0}^{\infty} \frac{(\delta^2/2)^j}{j!\Gamma((p+2j)/2)} \end{aligned} \tag{5.59}$$

Hence

$$E\left(\frac{\mu'X}{X'X}\right) = \delta E\left(\frac{Y_1}{Y'Y}\right)$$

$$= \exp\left(-\frac{1}{2}\delta^2\right) \sum_{j=0}^{\infty} \frac{(\delta^2/2)^j}{j!} \frac{2j}{p-2+2j}$$

$$= E\left(\frac{2\lambda}{p-2+2\lambda}\right).$$

(b) Since $X'X$ is distributed as gamma $G(\frac{1}{2}, \frac{1}{2}+\lambda)$, where λ is Poisson random variable with mean $\frac{1}{2}\delta^2$, we can easily show as in (a) that

$$E\left(\frac{p-2}{X'X}\right) = E\left(\frac{p-2}{p-2+2\lambda}\right).$$

Q.E.D.

From (5.51) and Lemma 5.3.1 we get

$$R(\mu, d_1) - R(\mu, d_2) = (p-2)^2 E\left(\frac{1}{p-2+2\lambda}\right) > 0$$

if $p \geq 3$. In other words, X is inadmissible for μ for $p \geq 3$.

Note

a. Let $X = (X_1, \ldots, X_p)'$ be such that $E(X) = \theta$ and the components $X_1, \ldots, X_p$ are independent with variance σ^2, the James-Stein estimator is

$$\left(1 - \frac{(p-2)\sigma^2}{X'X}\right)X. \tag{5.60}$$

b. If we choose μ_0 instead of 0 to be origin, the James-Stein estimator is

$$\mu_0 + \left(1 - \frac{(p-2)\sigma^2}{(X-\mu_0)'(X-\mu_0)}\right)(X - \mu_0). \tag{5.61}$$

The next theorem gives a more general result, due to James and Stein (1961).

Estimators of Parameters and Their Functions

Theorem 5.3.2. Let X be distributed as $N_p(\mu, I)$ and $L(\mu, d) = (\mu - d)'(\mu - d)$. Then

$$d_a(X) = \left(1 - \frac{a}{X'X}\right)X \tag{5.62}$$

dominates X for $0 < a < 2(p-1), p \geq 3$ and $d_{p-2}(X)$ is the uniformly best estimator of μ in the class of estimators $d_a(X), 0 < a < 2(p-1)$.

Proof. Using Lemma 5.3.1

$$R(\mu, d_a(X)) = E\left((X - \mu)'(X - \mu) - \frac{2aX'(X - \mu)}{X'X} + a^2 \frac{1}{X'X}\right)$$

$$= p + a^2 E\left(\frac{1}{p - 2 + 2\lambda}\right) + 2a\left[E\left(\frac{2\lambda}{p - 2 + 2\lambda}\right) - 1\right]$$

$$= p + [a^2 - 2a(p - 2)]E\left(\frac{1}{p - 2 + 2\lambda}\right).$$

Since $a^2 - 2a(p - 2) < 0$ for $0 < a < 2(p - 2)$ and $a^2 - 2a(p - 2)$ is minimum at $a = p - 2$ we get the results. Q.E.D.

Note

(a)

$$R(0, d_{p-2}(X)) = p + (p - 2)^2 E\left(\frac{1}{X'X}\right) - 2(p - 2)$$

$$= p + (p - 2) - 2(p - 2) = 2.$$

(b) Since the James-Stein estimator has smaller risk than that of X and X is minimax, the James-Stein estimator is minimax.

5.3.2. Interpretation of James-Stein Estimator

The following geometric interpretation of the above estimators is due to Stein (1962). Let $X = (X_1, \ldots, X_p)'$ be such that $E(X) = \mu$ and the components $X_1, \ldots, X_p$ are independent with the same variance σ^2. Since

$$E(X - \mu)'\mu = 0$$

$X - \mu, \mu$ are expected to be orthogonal especially when $\mu'\mu$ is large. Because

$$E(X'X) = E(\text{tr } X'X) = E(\text{tr}(X - \mu)(X - \mu)' + \mu'\mu$$
$$= p + \mu'\mu,$$

it appears that, as an estimator of μ, X might be too long. A better estimator of μ may be given by the projection of μ on X. Let the projection of μ on X be $(1 - a)X$.

Since the projection of μ on X depends on μ, in order to approximate it we may assume

$$X'X = p\sigma^2 + \mu'\mu$$

and $X - \mu$ is orthogonal to μ.

Thus we obtain ($\hat{a}$ being an estimate of a)

$$Y'Y = (X - \mu)'(X - \mu) - \hat{a}^2 X'X = p\sigma^2 - \hat{a}^2 X'X,$$

$$Y'Y = \mu'\mu - (1 - \hat{a})^2 X'X.$$

Hence

$$p\sigma^2 - \hat{a}^2 X'X = X'X - p\sigma^2 - (1 - \hat{a})^2 X'X.$$

Or

$$(1 - 2\hat{a})X'X = X'X - 2p\sigma^2$$

which implies that $\hat{a} = p\sigma^2 (X'X)^{-1}$. Thus the appropriate estimate of μ is

$$(1 - \hat{a})X = \left(1 - \frac{p\sigma^2}{X'X}\right)X.$$

A second interpretation of James-Stein estimator is a Bayes estimator given in (5.45–5.46) where X is distributed as $N_p(\mu, I)$ and the prior density of μ is $N_p(0, bI)$ with b unknown. The Bayes estimator of μ in this setup is

$$\frac{b}{b+1} X = \left(1 - \frac{1}{1+b}\right) X$$

To estimate b we proceed as follows. Since $X - \mu$ given μ is $N_p(0, (1+b)I)$, which implies that $(1+b)^{-1} X'X$ is distributed as χ_p^2. Hence $E((1+b)/X'X) = 1/p - 2$, provided $p > 2$. Thus a reasonable estimate of $(1+b)^{-1}$ is $(p-2)/X'X$. So the Bayes estimator of μ is

$$\left(1 - \frac{p-2}{X'X}\right) X$$

Estimators of Parameters and Their Functions

which is the James-Stein estimator of μ. If $\text{cov}(X) = \sigma^2 I$, the Bayes estimator of μ is $(1 - ((p-2)\sigma^2/X'X)X$.

5.3.3. Positive Part of James-Stein Estimator

The James-Stein Estimator has the disadvantage that for smaller values of $(X'X)$, the multiplicative factor of the James-Stein Estimator can be negative. In other words this estimator can be in the direction from the origin opposite to that of X. To remedy this situation a "shrinkage estimator" in terms of the positive part of the James-Stein estimator has been introduced and it is given by

$$d_2^+ = \left(1 - \frac{p-2}{X'X}\right)^+ X \tag{5.63}$$

where, by definition, for any function

$$g(t) = \left(1 - \frac{p-2}{t}\right),$$

$$g^+(t) = \begin{cases} g(t), & \text{if } g(t) > 0, \\ 0, & \text{otherwise.} \end{cases} \tag{5.64}$$

The theorem below will establish that the positive part of the James-Stein estimator has smaller risk than the James-Stein estimator. This implies that the James-Stein estimator is not admissible. It is known that the positive part of the James-Stein estimator is also not admissible. However one can obtain smoother shrinkage estimator than the positive part of James-Stein estimator. We will deal with it in Theorem 5.3.3.

Theorem 5.3.3. *The estimator d_2^+ has smaller risk than d_2 and is minimax.*

Proof. Let

$$h(X'X) = \left(1 - \frac{p-2}{X'X}\right);$$

$$h^+(X'X) = \begin{cases} h(X'X), & \text{if } h(X'X) > 0, \\ 0, & \text{otherwise} \end{cases}$$

Then

$$\begin{aligned}
R(\mu, d_2) &- R(\mu, d_2^+)) \\
&= E(h(X'X)X - \mu)'(h(X'X)X - \mu) \\
&\quad - E(h^+(X'X)X - \mu)'(h^+(X'X))X - \mu) \\
&= E[(h(X'X))^2(X'X) - ((h^+(X'X))^2 X'X)] \\
&\quad - 2E[\mu'Xh(X'X)) - \mu'X(h^+(X'X)].
\end{aligned}$$

Using (5.54)

$$E[\mu'X(h(X'X)) - \mu'X(h^+(X'X))]$$

$$= \frac{\delta}{(2\pi)^{p/2}} \int_{-\infty}^{\infty} y_1[h(y'y) - h^+(y'y)]$$

$$\times \exp\left\{-\frac{1}{2}\left(\sum_1^p y_i^2 - 2y_1\delta + \delta^2\right)\right\} dy$$

$$= \delta \exp\left\{-\frac{1}{2}\delta^2\right\} \int_{-\infty}^{\infty} \cdots \int_{-\infty}^{\infty} y_1[h(y'y) - h^+(y'y)] \exp\{y_1\delta\}$$

$$\times \frac{1}{(2\pi)^{p/2}} \exp\left\{-\frac{1}{2}\sum_{i=1}^p y_i^2\right\} dy = \delta \exp\left\{-\frac{1}{2}\delta^2\right\} \int_{-\infty}^{\infty} \cdots \int_{-\infty}^{\infty}$$

$$\left[y_1((h(y'y)) - h^+(y'y)) \times (\exp(y_1\delta) - \exp(-y_1\delta))\right.$$

$$\left. \times \exp\left\{-\frac{1}{2}\sum_{i=1}^p y_i^2\right\}\right] dy \leq 0.$$

Since

$$E((h(X'X))^2(X'X) - (h^+(X'X)^2(X'X))) \geq 0,$$

we conclude that

$$R(\mu, d_2) - R(\mu, d_2^+) \geq 0.$$

Q.E.D.

5.3.4. Unbiased Estimation of Risk

As stated earlier we have used the technique of the Poisson approximation of noncentral chi-square to evaluate the risk function. Stein, during 1976, (published in Stein (1981)) gave a new technique based on unbiased estimation of risk for the evaluation of the risk function which simplifies the computations considerably. His technique depends on this lemma.

Lemma 5.3.2. *Let X be distributed as $N(\mu, 1)$ and let $h(X)$ be a function of X such that for all $a < b$*

$$h(b) - h(a) = \int_a^b h'(x)dx.$$

Assume that $E(|h'(X)|) < \infty$. Then

$$\mathrm{cov}(h(X), (X - \mu)) = Eh(X)(X - \mu) = E(h'(X)).$$

Proof.

$$E(h(X)(X - \mu)) = \int_{-\infty}^{\infty} h(x)(x - \mu) \frac{1}{\sqrt{2\pi}} \exp\left\{-\frac{1}{2}(x - \mu)^2\right\} dx$$

$$= \int_{\mu}^{\infty} (h(x) - h(\mu))(x - \mu) \frac{1}{\sqrt{2\pi}} \exp\left\{-\frac{1}{2}(x - \mu)^2\right\} dx$$

$$+ \int_{-\infty}^{\mu} (h(x) - h(\mu))(x - \mu) \frac{1}{\sqrt{2\pi}} \exp\left\{-\frac{1}{2}(x - \mu)^2\right\} dx$$

$$= \int_{\mu}^{\infty} \int_{\mu}^{x} h'(y)(x - \mu) \frac{1}{\sqrt{2\pi}} \exp\left\{-\frac{1}{2}(x - \mu)^2\right\} dy\, dx$$

$$- \int_{-\infty}^{\mu} \int_{x}^{\mu} h'(y)(x - \mu) \frac{1}{\sqrt{2\pi}} \exp\left\{-\frac{1}{2}(x - \mu)^2\right\} dy\, dx.$$

Interchanging the order of integration which is permitted by Fubini's theorem we can write

$$E(h(X)(X - \mu)) = \int_\mu^\infty \int_y^\infty h'(y)(x - \mu) \frac{1}{\sqrt{2\pi}} \exp\left\{-\frac{1}{2}(x - \mu)^2\right\} dx\, dy$$

$$- \int_{-\infty}^\mu \int_{-\infty}^y h'(y)(x - \mu) \frac{1}{\sqrt{2\pi}} \exp\left\{-\frac{1}{2}(x - \mu)^2\right\} dx\, dy$$

$$= \int_{-\infty}^\infty h'(y) \frac{1}{\sqrt{2\pi}} \exp\left\{-\frac{1}{2}(y - \mu)^2\right\} dy$$

$$= E(h'(X)).$$

Q.E.D.

5.3.5. Smoother Shrinkage Estimator of Mean

We now turn to the problem of finding a smoother shrinkage factor than the positive part of the James-Stein estimator. The following Lemma, due to Baranchik (1970), allows us to obtain such smoother factors.

Lemma 5.3.3. *Let $X = (X_1, \ldots, X_p)'$ be distributed as $(N_p(\mu, I)$.*

The estimator

$$\left(1 - \frac{r(X'X)}{X'X}\right) X \tag{5.65}$$

is minimax with loss (5.30) provided that $0 \leq r(X'X) \leq 2(p - 2)$ and $r(X'X)$ is monotonically increasing in $(X'X)$.

Proof. Using Lemma 5.3.2

$$E\left[(X - \mu)'X \frac{r(X'X)}{X'X}\right]$$

$$= pE\left(\frac{r(X'X)}{X'X}\right) - 2E\left(\frac{r(X'X)}{X'X}\right) + 2E(r'(X'X))$$

$$\geq (p - 2)E\left(\frac{r(X'X)}{X'X}\right),$$

if $r(X'X)$ is monotonically increasing in $(X'X)$.

Estimators of Parameters and Their Functions

Now the risk of

$$\left(1 - \frac{r(X'X)}{X'X}\right)X$$

is

$$E\left(\left(1 - \frac{r(X'X)}{X'X}\right)X - \mu\right)'\left(\left(1 - \frac{r(X'X)}{X'X}\right)X - \mu\right)$$

$$= E(X - \mu)'(X - \mu) + E\left[\frac{r^2(X'X)}{X'X} - 2\frac{(X - \mu)'Xr(X'X)}{X'X}\right]$$

$$\leq p + [2(p - 2) - 2(p - 2)]E\left(\frac{r(X'X)}{X'X}\right)$$

$$= p = \text{Risk of } (X).$$

Since X is minimax for μ we conclude that $(1 - (r(X'X)/X'X))X$ is also minimax for μ. Q.E.D.

This lemma gives smoother shrinkage factor than the positive part of James-Stein estimator. We shall now show that such smoother shrinkage estimators can be obtained as Bayes estimators.

Let $X = (X_1, \ldots, X_p)'$ be distributed as $N_p(\mu, I)$ and let us consider a two-stage prior density of μ such that at the first stage the prior density $\Pi(\mu|\lambda)$ of μ given λ is $N_p(0, (\lambda/1 - \lambda)I)$ and at the second stage the prior density of λ is

$$\pi(\lambda) = (1 - b)\lambda^{-b}$$

with $b < 1, 0 \leq \lambda \leq 1$. With the loss (5.30) the Bayes estimator (using Exercise 4.15)

$$E(\mu|X) = E(E(\mu|\lambda, X)|X)$$

$$= E\left(\left(1 - \frac{1}{1 + [(1 - \lambda)\lambda^{-1}]}\right)X\Big|X\right) \qquad (5.66)$$

$$= (1 - E(\lambda|X))X,$$

where

$$E(\lambda|X) = (X'X)^{-1}\left[p+2-2b-2\left(\int_0^1 \lambda^{\frac{1}{2}p-b}\exp\left(\frac{1}{2}(1-\lambda)X'X\right)d\lambda\right)^{-1}\right]$$

$$= \frac{r(X'X)}{X'X}$$

where $r(X'X)$ is the expression inside the brackets. Since $r(X'X) \leq p+2-2b$ and $\int_0^1 \lambda^{\frac{1}{2}p-b}\exp\{\frac{1}{2}(1-\lambda)X'X\}d\lambda$ is increasing in $X'X$, using Lemma 5.3.3, we conclude that the Bayes estimator $(1-(r(X'X)/X'X)X$ is minimax if $p+2-2b \leq 2(p-2)$ or $b \geq (6-p)/2$.

Since $b < 1$, this implies $p \geq 5$. Hence we get the following theorem.

Theorem 5.3.4. *The Bayes estimator (5.66) is minimax for $p \geq 5$ if $\frac{1}{2}(6-p) \leq b < 1$.*

Strawderman (1972) showed that no Bayes minimax estimator exists for $p < 5$.

5.3.6. Estimation of Mean with Σ Unknown

Let $X^\alpha = (X_{\alpha 1}, \ldots, X_{\alpha p})'$, $\alpha = 1, \ldots, N(N > p)$ be a sample from a p-variate normal population with unknown mean μ and unknown covariance matrix Σ. Let $\bar{X} = 1/N \sum_{\alpha=1}^N X^\alpha$, $S = \sum_{\alpha=1}^N (X^\alpha - \bar{X})(X^\alpha - \bar{X})'$. We have proved that $\bar{X}$ is minimax for estimating μ with the loss function

$$L(\theta, d) = N(\mu - d)'\Sigma^{-1}(\mu - d)$$

with $\theta = (\mu, \Sigma)$ and Σ is unknown. James and Stein (1961) considered the estimator

$$\bar{d} = \left(1 - \frac{p-2}{N(N-p+2)\bar{X}'S^{-1}\bar{X}}\right)\sqrt{N}\bar{X} \qquad (5.67)$$

for estimating $\sqrt{N}\mu$. Using the fact that $N\bar{X}'S^{-1}\bar{X}$ is distributed as the ratio $\chi_p^2(N\mu'\Sigma^{-1}\mu)/\chi_{N-p}^2$ of two independent chi-squares (see Section 6.8) and with arguments similar to that above we can derive

$$R(\theta, \bar{d}) = p - \left(\frac{N-p}{N-p+2}\right)(p-2)^2 E\left(\frac{1}{p-2+2K}\right) \qquad (5.68)$$

Estimators of Parameters and Their Functions

where K is a Poisson random variable with parameter

$$\frac{N}{2}\mu'\Sigma^{-1}\mu \quad \text{and} \quad \theta = (\sqrt{N}, \mu, \Sigma).$$

Hence

$$R(\theta, \sqrt{N}\bar{X}) - R(\theta, \bar{d})$$

$$= \frac{(p-2)^2(N-p)}{N-p+2} E\left(\frac{1}{p-2+2K}\right) \geq 0, \quad \text{if} \quad p \geq 3.$$

The problem of the determination of the confidence region for μ is discussed in Chapter 6.

5.3.7. Estimation of Covariance Matrix

Let $X^\alpha = (X_{\alpha 1}, \ldots, X_{\alpha p})'$, $\alpha = 1, \ldots, N(N > p)$ be a sample of size N from a p-variate normal population with mean μ and covariance matrix Σ and let $N\bar{X} = \sum_{\alpha=1}^{N} X^\alpha$, $S = \sum_{\alpha=1}^{N}(X^\alpha - \bar{X})(X^\alpha - \bar{X})'$. We will show in Section 6.1 that S has a central Wishart distribution $W_p(n, \Sigma)$ with $n = N - 1$ degrees of freedom and $E(S) = n\Sigma$. We will use here some more results on Wishart distrituion which will be proved in Chapter 6.

We consider here the problem of estimating Σ by $\hat{\Sigma} = \delta(S)$, a $p \times p$ positive definite matrix with elements that are functions of S. The performance of any estimator is evaluated in terms of the risk function of a given loss function. Two commonly used loss functions are

$$L_1(\Sigma, \delta) = \text{tr } \Sigma^{-1}\delta - \log \det(\Sigma^{-1}\delta) - p,$$

$$L_2(\Sigma, \delta) = \text{tr}(\Sigma^{-1}\delta - I)^2 = \text{tr}(\delta - \Sigma)\Sigma^{-1}(\delta - \Sigma)\Sigma^{-1}.$$

(5.69)

They are non negative and are zero when $\delta = \Sigma$. There are other loss functions with these properties but these two are relatively easy to work with. The loss function L_1 is known as Stein's loss and was first considered by Stein (1956) and James and Stein (1961). The estimation of Σ using L_2 was considered by Olkin and Seillah (1977) and Haff (1980). Begining with the works of Stein (1956) and James and Stein (1961) the problem of estimating the matrix of regression coefficients in the normal case has been considered by Robert (1994), Kubokawa (1998), Kubokawa and Srivastava (1999) among others. If we restrict our attention to the estimators of the form αS, where α is a scalar, the following theorem will show that the unbiased estimator S/n of Σ is the best in the sense that it has the minimum risk for the loss function L_1 and the estimator $S/(n + p + 1)$ of Σ is the best for the loss function L_2.

Theorem 5.3.5. *Among all estimators of the form αS of Σ, the unbiased estimator S/n is the best for the loss L_1 and the estimator $S/(n+p+1)$ is the best for the loss L_2.*

Proof. For the loss L_1 the risk of αS is

$$R_1(\Sigma, \alpha S) = E[\alpha \text{tr}(\Sigma^{-1}S) - \log \det(\alpha \Sigma^{-1}S) - p]$$

$$= \alpha \text{tr } \Sigma^{-1}E(S) - p \log \alpha - E\left[\log \frac{\det S}{\det \Sigma}\right] - p.$$

Using Theorem 6.6.1 we get

$$R_1(\Sigma, \alpha S) = \alpha p n - p \log \alpha - E \log \left[\prod_{i=1}^{p} \chi^2_{n+1-i}\right] - p$$

$$= \alpha p n - n \log \alpha - \sum_{i=1}^{p} E(\log \chi^2_{n+1-i}) - p$$

and $R_1(\Sigma, \alpha S)$ is minimum when $\alpha = 1/n$.

Since $\Sigma > 0$, there exists $g \in G_\ell(p)$ such that $g\Sigma g' = I$. Let $gSg' = S^* = (S^{**}_{ij})$. Hence with $\delta = \alpha S$, using Section 6.3, we get

$$R_2(\Sigma, \alpha S) = E_I L_2(I, \alpha S^*)$$

$$= E_I \text{tr}(\alpha S^* - I)^2$$

$$= E_I \left(\alpha^2 \sum_{i,j=1}^{p} S^{*2}_{ij} - 2\alpha \sum_{i=1}^{p} S^{**}_{ii} + p\right).$$

Since $S^{**}_{ii} \simeq \chi^2_n$ and $S^{*2}_{ij} \simeq \chi^2_1$ $(i \neq j)$ we get

$$E_I(S^{*2}_{ii}) = 2n + n^2 = n(n+2), E_I(S^{*2}_{ij}) = 1 (i \neq j).$$

Hence

$$R_2(\Sigma, \alpha S) = \alpha^2[n(n+2)p + np(p-1)] - 2\alpha n p + p$$

which is minimum when $\alpha = (n+p+1)^{-1}$.

Hence we get the Theorem. Q.E.D.

The minimum value of $R_1(\Sigma, \alpha S)$ is $p \log n - \sum_{i=1}^{p} E(\log \chi^2_{n+1-i})$ and the minimum value of $R_2(\Sigma, \alpha S)$ is $p(p+1)(n+p+1)^{-1}$.

Stein (1975) pointed out that the characteristic roots of S/n spread out more than the corresponding characteristic roots of Σ and the problem gets more

Estimators of Parameters and Their Functions 173

serious when $\Sigma \simeq I_p$. This fact suggests that S/n should be shrunk towards a middle value. A similar phenomenon exists in the case of Stein type estimation of the multivariate normal mean. We refer to Stein (1975, 1977a,b), Young and Berger (1994) and the cited references therein. The problem of minimax estimation of Σ was first considered by James and Stein (1961). They utilised a result of Kiefer (1957) which states that if an estimator is minimax in the class of equivariant estimators with respect to a group of transformations which is solvable then it is minimax among all estimators. Here the group $G_\ell(p)$, full linear group of $p \times p$ nonsingular matrices, is not solvable but the subgroup $G_T(p)$ of $p \times p$ nonsingular lower triangular matrices and the subgroup $G_{UT}(p)$ of $p \times p$ nonsingular upper triangular matrices, are solvable. If $\delta(S)$ is an estimator of Σ and $g \in G_\ell(p)$ then δ should satisfy

$$\delta(gSg') = g\delta(S)g'.$$

Because gSg' has the Wishart distribution $W_p(n, g\Sigma g')$, gSg' estimates $g\Sigma g'$ as does $gS(S)g'$. If this holds for all $g \in G_\ell(p)$ then $\delta(S) = \alpha S$ for some scalar α. Since $G_\ell(p)$ is not solvable we can not assert the minimax property of αS from Theorem 5.3.5.

In the approach of James and Stein (1961) we consider $G_T(p)$ instead of $G_\ell(p)$ and find the best estimator $\delta(S)$ satisfying

$$\delta(gSg') = g\delta(S)g' \qquad (5.70)$$

for all $g \in G_T(p)$. It may be remarked that $\delta(S) = \alpha S$ satisfies (5.70). Since $G_T(p)$ is solvable. The best estimator will be minimax.

Since (5.70) holds for all $S(> 0)$, taking $S = I$, in particular, we get

$$\delta(gg') = g\delta(I)g'. \qquad (5.71)$$

Now, let g be a diagonal matrix with diagonal elements ± 1. Obviously $g \in G_T$ and $gg' = I$. From (5.71) we get $\delta(I) = g\delta(I)g'$ for all such g. This implies that $\delta(I)$ is a diagonal matrix Δ with diagonal elements $\delta_1, \ldots, \delta_p$ (say).

Write $S = TT'$ where $T = (T_{ij}) \in G_T(p)$ with positive diagonal elements (which we need to impose for the uniqueness). From (5.71) we get

$$\delta(S) = \delta(TT') = T\delta(I)T' = T\Delta T'. \qquad (5.72)$$

Theorem 5.3.6. *The best estimator of Σ in the class of all estimators satisfying (5.70) (hence minimax) is $\delta(S) = T\Delta T'$ where $S = TT'$ with T a $p \times p$ lower triangular nonsingular matrix with positive diagonal elements and Δ is a diagonal matrix with diagonal elements $\delta_1, \ldots, \delta_p$, given by*

a. $\delta_i = (n + 1 + p - 2i)^{-1}, i = 1, \ldots, p$, *when the loss function is L_1;*

b. $(\delta_1, \ldots, \delta_p)' = A^{-1}b$, where $A = (a_{ij})$ is a $p \times p$ symmetric matrix with
$$a_{ii} = (n+p-2i+1)(n+p-2i+3)$$
$$a_{ij} = n+p-2j+1, \quad i < j$$
$$b = (b_1, \ldots, b_p)' \quad \text{with} \quad b_i = n+p-2i+1,$$
when the loss function is L_2.

Proof. (a) For the loss function L_1, using (6.32),
$$E_\Sigma(L_1(\Sigma, \delta(S)))$$
$$= E_\Sigma(\operatorname{tr} \Sigma^{-1}\delta(S) - \log \det \Sigma^{-1}\delta(S) - p)$$
$$= (\det \Sigma)^{-n/2} c_{n,p}$$
$$\times \left\{ \int [\operatorname{tr} \Sigma^{-1}\delta(s) - \log \det \Sigma^{-1}\delta(s) - p] \right.$$
$$\left. \times \exp\left\{ -\frac{1}{2} \operatorname{tr} \Sigma^{-1} s \right\} (\det s)^{(n-p-1/2)} ds \right\}.$$

Let $\Sigma^{-1} = g'g$, $g \in G_T(p)$ by (5.70)
$$\operatorname{tr} \Sigma^{-1}\delta(S) - \log \det \Sigma^{-1}\delta(S) - p$$
$$= \operatorname{tr} g'g\delta(S) - \log \det g'g\delta(S) - p$$
$$= \operatorname{tr} g\delta(S)g' - \log \det g\delta(S)g' - p$$
$$= \operatorname{tr} \delta(gSg') - \log \det \delta(gSg') - p.$$

Transform $s \to u = gsg'$. From Theorem 2.4.10 the Jacobian of this transformation is $(\det g)^{-(p+1)}$. Hence
$$E_\Sigma L_1(\Sigma, \delta(S)) = c_{n,p}[\det gg']^{n/2}$$
$$\times \int [\operatorname{tr} \delta(gsg') - \log \det \delta(gsg') - p]$$
$$\times \exp\left\{ -\frac{1}{2} \operatorname{tr} gsg' \right\} (\det s)^{n-p-1/2} ds \quad (5.73)$$
$$= c_{n,p} \int [\operatorname{tr} \delta(u) - \log \det \delta(u) - p]$$
$$\times \left\{ -\frac{1}{2} \operatorname{tr} u \right\} (\det u)^{n-p-1/2} du$$
$$= E_I L_1(I, \delta(S)).$$

Estimators of Parameters and Their Functions 175

Hence the risk $R_1(\Sigma, \delta) = E_\Sigma(L_1(\Sigma, \delta(S))$ does not depend on Σ. Now, using Theorem 6.6.1 and results of Section 6.7, the T_{ij} in (5.72) are independent. T_{ii}^2 is distributed as χ^2_{n-i+1} and $T_{ij}^2 (i \neq j)$ is distributed as χ_1^2. Hence

$$R_1(I, \delta) = E_I(\text{tr } \delta(S) - \log \det \delta(S) - p)$$
$$= E_I(\text{tr } T\Delta T' - \log \det T\Delta T' - p)$$
$$= \sum_{i=1}^{p} \delta_i E(T_{ii}^2) + \sum_{\substack{i,j=1 \\ j<i}}^{p} \delta_i E(T_{ij}^2)$$
$$- \sum_{i=1}^{p} E(\log \chi^2_{n+1-i}) - \log \det \Delta - p$$
$$= \sum_{i=1}^{p} \delta_i(n+1+p-2i) + \sum_{\substack{i,j=1 \\ j<i}}^{p} \delta_i$$
$$- \sum_{i=1}^{p} E \log \chi^2_{n+1-i} - \log \det \Delta - p$$
$$= \sum_{i=1}^{p} \delta_i(n+1+p-2i) - \sum_{i=1}^{p} \log \delta_i - \sum_{i=1}^{p} E \log \chi^2_{n+1-i} - p.$$

This attains its minimum value at $\delta_i = (n+1+p-2i)^{-1}, i = 1, \ldots, p$ and the minimum risk is $\sum_{i=1}^{p} \log(n+1+p-2i) - \sum_{i=1}^{p} E \log \chi^2_{n+1-i}$.

(b) $R_1(\Sigma, \delta) = E_\Sigma L_2(\Sigma, \delta(S))$ does not depend on Σ. Now using (5.72)–(5.73) and the above arguments we get

$$E_I(L_2(I, \delta(S)) = E_I \text{tr}(\delta(S) - I)^2$$
$$= E_I \text{tr}(T\Delta T' - I)^2$$
$$= E_I \text{tr}(T\Delta TT\Delta T' - 2T\Delta T' + I)$$
$$= E_I \left(\sum_{i,j,k,\ell=1}^{p} T_{ij} \delta_j T_{kj} T_{k\ell} \delta_\ell T_{i\ell} \right) \quad (5.74a)$$
$$- 2E_I \left(\sum_{i,j=1}^{p} T_{ij}^2 \delta_j \right) + p$$
$$= (\delta_1, \ldots, \delta_p) A (\delta_1, \ldots, \delta_p)'$$
$$- 2b'(\delta_1, \ldots, \delta_p)' + p.$$

Since $(\delta_1,\ldots,\delta_p)A(\delta_1,\ldots,\delta_p)' = E(\text{tr}(T\Delta T')^2) > 0$, Δ is positive definite and (5.74a) has a unique minimum at $(\delta_1,\ldots,\delta_p)' = A^{-1}b$. The minimum value of $E_1 L_2(I, \delta(S)) = p - b'A^{-1}b$. Q.E.D.

5.3.8. Estimation of Parameters in $CN_p(\alpha, \Sigma)$

Let $Z^i = (Z_{i1},\ldots,Z_{ip})'$, $i = 1,\ldots,N$ be independently and identically distributed as $CN_p(\alpha, \Sigma)$ and let

$$\bar{Z} = \frac{1}{N}\sum_{i=1}^N Z^i, \quad A = \sum_{i=1}^N (Z^i - \bar{Z})(Z^i - \bar{Z})^*.$$

To find the maximum likelihood estimator of α, Σ we need the following Lemma.

Lemma 5.3.4. *Let $f(\Sigma) = c(\det \Sigma)^n \exp\{-\text{tr }\Sigma\}$ where Σ is positive definite Hermitian and c is a positive constant.*
Then $f(\Sigma)$ is maximum at $\Sigma = \hat{\Sigma} = nI$.

Proof. By Theorem 1.8.1. there exists a unitary matrix U such that $U^*\Sigma U$ is a diagonal matrix with diagonal elements $\lambda_1,\ldots,\lambda_p$, the characteristic roots of Σ and $\lambda_i > 0$ for all i. Hence

$$f(\Sigma) = c(\det(U^*\Sigma U))^n \exp\{-\text{tr}(U^*\Sigma U)\}$$

$$= c\prod_{i=1}^p (\lambda_i^n e^{-\lambda_i}) = c\prod_{i=1}^p (\lambda_i^n e^{-\lambda_i/n})^n,$$

which is maximum if $\lambda_i = n, i = 1,\ldots,p$. Hence $f(\Sigma)$ is maximum when $\Sigma = nI$. Q.E.D.

Theorem 5.3.7. *The maximum likelihood estimates of $\hat{\alpha}, \hat{\Sigma}$ of α, Σ respectively are given by $\hat{\alpha} = \bar{Z}, \hat{\Sigma} = A/N$.*

Proof. From (4.18) the likelihood of $z_i, i = 1,\ldots,N$ is

$$L(z^1,\ldots,z^N) = \pi^{-Np}(\det \Sigma)^{-N}$$

$$\exp\left\{-\text{tr }\Sigma^{-1}\left(\sum_{i=1}^N (z^i - \alpha)(z^i - \alpha)^*\right)\right\}$$

$$= \pi^{-Np}(\det \Sigma)^{-N}\exp\{-\text{tr }\Sigma^{-1}(A + N(\bar{z} - \alpha)(\bar{z} - \alpha)^*)\}.$$

Estimators of Parameters and Their Functions

Hence

$$\max_{\alpha,\Sigma} L(z^1,\ldots,z^N) = \max_{\Sigma}[\pi^{-Np}(\det \Sigma)^{-N} \exp\{-\text{tr } \Sigma^{-1}A\}]$$

and the maximum likelihood estimate of α is $\hat{\alpha} = \bar{z}$. Let us assume that A is positive definite Hermitian, which we can do with probability one if $N > p$. By Theorem 1.8.3 there exists a Hermitian nonsingular matrix B such that $A = BB^*$, so that

$$\max_{\alpha,\Sigma} L(z^1,\ldots,z^N) = \max_{\Sigma}[\pi^{-Np}(\det \Sigma)^{-N} \exp\{-\text{tr } B^*\Sigma^{-1}B\}]$$

$$= \max_{\Sigma}[\pi^{-Np}(\det(BB^*))^{-N}(\det(B^*\Sigma^{-1}B))^N \exp\{-\text{tr } B^*\Sigma^{-1}B\}]$$

By Lemma 5.3.4 the maximum likelihood estimate of Σ is $\hat{\Sigma} = BB^*/N = A/N$. Q.E.D.

Theorem 5.3.8. *$\sqrt{N}\bar{Z}, A$ are independent in distribution. $\sqrt{N}\bar{Z}$ has a p-variate complex normal distribution with mean $\sqrt{N}\alpha$ and complex covariance Σ; A is distributed as $\sum_{i=2}^{N} \xi^i \xi^{i*}$ where $\xi^i, i = 2,\ldots,N$ are independently and identically distributed $CN_p(0,\Sigma)$.*

Proof. Let $U = (u_{ij})$ be a $N \times N$ unitary matrix with first row $(N^{-1/2},\ldots,N^{-1/2})$. Consider the transformations from $(Z^1,\ldots,Z^N)$ to $(\xi^1,\ldots,\xi^N)$, given by

$$\xi^1 = N^{1/2}\bar{Z},$$

$$\xi^i = \sum_{j=1}^{N} \mathbf{u}_{ij} Z^j, \quad i \geq 1$$

It may be verified that

$$E(\xi^1) = N^{1/2}\alpha, E(\xi^i) = 0, i \geq 2;$$

$$\text{cov}(\xi^i, \xi^j) = \begin{cases} \Sigma, & \text{if } i = j, \\ 0, & \text{if } i \neq j, \end{cases}$$

$$A = \sum_{i=2}^{N} \xi^i \xi^{i*}.$$

By Theorem 4.2.3 and Theorem 4.2.5 $\xi^i, i = 1,\ldots,N$ are independently distributed p-variate complex normals. Hence we get the theorem. Q.E.D.

The distribution of A is known as the complex Wishart distribution with parameter Σ and degrees of freedom $N-1$ with pdf given by (Goodman, 1963)

$$f(A) = \frac{(\det A)^{N-p-1}}{I(\Sigma)} \exp\{-\operatorname{tr} \Sigma^{-1} A\} \qquad (5.75)$$

where $I(\Sigma) = \pi^{p(p-1)/2} (\det \Sigma)^{N-1} \prod_{i=1}^{p} \Gamma(N-i)$. From Theorem 5.3.8 $E(A) = (N-1)\Sigma$. Since

$$L(z^1, \ldots, z^N) = (2\pi)^{-Np} (\det \Sigma)^{-N} \exp\{-\operatorname{tr} \Sigma^{-1} (A + N(\bar{z} - \alpha)(z - \alpha)^*)\}$$

$(\bar{Z}, A)$ is sufficient for (α, Σ) (Halmos and Savage (1949)).

5.3.9. Estimation of Parameters in Symmetrical Distribution and Related Results

Let $X = (X_1, \ldots, X_p)'$ be distributed as $E_p(\mu, \Sigma)$ with pdf

$$f_X(x) = (\det \Sigma)^{-1/2} q((x-\mu)' \Sigma^{-1}(x-\mu))$$

where q is a function on $[0, \infty)$ such that $\int q(y'y) dy = 1$.

Theorem 5.3.9. *Let q be such that $u^{p/2} q(u)$ has finite positive maximum u_q. Suppose that on the basis of one observation x from*

$$f_X(x) = (\det \Sigma)^{-1/2} q((x-\mu)' \Sigma^{-1}(x-\mu))$$

the maximum likelihood estimators $(\hat{\mu}, \hat{\Sigma})$, under $N_p(\mu, \Sigma)$ exist and are unique and that $\hat{\Sigma} > 0$ with probability one. Then the maximum likelihood estimators $(\bar{\mu}, \bar{\Sigma})$ of (μ, Σ) for a general q are given by

$$\bar{\mu} = \hat{\mu}, \quad \bar{\Sigma} = \frac{p}{u_q} \hat{\Sigma}.$$

Proof. Let $\Delta = \Sigma/(\det \Sigma)^{1/p}$ and let $d = (x-\mu)' \Sigma^{-1}(x-\mu)$.

Then

$$d = (x-\mu)' \Delta^{-1} (x-\mu)/(\det \Sigma)^{1/p} \text{ and } \det \Delta = 1.$$

Hence

$$f_X(x) = [(x-\mu)' \Delta^{-1} (x-\mu)]^{-p/2} d^{p/2} q(d). \qquad (5.76)$$

Under $N_p(\mu, \Sigma)$, $q(d) = (2\pi)^{-p/2} \exp\{-\frac{1}{2} d\}$ and the maximum of (5.76) is attained at $\mu = \hat{\mu}$, $\Delta = \hat{\Delta} = (\hat{\Sigma}/(\det \hat{\Sigma})^{1/p})$ and $d = p$. For a general q the maximum of (5.76) is attained at $\mu = \bar{\mu} = \hat{\mu}$, $\Delta = \bar{\Delta} = \hat{\Delta}$ and $d = \bar{d} = u_q$. Hence the maximum likelihood estimator of Σ is $\bar{\Sigma} = (\det \bar{\Sigma})^{1/p} \hat{\Delta} = ((\det \bar{\Sigma})^{1/p}/(\det \hat{\Sigma})^{1/p}) \hat{\Sigma}$.

Since

$$p = \frac{(x - \hat{\mu})'\hat{\Delta}^{-1}(x - \hat{\mu})}{(\det \hat{\Sigma})^{1/p}}, \quad u_q = \frac{(x - \bar{\mu})'\bar{\Delta}^{-1}(x - \bar{\mu})}{\det \bar{\Sigma}^{1/p}}$$

$$= \frac{(x - \hat{\mu})'\hat{\Delta}^{-1}(x - \hat{\mu})}{\det \bar{\Sigma}^{1/p}} = \frac{p(\det \hat{\Sigma})^{1/p}}{(\det \bar{\Sigma})^{1/p}}$$

we get

$$\frac{p}{u_q} = \frac{(\det \bar{\Sigma})^{1/p}}{(\det \hat{\Sigma})^{1/p}}.$$

Hence the maximum likelihood estimators under a general q are $\bar{\mu} = \hat{\mu}, \bar{\Sigma} = p/u_q \hat{\Sigma}$. Q.E.D.

Let $X = (X_1, \ldots, X_N)'$, where $X_i = (X_{i1}, \ldots, X_{ip})'$, be a $N \times p$ random matrix having an elliptically symmetric distribution with *pdf*, as given in (4.36) with $\mu_i = \mu, \Sigma_i = \Sigma$ for all i. Define $\bar{X} = 1/N \sum_{i=1}^{N} X_i, S = \sum_{i=1}^{N}(X_i - \bar{X})(X_i - \bar{X})'$. Using Theorem 5.3.6 we get the maximum likelihood estimators $\bar{\mu}, \bar{\Sigma}$ of μ, Σ are given by

$$\bar{\mu} = \bar{X}, \bar{\Sigma} = \frac{p}{u_q} S \qquad (5.77)$$

where u_q is the maximum value of $u^{Np/2} q(u)$.

Let $X = (X_1, \ldots, X_p)'$ be distributed as a contaminated normal with *pdf*

$$f_X(x) = \int (2\pi)^{-p/2} \sigma^{-p} \exp\left\{-\frac{1}{2\sigma^2}(x - \mu)'(x - \mu)\right\} dG(\sigma)$$

where $G(\cdot)$ is a known distribution function. To estimate μ with loss (5.30) Strawderman (1974) showed that the estimator

$$\left(1 - \frac{a}{X'X}\right) X$$

has smaller risk than X provided $0 < a < 2/E(X'X)^{-1}$.

To estimate μ with loss (5.30) in spherically symmetric distribution with *pdf*

$$f_X(x) = q((x - \mu)'(x - \mu))$$

where q is on $[0, \infty]$ and $E(X'X), E(X'X)^{-1}$ are both finite Brandwein (1979) showed that for $p \geq 4$ the estimator $(1 - (a/X'X))X$ has smaller risk than X provided

$$0 < a < 2(p - 2)p^{-1}[E(X'X)]^{-1}.$$

Let $X = (X_1, \ldots, X_N)'$ be an $N \times p$ matrix having elliptically symmetric distribution with parameters μ, Σ. Let $\bar{X} = N^{-1} \sum_{\alpha=1}^{N} X_\alpha, S = \sum_{\alpha=1}^{N}(X_i - \bar{X})(X_i - \bar{X})'$.

Srivastava and Bilodeau (1989) have shown that for estimating μ with Σ unknown with loss (5.30) the estimator $(1 - (\kappa/N\bar{X}'S^{-1}\bar{X}))\bar{X}$ has smaller risk than $\bar{X}$ provided that $0 < \kappa < 2(p-2)(N-p+2)^{-1}$ and $p \geq 3$. Kubakawa and Srivastava (1999) have shown that the minimax estimator of the covariance matrix Σ obtained under the multivariate normal model remains robust under the elliptically symmetric distributions. For non negative estimation of multivariate components of variance we refer to Srivastava and Kubokawa (1999). The determination of the confidence region of μ is discussed in Chapter 7.

Example 5.3.1. Observations were made in the Indian Agricultural Research Institute, New-Delhi, India, on six different characters:

X_1 plant height at harvesting (cm)
X_2 number of effective tillers
X_3 length of ear (cm)
X_4 number of fertile spikelets per 10 ears
X_5 number of grains per 10 ears
X_6 weight of grains per 10 ears

for 27 randomly selected plants of Sonalika, a late-sown variety of wheat in two consecutive years (1971, 1972). The observations are recorded in Table 5.1.

Assuming that each year's data constitute a sample from a six-variate normal distribution with mean μ and covariance matrix Σ, we obtain the following maximum likelihood estimates.

For 1971

i.
$$\hat{\mu} = \begin{pmatrix} 84.8911 \\ 186.2963 \\ 9.7411 \\ 13.4593 \\ 304.3701 \\ 13.6259 \end{pmatrix}$$

ii. $\hat{\Sigma} = s/27$

	X_1	X_2	X_3	X_4	X_5	X_6
X_1	12.247					
X_2	14.389	353.293				
X_3	−0.245	2.703	0.191			
X_4	0.209	3.278	0.155	0.519		
X_5	14.000	173.400	8.519	7.003	1130.456	
X_6	−0.464	8.403	0.465	0.313	40.970	2.383

Estimators of Parameters and Their Functions

Table 5.1.

Plant No.	X_1 1971	X_1 1972	X_2 1971	X_2 1972	X_3 1971	X_3 1972	X_4 1971	X_4 1972	X_5 1971	X_5 1972	X_6 1971	X_6 1972
1	82.85	74.35	150	162	8.97	9.76	12.6	12.2	261	337	11.8	13.7
2	79.10	66.05	163	145	10.19	10.10	13.1	12.5	320	351	14.3	13.9
3	86.95	80.30	181	156	9.63	10.71	13.5	13.8	339	424	15.4	17.7
4	83.31	77.60	205	148	9.47	10.75	13.8	13.0	287	379	12.7	17.3
5	88.90	80.45	187	142	9.59	9.56	13.3	12.4	308	327	14.3	13.8
6	83.10	81.00	182	200	9.19	10.48	12.8	13.9	314	378	13.9	15.7
7	89.50	85.05	152	163	9.60	10.90	13.5	13.3	311	367	13.5	16.3
8	86.50	80.75	188	170	9.30	10.65	12.5	13.0	281	372	12.9	15.1
9	87.30	80.95	170	165	9.00	10.57	12.7	13.8	264	357	11.8	14.6
10	88.75	64.40	193	142	9.78	10.21	13.4	12.2	293	352	14.0	14.8
11	84.60	75.90	188	157	10.43	10.79	14.2	13.6	346	357	16.7	13.8
12	83.60	69.00	164	170	9.58	8.61	13.9	9.8	290	258	13.0	9.4
13	86.60	82.25	193	156	10.43	11.06	15.3	13.8	336	404	15.7	17.5
14	84.55	80.75	200	156	9.07	11.14	12.6	14.7	237	412	10.4	17.1
15	87.95	82.25	202	164	9.31	10.30	14.4	13.3	287	390	11.7	17.2
16	85.50	79.55	225	174	10.32	10.75	12.9	13.4	355	400	16.5	16.9
17	86.30	81.90	184	163	9.50	10.75	12.7	13.4	300	355	13.4	16.5
18	86.10	83.55	198	182	9.73	11.43	13.7	14.3	295	406	12.8	15.0
19	81.80	65.45	203	147	10.41	9.55	13.9	11.3	314	300	13.1	12.0
20	75.20	68.00	185	156	10.10	9.88	13.4	11.7	320	330	15.9	13.2
21	78.60	66.85	174	194	9.77	9.56	12.6	11.9	310	304	13.8	9.3
22	85.20	81.45	159	192	9.92	11.12	13.5	14.2	286	384	12.1	17.8
23	81.05	75.65	189	191	9.74	10.93	15.0	13.7	307	380	13.6	13.0
24	86.65	77.30	198	170	10.22	11.09	14.0	14.1	324	404	13.4	16.3
25	89.30	81.35	212	186	9.90	10.41	13.5	13.3	323	340	14.4	12.5
26	84.50	79.45	173	165	9.86	10.79	13.0	13.6	282	384	12.2	14.4
27	88.30	81.35	212	198	10.08	10.53	13.6	13.4	328	310	15.0	13.8

iii. The matrix of sample correlation coefficients $R = (r_{ij})$ is

	X_1	X_2	X_3	X_4	X_5	X_6
X_1	1.000					
X_2	0.219	1.000				
X_3	−0.160	0.329	1.000			
X_4	0.083	0.242	0.491	1.000		
X_5	0.119	0..274	0.579	0.289	1.000	
X_6	−0.086	0.290	0.689	0.281	0.789	1.000

iv. The maximum likelihood estimate of the regression of X_6 on $X_1 = x_1, \ldots, X_5 = x_5$ is

$$\hat{E}(X_6|X_1 = x_1, \ldots, X_5 = x_5) = 3.39768 - 0.03721x_1 - 0.00008x_2 \\ - 0.14427x_3 - 0.15360x_4 + 0.05544x_5.$$

v. The maximum likelihood estimate of the square of the multiple correlation coefficient of X_6 on $(X_1, \ldots, X_5)$ is

$$r^2 = 0.85358.$$

vi. The maximum likelihood estimates of some of the partial correlation coefficients are

$$
\begin{aligned}
r_{23.5} &= 0.0156 \\
r_{23.1} &= -0.2130 \\
r_{23.15} &= -0.2363 \\
r_{23.16} &= -0.1982 \\
r_{23.456} &= 0.0100 \\
r_{23.45} &= -0.0063 \\
r_{23.46} &= -0.1823 \\
r_{23.145} &= -0.2252 \\
r_{23.146} &= -0.1906 \\
r_{23.14} &= -0.2000 \\
r_{23.56} &= 0.0328 \\
r_{23.156} &= -0.2074 \\
r_{23.1456} &= -0.1999
\end{aligned}
$$

For 1972

i.

$$\hat{\mu} = \begin{pmatrix} 77.1444 \\ 167.1852 \\ 10.4585 \\ 13.0963 \\ 361.5553 \\ 14.7630 \end{pmatrix}$$

ii. $\hat{\Sigma} = s/27$

	X_1	X_2	X_3	X_4	X_5	X_6
X_1	38.829					
X_2	32.872	299.772				
X_3	2.603	2.102	0.404			
X_4	4.899	4.997	0.632	1.138		
X_5	141.722	−11.799	21.470	35.591	1553.795	
X_6	9.027	−6.689	1.090	1.779	76.664	5.318

iii. The matrix of sample correlation coefficients $R = (r_{ij})$ is

	X_1	X_2	X_3	X_4	X_5	X_6
X_1	1.000					
X_2	0.305	1.000				
X_3	0.657	0.191	1.000			
X_4	0.737	0.271	0.932	1.000		
X_5	0.577	−0.017	0.857	0.846	1.000	
X_6	0.628	−0.168	0.743	0.723	0.843	1.000

iv. The maximum likelihood estimate of the regression of X_6 on $X_1 = x_1, \ldots, X_5 = x_5$ is

$$\hat{E}(X_6|X_1 = x_1, \ldots, X_5 = x_5) = -4.82662 + 0.12636x_1 - 0.03436x_2 + 0.61897x_3 - 0.28526x_4 + 0.03553x_5.$$

v. The maximum likelihood estimate of the square of the multiple correlation coefficient of X_6 on $(X_1, \ldots, X_5)$ is

$$r^2 = 0.80141.$$

vi. The maximum likelihood estimates of some of the partial correlation coefficients are

$r_{23.4} = 0.7234$ $\quad r_{23.46} = 0.3539$ $\quad r_{23.456} = 0.3861$
$r_{23.5} = 0.2776$ $\quad r_{23.14} = 0.4887$ $\quad r_{23.145} = 0.4578$
$r_{23.6} = 0.2042$ $\quad r_{23.56} = 0.2494$ $\quad r_{23.146} = 0.4439$
$r_{23.1} = 0.3226$ $\quad r_{23.15} = 0.3194$ $\quad r_{23.156} = 0.3795$
$r_{23.45} = 0.4576$ $\quad r_{23.16} = 0.3709$ $\quad r_{23.1456} = 0.4532$

5.4. EQUIVARIANT ESTIMATION UNDER CURVED MODELS

In the recent years some attention has been focused on the estimation of multivariate mean with constrain. This problem was originally considered by R.A. Fisher a long time ago. It is recently focused again in the works of Efron (1978), Cox and Hinkley (1977), Kariya (1989), Kariya, Giri and Perron (1988), Perron and Giri (1990), Marchand and Giri (1993), Marchand (1994), Fourdrinier and Strawderman (1996), Fourdrinier and Onassou (2000) among others. The motivation behind it is primarily based on the observed fact that in the univariate normal population with mean μ and variance σ^2, σ becomes large proportionally to μ so that $|\mu|/\sigma$ remains constant. This is also evident in the multivariate observations. But in the multivariate case no well accepted measure of variation between the mean vector μ and the covariance matrix Σ is available.

Let

$$\lambda = \mu'\Sigma^{-1}\mu, \quad \nu = \Sigma^{-\frac{1}{2}}\mu \qquad (5.78)$$

where $\Sigma^{\frac{1}{2}}$ is a $p \times p$ lower triangular matrix with positive diagonal elements such that $\Sigma^{\frac{1}{2}}\Sigma^{\frac{1}{2}} = \Sigma$. Kariya, Giri and Perron (1990) considered the problem of estimating μ with either λ or ν known under the loss (5.30) in the context of curved models. In all cases the best invariant estimators (BEE) are obtained as infinite series which in some special cases can be expressed as a finite series. They also proved that the BEE improves uniformly on the maximum likelihood estimator (MLE). Marchand (1994) gave a explicit expression for BEE and proved that the BEE dominates the MLE and the best linear estimator (BLE). Marchand and Giri (1993) obtained an optimal estimator within the class of James-Stein type estimators when the underlying distribution is that of a variance mixture of normals and when the norm $\|\mu\|$ is known. When the norm is restricted to a known interval, typically no optimum James-Stein type estimator exists.

Estimators of Parameters and Their Functions

When μ is restricted, the most usual constraint is a ball, that is a set for which $\|\mu\|$ is bounded by some constant m. By an invariance argument and analyticity considerations Bickel (1981) noted that the minimax estimator is Bayes with respect to a unique spherically symmetric least favorable prior distribution concentrating on a finite number of spherical shells, that is $\|\mu\|$ is constant. More recently Berry (1990) specified that when m is small enough, the corresponding prior is supported by a single spherical shell. This result is related to a more general class of models where Das Gupta (1985) showed that, when the parameter is restricted to a arbitrary bounded convex set in R^p, the Bayes estimator against the least favorable prior on the boundary of the parameter space is minimax.

Let $(\mathcal{X}, \mathcal{A})$ be a measure space and let Ω be the parametric space of θ. Denote by $\{P_\theta, \theta \in \Omega\}$ the set of probability distribution on $\mathcal{X}$. Let G be a group of transformations operating on $\mathcal{X}$ such that $g \in G, g : \mathcal{X} \to \mathcal{X}$ (sample space) is one to one onto (bijective). Let $\bar{G}$ be the corresponding group of induced transformations $\bar{g}$ on Ω. Assume

a. for $\theta_i \in \Omega, i = 1, 2, \theta_1 \neq \theta_2, P_{\theta_1} \neq P_{\theta_2}$,
b. $P_\theta(A) = P_{\bar{g}\theta}(gA), A \in \mathcal{A}, g \in G, \bar{g} \in \bar{G}$.

Let $\lambda(\theta)$ be a maximal invariant on Ω under $\bar{G}$ and let

$$\Omega^* = \{\theta | \theta \in \Omega \text{ with } \lambda(\theta) = \lambda_0\} \tag{5.79}$$

where λ_0 is known. We assume that $\mathcal{N}$ is the space of minimal sufficient statistic for θ. A point estimator $\hat{\theta}(X), X \in \mathcal{X}$ is equivariant if $\hat{\theta}(gX) = g\hat{\theta}(X), g \in G$. For notational simplification we take G to be the group of transformations on $\hat{\theta}(X)$.

Let $T(X), X \in \mathcal{X}$ be a maximal invariant under G (definition 3.2.4). Since the distribution of $T(X)$ depends on $\theta \in \Omega$ only through $\lambda(\theta)$, given $\lambda(\theta) = \lambda_0$, $T(X)$ is an ancillary statistic.

Definition 5.4.1. *Ancillary Statistic.* It is defined to be a part of the minimal sufficient statistic whose marginal distribution is parameter free.

Such models are assumed to be generated as an orbit under the induced group $\bar{G}$ on Ω and the ancillary statistic is realized as the maximal invariant on $\mathcal{X}$ under G.

Definition 5.4.2. *Curved Model.* A model with admits an ancillary statistic is called a curved model.

5.4.1. Best Equivariant Estimation of μ with λ Known

Let $X_1, \ldots, X_N (N > p)$ be independently and identically distributed $N_p(\mu, \Sigma)$. We want to estimate μ with loss function

$$L(\mu, d) = (d - \mu)' \Sigma^{-1} (d - \mu) \qquad (5.80)$$

when $\lambda = \mu' \Sigma^{-1} \mu = \lambda_0$ (known). Let $N\bar{X} = \sum_{i=1}^{N} X_i$, $S = \sum_{i=1}^{N} (X_i - \bar{X})(X_i - \bar{X})'$. The minimal sufficient statistic for (μ, Σ) is $(\bar{X}, S)$ and $\sqrt{N}\bar{X}$ is distributed independently of S as $N_p(\sqrt{N}\mu, \Sigma)$ and S is distributed as Wishart $W_p(N - 1, \Sigma)$ with $N - 1$ degrees of freedom and parameter Σ (see (6.32)). Under the loss function (5.80) this problem remains invariant under the full linear group $G_\ell(p)$ of $p \times p$ nonsingular matrices g transforming $(\bar{X}, S) \to (g\bar{X}, gSg')$. The corresponding group $\bar{G}$ of induced transformations $\bar{g}$ on the parametric space Ω transforms $\theta = (\mu, \Sigma) \to \bar{g}\theta = (g\mu, g\Sigma g')$. In Chapter 7 we will show that $T^2 = N(N-1)\bar{X}'S^{-1}\bar{X}$ is a maximal invariant in the space of $(\sqrt{N}\bar{X}, S)$ and the corresponding maximal invariant in the parametric space Ω in $\lambda = \mu' \Sigma^{-1} \mu$ and the distribution of T^2 depends on the parameters only through λ. Hence, given $\lambda = \lambda_0$, T^2 is an ancillary statistic. Since for any equivariant estimator $\bar{\mu}(X)$ of μ, the risk

$$\begin{aligned}
R(\theta, \bar{\mu}) &= E_\theta (\bar{\mu} - \mu)'(\bar{\mu} - \mu) \\
&= E_\theta (g\bar{\mu} - g\mu)'(g\Sigma g')^{-1}(g\bar{\mu} - g\mu) \\
&= E_\theta (\bar{\mu}(gX) - g\mu)'(g\Sigma g')^{-1}(\bar{\mu}(X) - g\mu) \qquad (5.81) \\
&= E_{\bar{g}\theta}(\bar{\mu}(X) - g\mu)'(g\Sigma g')^{-1}(\bar{\mu}(X) - g\mu) \\
&= R(\bar{g}\theta, \bar{\mu})
\end{aligned}$$

for $g \in G_\ell(p)$ and $\bar{g}$ is the induced transformation on Ω corresponding to g on $\mathcal{X}$. Since $\bar{G}_\ell(p)$ acts transitively on Ω we conclude from (5.80) that the risk $R(\theta, \bar{\mu})$ for any equivariant estimator $\bar{\mu}$ is a constant for all $\theta \in \Omega$. Taking $\lambda_0 = 1$ without any loss of generality and using the fact that $R(\theta, \bar{\mu})$ is a constant. Theorem 1.6.6 allows us to choose $\mu = e = (1, 0, \ldots, 0)'$ and $\Sigma = I$. To find the BEE which minimizes $R(\theta, \bar{\mu})$ among all equivariant estimators $\bar{\mu}$ satisfying

$$\bar{\mu}(g\bar{X}, gSg') = g\bar{\mu}(\bar{X}, S)$$

we need to characterize $\bar{\mu}$. Let $G_T(p)$ be the subgroup of $G_\ell(p)$ containing all $p \times p$ lower triangular matrices with positive diagonal elements. Since S is positive definite with probability one because of the assumption $N > p$ we can

Estimators of Parameters and Their Functions

write $S = WW'$, $W \in G_T(p)$. Let

$$V = W^{-1}Y, \qquad Q = \frac{V}{\|V\|}$$

where $Y = \sqrt{N}\bar{X}$, $\|V\|^2 = V'V = T^2/(N-1)$.

Theorem 5.4.1. *If $\bar{\mu}$ is an equivariant estimator of μ under $G_\ell(p)$ then*

$$\bar{\mu}(Y, S) = K(U)WQ \qquad (5.82)$$

where $K(U)$ is a measurable function of $U = T^2/(N-1)$.

Proof. Since $\bar{\mu}$ is equivariant under $G_\ell(p)$ we get for $g \in G_\ell(p)$

$$\bar{g}\bar{\mu}(Y, S) = \bar{\mu}(gY, gSg'). \qquad (5.83)$$

Replacing Y by $W^{-1}Y$, g by W and S by I in (5.83) we get

$$\bar{\mu}(Y, S) = W\bar{\mu}(V, I). \qquad (5.84)$$

Let O be a $p \times p$ orthogonal matrix with $Q = \dfrac{V}{\|V\|}$ as its first column. Then

$$\bar{\mu}(V, I) = \bar{\mu}(00'V, 00')$$
$$= 0\bar{\mu}(\sqrt{U}e, I).$$

Since the columns of O except the first one are arbitrary as far as they are orthogonal to Q, all components of $\mu(\sqrt{U}e, I)$, except the first component $\mu_1(\sqrt{U}e, I)$, are zero. Hence

$$\bar{\mu}(V, I) = Q\mu_1(\sqrt{U}e, I).$$

Q.E.D.

Theorem 5.4.2. *Under the loss function (5.79) the unique BEE of μ is*

$$\mu = \hat{K}(U)WQ \qquad (5.85)$$

where

$$\hat{K}(U) = E(Q'W'e|U)/E(Q'W'WQ|U). \qquad (5.86)$$

Proof. From Theorem 5.4.1 the risk function of an equivariant estimator μ, given that $\lambda_0 = 1$ is

$$R(\theta, \bar{\mu}) = R((e, I), \bar{\mu})$$
$$= E(K(U)WQ - e)'(K(U)WQ - e).$$

Since U is ancillary a unique BEE is $\tilde{\mu} = \hat{K}(U)WQ$, where $\hat{K}(U)$ minimizes the conditional risk given U

$$E((K(U)WQ - e)'(K(U)WQ - e)|U).$$

Using results of Section 1.7 we conclude that $\hat{K}(U)$ is given by (5.84). Q.E.D.

Maximum likelihood estimators

The maximum likelihood estimators $\hat{\mu}, \hat{\Sigma}$ of μ, Σ respectively under the restriction $\lambda_0 = 1$ are obtained by maximizing

$$-\frac{N}{2}\log\det\Sigma - \frac{1}{2}\operatorname{tr} S\Sigma^{-1}$$
$$-\frac{N}{2}(\bar{x} - \mu)'\Sigma^{-1}(\bar{x} - \mu) - \frac{\gamma}{2}(\mu'\Sigma^{-1}\mu - 1) \tag{5.87}$$

with respect to μ and Σ where γ is the Lagrange multiplier. Maximizing (5.85) we obtain

$$\hat{\mu} = \frac{N\bar{X}}{N + \gamma}, \qquad \hat{\Sigma} = \frac{S}{N} + \gamma \bar{X}\bar{X}'(\gamma + N)^{-1}.$$

Since $\hat{\mu}'\hat{\Sigma}^{-1}\hat{\mu} = 1$ we obtain

$$\hat{\mu} = \left(\frac{U - \sqrt{U(4 + 5U)}}{2U}\right)\bar{X},$$
$$\hat{\Sigma} = \frac{S}{N} + \frac{\sqrt{U(1 + 5U)}}{2U}\bar{X}\bar{X}'. \tag{5.88}$$

The maximum likelihood estimators are equivariant estimators. Hinkley (1977) investigated some properties of the model associated with Fisher information. Amari (1982a,b) proposed through a geometric approach what he called the dual *mle* which is also equivariant.

5.4.2. A Special Case

As a special case of λ constant we consider here the case $\Sigma = (\mu'\mu/C^2)I$ where C^2 is known. Let $Y = \sqrt{N}\bar{X}$ and $W = \operatorname{tr} S$. Then (Y, W) is a sufficient statistic and $C^2 W/(\mu'\mu)$ is distributed as $\chi^2_{(N-1)p}$. We are interested here to estimate μ with respect to the loss function.

$$L(\mu, d) = \frac{(d - \mu)'(d - \mu)}{(\mu'\mu)}. \tag{5.89}$$

Estimators of Parameters and Their Functions

The problem of estimating μ remains invariant under the group $G = R_+ \times 0(p)$, R_+ being the multiplicative group of positive reals and $0(p)$ being the multiplicative group of $p \times p$ orthogonal matrices transforming

$$X_i \to b\Gamma X_i, \quad i = 1, \ldots, N$$
$$d \to b\Gamma d$$
$$\left(\mu, \left(\frac{\mu'\mu}{C^2}\right)I\right) \to \left(b\Gamma\mu, b^2\left(\frac{\mu'\mu}{C^2}\right)I\right)$$

where $(b, \Gamma) \in G$ with $b \in R_+, \Gamma \in 0(p)$. The transformation induced by G on (Y, W) is given by

$$(Y, W) \to (b\Gamma y, bW).$$

A representation of an equivariant estimator under G is given in the following Theorem.

Theorem 5.4.3. *An estimator $d(Y, W)$ is equivariant if and only if there exists a measurable function $h : R_+ \to R$ such that*

$$d(Y, W) = h\left(\frac{Y'Y}{W}\right)Y$$

for all $(Y, W) \in R^p \times R_+$.

Proof. If h is a measurable function from $R_+ \to R$ and $d(Y, W) = h(Y'Y/W)Y$ then clearly d is equivariant under G. Conversely if d is equivariant under G, then

$$d(Y, W) = b\Gamma d\left(\frac{\Gamma'Y}{b}, \frac{W}{b^2}\right) \qquad (5.90)$$

for all $\Gamma \in 0(p), Y \in R^p, b > 0, W > 0$. We may assume without any loss of generality that $Y'Y > 0$.

Let Y and W be fixed and

$$d = \begin{pmatrix} d_1 \\ d_2 \end{pmatrix}$$

where d_1 is the first component of the p-dimensional vector d. Let $A \in 0(p)$ be a fixed $p \times p$ matrix such that $Y'A = (\|Y\|, 0, \ldots, 0)$ (see Theorem 1.6.6).

We now partition the matrix $A = (A_1, A_2)$ where $A_1 = \|Y\|^{-1}Y$, and choose $\Gamma = (A_1, A_2B)$ with $B \in O(p-1)$ and $b = \|Y\|$. From (5.86) we get

$$d(Y, W) = d_1\left((1, 0, \ldots, 0), \frac{W}{Y'Y}\right)Y$$
$$+ \|Y\| A_2 B d_2\left((1, 0, \ldots, 0), \frac{W}{Y'Y}\right). \tag{5.91}$$

Since the result holds for any choice of $B \in O(p-1)$ we must have

$$d(Y, W) = d_1\left((1, 0, \ldots, 0), \frac{W}{Y'Y}\right)Y.$$

Q.E.D.

It may be verified that a maximal invariant under G in the space of sufficient statistic is $V = W^{-1}(Y'Y)$ and a corresponding invariant in the parametric space is

$$\mu'\left(\frac{(\mu'\mu)I}{C^2}\right)^{-1}\mu = C^2.$$

As the group G acts transitively on the parametric space the risk function

$$R(\mu, d) = E_\mu(L(\mu, d))$$

of any equivariant estimator d is constant. Hence we can take $\mu = \mu_0 = (C, 0, \ldots, 0)'$. Thus the risk of any equivariant estimator d can be written as

$$R(\mu_0, d) = E_{\mu_0}\left(L\left(\mu_0, h\left(\frac{Y'Y}{W}\right)Y\right)\right)$$
$$= E_{\mu_0}\left(E\left(L\left(\mu_0, h\left(\frac{Y'Y}{W}\right)Y\right)|V = v\right)\right).$$

To find a BEE we need the function h_0 satisfying

$$E_{\mu_0}(L(\mu_0, h_0(V)Y)|V = v) \leq E_{\mu_0}(L(\mu_0, h(V)Y)|V = v)$$

for all $h : R_+ \to R$ measurable functions and for all values v of V. Since

$$E_{\mu_0}(L(\mu_0, h_0(v)Y)|V = v)$$
$$= h^2(v)E_{\mu_0}(Y'Y|V = v) - 2h(v)E_{\mu_0}(Y_1|V = v) + 1$$

Estimators of Parameters and Their Functions

where $Y = (Y_1, \ldots, Y_p)'$, we get

$$h_0(v) = \frac{E_{\mu_0}(Y_1|V=v)}{E_{\mu_0}(Y'Y|V=v)}. \qquad (5.92)$$

Theorem 5.4.4. *The BEE $d_0(X_1, \ldots, X_N; C) = d_0(Y, W)$ is given by*

$$d_0(Y, W) = \frac{NC^2}{2} \left[\frac{\sum_{i=1}^{\infty} \frac{\Gamma(\frac{1}{2}Np + i + 1)}{\Gamma(\frac{1}{2}p + i + 1)i!} \left(\frac{NC^2}{2}\right)^i}{\sum_{i=1}^{\infty} \frac{\Gamma(\frac{1}{2}Np + i + 1)}{\Gamma(\frac{1}{2}p + i)i!} \left(\frac{NC^2 t}{2}\right)^i} \right] \bar{X}$$

where $t = v(1+v)^{-1}$.

Proof. The joint probability density function of Y and W under the assumption that $\mu = \mu_0$ is

$$f_{Y,W}(y, w) = \begin{cases} \dfrac{\exp\{-(C^2/2)(y'y - 2\sqrt{N}y_1 + N + w)\}w^{((N-1)p-1)/2}}{2^{Np/2}(C^2)^{-Np/2}(\Gamma(\frac{1}{2}))^p \Gamma((N-1)p/2)}, & \text{if } w > 0 \\ 0, & \text{otherwise.} \end{cases}$$

Changing $(Y, W) \to (Y, V^{-1}Y'Y)$, the joint probability density function of Y and V is

$$f_{Y,V}(y, v) = \begin{cases} \dfrac{\exp\{C^2/2[((1+v)/v)(y'y + N]\}}{2^{Np/2}(C^2)^{-Np/2}[\Gamma(\frac{1}{2})]^p \Gamma((N-1)p/2)} \\ \quad \times \exp\{\sqrt{N}C^2 y_1\}(y'y)^{(N-1)p/2}(v)^{(-N-1)p/2-1}, & \text{if } v > 0 \\ 0, & \text{otherwise.} \end{cases}$$

Hence, with $t = (1+v)/v$, we get

$$h_0(v) = \frac{\int_{R^p} y_1 f_{Y,V}(y,v) dy}{\int_{R^p} (y'y) f_{Y,V}(y,v) dy}$$

$$= \sqrt{N} \frac{C^2}{2} \frac{\sum_{i=1}^{\infty} \frac{G(Np/2+i+1)}{\Gamma(p/2+i+1)i!} \left(\frac{NC^2 t}{2}\right)^i}{\sum_{i=0}^{\infty} \frac{\Gamma(Np/2+i+1)}{\Gamma(p/2+i)i!} \left(\frac{NC^2 t}{2}\right)^i}.$$

Q.E.D.

Theorem 5.4.5. *If $m = (N-1)p/2$ is an integer, the BEE is given by*

$$d_0(Y, W) = \frac{NC^2}{2} h_0(v) = g(t) \bar{X} \tag{5.93}$$

with $g(t) = u(t)/w(t)$ where

$$u(t) = \sum_{i=0}^{m+1} \frac{\binom{i}{m+1}\binom{m+1}{i}}{\Gamma(p/2+i)} \left(\frac{NC^2}{2}\right)^i t^i,$$

$$w(t) = \sum_{i=0}^{m+1} \frac{\binom{m+1}{i}\left(\frac{NC^2}{2}\right)^i}{\Gamma(p/2+i)} t^{i+1}.$$

Proof. Let Y_k be distributed as χ_k^2. Then

$$E(Y_k^\alpha) = 2^\alpha \frac{\Gamma(k/2+\alpha)}{\Gamma(k/2)} \quad \text{if} \quad \alpha > -\frac{k}{2}.$$

Hence with m as integer

$$h_0(v) = \sqrt{N} C^2 \frac{\sum_{i=0}^{\infty} E(Y_{p+2i+2}^m) \exp\left\{-\frac{NC^2 t}{2}\right\} \left(\frac{NC^2 t}{2}\right)^i \frac{1}{i!}}{\sum_{i=0}^{\infty} E(Y_{p+2i}^{m+1}) \exp\left\{-\frac{NC^2 t}{2}\right\} \left(\frac{NC^2 t}{2}\right)^i \frac{1}{i!}} \tag{5.94}$$

$$= \sqrt{N} C^2 \frac{E(V_1^m)}{E(V_2^{m+1})},$$

Estimators of Parameters and Their Functions

where V_1 is distributed as noncentral $\chi^2_{p+2}(NC^2 t)$ and V_2 is distributed as noncentral $\chi^2_p(NC^2 t)$. For $V = \chi^2_\nu(\delta^2)$ and r integer

$$E(V^r) = 2^r \sum_{k=0}^{r} \frac{\Gamma(\nu/2 + r)}{\Gamma(\nu/2 + k)} \binom{r}{k} \left(\frac{\delta^2}{r}\right)^k. \qquad (5.95)$$

From (5.94) and (5.95) we get (5.93). Q.E.D.

It may be verified that $g(t)$ is a continuous function of t and

$$\lim_{t \to 0^+} g(t) = \frac{NC^2}{p},$$

$$\lim_{t \to 1} g(t) = g(1) < 1,$$

and $g(t) > 0$ for all $t > 0$ (see Perron and Giri (1990) for details). Thus when $Y'Y$ is large the BEE is less than $\bar{X}$.

We can also write $d_0 = (1 - (\tau(\nu)/\nu))\bar{X}$ where $\tau(\nu)/\nu = 1 - g(t)$. This form is very popular in the literature. Perron and Giri (1990) have shown that $g(t)$ is a strictly decreasing function of t and $\tau(\nu)$ is strictly increasing in ν. The result that $g(t)$ is strictly decreasing in t tells what one may intuitively do if he has an idea of the true value of C and observe many large values concentrated. Normally one is suspicious of their effects on the sample mean and they have the tendency to shrink the sample mean towards the origin. That is what our estimator does. The result that $\tau(\nu)$ is strictly increasing in ν relates the BEE of the mean for C known with the class of minimax estimators of the mean for C unknown. Efron and Morris (1973) have shown that a necessary and sufficient condition for an equivariant estimator of the form $g(t)\bar{X}$ to be minimax is $g(t) \to 1$ as $t \to 1$. So our estimator fails to be minimax if we do not know the value of C. On the other hand they have shown that an estimator of the form $d = (1 - (\tau(\nu)/\nu))\bar{X}$ is minimax if (i) τ is an increasing function, (ii)

$$0 \leq \tau(\nu) \leq (p-2)/(n-1) + 2 \quad \text{for} \quad \nu \in (0, \infty).$$

Thus our estimator satisfies (i) but fails to satisfy (ii). So a truncated version of our estimator could be a compromise solution between the best when one knows the value of C and the worst, one can do by using the incorrect value of C.

5.4.3. Maximum Likelihood Estimator (mle)

The likelihood of $x_1, \ldots, x_N$ with C known is given by

$$L(x_1, \ldots, x_N|\mu) =$$

$$\left(\frac{2N}{C^2}\right)^{-Np/2} (\mu'\mu)^{Np/2} \exp\left\{\frac{C^2}{2\mu'\mu}(w + y'y - 2\sqrt{N}y'\mu + N\mu'\mu)\right\}.$$

Thus the *mle* $\hat{\mu}$ of μ (if it exists) is given by

$$[Np/c^2](\hat{\mu}'\hat{\mu}) - [w + y'y - 2\sqrt{N}y'\hat{\mu}|\hat{\mu} = \sqrt{N}\hat{\mu}'\hat{\mu}y. \qquad (5.96)$$

If this equation in $\hat{\mu}$ has a solution it must be collinear with y and hence

$$k[(Np/C^2)(y'y)k^2 + \sqrt{N}y'yk - (y'y + w)] = 0$$

Two nonzero solutions of k are

$$k_1 = \frac{-1 - \left(1 + \frac{4p}{C^2}\left(\frac{1+v}{v}\right)\right)^{\frac{1}{2}}}{2\sqrt{N}p/C^2}, \quad k_2 = \frac{-1 + \left(1 + \frac{4p}{C^2}\left(\frac{1+v}{v}\right)\right)^{\frac{1}{2}}}{2\sqrt{N}p/C^2}.$$

To find the value of k which maximizes the likelihood we compute the matrix of mixed derivatives

$$\left.\frac{\partial^2(-\log L)}{\partial \mu' \partial \mu}\right|_{\mu=ky} = \frac{C^2}{k^4(y'y)^2}\left[\sqrt{N}k(y'y)I + \frac{2Np}{C^2}k^2yy'\right]$$

and assert that matrix should be positive definite. The characteristic roots of this matrix are given by

$$\lambda_1 = \frac{\sqrt{N}C^2}{k^3 y'y}, \quad \lambda_2 = \frac{\sqrt{N}C^2 + 2Npk}{k^2 y'y}.$$

If $k = k_1$, then $\lambda_1 < 0$ and $\lambda_2 < 0$. But if $k = k_2$, then $\lambda_1 > 0$, $\lambda_2 > 0$, hence the *mle* $\hat{\mu} = d_1(x_1, \ldots, x_N, C)$ is given by

$$d_1(x_1, \ldots, x_N, C) = \left[\frac{(1 + 4p/C^2 t)^{\frac{1}{2}} - 1}{2p}\right] C^2 \bar{x}.$$

Since the maximum likelihood estimator is equivariant and it differs from the BEE d_0, the *mle* d_1 is inadmissible. The risk function of d_0 depends on C. Perron and Giri (1990) computed the relative efficiency of d_0 when compared with d_1, the James-Stein estimator d_2, the positive part of the James-Stein estimator d_3, and the sample mean $\bar{X}$ (d_4) for different values of C, N, and p. They have concluded that when the sample size N increases for a given p and C the relative

Estimators of Parameters and Their Functions

efficiency of d_0 when compared with d_i, $i = 1, \ldots, 4$ does not change significantly. This phenomenon changes markedly when C varies. When C is small, d_0 is markedly superior to others. On the other hand, when C is large all five estimators are more or less similar. These conclusions are not exact as the risk of d_0, d_1 are evaluated by simulation. Nevertheless, it gives us significant indication that for small value of C the use of BEE is advantageous.

5.4.4. An Application

An interesting application of this model is given by Kent, Briden and Mardia (1983). The natural remanent magnetization (NRM) in rocks is known to have, in general, originated in one or more relatively short time intervals during rock forming or metamorphic events during which NRM is frozen in by falling temperature, grain growth, etc. The NRM acquired during each such event is a single vector magnetization parallel to the then-prevailing geometric field and is called a component of NRM. By thermal, alternating fields or chemical demagnetization in stages these components can be identified. Resistance to these treatments is known as "stability of remanence". At each stage of the demagnetization treatment one measures the remanent magnetization as a vector in 3-dimensional space. These observations are represented by vectors $X_1, \ldots, X_N$ in R^3. They considered the model given by $X_i = \alpha_i + \beta_i + e_i$ where α_i denotes the true magnetization at the ith step, β_i represents the model error, and e_i represents the measurement error. They assumed that β_i and e_i are independent, β_i is distributed as $N_3(0, \tau^2(\alpha_i)I)$, and e_i is distributed as $N_3(0, \sigma^2(\alpha_i)I)$. The α_i are assumed to possess some specific structures, like collinearity etc., which one attempts to determine. Sometimes the magnitude of model error is harder to ascertain and one reasonably assumes $\tau^2(\alpha_i) = 0$. In practice $\sigma^2(\alpha_i)$ is allowed to depend on α_i and plausible model for $(\sigma^2(\alpha_i)$ which fits many data reasonably well is $\sigma^2(\alpha) = a(\alpha'\alpha) + b$ with $a > 0, b > 0$. When $\alpha'\alpha$ large, b is essentially 0 and a is unknown.

5.4.5. Best Equivariant Estimation in Curved Covariance Models

Let $X_1, \ldots, X_N$ ($N > p > 2$) be independently and identically distributed $N_p(\mu, \Sigma)$. Let Σ and $S = \sum_{i=1}^{N}(X_i - \bar{X})(X_i - \bar{X})'$ be partitioned as

$$\Sigma = \frac{1}{p-1}\begin{pmatrix} 1 & p-1 \\ \Sigma_{11} & \Sigma_{12} \\ \Sigma_{21} & \Sigma_{22} \end{pmatrix}, \quad S = \frac{1}{p-1}\begin{pmatrix} 1 & p-1 \\ S_{11} & S_{12} \\ S_{21} & S_{22} \end{pmatrix}$$

where $N\bar{X} = \sum_{i=1}^{N} X_i$. We are interested to find the BEE of $\beta = \Sigma_{22}^{-1}\Sigma_{12}$ on the basis of N observations $x_1, \ldots, x_N$ when one knows the value of the multiple correlation coefficient $\rho^2 = \Sigma_{11}^{-1}\Sigma_{12}\Sigma_{22}^{-1}\Sigma_{21}$. If the value of ρ^2 is significant one would naturally be interested to estimate β for the prediction purpose and also to estimate Σ_{22} to ascertain the variability of the prediction variables.

Let H_1 be the subgroup of the full linear group $G_\ell(p)$, define by

$$H_1 = \left\{ h \in G_\ell(p) : h = \begin{pmatrix} h_{11} & 0 \\ 0 & h_{22} \end{pmatrix} \text{ with } h_{11} \text{ is } 1 \times 1 \right\}$$

and let H_2 be the additive subgroup in R^p. Define $G = H_1 \oplus H_2$, the direct sum of H_1 and H_2. The transformation $g = (h_1, h_2) \in G$ transforms

$$X_i \to h_1 X_i + h_2, \quad i = 1, \ldots, N$$
$$(\mu, \Sigma) \to (h_1 \mu + h_2, h_1 \Sigma h_1').$$

The corresponding transformation on the sufficient statistic $(\bar{X}, S)$ is given by $(\bar{X}, S) \to (h_1 \bar{X} + h_2, h_1 S h_1')$. A maximal invariant in the space of $(\bar{X}, S)$ under G is

$$R^2 = S_{11}^{-1} S_{12} S_{22}^{-1} S_{21}$$

and a corresponding maximal invariant in the parametric space of (μ, Σ) is ρ^2 (see Section 8.3.1).

5.4.6. Charcterization of Equivariant Estimators of Σ

Let S_p, be the space of all $p \times p$ positive definite matrices and let $G_T^+(p)$ be the group of $p \times p$ lower triangular matrices with positive diagonal elements. An equivariant estimator $d(\bar{X}, S)$ of Σ with respect to the group of transformations G is a measurable function $d(\bar{X}, S)$ on $S_p \times R^p$ to S_p satisfying

$$d(h\bar{X} + \xi, hSh') = hd(\bar{X}, S)h'$$

for all $S \in S_p$, $h \in H_1$ and $\bar{X}, \xi \in R^p$. From this definition it is easy to conclude that if d is equivariant with respect to G then $d(\bar{X}, S) = d(0, S)$ for all $\bar{X} \in R^p, S \in S^p$. Thus without any loss of generality we can assume that d is a mapping from S_p to S_p. Furthermore, if u is a function mapping S_p into another space Y (say) then d^* is an equivariant estimator of $u(\Sigma)$ if and only if $d^* = u \cdot d$ for some equivariant estimator d of Σ.

Let

$$\Theta_\rho = \{(\mu, \Sigma) : \Sigma_{11}^{-1}\Sigma_{12}\Sigma_{22}^{-1}\Sigma_{21} = \rho^2\}$$

and let $\bar{G}$ be the group of induced transformations corresponding to G on Θ_ρ.

Estimators of Parameters and Their Functions

Theorem 5.4.6. $\bar{G}$ *acts transitively on* Θ_ρ.

Proof. It is sufficient to show that there exists a $h = (h, \xi) \in G$ with $h \in H_1, \xi \in R^p$ such that

$$(h\mu + \xi, h\Sigma h') = \left(0, \begin{pmatrix} 1 & \rho & 0 \\ \rho & 1 & 0 \\ 0 & 0 & I \end{pmatrix}\right) \tag{5.97}$$

with $I = I_{p-2}$. If $\rho = 0$, i.e. $\Sigma_{12} = 0$, we take $h_{11} = \Sigma_{11}^{-1/2} \cdot \xi = -h\mu$ to obtain (5.97). If $\rho \neq 0$, choose $h_{11} = \Sigma_{11}^{-1/2}, h_{22} = \Gamma \Sigma_{22}^{-1/2}$ where Γ is a $(p-1) \times (p-1)$ orthogonal matrix such that

$$\Sigma_{11}^{-\frac{1}{2}} \Sigma_{12} \Sigma_{22}^{-\frac{1}{2}} \Gamma = (\rho, 0, \ldots, 0),$$

and $\xi = -h\mu$ to get (5.97). Q.E.D.

The Theorem below gives a characterization of the equivariant estimator $d(S)$ of Σ.

Theorem 5.4.7. *An estimator d of Σ is equivariant if and only if it admits the decomposition*

$$d(S) = \begin{pmatrix} a_{11}(R) & a_{12}(R)R^{-1}S_{12} \\ a_{12}(R)R^{-1}S_{21} & R^{-2}a_{22}(R)S_{21}S_{11}^{-1}S_{12} + C(R)(S_{22} - S_{21}S_{11}^{-1}S_{12}) \end{pmatrix} \tag{5.98}$$

where $C(R) > 0$ and

$$A(R) = \begin{pmatrix} a_{11}(R) & a_{12}(R) \\ a_{21}(R) & a_{22}(R) \end{pmatrix}$$

is a 2×2 positive definite matrix. Furthermore

$$d_{11}^{-1} d_{12} d_{22}^{-1} d_{21} = \rho^2 = \Sigma_{11}^{-1} \Sigma_{12} \Sigma_{22}^{-1} \Sigma_{21}$$

if and only if $a_{11}^{-1} a_{12} a_{22}^{-1} a_{21} = \rho^2$.

Note The d_{ij} are submatrices of d as partitioned in (5.98) and $a_{ij} = a_{ij}(R)$.

Proof. The sufficiency part of the proof is computational. It consists in verifying $d(hSh') = hd(S)h'$ for all $h \in H$, $S \in S_p$ and $d_{11}^{-1}d_{12}d_{22}^{-1}d_{21} = a_{11}^{-1}a_{12}a_{22}^{-1}a_{21}$. It can be obtained in a straightforward way from the computations presented in the necessary part.

To prove the necessary part we observe that if

$$P = \begin{pmatrix} 1 & R \\ R & 1 \end{pmatrix}, \qquad R > 0$$

and d satisfies

$$d\begin{pmatrix} P & 0 \\ 0 & I_{p-2} \end{pmatrix} = \begin{pmatrix} I_2 & 0 \\ 0 & \Gamma \end{pmatrix} d\begin{pmatrix} P & 0 \\ 0 & I_{p-2} \end{pmatrix} \begin{pmatrix} I_2 & 0 \\ 0 & \Gamma' \end{pmatrix}$$

for all $\Gamma \in O(p-2)$, then

$$d\begin{pmatrix} P & 0 \\ 0 & I_{p-1} \end{pmatrix} = \begin{pmatrix} A(R) & 0 \\ 0 & C(R)I_{p-2} \end{pmatrix}$$

with $C(R) > 0$. In general, S has a unique decomposition of the form

$$S = \begin{pmatrix} S_{11} & S_{12} \\ S_{21} & S_{22} \end{pmatrix}$$

$$= \begin{pmatrix} T_1 & 0 \\ 0 & T_2 \end{pmatrix} \begin{pmatrix} 1 & U' \\ U & I_{p-1} \end{pmatrix} \begin{pmatrix} T_1' & 0 \\ 0 & T_2' \end{pmatrix}$$

where $T_1 \in G_T^+(1)$, $T_2 \in G_T^+(p-1)$, and $U = T_2^{-1} S_{21} T_1^{-1}$.

Without any loss of generality we may assume that $U \neq 0$. Corresponding to U there exists a $B \in 0(p-1)$ such that $U'B = (R, 0, \ldots, 0)$ with $R = \|U\| = (S_{11}^{-1} S_{12} S_{22}^{-1} S_{21})^{1/2} > 0$. For $p > 2$, B is not uniquely determined but its first column is $R^{-1}U$. Using such a B we have the decomposition

$$\begin{pmatrix} 1 & U' \\ U & I_{p-1} \end{pmatrix} = \begin{pmatrix} 1 & 0 \\ 0 & B \end{pmatrix} \begin{pmatrix} P & 0 \\ 0 & I_{p-2} \end{pmatrix} \begin{pmatrix} 1 & 0 \\ 0 & B' \end{pmatrix}$$

Estimators of Parameters and Their Functions

and

$$d(S) = \begin{pmatrix} T_1 & 0 \\ 0 & T_2 \end{pmatrix} \begin{pmatrix} 1 & 0 \\ 0 & B \end{pmatrix} \begin{pmatrix} A(R) & 0 \\ 0 & I_{p-2} \end{pmatrix} \begin{pmatrix} 1 & 0 \\ 0 & B' \end{pmatrix} \begin{pmatrix} T'_1 & 0 \\ 0 & T'_2 \end{pmatrix}$$

$$= \begin{pmatrix} T_1 & 0 \\ 0 & T_2 \end{pmatrix}$$

$$\times \begin{pmatrix} a_{11}(R) & R^{-1}a_{12}(R)U' \\ R^{-1}a_{21}(R)U & R^{-2}a_{22}(R)UU' + C(R)(I_{p-1} - UU') \end{pmatrix}$$

$$\times \begin{pmatrix} T'_1 & 0 \\ 0 & T'_2 \end{pmatrix}$$

$$= \begin{pmatrix} a_{11}(R)S_{11} & R^{-1}a_{12}(R)S_{12} \\ R^{-1}a_{21}(R)S_{21} & R^{-2}(R)S_{21}S_{11}^{-1}S_{12} + C(R)(S_{22} - S_{21}S_{11}^{-1}S_{12}) \end{pmatrix}$$

which proves the necessary part of the Theorem.   Q.E.D.

5.4.7. Characterization of Equivariant Estimators of β

The following Theorem gives a characterization of the equivariant estimator of β.

Theorem 5.4.8. *If d^* is an equivariant estimator of β then $d^*(S)$ has the form*

$$d^*(S) = R^{-1}a(R)S_{22}^{-1}S_{21}$$

where $a(R) : R_+ \to R^1$.

Proof. Define $u : S^p \to R^{p-1}$ by $u(\Sigma) = \beta = \Sigma_{22}^{-1}\Sigma_{21}$. If $d^*(S)$ is equivariant, from Theorem 5.4.7, we get

$$d^*(S) = (R^{-2}a_{22}(R)S_{21}S_{11}^{-1}S_{12}$$
$$+ C(R)(S_{22} - S_{21}S_{11}^{-1}S_{12}))^{-1}S_{21}R^{-1}a_{21}(R)$$
$$= (T_2(R^{-2}a_{22}(R)UU' + C(R)(I_{p-1} - UU')T'_2)^{-1}S_{21}R^{-1}a_{21}(R)$$
$$= R^{-1}(a_{22}(R) + (1 - R^2)C(R))^{-1}a_{21}(R)S_{22}^{-1}S_{21}$$
$$= R^{-1}a(R)S_{22}^{-1}S_{21}$$

Q.E.D.

The risk function of an equivariant estimator d^* of β is given by

$$R(\beta, d^*) = E_\Sigma(L(\beta, d^*))$$
$$- E_\Sigma\{S_{11}^{-1}(R^{-1}a(R)S_{12}S_{22}^{-1} - \beta)'S_{22}(R^{-1}a(R)S_{22}^{-1}S_{21} - \beta)\} \quad (5.99)$$
$$= E_\Sigma\{a^2(R) - 2R^{-1}a(R)S_{11}^{-1}S_{12}\beta + S_{11}^{-1}\beta'S_{22}\beta\}.$$

Theorem 5.4.9. *The best equivariant estimator of β given ρ, under the loss function L, is given by*

$$R^{-1}a^*(R)S_{22}^{-1}S_{21} \quad (5.100)$$

where

$$a^*(R) = R\rho^2 \frac{\sum_{i=0}^\infty \Gamma\left(\frac{N+1}{2}+i\right)\Gamma\left(\frac{N-1}{2}+i\right)(R^2\rho^2)^i/i!\Gamma\left(\frac{p+1}{2}+i\right)}{\sum_{j=0}^\infty \Gamma^2\left(\frac{N-1}{2}+j\right)(\rho^2R^2)^j/j!\Gamma\left(\frac{p-1}{2}+j\right)}.$$

$$(5.101)$$

Proof. From (5.94), the minimum of $R(\beta, \delta^*)$ is attained when $a(R) = a^*(R) = E_\Sigma(S_{11}^{-1}S_{12}\beta R^{-1}/R)$. Since the problem is invariant and d^* is equivariant we may assume, without any loss of generality, that

$$\Sigma = \Sigma_\rho = \begin{pmatrix} C(\rho) & 0 \\ 0 & I_{p-2} \end{pmatrix}$$

with

$$C(\rho) = \begin{pmatrix} 1 & \rho \\ \rho & 1 \end{pmatrix}.$$

To evaluate $a^*(R)$ we write

$$S_{22} = TT', T \in G_T^+(p-1), T = (t_{ij}),$$
$$S_{21} = RTW, 0 < R < 1, W \in R^{p-1},$$
$$S_{11} = W'W.$$

Estimators of Parameters and Their Functions

The joint probability density function of (R, W, T) (see Chapter 6) is given by

$$f_{R,W,T}(r, w, t) = K^{-1} r^{p-2} (1 - r^2)^{(N-p)/2 - 1} (w'w)^{(N-p)/2}$$

$$\times \exp\left\{ -\frac{1}{2} \sum_{i=2}^{p-1} \sum_{j=1}^{i} t_{ij}^2 \right\}$$

$$\times \exp\left\{ -\frac{1}{2(1 - \rho^2)} (w'w + t_{11}^2 - 2r\rho t_{11} w_1) \right\} \quad (5.102)$$

$$\times \prod_{i=1}^{p-1} (t_{ii})^{N-i-1}$$

where

$$K = (1 - \rho^2)^{(N-1)/2} \pi^{p(p-1)/4} 2^{(N-3)p/2} \prod_{i=1}^{p} \Gamma\left(\frac{N-i}{2}\right).$$

A straightforward computation gives (5.100). Q.E.D.

The following Lemma reduces the last expression in (5.101) into a rational polynomial when $(N - p)/2$ is an integer.

Lemma 5.4.1. *Let $\beta > 0$, $\gamma \in (0, 1)$ and $m \in \mathbb{N}$. Then*

$$\sum_{i=0}^{\infty} \frac{\Gamma(\alpha + m + i) \Gamma(\beta + i)}{\Gamma(\alpha + i) i!} \gamma^i = (1 - \gamma)^{-\beta}$$

$$\times \sum_{j=0}^{m} \binom{m}{j} \frac{\Gamma(\alpha + m) \Gamma(\beta + j)}{\Gamma(\alpha + j)} \left(\frac{\gamma}{1 - \gamma} \right)^j$$

Proof.

$$\sum_{i=0}^{\infty} \frac{\Gamma(\alpha+m+i)}{\Gamma(\alpha+i)} \frac{\Gamma(\beta+i)}{i!} \gamma^i$$

$$= \frac{d^m}{dt^m} t^{\alpha+m-1}(1-\gamma t)^{-\beta}\Gamma(\beta)|_{t=1}$$

$$= (1-\gamma)^{-\beta}\Gamma(\beta)\frac{d^m}{du^m}(1+u)^{\alpha+m-1}\left(-\frac{\gamma}{1-\gamma}u\right)^{-\beta}\Bigg|_{u=0}$$

$$= (1-\gamma)^{-\beta}\sum_{j=0}^{m}\binom{m}{j}\frac{\Gamma(\alpha+m)}{\Gamma(\alpha+j)}\Gamma(\beta+j)\left(\frac{\gamma}{1-\gamma}\right)^j.$$

Q.E.D.

If $(N-p)$ is even then with $m = (N-p)/2$,

$$a^*(R) = \frac{(N-1)}{2}R\rho^2 \frac{\sum_{i=1}^{\infty}\binom{m}{i}\left[\Gamma\left(\frac{N-1}{2}+i\right)/\Gamma\left(\frac{p+1}{2}+i\right)\right]\left(\frac{R^2\rho^2}{1-R^2\rho^2}\right)^i}{\sum_{j=0}^{m}\binom{m}{j}\left[\Gamma\left(\frac{N-1}{2}+j\right)/\Gamma\left(\frac{p-1}{2}+j\right)\right]\left(\frac{R^2\rho^2}{1-R^2\rho^2}\right)^j}.$$

If the value of ρ^2 is such that terms of order $(R\rho)^2$ and higher can be neglected, the BEE of β is approximately equal to $\rho^2(N-1)(p-1)^{-1}S_{22}^{-1}S_{21}$. The *mle* of β is $S_{22}^{-1}S_{21}$.

For the BEE of Σ_{22} we refer to Perron and Giri (1992).

EXERCISES

1 The data in Table 5.2 were collected in an experiment on jute in Bishnupur village of West Bengal, India, in which the weights of green jute plants (X_2) and their dry jute fibers (X_1) were recorded for 20 randomly selected individual plants. Assume that $X = (X_1, X_2)'$ is normally distributed with mean $\mu = (\mu_1, \mu_2)'$ and positive defiite covariance matrix Σ.
   (a) Find maximum likelihood estimates of μ, Σ.
   (b) Find the maximum likelihood estimate of the coefficient of correlation ρ between the components.
   (c) Find the maximum likelihood estimate of $E(X_1|X_2 = x_2)$.
2 The variability in the price of farmland per acre is to be studied in relation to three factors which are assumed to have major influence in determining the selling price. For 20 randomly selected farms, the price (in dollars) per acre

Table 5.2.

Plant No.	X_1	X_2	Plant No.	X_1	X_2
1	68	971	11	33	462
2	63	892	12	27	352
3	70	1125	13	21	305
4	6	82	14	5	84
5	65	931	15	14	229
6	9	112	16	27	332
7	10	162	17	17	185
8	12	321	18	53	703
9	20	315	19	62	872
10	30	375	20	65	740

(Weight (gm) column headers above both X_1, X_2 groups.)

(X_1), the depreciated cost (in dollars) of building per acre (X_2), and the distance to the nearest shopping center (in miles) (X_3) are recorded in Table 5.3. Assuming that $X = (X_1, X_2, X_3)'$ has three-variate normal distribution, find the maximum likelihood estimates of the following:

(a) $E(X_1|X_2 = x_2, X_3 = x_3)$;
(b) the partial correlation coefficient between X_1 and X_3 when X_2 is kept fixed;
(c) the multiple correlation coefficient between X_1 and (X_2, X_3).

Table 5.3.

Farm	X_1	X_2	X_3	Farm	X_1	X_2	X_3
1	75	15	6.0	11	13.5	13	0.5
2	156	6	2.5	12	175	12	2.5
3	145	60	0.5	13	240	7	2.0
4	175	24	3.0	14	175	27	4.0
5	70	5	2.0	15	197	16	6.0
6	179	8	1.5	16	125	6	5.0
7	165	14	4.0	17	227	13	5.0
8	134	13	4.0	18	172	13	11.0
9	137	7	1.5	19	170	34	2.0
10	175	19	2.5	20	172	19	6.5

3 Let $X^\alpha = (X_{\alpha 1}, \ldots, X_{\alpha p})'$, $\alpha = 1, \ldots, N$, be a random sample of size N from a p-variate normal distribution with mean μ and positive definite covariance matrix Σ. Show that the distribution of $\bar{X} = (1/N) \sum_{\alpha=1}^{N} X^\alpha$ is complete for given Σ.

4 Prove the equivalence of the three criteria of stochastic convergence of a random matrix as given in (5.5).

5 Let $X^\alpha = (X_{\alpha 1}, \ldots, X_{\alpha 1})'$, $\alpha = 1, \ldots, N$, be a random sample of size N from a p-dimensional normal distribution with mean μ positive definite covariance matrix Σ.

(a) Let $\mu = (\mu, \ldots, \mu)'$,

$$\Sigma = \begin{pmatrix} 1 & \rho & \rho & \cdots & \rho \\ \rho & 1 & \rho & \cdots & \rho \\ \vdots & \vdots & \vdots & & \vdots \\ \rho & \rho & \rho & \cdots & 1 \end{pmatrix} \sigma^2,$$

with $-1/(p-1) < \rho < 1$. Find the maximum likelihood estimators of ρ, σ^2, and μ.

(b) Let $\mu = (\mu_1, \ldots, \mu_p)'$,

$$\Sigma = \begin{pmatrix} \sigma_1^2 & \rho\sigma_1\sigma_2 & \rho\sigma_1\sigma_3 & \cdots & \rho\sigma_1\sigma_p \\ \rho\sigma_1\sigma_2 & \sigma_2^2 & \rho\sigma_2\sigma_3 & \cdots & \rho\sigma_2\sigma_p \\ \vdots & \vdots & \vdots & & \vdots \\ \rho\sigma_1\sigma_p & \rho\sigma_2\sigma_p & \rho\sigma_3\sigma_p & \cdots & \sigma_p^2 \end{pmatrix}$$

with $-1/p - 1 < \rho < 1$. Find the maximum likelihood estimator of $\mu, \rho, \sigma_1^2, \ldots, \sigma_p^2$.

6 Find the maximum likelihood estimators of the parameters of the multivariate log-normal distribution and of the multivariate Student's t-distribution as defined in Exercise 4.

7 Let $Y = (Y_1, \ldots, Y_N)'$ be normally distributed with

$$E(Y) = X\beta, \qquad \text{cov}(Y) = \sigma^2 I$$

where $X = (x_{ij})$ is an $N \times p$ matrix of known constants x_{ij}, and $\beta = (\beta_1, \ldots, \beta_p)'$, σ^2 are unknown constants.

(a) Let the rank of X be p. Find the maximum likelihood estimators $\hat{\beta}, \hat{\sigma}^2$ of β, σ^2. Show that $\hat{\beta}, \hat{\sigma}^2$ are stochastically independent and $N\hat{\sigma}^2/\sigma^2$ is distributed as chi-square with $N - p$ degrees of freedom.

(b) A linear parametric function $L'\beta, L = (l_1, \ldots, l_p)' \neq 0$, is called estimable if there exists a linear estimator $b'Y, b = (b_1, \ldots, b_N)' \neq 0$,

Estimators of Parameters and Their Functions

such that

$$E(b'Y) = L'\beta.$$

Let the rank of X be less than p and let the linear parametric function $L'\beta$ be estimable. Find the unique minimum variance linear unbiased estimator of $L'\beta$.

8 [*Inverted Wishart distribution*—$W_p^{-1}(A, N)$] A $p \times p$ symmetric random matrix V has an inverted Wishart distribution with parameter A (symmetric positive definite matrix) and with N degrees of freedom if its probability density function is given by

$$c(\det A)^{(N-p-1)/2}(\det V^{-1})^{N/2} \exp\left\{-\frac{1}{2}\operatorname{tr} V^{-1}A\right\}$$

where

$$c^{-1} = 2^{(N-p-1)p/2}\pi^{p(p-1)/4}\Pi_{i=1}^{p}(N-p-i)/2,$$

provided $2p < N$ and V is positive definite, and is zero otherwise.
  (a) Show that if a $p \times p$ random matrix S has a Wishart distribution as given in (5.2), then S^{-1} has an inverted Wishart distribution with parameters Σ^{-1} and with $N+p$ degrees of freedom.
  (b) Show that $E(S^{-1}) = \Sigma^{-1}/(N-p-1)$.

9 Let $X^\alpha = (X_{\alpha 1}, \ldots, X_{\alpha p})'$, $\alpha = 1, \ldots, N_1$, be a random sample of size N_1 from a p-dimensional normal distribution with mean $\mu = (\mu_1, \ldots, \mu_p)'$ and positive definite covariance matrix Σ, and let $Y^\alpha = (Y_{\alpha 1}, \ldots, Y_{\alpha p})'$, $\alpha = 1, \ldots, N_2$, be a random sample of size N_2 (independent of $X^\alpha, \alpha = 1, \ldots, N_1$) from a normal distribution with mean $\nu = (\nu_1, \ldots, \nu_p)'$ and the same covariance matrix Σ.
  (a) Find the maximum likelihood estimators of $\hat{\mu}, \hat{\nu}, \hat{\Sigma}$ of μ, ν and Σ, respectively.
  (b) Show that $\hat{\mu}, \hat{\nu}, \hat{\Sigma}$ are stochastically independent and that $(N_1 + N_2)\hat{\Sigma}$ is distributed as $\sum_{\alpha=1}^{N_1+N_2-2} Z^\alpha Z^{\alpha'}$, where $Z^\alpha = (Z_{\alpha 1}, \ldots, Z_{\alpha p})'$, $\alpha = 1, \ldots, N_1 + N_2 - 2$, are independently distributed p-variate normal random variables with mean 0 and the same covariance matrix Σ.

10 [Giri (1965); Goodman (1963)] Let $\xi^\beta = (\xi_{\beta_1}, \ldots, \xi_{\beta p})'$, $\beta = 1, \ldots, N$, be N independent and identically distributed p-variate complex Gaussian random variables with the same mean $E(\xi^\beta) = \alpha$ and with the same Hermitian positive definite complex covariance matrix $\Sigma = E(\xi^\beta - \alpha)(\xi^\beta - \alpha)^*$, where $(\xi^\beta - \alpha)^*$ is the adjoint of $(\xi^\beta - \alpha)$.
  (a) Show that, if α is known, the maximum likelihood estimator of Σ is $\hat{\Sigma} = 1/N \sum_{\beta=1}^{N}(\xi^\beta - \alpha)(\xi^\beta - \alpha)^*$. Find $E(\hat{\Sigma})$.

(b) Show that, if α, Σ are unknown, $(\hat{\xi}, \hat{\Sigma})$ where

$$\hat{\xi} = \frac{1}{N}\sum_{\beta=1}^{N} \xi^{\beta}, \qquad \hat{\Sigma} = \frac{1}{N}\sum_{\beta=1}^{N}(\xi^{\beta} - \bar{\xi})(\xi^{\beta} - \bar{\xi})^{*}.$$

is sufficient for (α, Σ).

11 Let $X = (X_1, \ldots, X_p)'$ be distributed as $N_p(\mu, \gamma^2 \Sigma)$ and let A be a $p \times p$ positive definite matrix. Show that (a)

$$E\left(\frac{X(X-\mu)'}{X'AX}\right) = \gamma^2 E\left(\frac{1}{X'AX}\left(I - \frac{2XX'A}{X'AX}\right)\right)\Sigma,$$

and (b)

$$E\left(\frac{X'(X-\mu)}{X'AX}\right) = \gamma^2(p-2)E\left(\frac{1}{X'AX}\right), \quad \text{if } \Sigma = I$$

12 Prove (5.68) and (5.71).

13 Show that

$$\int_{\mu}^{\infty}\int_{y}^{\infty} h'(y)(x-\mu)\frac{1}{\sqrt{2\pi}\sigma}\exp\left\{-\frac{1}{2}\left(\frac{x-\mu}{\sigma}\right)^2\right\}dx\,dy$$

$$= \int_{\mu}^{\infty} |h'(y)| \frac{1}{\sqrt{2\pi}\sigma} e^{-\frac{1}{2}\left(\frac{(x-\mu)^2}{\sigma^2}\right)} dy.$$

14 Let L be a class of nonnegative definite symmetric $p \times p$ matrices, and suppose J is a fixed nonsingular member of L. If $J^{-1}B$ (over B in L) is maximized by J, then $\det B$ is also maximized by J. Conversely, if L is convex and J maximizes $\det B$ then $J^{-1}B$ is maximized by $B = J$.

15 In Theorem 5.3.6 for $p = 2$, compute δ_1, δ_2 and the risk of the minimax estimator. Show that the risk is $2(2n^2 + 5n + 4)(n^3 + 5n^2 + 6n + 4)^{-1}$.

REFERENCES

Amari, S. (1982(a)). Differential geometry of curved exponential families—curvature and information loss. *Ann. Statist.* 10:357–385.

Amari, S. (1982(b)). Geometrical theory of asymptotic ancillary and conditional inference. *Biometrika* 69:1–17.

Bahadur, R. R. (1955). Statistics and subfields. *Ann. Math. Statist.* 26:490–497.

Bahadur, R. R. (1960). On the asymptotic efficiency of tests and estimators. *Sankhya* 22:229–252.

Baranchik, A. J. (1970). A family of minimax estimator of the mean of a multivariate normal distribution. *Ann. Math. Statist.* 41:642–645.

Basu, D. (1955). An inconsistency of the method of maximum likelihood. *Ann. Math. Statist.* 26:144–145.

Berry, P. J. (1990). Minimax estimation of a bounded normal mean vector. *Jour. Mult. Anal.* 35:130–139.

Bickel, P. J. (1981). Minimax estimation of the mean of a normal distribution when the parameter space is restricted. *Ann. Statist.* 9:1301–1309.

Brandwein, A. C., Strawderman, W. E. (1990). Stein estimation: the spherically symmetric case. *Statistical Science* 5:356–369.

Brandwein, A. C. (1979). Minimax estimation of the mean of spherically symmetric distribution under general quadratic loss. *J. Mult. Anal.* 9:579–588.

Berger, J. (1980). A robust generalized Bayes estimator and confidence region for a multivariate normal mean. *Ann. Statist.* 8:716–761.

Das Gupta, A. (1985). Bayes minimax estimation in multiparameter families when the parameter is restricted to a bounded convex set. *Sankhya A*, 47:281–309.

Cox, D. R., Hinkley, D. V. (1977). *Theoritical Statistics*. London: Chapman and Hall.

Dykstra, R. L. (1970). Establishing the positive definiteness of the sample covariance matrix. *Ann. Math. Statist.* 41:2153–2154.

Eaton, M. L., Pearlman, M. (1973). The nonsingularity of generalized sample covariance matrix. *Ann. Statist.* 1:710–717.

Efron, B. (1978). The geometry of exponential families. *Ann. Statist.* 6:362–376.

Efron, B. Morris, C. (1973) Stein's estimation rule and its competitors. An empirical Bayes approach. *J. Amer. Statist. Assoc.* 68:117–130.

Ferguson, T. S. (1967). *Mathematical Statistics, A Decision Theoritic Approach*. New York: Academic Press.

Fisher, R. A. (1925). Theory of statistical estimation. *Proc. Cambridge Phil. Soc.* 22:700–715.

Fourdrinier, D., Strawderman, W. E. (1996). A paradox concerning shrinkage estimators: should a known scale parameter be replaced by an estimated value in the shrinkage factor? *J. Mult. Anal.* 59:109–140.

Fourdrinier, D., Ouassou, Idir (2000). Estimation of the mean of a spherically symmetric distribution with constraints on the norm. *Can. J. Statistics* 28:399–415.

Giri, N. (1965). On the complex analogues of T^2- and R^2-tests. *Ann. Math. Statist.* 36:664–670.

Giri, N. (1975). *Introduction to Probability and Statistics, Part 2, Statistics*. New York: Dekker.

Giri, N. (1993). *Introduction to Probability and Statistics*, (revised and expanded edition). New York: Dekker.

Goodman, N. R. (1963). Statistical analysis based on a certain multivariate Gaussian distributions (an introduction). *Ann. Math. Statist.* 34:152–177.

Haff, L. R. (1980). Empirical Bayes estimation of the multivariate normal covariance matrix. *Ann. Statist.* 8:586–597.

Halmos, P. L., Savage, L. J. (1949). Application of Radon-Nikodyn theorem of the theory of sufficient statistics. *Ann. Math. Statist.* 20:225–241.

Hinkley, D. V. (1977). Conditional inference about a normal mean with known coefficient of variation. *Biometrika* 64:105–108.

James, W., Stein, C. (1961). Estimation with quadratic loss. *Barkeley Symp. Math. Statist. Prob. 2*, 4:361–379.

Kariya, T. (1989). Equivariant estimation in a model with ancillary statistic. *Ann. Statist.* 17:920–928.

Kariya, T., Giri, N., Perron. F. (1988). Equivariant estimation of a mean vector μ of $N(\mu, \Sigma)$ with $\mu'\Sigma^{-1}\mu = 1$ or $\Sigma^{-1/2}\mu = c$ or $\Sigma = \sigma^2(\mu'\mu)I$. *J. Mult. Anal.* 27:270–283.

Kent, B., Briden, C., Mardia, K. (1983). Linear and planar structure in ordered multivariate data as applied to progressive demagnetization of palaemagnetic remanance. *Geophys. J. Roy. Astron. Soc.* 75:593–662.

Kiefer, J. (1957). Invariance, minimax and sequential estimation and continuous time processes. *Ann. Math. Statist.* 28:573–601.

Kiefer, J., Wolfowitz, J. (1956). Consistency of maximum likelihood estimator in the presence of infinitely many incident parameters. *Ann. Math. Statist.* 27:887–906.

Kubokawa, T. (1998). The Stein phenomenon in simultaneous estimation, a review. In: Ahmad, S.E., Ahsanullah, M., Sinha, B. K., eds. *Applied Statistical Sciences, 3*. New York: Nova, pp. 143–173.

Estimators of Parameters and Their Functions 209

Kubokawa, T., Srivastava, M. S. (1999). Robust improvement in estimation of a covariance matrix in an elliptically contoured distribution. *Ann. Statist.* 27:600–609.

LeCam, L. (1953). On some asymptotic properties of the maximum likelihood estimates and related Bayes estimates. *Univ. California Publ. Statist.* 1:277–330.

Lehmann, E. L. (1959). *Testing Statistical Hypotheses*. New York: Wiley.

Lehmann, E. L., Scheffie, H. (1950). Completeness, similar regions and unbiased estimation, part 1. *Sankhya* 10:305–340.

Marchand, E., Giri, N. (1993). James-Stein estimation with constraints on the norm. *Commun. Statist., Theory Method* 22:2903–2924.

Marchand, E. (1994). On the estimation of the mean of a $N_p(\mu, \Sigma)$ population with $\mu'\Sigma^{-1}\mu$ known. *Statist. Probab. Lett.* 21:6975.

Neyman, J., Scott, E. L. (1948). Consistent estimates based on partially consistent observations. *Econometrika* 16:1–32.

Olkin, I., Selliah, J. B. (1977). Estimating covariance in a multivariate normal distribution. I: Gupta, S. S., Moore, D. S., eds. *Statistical Decision Theory and Related Topics*, Vol II, pp. 312–326.

Pearson, K. (1896). Mathematical contribution to the theory of evolution III, regression, heridity and panmixia. *Phil. Trans. A.* 187:253–318.

Perron, F., Giri, N. (1990). On the best equivariant estimation of mean of a multivariate normal population. *J. Mult. Anal.* 32:1–16.

Perron, F., Giri, N. (1992). Best equivariant estimator in curved covariance models. *J. Mult. Anal.* 44:46–55.

Press, S. J. (1972). *Applied Multivariate Analysis*. New York: Holt.

Raiffa, H., Schlaifer, R. (1961). *Applied Statistical Decision Theory*. Cambridge, Massachusetts: Harvard University Press.

Robert, C. P. (1994). *The Bayesian Choice a Decision Theoritic Motivation*. N.Y.: Springer.

Rao, C. R. (1965). *Linear Statistical Inference and its Applications*. New York: Wiley.

Srivastava, M. S., Bilodeau, M. (1989). Stein estimation under elliptical distributions. *J. Mult. Anal.* 28:247–259.

Strawderman, W. E. (1972). On the existance of proper Bayes minimax estimators of the mean of a multivariate normal distribution. *Proc. Barkeley Symp. Math. Stat. Prob. 1*, 6th:51–55.

Strawderman, W. E. (1974). Minimax estimation of location parameters of certain spherically symmetric distributions. *J. Mult. Anal.* 4:255–264.

Stein, C. (1956). Inadmissibility of the usual estimator for the mean of a multivariate normal distribution. *Barkeley Symp., Math. Stat. Prob. 3rd*, 5:196–207.

Stein, C. (1962). Confidence sets for the mean of a multivariate normal distribution. *J. Roy Statist. Soc. Ser. B* 24:265–285.

Stein, C. (1969). Multivariate analysis I (notes recorded by M. L. Eaton). Tech. Rep. No. 42, Statist. Dept., Stanford Univ., California.

Stein, C. (1981). Estimation of the mean of a multivariate normal distribution. *Ann. Statist.* 9:1135–1151.

Stein, C. (1975). Estimation of a Covariance Matrix. Rietz Lecture, 39th IMS Annual Meeting, Atlanta, Georgia.

Stein (1977a). Estimating the Covariance Matrix. Unpublished manuscript.

Stein (1977b). Lectures on the theory of many parameters. In: Ibrogimov, I. A., Nikulin, M. S., eds. *Studies in the Statistical Theory of Estimation. I.* Proceedings of Scientific Seminars of the Steklov Institute, Leningrad Division, 74, pp. 4–65 (In Russian).

Young, R., Bergen, J. O. (1994). Estimation of a covariance matrix using reference prior. *Ann. Statist.* 22:1195–1211.

Wald, A. (1943). Tests of statistical hypotheses concerning several parameters when the number of observations is large. *Trans. Am. Math. Soc.* 54:426–482.

Wolfowitz, J. (1949). On Wald's proof of the consistency of the maximum likelihood estimate. *Ann. Math. Statist.* 20:601–602.

Zacks, S. (1971). *The Theory of Statistical Inference*. New York: Wiley.

Zehna, P. W. (1966). Invariance of maximum likelihood estimation. *Ann. Math. Statist.* 37:744.

6

Basic Multivariate Sampling Distributions

6.0. INTRODUCTION

This chapter deals with some basic distributions connected with multivariate distributions. We discuss first the basic distributions connected with multivariate normal distributions. Then we deal with distributions connected with multivariate complex normal and basic distributions connected with elliptically symmetric distributions. The distributions of other multivariate test statistics needed for testing hypotheses concerning the parameters of multivariate populations will be derived where relevant. For better understanding and future reference we will also describe briefly the noncentral chi-square, noncentral Student's t, and noncentral F-distributions. For derivations of these noncentral distributions the reader is referred to Giri (1993).

6.1. NONCENTRAL CHI-SQUARE, STUDENT'S t-, F-DISTRIBUTIONS

6.1.1. Noncentral Chi-square

Let $X_1, \ldots, X_N$ be independently distributed normal random variables with $E(X_i) = \mu_i$, $\text{var}(X_i) = \sigma_i^2$, $i = 1, \ldots, N$. Then the random variable

$$Z = \sum_{i=1}^{N} \frac{X_i^2}{\sigma_i^2}$$

has the probability density function given by

$$f_Z(z|\delta^2) = \begin{cases} \dfrac{\exp\{-\frac{1}{2}(\delta^2+z)\}z^{(\frac{N}{2}-1)}}{\sqrt{\pi}\,2^{N/2}} \sum_{j=0}^{\infty} \dfrac{(\delta^2)^j z^j \Gamma(j+1/2)}{(2j)!\,\Gamma(N/2+j)}, & z \geq 0; \\ 0, & \text{otherwise}, \end{cases} \quad (6.1)$$

where $\delta^2 = \sum_i^N (\mu_i^2/\sigma_i^2)$. This is called the noncentral chi-square distribution with N degrees of freedom and with the noncentrality parameter δ^2. The random variable Z is often written as $\chi_N^2(\delta^2)$. The characteristic function of Z is (t real)

$$\phi_Z(t) = E(e^{itZ}) = (1-2it)^{-N/2}\exp\{it\,\delta^2/(1-2it)\} \quad (6.2)$$

with $i = (-1)^{1/2}$. From this it follows that if $Y_1, \ldots, Y_k$ are independently distributed noncentral chi-square random variables $\chi_{N_i}^2(\delta_i^2), i = 1, \ldots, k$ then $\sum_1^k Y_i$ is distributed as $\chi_{\sum_1^k N_i}^2(\sum_1^k \delta_i^2)$. Furthermore,

$$E(\chi_N^2(\delta^2)) = N + \delta^2, \ \text{var}(\chi_N^2(\delta^2)) = 2N + 4\delta^2. \quad (6.3)$$

Since for any integer k

$$\Gamma(2k+1)\sqrt{\pi} = 2^{2k}\Gamma\left(k+\frac{1}{2}\right)\Gamma(k+1), \quad (6.4)$$

we can write $f_Z(z|\delta^2)$ as

$$f_Z(z|\delta^2) = \sum_{k=0}^{\infty} p_K(k) f_{\chi_{N+2k}^2}(z) \quad (6.5)$$

where $p_K(k)$ is the probability mass function of the Poisson random variable K with parameter $\frac{1}{2}\delta^2$ and $f_{\chi_{N+2k}^2}(z)$ is the probability density function of the central chi-square random variable with $N + 2k$ degrees of freedom.

6.1.2. Noncentral Students's t

Let the random variable X, distributed normally with mean μ and variance σ^2, and the random variable Y such that Y/σ^2 has a chi-square distribution with n degrees of freedom, be independent and let $t = \sqrt{n}X/\sqrt{Y}$. The probability

Basic Multivariate Sampling Distributions

density function of t is given by

$$f_t(t|\lambda) = \begin{cases} \dfrac{n^{n/2}\exp\{-\frac{1}{2}\lambda^2\}}{(n+t^2)^{(n+1)/2}} \sum_{j=0}^{\infty} \dfrac{\Gamma((n+j+1)/2)\lambda^j}{j!}\left(\dfrac{2t^2}{n+t^2}\right)^{j/2}, & -\infty < t < \infty; \\ 0 & \text{otherwise,} \end{cases}$$

(6.6)

where $\lambda = \mu/\sigma$. The distribution of t is known as the noncentral t-distribution with n degrees of freedom and the noncentrality parameter λ.

6.1.3. Noncentral F-Distribution

Let the random variable X, distributed as $\chi_m^2(\delta^2)$, and the random variable Y, distributed as χ_n^2, be independent and let

$$F = \frac{n}{m} \frac{\chi_m^2(\delta^2)}{\chi_n^2}$$

The distribution F is known as the noncentral F-distribution and its probability density function is given by

$$f_F(z) =$$

$$\begin{cases} \dfrac{m}{n}\exp(-\frac{1}{2}\delta^2) \sum_{j=0}^{\infty} \dfrac{(\delta^2/2)^j \Gamma((m+n)/2+j)((m/n)z)^{m/2+j-1}}{\Gamma(m/2+j)\Gamma(n/2)(1+(m/n)z)^{(m+n)/2+j}}, & z \geq 0; \\ 0, & \text{otherwise.} \end{cases}$$

(6.7)

6.2. DISTRIBUTION OF QUADRATIC FORMS

Theorem 6.2.1. Let $X = (X_1, \ldots, X_p)'$ be normally distributed with mean μ and symmetric positive definite covariance matrix Σ, and let

$$X'\Sigma^{-1}X = Q_1 + \cdots + Q_k, \tag{6.8}$$

where $Q_i = X'A_i X$ and the rank of A_i is p_i, $i = 1, \ldots, k$. Then the Q_i are independently distributed as noncentral chi-square $\chi_{p_i}^2(\mu'A_i\mu)$ with p_i degrees of freedom and the noncentrality parameter $\mu'A_i\mu$ if and only if $\Sigma_1^k p_i = p$, in which case $\mu'\Sigma^{-1}\mu = \Sigma_1^k \mu'A_i\mu$.

Proof. Since Σ is symmetric and positive definite there exists a nonsingular matrix C such that $\Sigma = CC'$. Let $Y = C^{-1}X$. Obviously Y has a p-variate normal distribution with mean $v = C^{-1}\mu$ and covariance matrix I (identity matrix). From (6.8) we get

$$Y'Y = Y'B_1Y + \cdots + Y'B_kY, \tag{6.9}$$

where $B_i = C'A_iC$. Since C is nonsingular, rank (A_i) = rank(B_i), $i = 1, \ldots, k$. Obviously the theorem will be proved if we show that $Y'B_iY$, $i = 1, \ldots, k$, are independently distributed noncentral chi-squares $\chi^2_{p_i}(v'B_iv)$ if and only if $\Sigma^k_1 p_i = p$, in which case $v'v = \Sigma^k_{i=1} v'B_iv$. Let us suppose that $Y'B_iY$, $i = 1, \ldots, k$, are independently distributed as $\chi^2_{p_i}(v'B_iv)$. Then $\Sigma^k_{i=1} Y'B_iY$ is distributed as noncentral chi-square $\chi^2_{\Sigma^k_{i=1} p_i}(\Sigma^k_{i=1} v'B_iv)$. Since $Y'Y$ is distributed as $\chi^2_p(v'v)$ and (6.9) holds, it follows from the uniqueness of the characteristic function that $\Sigma^k_1 p_i = p$ and $v'v = \Sigma^k_{i=1} v'B_iv$, which proves the necessity part of the theorem. To prove the sufficiency part of the theorem let us assume that $\Sigma^k_1 p_i = p$. Since Q_i is a quadratic form in Y of *rank p_i* (rank of B_i) by Theorem 1.5.8, Q_i can be expressed as

$$Q_i = \sum_{j=1}^{p_i} \pm Z_{ij}^2 \tag{6.10}$$

where the Z_{ij} are linear functions of $Y_1, \ldots, Y_p$. Let

$$Z = (Z_{11}, \ldots, Z_{1p_1}, \ldots, Z_{k1}, \ldots, Z_{kp_k})'$$

be a vector of dimension $\Sigma^k_1 p_i = p$. Then

$$Y'Y = \sum_1^k Q_i = Z'\Delta Z, \tag{6.11}$$

where Δ is a diagonal matrix of dimension $p \times p$ with diagonal elements $+1$ or -1. Let $Z = AY$ be the linear transformation that transforms the positive definite quadratic form $Y'Y$ to $Z'\Delta Z$. Since

$$Y'Y = Z'\Delta Z = Y'A'\Delta AY \tag{6.12}$$

for all values of Y we conclude that $A'\Delta A = I$. In other words, A is nonsingular. Thus $Z'\Delta Z$ is positive definite and hence $\Delta = I, A'A = I$. Since A is orthogonal and Y has a p-variate normal distribution with mean v and covariance matrix I, the components of Z are independently normally distributed with unit variance. So $Q_i (i = 1, \ldots, k)$ are independently distributed chi-square random variables with p_i degrees of freedom and noncentrality parameter $v'B_iv$, $i = 1, \ldots, k$ (see

Basic Multivariate Sampling Distributions

Exercise 6.1). But $Y'Y$ is distributed as $\chi_p^2(v'v)$. Therefore

$$v'v = \sum_1^k v'B_i v.$$

Q.E.D.

Theorem 6.2.2. Let $X = (X_1, \ldots, X_p)'$ be normally distributed with mean μ and positive definite covariance matrix Σ. Then $X'AX$ is distributed as a noncentral chi-square with k degrees of freedom if and only if ΣA is an idempotent matrix of rank k, i.e., $A\Sigma A = A$.

Proof. Since Σ is positive definite there exists a nonsingular matrix C such that $\Sigma = CC'$. Let $X = CY$. Then Y has a p-variate normal distribution with mean $v = C^{-1}\mu$ and covariance matrix I, and

$$X'AX = Y'BY \tag{6.13}$$

where $B = C'AC$ and rank(A) = rank(B). The theorem will now be proved if we show that $Y'BY$ has a noncentral chi-square distribution $\chi_k^2(v'Bv)$ if and only if B is an idempotent matrix of rank k. Let us assume that B is an idempotent matrix of rank k. Then there exists an orthogonal matrix θ such that $\theta B \theta'$ is a diagonal matrix

$$D = \begin{pmatrix} I & 0 \\ 0 & 0 \end{pmatrix}$$

where I is the identity matrix of dimension $k \times k$ (see Chapter 1). Write $Z = (Z_1, \ldots, Z_p)' = \theta Y$. Then

$$Y'BY = Z'DZ = \sum_{i=1}^k Z_i^2 \tag{6.14}$$

is distributed as chi-square $\chi_k^2(v'Bv)$ (see Exercise 6.1). To prove the necessity of the condition let us assume that $Y'BY$ is distributed as $\chi_k^2(v'Bv)$. If B is of rank m, there exists an orthogonal matrix θ such that $\theta B \theta'$ is a diagonal matrix with m nonzero diagonal elements $\lambda_1, \ldots, \lambda_m$, the characteristic roots of B (we can without any loss of generality assume that the first m diagonal elements are nonzero). Let $Z = \theta Y$. Then

$$Y'BY = \sum_{i=1}^m \lambda_i Z_i^2. \tag{6.15}$$

Since the Z_i^2 are independently distributed each as noncentral chi-square with one degree of freedom and $Y'BY$ is distributed as non-central $\chi_k^2(v'Bv)$, it follows from the uniqueness of the characteristic function that $m = k$ and $\lambda_i = 1, i = 1, \ldots, k$. In other words, $\theta B \theta'$ is a diagonal matrix with k diagonal elements each equal to unity and the rest are zero. This implies that B is an idempotent matrix of rank k.   Q.E.D.

From this theorem it follows trivially that

(a) $X'\Sigma^{-1}X$ is distributed as noncentral chi-square $\chi_p^2(\mu'\Sigma^{-1}\mu)$;
(b) $(X - \mu)'\Sigma^{-1}(X - \mu)$ is distributed as χ_p^2;
(c) for any vector $\alpha = (\alpha_1, \ldots, \alpha_p)'$, $(X - \alpha)'\Sigma^{-1}(X - \alpha)$ is distributed as $\chi_p^2((\mu - \alpha)'\Sigma^{-1}(\mu - \alpha))$.

Theorem 6.2.3. *Let $X = (X_1, \ldots, X_p)'$ be a normally distributed p-vector with mean μ and positive definite covariance matrix Σ and let B be an $m \times p$ matrix of rank $m(< p)$. Then the quadratic form $X'X$ is distributed independently of the linear form BX if $B\Sigma A = 0$.*

Proof. Since Σ is positive definite there exists a nonsingular matrix C such that $\Sigma = CC'$. Write $X = CY$. Obviously Y is normally distributed with mean $v = C^{-1}\mu$ and covariance matrix I. Now

$$X'AX = Y'DY, \quad BX = EY \qquad (6.16)$$

where $D = C'AC$, $E = BC$. To prove the theorem we need to show that $Y'DY$, EY are independently distributed if $ED = 0$. Assume that $ED = 0$ and that the rank of D is k ($< p$). There exists an orthogonal matrix θ such that $\theta D \theta'$ is a diagonal matrix

$$\begin{pmatrix} D_1 & 0 \\ 0 & 0 \end{pmatrix}$$

where D_1 is a diagonal matrix of dimension $k \times k$ with nonzero diagonal elements. Now

$$Y'DY = Z'_{(1)}D_1 Z_{(1)}, \quad EY = E\theta'\theta Y = E^*Z$$

where $Z = \theta Y = (Z_1, \ldots, Z_p)'$, $Z_{(1)} = (Z_1, \ldots, Z_k)'$, and

$$E^* = E\theta' = \begin{pmatrix} E^*_{11} & E^*_{12} \\ E^*_{21} & E^*_{22} \end{pmatrix}$$

Basic Multivariate Sampling Distributions

with E_{11}^* a $k \times k$ submatrix of E^*. Since $ED = 0$ implies that $E^*\theta D\theta' = 0$, we get $E_{11}^* D_1 = E_{21}^* D_1 = 0$, and hence

$$E^* = \begin{pmatrix} 0 & E_{12}^* \\ 0 & E_{22}^* \end{pmatrix} = (0 \quad E_2^*) \quad (say),$$

and EY is distributed as $E_2^* Z_{(2)}$, where $Z_{(2)} = (Z_{k+1}, \ldots, Z_p)'$. Since $Y_1, \ldots, Y_p$ are independently distributed normal random variables and θ is an orthogonal matrix we conclude that $Y'DY$ is independent of EY. Q.E.D.

Theorem 6.2.4. *Cochran's Theorem. Let $X^\alpha = (X_{\alpha 1}, \ldots, X_{\alpha p})'$, $\alpha = 1, \ldots, N$, be a random sample of size N from a p-variate normal distribution with mean 0 and positive definite covariance matrix Σ. Assume that*

$$\sum_{\alpha=1}^{N} (X^\alpha)(X^\alpha)' = Q_1 + \cdots + Q_k, \tag{6.17}$$

where $Q_i = \Sigma_{\alpha,\beta=1}^{N}(X^\alpha) a_{\alpha\beta}^i (X^\beta)'$ with $A_i = (a_{\alpha\beta}^i)$ of rank N_i, $i = 1, \ldots, k$. Then the Q_i independently distributed as

$$\sum_{\alpha=N_1+\cdots+N_{i-1}+1}^{N_1+\cdots+N_i} (Z^\alpha)(Z^\alpha)'. \tag{6.18}$$

where $Z^\alpha = (Z_{\alpha 1}, \ldots, Z_{\alpha p})'$, $\alpha = 1, \ldots, \Sigma_1^k N_i$, are independently distributed normal p-vectors with mean 0 and covariance matrix Σ if and only if $\Sigma_1^k N_i = N$.

Proof. Suppose that the Q_i are independently distributed as in (6.18). Hence $\Sigma_1^k Q_i$ is distributed as

$$\sum_{\alpha=1}^{N_1+\cdots+N_k} (Z^\alpha)(Z^\alpha)' \tag{6.19}$$

From (6.17) and (6.19) and the uniqueness of the characteristic function we conclude that $\Sigma_1^k N_i = N$. To prove the sufficiency part of the theorem let us assume that $\Sigma_1^k N_i = N$. In the same way as in Theorem 6.2.1 we can assert that there exists an orthogonal matrix B

$$B = \begin{pmatrix} B_1 \\ \vdots \\ B_k \end{pmatrix} \quad \text{with} \quad A_i = B_i B_i'.$$

Since $B = (b_{\alpha\beta})$ is orthogonal,

$$Z^\alpha = \sum_{\beta=1}^{N} b_{\alpha\beta} X^\beta, \ \alpha = 1, \ldots, N,$$

are independently distributed normal p-vectors with mean 0 and covariance matrix Σ. It easy to see that for $i = 1, \ldots, k$,

$$Q_i = \sum_{\alpha,\beta=1}^{N} (X^\alpha) a^i_{\alpha\beta} (X^\beta)' = \sum_{\alpha=N_1+\cdots+N_{i-1}+1}^{N_1+\cdots+N_i} (Z^\alpha)(Z^\alpha)'.$$

Q.E.D.

This theorem is useful in generalizing the univariate analysis of variance results to multivariate analysis of variance problems. There is considerable literature on the distribution of quadratic forms and related results. The reader is referred to Cochran (1934), Hogg and Craig (1958), Ogawa (1949), Rao (1965), and Graybill (1961) for further references and details.

6.3. THE WISHART DISTRIBUTION

In Chapter 5 we remarked that a symmetric random matrix S of dimension $p \times p$ has a Wishart distribution with n degrees of freedom ($n \geq p$) and parameter Σ (a positive definite matrix) if S can be written as

$$S = \sum_{\alpha=1}^{n} X^\alpha (X^\alpha)'$$

where $X^\alpha = (X_{\alpha 1}, \ldots, X_{\alpha p})'$, $\alpha = 1, \ldots, n$ are independently distributed normal p-vectors with mean 0 and covariance matrix Σ. In this section we shall derive the Wishart probability density function as given in (5.2). In the sequel we shall need the following lemma.

Lemma 6.3.1. *Suppose X with values in the sample space $\mathcal{X}$ is a random variable with probability density function $f(t(x))$ with respect to a σ-finite measure μ on $\mathcal{X}$ where $t: \mathcal{X} \to \mathcal{Y}$ is measurable. For any measurable subset $B \in \mathcal{Y}$ define the measure ν by*

$$\nu(B) = \mu(t^{-1}(B)). \tag{6.20}$$

Then the probability density function of $Y = t(X)$ with respect to the measure ν is $f(y)$.

Basic Multivariate Sampling Distributions

Proof. It suffices to show that if $g : \mathcal{Y} \to R$ (real line), then

$$E(g(Y)) = \int_{\mathcal{Y}} g(y)f(y)dv(y).$$

From (6.20)

$$E(g(Y)) = Eg(t(X)) = \int_{\mathcal{X}} g(t(x))f(t(x))d\mu(x) = \int_{\mathcal{Y}} g(y)f(y)dv(y).$$

Q.E.D.

We shall assume that $n \geq p$ so that S is positive definite with probability 1. The joint probability density function of X^α, $\alpha = 1, \ldots, n$, is given by

$$f(x^1, \ldots, x^n) = (2\pi)^{-np/2}(\det \Sigma^{-1})^{n/2}$$

$$\times \exp\left\{-\frac{1}{2}tr\Sigma^{-1}\sum_{\alpha=1}^{n}x^\alpha(x^\alpha)'\right\}. \quad (6.21)$$

For any measurable set A in the space of S, the probability that S belongs to A depends on Σ and is given by

$$P_\Sigma(S \in A) = (2\pi)^{-np/2} \int_{\sum_{\alpha=1}^{n}x^\alpha(x^\alpha)'=s\in A} (\det \Sigma)^{-n/2} \exp\left\{-\frac{1}{2}tr\Sigma^{-1}s\right\}$$

$$\times \prod_{\alpha=1}^{n}dx^\alpha = (2\pi)^{-np/2} \int_{s\in A} (\det \Sigma)^{-n/2} \exp\left\{-\frac{1}{2}tr\Sigma^{-1}s\right\} dm(s)$$

(6.22)

where m is the measure corresponding to the measure v of (6.20). Let us now define the measure m^* on the space of S by

$$dm^*(s) = \frac{dm(s)}{(\det s)^{n/2}}. \quad (6.23)$$

Then

$$P_\Sigma(S \in A)$$

$$= (2\pi)^{-np/2} \int_A (\det(\Sigma^{-1}s))^{n/2} \exp\left\{-\frac{1}{2}tr\Sigma^{-1}s\right\} dm^*(s). \quad (6.24)$$

Obviously to find the probability density function of S it is sufficient to find $dm^*(s)$. To do this let us first observe the following: (i) Since Σ is positive definite there exists $C \in G_l(p)$, the multiplicative group of $p \times p$ nonsingular matrices, such that, $\Sigma = CC'$. (ii) Let

$$\tilde{S} = C^{-1}S(C^{-1})' = \sum_{\alpha=1}^{N}(C^{-1}X^{\alpha})(C^{-1}X^{\alpha})'. \tag{6.25}$$

Since $C^{-1}X^{\alpha}$ are independently normally distributed with mean 0 and covariance matrix I, $\tilde{S}$ is distributed as $W_p(n, I)$. Thus by (6.20)

$$P_{CC'}(S \in A) = P_I(C\tilde{S}C' \in A). \tag{6.26}$$

Now

$$P_{CC'}(S \in A) = (2\pi)^{-np/2} \int_A (\det((CC')^{-1}s))^{n/2}$$

$$\times \exp\left\{-\frac{1}{2}tr(CC')^{-1}s\right\} dm^*(s), \tag{6.27}$$

$$P_i(C\tilde{S}C' \in A) = (2\pi)^{-np/2} \int_{C\tilde{S}C' \in A} (\det(\tilde{s}))^{m/2} \exp\left\{-\frac{1}{2}tr\tilde{s}\right\} dm^*(\tilde{s})$$

$$= (2\pi)^{-np/2} \int_A (\det((CC')^{-1}s))^{n/2}$$

$$\times \exp\left\{-\frac{1}{2}tr((CC')^{-1}s)\right\} dm^*(C^{-1}sC'^{-1}). \tag{6.28}$$

Since (6.26) holds for all measurable sets A in the space of S we must then have

$$dm^*(s) = dm^*(CsC') \tag{6.29}$$

for all $C \in G_l(p)$ and all s in the space of S. This implies that for some positive constant k

$$dm^*(s) = \frac{kds}{(\det s)^{(p+1)/2}} \tag{6.30}$$

where ds stands for the Lebesgue measure $\prod_{i \leq j} ds_{ij}$ in the space of S. By Theorem 2.4.10 for the Jacobian of the transformation $s \to CsC', C \in G_l(p)$, is $(\det(CC'))^{(p+1)/2}$. Hence

$$dm^*(CsC') = \frac{kd(CsC')}{(\det(CsC'))^{(p+1)/2}} = \frac{kds}{(\det s)^{(p+1)/2}} \tag{6.31}$$

In other words, $dm^*(s)$ is an invariant measure on the space of S under the action of the group of transformations defined by $s \to CsC', C \in G_l(p)$. Now (6.30) follows from the uniqueness of invariant measures on homogeneous spaces (see Nachbin, 1965; or Eaton, 1972). From (6.24) and (6.30) the probability density

Basic Multivariate Sampling Distributions

function $W_p(n, \Sigma)$ of a Wishart random variable S with m degrees of freedom and parameter Σ is given by (with respect to the Lebesgue measure ds)

$$W_p(n, \Sigma) = \begin{cases} K(\det \Sigma)^{-n/2}(\det s)^{(n-p-1)/2} \exp\{-\tfrac{1}{2}tr\Sigma^{-1}s\}, & \text{if } s \text{ is positive definite,} \\ 0, & \text{otherwise} \end{cases}$$

(6.32)

where K is the normalizing constant independent of Σ. To specify the probability density function we need to evaluate the constant K. Since K is independent of Σ, we can in particular take $\Sigma = I$ for the evaluation of K. Since K is a function of n and p, we shall denote it by $C_{n,p}$. Let us partition $S = (S_{ij})$ as

$$S = \begin{pmatrix} S_{(11)} & S_{(12)} \\ S_{(21)} & S_{(22)} \end{pmatrix}$$

with $S_{(11)}$ a $(p-1) \times (p-1)$ submatrix of S, and let

$$Z = S_{(22)} - S_{(21)} S_{(11)}^{-1} S_{(12)}.$$

From (6.32)

$$1 = \int C_{n,p} (\det s_{(11)})^{(n-p-1)/2} (s_{(22)} - s_{(21)} s_{(11)}^{-1} s_{(12)})^{(n-p-1)/2}$$

$$\times \exp\left(-\frac{1}{2}tr(s_{(22)} + s_{(11)})\right) ds_{(11)} ds_{(12)} ds_{(22)}$$

$$= C_{n,p} \int (\det(s_{(11)}))^{(n-p-1)/2} \exp\left\{-\frac{1}{2}trs_{(11)}\right\}$$

$$\times \left(\int \exp\left(-\frac{1}{2} s_{(21)} s_{(11)}^{-1} s_{(12)}\right) ds_{(12)}\right) ds_{(11)}$$

(6.33)

$$\times \int_0^\infty z^{(n-p-1)/2} \exp\left\{-\frac{1}{2}z\right\} dz$$

$$= C_{n,p} 2^{(n-p+1)/2} \Gamma\left(\frac{n-p+1}{2}\right)(2\pi)^{(p-1)/2} \int (\det s_{(11)})^{(n-p)/2}$$

$$\times \exp\left\{-\frac{1}{2}trs_{(11)}\right\} ds_{(11)}$$

as

$$\int_0^\infty z^{(n-p-1)/2} \exp\left\{-\frac{1}{2}z\right\} dz = 2^{(n-p+1)/2} \Gamma\left(\frac{n-p+1}{2}\right),$$

$$\times \int \exp\left\{-\frac{1}{2} s_{(21)} s_{(11)}^{-1} s_{(12)}\right\} ds_{(12)} = (2\pi)^{(p-1)/2} (\det(s_{(11)}))^{1/2}.$$

Since $W_p(n, I)$ is a probability density function with the constant $K = C_{n,p}$, we obtain

$$\int (\det s_{(11)})^{(n-p)/2} \exp\left\{-\frac{1}{2} \operatorname{tr} s_{(11)}\right\} ds_{(11)} = (C_{n,p-1})^{-1}. \tag{6.34}$$

From (6.33) and (6.34) we get

$$C_{n,p} = \left[\Gamma\left(\frac{n-p+1}{2}\right) 2^{n/2} \pi^{(p-1)/2}\right]^{-1} C_{n,p-1}$$

$$= \left[\Gamma\left(\frac{n-p+1}{2}\right) 2^{n/2} \pi^{(p-1)/2}\right]^{-1} \cdots \left[\Gamma\left(\frac{n-1}{2}\right) 2^{n/2} \pi^{1/2}\right]^{-1} \tag{6.35}$$

$$\times C_{n,1}.$$

But $C_{n,1}$ is given by

$$C_{n,1} \int_0^\infty x^{(n-2)/2} \exp\left\{-\frac{1}{2}x\right\} dx = 1,$$

that implies

$$C_{n,1} = [\Gamma(n/2) 2^{n/2}]^{-1}. \tag{6.36}$$

From (6.35) and (6.36) we get

$$(C_{n,p})^{-1} = \left(\prod_{i=0}^{p-1} \Gamma\left(\frac{n-i}{2}\right)\right) 2^{np/2} \pi^{p(p-1)/4} = K^{-1}. \tag{6.37}$$

The derivation of the Wishart distribution, which is very fundamental in multivariate analysis, was a major breakthrough for the development of multivariate analysis. Several derivations of the Wishart distribution are available in the literature. The derivation given here involves a property of invariant measure and is quite short and simple in nature.

Alternate Derivation. Since the preceding derivation of the Wishart distribution involves some deep theoretical concepts, we will now give a straightforward derivation. S is distributed as $\sum_{\alpha=1}^{N-1} X^\alpha (X^\alpha)'$, where $X^\alpha, \alpha = 1, 2, \ldots, N-1$, are independently distributed normal p-vectors with

Basic Multivariate Sampling Distributions

mean 0 and positive definite covariance matrix Σ. Let

$$\Sigma = CC', \quad Y^\alpha = C^{-1}X^\alpha, \quad \alpha = 1, 2, \ldots, N-1,$$

where C is a nonsingular matrix. Let us first consider the distribution of

$$A = \sum_{\alpha=1}^{N-1} Y^\alpha (Y^\alpha)'.$$

Write $Y = (Y^1, \ldots, Y^{N-1})$. Then $A = YY'$. By the Gram-Schmidt orthogonalization process on the row vectors $Y_1, \ldots, Y_p$ of Y we obtain new row vectors $Z_1, \ldots, Z_p$ such that

$$ZZ' = I,$$

where

$$Z = \begin{pmatrix} Z_1 \\ \vdots \\ Z_p \end{pmatrix}$$

Let the transformation involved in transforming Y to Z be given by $Z = B^{-1}Y$. Obviously $B = (b_{ij})$ is a random lower triangular nonsingular matrix. Now

$$A = YY' = BZZ'B = BB',$$

where $B = (b_{ij})$ is a random lower triangular nonsingular matrix with positive diagonal elements satisfying $Y = BZ$. Thus we get

$$Y_i = \sum_{j=1}^{i} = b_{ij}Z_j, \quad i = 1, \ldots, p,$$

and $Z_j Y_i' = b_{ij}$. Hence with $A = (a_{ij})$,

$$a_{ii} = Y_i Y_i' = \sum_{j=1}^{i} b_{ij}^2, \quad b_{ii}^2 = a_{ii} - \sum_{j=1}^{i-1} b_{ij}^2.$$

In other words,

$$\begin{pmatrix} b_{i1} \\ \vdots \\ b_{i,i-1} \end{pmatrix} = \begin{pmatrix} Z_1 \\ \vdots \\ Z_{i-1} \end{pmatrix} Y_i'$$

Since $ZZ' = I$, the components of $Y^\alpha, \alpha = 1, \ldots, N-1$, are independently distributed normal variables with mean 0 and variance 1, and $Z_1, \ldots, Z_{i-1}$ are functions of $Y_1, \ldots, Y_{i-1}$, the conditional distributions of $b_{i1}, \ldots, b_{i,i-1}$, given

$Z_1, \ldots, Z_{i-1}$ are independent normal with mean 0 and variance 1,

$$b_{ii}^2 = Y_i Y_i' - \sum_{j=1}^{i-1} b_{ij}^2$$

is distributed as χ_{N-i}^2, and all $b_{ij}, j = 1, \ldots, i$, are independent. Since the conditional distributions of $b_{ij}, j = 1, \ldots, i$, do not involve $Z_1, \ldots, Z_{i-1}$, these conditional distributions are also the unconditional distributions of $b_{ij}, j = 1, \ldots, i-1$. Furthermore, $b_{i1}, \ldots, b_{ii}$ are distributed independently of $Y_1, \ldots, Y_{i-1}$ and hence of $b_{rs}, r, s = 1, \ldots, i-1 (r \geq s)$, and $Z_1, \ldots, Z_{i-1}$, which are functions of $Y_1, \ldots, Y_{i-1}$ only. Hence $b_{ij}, i, j = 1, \ldots, p(i > p)$, are independently distributed normal random variables with mean 0 and variance 1, and $b_{ii}^2, i = 1, \ldots, p$, are independently distributed (independently of the b_{ij}) as χ_{N-i}^2. From Theorem 2.4.6 the Jacobian of the transformations $B \to A = BB'$ is $2^p \prod_{i=1}^{p} (b_{ii})^{i-p-1}$. Hence the distribution of A is

$$f_A(a) = K(\det a)^{(N-p-2)/2} \exp\left\{-\frac{\operatorname{tr} a}{2}\right\}$$

provided a is positive definite, where K is a constant depending on N and p. By Theorem 2.4.9 the probability density function of $S = CAC'$ is given by (6.32).

The Wishart distribution was first derived by Fisher (1915) for $p = 2$. Wishart (1928) gave a geometrical derivation of this distribution for general p. Ingham (1933) derived this distribution from its characteristic function. Elfving (1947), Mauldan (1955), and Olkin and Roy (1954) used matrix transformations to derive the Bartlett decomposition of the Wishart matrix from sample observations and then derived the distribution of the Wishart matrix. Khirsagar (1959) used random orthogonal transformations to derive the Wishart distribution and the distribution of Bartlett decomposition. Sverdrup (1947) derived this distribution by straightforward integration over the sample space. Narain (1948) and Ogawa (1953) used the regression approach, Ogawa's approach being more elegant. Rasch (1948) and Khatri (1963) also gave alternative derivations of this distribution.

6.4. PROPERTIES OF THE WISHART DISTRIBUTION

This section deals with some important properties of the Wishart distribution which are often used in multivariate analysis.

Theorem 6.4.1. *Let*

$$S = \begin{pmatrix} S_{(11)} & S_{(12)} \\ S_{(21)} & S_{(22)} \end{pmatrix}$$

Basic Multivariate Sampling Distributions

where $S_{(11)}$ is the $q \times q$ left-hand corner submatrix of $S (q < p)$, be distributed as $W_p(n, \Sigma)$, and let Σ be similarly partitioned into

$$\Sigma = \begin{pmatrix} \Sigma_{(11)} & \Sigma_{(12)} \\ \Sigma_{(21)} & \Sigma_{(22)} \end{pmatrix}$$

Then

(a) $S_{(11)} - S_{(12)} S_{(22)}^{-1} S_{(21)}$ *is distributed as Wishart* $W_q(n - (p-q), \Sigma_{(11)} - \Sigma_{(12)} \Sigma_{(22)}^{-1} \Sigma_{(21)})$;
(b) $S_{(22)}$ *is distributed as Wishart* $W_{p-q}(n, \Sigma_{(22)})$;
(c) *the conditional distribution of* $S_{(12)} S_{(22)}^{-1}$ *given that* $S_{(22)} = s_{(22)}$ *is normal with mean* $\Sigma_{(12)} \Sigma_{(22)}^{-1}$ *and covariance matrix* $(\Sigma_{(11)} - \Sigma_{(12)} \Sigma_{(22)}^{-1} \Sigma_{(21)}) \otimes s_{(22)}^{-1}$;
(d) $S_{(11)} - S_{(12)} S_{(22)}^{-1} S_{(21)}$ *is independent of* $(S_{(12)}, S_{(22)})$.

Proof. Let $\Sigma^{-1} = \Lambda$ be partitioned into

$$\Lambda = \begin{pmatrix} \Lambda_{(11)} & \Lambda_{(12)} \\ \Lambda_{(21)} & \Lambda_{(22)} \end{pmatrix}$$

where $\Lambda_{(11)}$ is a $q \times q$ submatrix of Λ and let $(S_{(11)}, S_{(12)}, S_{(22)})$ be transformed to (W, U, V) where

$$W = S_{(11)} - S_{(12)} S_{(22)}^{-1} S_{(21)}, \ U = S_{(12)} S_{(22)}^{-1/2}, \ V = S_{(22)} \qquad (6.38)$$

or, equivalently,

$$S_{(11)} = W + UU', \ S_{(12)} = UV^{1/2}, \ S_{(22)} = V.$$

The Jacobian of this transformation is given by the absolute value of the determinant of the following matrix of partials:

$$\begin{array}{c} \\ S_{(11)} \\ S_{(12)} \\ S_{(22)} \end{array} \begin{pmatrix} W & U & V \\ I & - & - \\ 0 & A & - \\ 0 & 0 & I \end{pmatrix}$$

where the dash indicates some matrix which need not be known and A is the matrix of partial derivatives of the transformation $S_{(12)} \to UV^{1/2}$ (V fixed). By a result analogous to Theorem 2.4.1, the Jacobian is

$$|\det(A)| = |\det(V)|^{q/2} \qquad (6.39)$$

Now

$$\operatorname{tr} \Sigma^{-1} s = \operatorname{tr} \Lambda s$$

$$= tr(\Lambda_{(11)}s_{(11)} + \Lambda_{(12)}s_{(21)}) + tr(\Lambda_{(21)}s_{(12)} + \Lambda_{(22)}s_{(22)})$$

$$= tr(\Lambda_{(11)}s_{(11)} + \Lambda_{(12)}s_{(21)} + s_{(12)}\Lambda_{(21)} + \Lambda_{(11)}^{-1}\Lambda_{(12)}s_{(22)}\Lambda_{(21)})$$

$$+ \operatorname{tr}(\Lambda_{(22)} - \Lambda_{(21)}\Lambda_{(11)}^{-1}\Lambda_{(12)})s_{(22)}$$

$$= \operatorname{tr} \Lambda_{(11)}(w + uu') + 2\operatorname{tr} \Lambda_{(12)}v^{1/2}u' + \operatorname{tr} \Lambda_{(21)}\Lambda_{(11)}^{-1}\Lambda_{(12)}v \quad (6.40)$$

$$+ \operatorname{tr}(\Lambda_{(22)} - \Lambda_{(21)}\Lambda_{(11)}^{-1}\Lambda_{(12)})v$$

$$= \operatorname{tr} \Lambda_{(11)}w + \operatorname{tr} \Lambda_{(11)}uu' + 2\operatorname{tr} \Lambda_{(12)}v^{1/2}u'$$

$$+ \operatorname{tr} \Lambda_{(12)}v\Lambda_{(21)}\Lambda_{(11)}^{-1} + \operatorname{tr}(\Lambda_{(22)} - \Lambda_{(21)}\Lambda_{(11)}^{-1}\Lambda_{(12)})v$$

$$= \operatorname{tr} \Lambda_{(11)}w + \operatorname{tr} \Lambda_{(11)}(u + \Lambda_{(11)}^{-1}\Lambda_{12}v^{1/2})(u + \Lambda_{(11)}^{-1}\Lambda_{(12)}v^{1/2})'$$

$$+ \operatorname{tr}(\Lambda_{(22)} - \Lambda_{(21)}\Lambda_{(11)}^{-1}\Lambda_{(12)})v.$$

Since

$$\Lambda_{(11)} = (\Sigma_{(11)} - \Sigma_{(12)}\Sigma_{(22)}^{-1}\Sigma_{(21)})^{-1},$$

$$\Sigma_{(22)}^{-1} = (\Lambda_{(22)} - \Lambda_{(21)}\Lambda_{(11)}^{-1}\Lambda_{(12)}),$$

$$\Lambda_{(11)}^{-1}\Lambda_{(12)} = -\Sigma_{(12)}\Sigma_{(22)}^{-1},$$

$$\det(S) = \det(S_{(22)})\det(S_{(11)} - S_{(12)}S_{(22)}^{-1}S_{(21)}),$$

$$\det \Sigma = \det(\Sigma_{(22)})\det(\Sigma_{(11)} - \Sigma_{(12)}\Sigma_{(22)}^{-1}\Sigma_{(21)}),$$

from (6.39), (6.40), and (6.32), the joint probability density function of (W, U, V) can be written as

$$f_{W,U,V}(w, u, v) = f_W(w)f_{U|V}(u|v)f_V(v) \quad (6.41)$$

Basic Multivariate Sampling Distributions

where

$$f_W(w) = k_1(\det(\Sigma_{(11)} - \Sigma_{(12)}\Sigma_{(22)}^{-1}\Sigma_{(21)}))^{-(n-(p-q))/2}$$
$$\times (\det(w))^{(n-(p-q)-q-1)/2}$$
$$\times \exp\left\{-\frac{1}{2}tr(\Sigma_{(11)} - \Sigma_{(12)}\Sigma_{(22)}^{-1}\Sigma_{(21)})^{-1}w\right\},$$

$$f_{U|V}(u,v) = k_2(\det((\Sigma_{(11)} - \Sigma_{(12)}\Sigma_{(22)}^{-1}\Sigma_{(21)}) \otimes I_{p-q}))^{-1/2}$$
$$\times \exp\left\{-\frac{1}{2}tr((\Sigma_{(11)} - \Sigma_{(12)}\Sigma_{(22)}^{-1}\Sigma_{(21)}) \otimes I_{p-q})^{-1}\right.$$
$$\left.\times (u - \Sigma_{(12)}\Sigma_{(22)}^{-1}v^{1/2})'(u - \Sigma_{(12)}\Sigma_{(22)}^{-1}v^{1/2})\right\},$$

$$f_V(v) = k_3(\det \Sigma_{(22)})^{-n/2}(\det v)^{(n-(p-q)-1)/2}\exp\left\{-\frac{1}{2}tr\Sigma_{(22)}^{-1}v\right\},$$

where k_1, k_2, k_3 are normalizing constants independent of Σ. Thus $S_{(11)} - S_{(12)}S_{(22)}^{-1}S_{(21)}$ is distributed as Wishart

$$W_q(n - (p-q), \Sigma_{(11)} - \Sigma_{(12)}\Sigma_{(22)}^{-1}\Sigma_{(21)})$$

and is independent of $(S_{(12)}, S_{(22)})$. The conditional distribution of $S_{(12)}S_{(22)}^{-1/2}$ given $S_{(22)} = s_{(22)}$, is normal (in the sense of Example 4.3.6.0 with mean $\Sigma_{(12)}\Sigma_{(22)}^{-1}s_{(22)}^{1/2}$ and covariance matrix $(\Sigma_{(11)} - \Sigma_{(12)}\Sigma_{(22)}^{-1}\Sigma_{(21)}) \otimes I_{p-q}$. Multiplying $S_{(12)}S_{(22)}^{-1/2}$ by $S_{(22)}^{-1/2}$ we conclude that the conditional distribution of $S_{(12)}S_{(22)}^{-1/2}$ given $S_{(22)} = s_{(22)}$ is normal in the sense of Example 4.3.6.0 with mean $\Sigma_{(12)}\Sigma_{(22)}^{-1}$ and covariance matrix $(\Sigma_{(11)} - \Sigma_{(12)}\Sigma_{(22)}^{-1}\Sigma_{(21)}) \otimes s_{(22)}^{-1}$. Finally, $S_{(22)}$ is distributed as Wishart $W_{(p-q)}(n, \Sigma_{(22)})$. Q.E.D.

Theorem 6.4.2. *If S is distributed as $W_p(n, \Sigma)$ and C is a nonsingular matrix of dimension $p \times p$, then CSC' is distributed as $W_p(n, C\Sigma C')$.*

Proof. Since S is distributed as $W_p(n, \Sigma)$, S can be written as

$$S = \sum_{\alpha=1}^{n} Y^{\alpha}(Y^{\alpha})'$$

where $Y^{\alpha} = (Y_{\alpha 1}, \ldots, Y_{\alpha p})'$, $\alpha = 1, \ldots, n$, are independently distributed normal p-vectors with mean 0 and the same covariance matrix Σ. Hence CSC' is distributed as

$$\sum_{\alpha=1}^{n}(CY^{\alpha})(CY^{\alpha})' = \sum_{\alpha=1}^{n} Z^{\alpha}(Z^{\alpha})',$$

where $Z^{\alpha} = (Z_{\alpha 1}, \ldots, Z_{\alpha p})'$, $\alpha = 1, \ldots, n$, are independently and identically distributed normal p-vectors with mean 0 and covariance matrix $C\Sigma C'$ and hence the theorem. Q.E.D.

Theorem 6.4.3. *Let the $p \times p$ symmetric random matrix $S = (S_{ij})$ be distributed as $W_p(n, \Sigma)$. The characteristic function of S (i.e., the characteristic function of $S_{11}, S_{22}, \ldots, S_{pp}, 2S_{12}, \ldots, 2S_{p-1,p}$) is given by*

$$E(\exp(itr\,\theta S)) = (\det(I - 2i\Sigma\theta))^{-n/2} \qquad (6.42)$$

where $\theta = (\theta_{ij})$ is a real symmetric matrix of dimension $p \times p$.

Proof. S is distributed as $\sum_{\alpha=1}^{n} Y^{\alpha}(Y^{\alpha})'$ where the Y^{α}, $\alpha = 1, \ldots, n$, have the same distribution as in Theorem 6.4.2. Hence

$$E(\exp(itr\,\theta S)) = E\left(\exp\left(itr\,\theta \sum_{\alpha=1}^{n} Y^{\alpha}(Y^{\alpha})'\right)\right)$$

$$= \prod_{\alpha=1}^{n} E(\exp(itr\,\theta Y^{\alpha}(Y^{\alpha})')) \qquad (6.43)$$

$$= E(\exp(itr\,\theta Y(Y)'))^{n},$$

where Y has p-dimensional normal distribution with mean 0 and covariance matrix Σ. Since θ is real and Σ is positive definite there exists a nonsingular matrix C such that

$$C'\Sigma^{-1}C = I \quad \text{and} \quad C'\theta C = D,$$

Basic Multivariate Sampling Distributions

where D is a diagonal matrix of diagonal elements d_{ii}. Let $Y = CZ$. Then Z has a p-dimensional normal distribution with mean 0 and covariance matrix I. Hence

$$E(\exp(itrY'\theta Y)) = E(\exp(itrZ'DZ))$$

$$= \prod_{j=1}^{p} E(\exp(id_{jj}Z_j^2)) = \prod_{j=1}^{p}(1 - 2id_{jj})^{-1/2}$$

$$= (\det(I - 2iD))^{-1/2} = (\det(I - 2iC'\theta C))^{-1/2}$$

$$= (\det(C'C))^{-1/2}(\det(\Sigma^{-1} - 2i\theta))^{-1/2}$$

$$= (\det(I - 2i\theta\Sigma))^{-\frac{1}{2}}.$$

Hence

$$E(\exp(itr\theta S)) = (\det(I - 2i\theta\Sigma))^{-n/2}$$

Q.E.D.

From this it follows that

$$E(S) = n\Sigma, \quad cov(S) = 2n\Sigma \otimes \Sigma. \tag{6.44}$$

Theorem 6.4.4. *If $S_i, i = 1, \ldots, k$ are independently distributed as $W_p(n_i, \Sigma)$ then $\sum_1^k S_i$ is distributed as $W_p(\sum_1^k n_i, \Sigma)$.*

Proof. Since $S_i, i = 1, \ldots, k$, are independently distributed as Wishart we can write

$$S_i = \sum_{\alpha = n_1 + \cdots + n_{i-1}+1}^{n_1+\cdots+n_i} Y^\alpha (Y^\alpha)', i = 1, \ldots, k,$$

where $Y^\alpha = (Y_{\alpha 1}, \ldots, Y_{\alpha p})', \alpha = 1, \ldots, \sum_1^k n_i$, are independently distributed p-dimensional normal random vectors with mean 0 and covariance matrix Σ. Hence

$$\sum_1^k S_i = \sum_{\alpha=1}^{n_1+\cdots+n_k} Y^\alpha (Y^\alpha)'$$

is distributed as $W_p(\sum_1^k n_i, \Sigma)$.

Q.E.D.

Theorem 6.4.5. Let S be distributed as $W_p(n, \Sigma)$ and let B be a $k \times p$ matrix of rank $k \leq p$. Then $(BS^{-1}B')^{-1}$ is distributed as

$$W_k(n - p + k, (B\Sigma^{-1}B')^{-1}).$$

Proof. Let $A = \Sigma^{-1/2} S \Sigma^{-1/2}$ where $\Sigma^{1/2}$ is the symmetric positive definite matrix such that $\Sigma = \Sigma^{1/2}\Sigma^{1/2}$. From Theorem 6.4.2 A is distributed as $W_p(n, I)$. Now

$$\begin{aligned}(BS^{-1}B')^{-1} &= (B\Sigma^{-1/2}A^{-1}\Sigma^{-1/2}B')^{-1} \\ &= (MA^{-1}M')^{-1},\end{aligned} \quad (6.45)$$

and

$$(B\Sigma^{-1}B')^{-1} = (MM')^{-1} \quad (6.46)$$

where $M = B\Sigma^{-1/2}$. Since M is a $k \times p$ matrix of rank k, by Theorem 1.5.14 there exist a $k \times k$ nonsingular matrix C and a $p \times p$ orthogonal matrix θ such that

$$M = C(I_k, 0)\theta. \quad (6.47)$$

Let $D = \theta A \theta'$. Then

$$(MA^{-1}M')^{-1} = (C')^{-1}[(I_k, 0)D^{-1}(I_k, 0)']^{-1}C^{-1} \quad (6.48)$$

and D is distributed $W_p(n, I)$. Write

$$D^{-1} = H = \begin{pmatrix} H_{11} & H_{12} \\ H_{21} & H_{22} \end{pmatrix}, \quad D = \begin{pmatrix} D_{11} & D_{12} \\ D_{21} & D_{22} \end{pmatrix} \quad (6.49)$$

where H_{11}, D_{11} are left-hand corner submatrices of order $k \times k$. From (6.48) and (6.49)

$$(MA^{-1}M')^{-1} = (C')^{-1}H_{11}^{-1}C^{-1},$$

$$H_{11}^{-1} = D_{11} - D_{12}D_{22}^{-1}D_{21}.$$

By Theorem 6.4.1 (since D is $W_p(n, I)$) H_{11}^{-1} is distributed as a $W_k(n - p + k, I_k)$. Hence $(C')^{-1}H_{11}^{-1}C^{-1}$ is distributed as $W_k(n - p + k, (CC')^{-1})$ and, from (6.47), $(CC')^{-1} = (MM')^{-1} = (B\Sigma^{-1}B')^{-1}$. Q.E.D.

Basic Multivariate Sampling Distributions

Theorem 6.4.6. *(Inverted Wishart). Let S be distributed as $W_p(n, \Sigma)$. Then the distribution of $A = S^{-1}$ is given by*

$$f_A(a) = \begin{cases} K(\det \Sigma^{-1})^{n/2}(\det a)^{-\frac{(n+p-1)}{2}} e^{-\frac{1}{2}\mathrm{tr}\Sigma^{-1}a^{-1}}, & \text{if a is positive definite,} \\ 0, & \text{otherwise} \end{cases} \qquad (6.50)$$

where K is given in (6.32).

Proof. From Theorem 2.4.11 the Jacobian of the transformation $S \to A = S^{-1}$ is $|A|^{2p}$. Using (6.32) the distribution of A is given by (6.50). Q.E.D.

6.5. THE NONCENTRAL WISHART DISTRIBUTION

Let $X^\alpha = (X_{\alpha 1}, \ldots, X_{\alpha p})'$, $\alpha = 1, \ldots, N$, be independently distributed normal p-vectors with mean $\mu^\alpha = (\mu_{\alpha 1}, \ldots, \mu_{\alpha p})'$ and the same covariance matrix Σ. Let

$$X = (X^1, \ldots, X^N), D = XX', M = (\mu^1, \ldots, \mu^N).$$

The probability density function of X is given by

$$f_X(x) = (2\pi)^{Np/2}(\det \Sigma)^{-N/2} \exp\left\{-\frac{1}{2}\mathrm{tr}\Sigma^{-1}(x-M)(x-M)'\right\}. \qquad (6.51)$$

The distribution of D is called the noncentral Wishart distribution. Its probability density function in its most general form was first derived by James, 1954, 1955, 1964; see also Constantine, 1963; Anderson, 1945, 1946, and it involves the characteristic roots of $\Sigma^{-1}MM'$. The noncentral Wishart distribution is said to belong to the linear case if the rank of M is 1, and to the planar case if the rank of M is 2. In particular, if $\Sigma = I$, the probability density function of D can be written as (with respect to the Lebesgue measure)

$$f_D(d) = 2^{-Np/2} \pi^{-p(p-1)/4} \left[\prod_{i=1}^p \Gamma\left(\frac{N-i+1}{2}\right)\right]^{-1}$$

$$\times \exp\left\{-\frac{1}{2}(\mathrm{tr}MM' + \mathrm{tr}d)\right\}(\det d)^{(N-p-1)/2} \qquad (6.52)$$

$$\times \int_{O(N)} \exp\{-\mathrm{tr}M'x\theta\} d\theta$$

where $O(N)$ is the group of $N \times N$ orthogonal matrices θ and $d\theta$ is the Lebesgue measure in the space of $O(N)$. In particular, if $\Sigma = I$ and

$$M = \begin{pmatrix} \mu_1 & \cdots & \mu_N \\ 0 & \cdots & 0 \\ 0 & \cdots & 0 \end{pmatrix}$$

then the distribution of $D = (D_{ij})$ is given by $[d = (d_{ij})]$

$$f_D(d) = \exp\left\{-\frac{1}{2}\lambda^2\right\} \sum_{\alpha=0}^{\infty} \frac{(\lambda^2/2)^\alpha}{\alpha!} \frac{\Gamma(N/2)}{\Gamma(N/2 + \alpha)} \left(\frac{d_{11}}{2}\right)^\alpha \quad (6.53)$$

where $\lambda^2 = \sum_1^N \mu_i^2$. This is called the canonical form of the noncentral Wishart distribution in the linear case.

6.6. GENERALIZED VARIANCE

For the p-variate normal distribution with mean μ and covariance matrix Σ, $\det \Sigma$ is called the generalized variance of the distribution (see Wilks, 1932). Its estimate, based on sample observations $x^\alpha = (X_{\alpha 1}, \ldots, X_{\alpha p})'$, $\alpha = 1, \ldots, N$,

$$\det\left(\frac{1}{N-1} \sum_{\alpha=1}^N (x^\alpha - \bar{x})(x^\alpha - \bar{x})'\right) = \frac{1}{(N-1)^p} \det(s)$$

is called the sample generalized variance or the generalized variance of the sample observations x^α, $\alpha = 1, \ldots, N$. The sample generalized variance occurs in many test criteria of statistical hypotheses concerning the means and covariance matrices of multivariate distributions. We will now consider the distribution of $\det S$ where S is distributed as $W_p(n, \Sigma)$.

Theorem 6.6.1. Let S be distributed as $W_p(n, \Sigma)$. Then $\det S$ is distributed as $(\det \Sigma) \prod_{i=1}^p \chi_{n+1-i}^2$, where χ_{n+1-i}^2, $i = 1, \ldots, p$, are independent central chi-square random variables.

Note. $W_1(n, 1)$ is a central chi-square random variable with n degrees of freedom.

Proof. Since Σ is positive definite there exists a nonsingular matrix C such that $C\Sigma C' = I$. Let $S^* = CSC'$. Then S^* is distributed as Wishart $W_p(n, I)$. Now

Basic Multivariate Sampling Distributions 233

$\det S^* = (\det \Sigma^{-1})(\det S)$. Hence $\det S$ is distributed as $(\det \Sigma)(\det S^*)$. Write $S^* = (S^*_{ij})$ as

$$S^* = \begin{pmatrix} S^*_{(11)} & S^*_{(12)} \\ S^*_{(21)} & S^*_{(pp)} \end{pmatrix}$$

where $S^*_{(pp)}$ is 1×1. Then $\det S^* = S^*_{(pp)} \det(S^*_{(11)} - S^*_{(12)} S^{*-1}_{(pp)} S^*_{(21)})$. By Theorem 6.4.1 $S^*_{(pp)}$ is distributed as $W_1(n, 1)$ independently of $(S^*_{(11)} - S^*_{(21)} S^{*-1}_{(pp)} S^*_{(12)})$ and $S^*_{(11)} - S^*_{(21)} S^{*-1}_{(pp)} S^*_{(12)}$ is distributed as $W_{p-1}(n - 1, I_{p-1})$, where I_{p-1} is the identity matrix of dimension $(p - 1) \times (p - 1)$. Thus $\det(W_p(n, I_p))$ is distributed as the product of χ_n^2 and $\det(W_{p-1}(n - 1, I_{p-1}))$ where χ_n^2 and $W_{p-1}(n - 1, I_{p-1})$ are independent. Repeating this argument $p - 1$ times we conclude that $\det S^*$ is distributed as $\prod_{i=1}^{p} \chi^2_{n+1-i}$, where χ^2_{n+1-i}, $i = 1, \ldots, p$, are independent chi-square random variables. Q.E.D.

6.7. DISTRIBUTION OF THE BARTLETT DECOMPOSITION (RECTANGULAR COORDINATES)

Let S be distributed as $W_p(n, \Sigma)$ and $n \geq p$. As we have observed earlier, S is positive definite with probability 1. Let $B = (B_{ij})$, $B_{ij} = 0, i < j$, be the unique lower triangular matrix with positive diagonal elements such that

$$S = BB' \tag{6.54}$$

(see Theorem 1.6.5). By Theorem 2.4.6 the Jacobian of the transformation $S \to B$ is given by $[s = (s_{ij}), b = (b_{ij})]$

$$\det\left(\frac{\partial s}{\partial b}\right) = 2^p \prod_{i=1}^{p} (b_{ii})^{p+1-i} \tag{6.55}$$

From (6.32), (6.54), and (6.55) the probability density function of B with respect to the Lebesgue measure db is given by

$$f_B(b) = K 2^p (\det \Sigma)^{-n/2} (\det b)^{n-p-1} \exp\left\{-\frac{1}{2} tr \Sigma^{-1} bb'\right\} \prod_{i=1}^{p} (b_{ii})^{p+1-i}$$

$$= K 2^p (\det \Sigma)^{-n/2} \prod_{i=1}^{p} (b_{ii}^{(n-i)}) \exp\left\{-\frac{1}{2} tr \Sigma^{-1} bb'\right\} \tag{6.56}$$

where K is given by (6.37).

Let $T = (T_{ij})$ be a nonsingular lower triangular matrix (not necessarily with positive diagonal elements). Then we can write $T = B\theta$ where θ is a diagonal matrix with diagonal entries ± 1. Since the Jacobian of the transformation $B \to T = B\theta$ is unity, from (6.56) the probability density function of T is given by

(with respect to the Lebesgue measure dt)

$$f_T(t) = K2^p(\det \Sigma)^{-n/2}(\det(tt'))^{(n-p-1)/2}$$
$$\times \exp\left\{-\frac{1}{2}tr\Sigma^{-1}tt'\right\} \prod_{i=1}^{p} |t_{ii}|^{p+1-i}, \quad (6.57)$$

$t = (t_{ij})$, where K is given by (6.37). If $\Sigma = I$, (6.57) reduces to

$$f_T(t) = K2^p \exp\left\{-\frac{1}{2}\sum_{i=1}^{p}\sum_{j=1}^{i} t_{ij}^2\right\} \prod_{i=1}^{p} (t_{ii}^2)^{(n-i)/2}. \quad (6.58)$$

From (6.58) it is obvious that in this particular case the T_{ij} are independently distributed and T_{ii}^2 is distributed as central chi-square with $n - i + 1$ degrees of freedom ($i = 1, \ldots, p$), and $T_{ij}(i \neq j)$ is normally distributed with mean 0 and variance 1.

6.8. DISTRIBUTION OF HOTELLING'S T^2

Let $X^\alpha = (X_{\alpha 1}, \ldots, X_{\alpha p})'$, $\alpha = 1, \ldots, N$, be independently distributed p-variate normal random variables with the same mean μ and the same positive definite covariance matrix Σ. Let

$$\bar{X} = \frac{1}{N}\sum_{1}^{N} X^\alpha, \quad S = \sum_{\alpha=1}^{N}(X^\alpha - \bar{X})(X^\alpha - \bar{X})'.$$

We have observed that $\sqrt{N}\bar{X}$ has a p-variate normal distribution with mean $\sqrt{N}\mu$ and covariance matrix Σ and that it is independent of S, which is distributed as $\sum_{\alpha=1}^{N-1} Y^\alpha(Y^\alpha)'$, where $Y^\alpha = (Y_{\alpha 1}, \ldots, Y_{\alpha p})'$, $\alpha = 1, \ldots, N-1$, are independently and identically distributed normal p-vectors with mean 0 and covariance matrix Σ. We will prove the following theorem (due to Bowker).

Theorem 6.8.1. $N\bar{X}'S^{-1}\bar{X}$ is distributed as

$$\frac{\chi_p^2(N\mu'\Sigma^{-1}\mu)}{\chi_{N-p}^2} \quad (6.59)$$

where $\chi_p^2(N\mu'\Sigma^{-1}\mu)$ and χ_{N-p}^2 are independent.

Proof. Since Σ is positive definite there exists a nonsingular matrix C such that $C\Sigma C' = I$. Define $Z = \sqrt{N}C\bar{X}$, $A = CSC'$, and $\nu = \sqrt{N}c\mu$. Then Z is normally distributed with mean ν and covariance matrix I, and A is distributed as

Basic Multivariate Sampling Distributions

$\sum_{\alpha=1}^{N-1} Z^{\alpha}(Z^{\alpha})'$ where $Z^{\alpha} = (Z_{\alpha 1}, \ldots, Z_{\alpha p})'$, $\alpha = 1, \ldots, N-1$, are independently and identically distributed normal p-vectors with mean 0 and covariance matrix I. Furthermore, A and Z are independent. Consider a random orthogonal matrix Q of dimension $p \times p$ whose first row is $Z'(Z'Z)^{-1/2}$ and whose remaining $p-1$ rows are defined arbitrarily. Let

$$U = (U_1, \ldots, U_p)' = QZ, B = (B_{ij}) = QAQ'.$$

Obviously,

$$U_1 = (Z'Z)^{1/2}, U_i = 0, i = 2, \ldots, p$$

and

$$N\bar{X}'S^{-1}\bar{X} = Z'A^{-1}Z = U'B^{-1}U = U_1^2/(B_{11} - B_{(12)}B_{(22)}^{-1}B_{(21)}) \qquad (6.60)$$

where

$$B = \begin{pmatrix} B_{11} & B_{(12)} \\ B_{(21)} & B_{(22)} \end{pmatrix}$$

Since the conditional distribution of B given Q is Wishart with $N-1$ degrees of freedom and parameter I, by Theorem 6.4.1, the conditional distribution of $B_{11} - B_{(12)}B_{(22)}^{-1}B_{(21)}$ given Q is central chi-square with $N-p$ degrees of freedom. As this conditional distribution does not depend on Q, the unconditional distribution of $B_{11} - B_{(12)}B_{(22)}^{-1}B_{(21)}$ is also central chi-square with $N-p$ degrees of freedom. By the results presented in Section 6.1, $Z'Z$ is distributed as a noncentral chi-square with p degrees of freedom and the noncentrality parameter $\nu'\nu = N\mu'\Sigma^{-1}\mu$. The independence of $Z'Z$ and $B_{11} - B_{(12)}B_{(22)}^{-1}B_{(21)}$ is obvious. Q.E.D.

We now need the following lemma to demonstrate the remaining results in this section.

Lemma 6.8.1. *For any p-vector $Y = (Y_1, \ldots, Y_p)'$ and any $p \times p$ positive definite matrix A*

$$Y'(A + YY')^{-1}Y = \frac{Y'A^{-1}Y}{1 + Y'A^{-1}Y} \qquad (6.61)$$

Proof. Let

$$(A + YY')^{-1} = A^{-1} + C.$$

Then

$$I = (A^{-1} + C)(A + YY') = I + A^{-1}YY' + CA + CYY'.$$

Since $(A + YY')$ is positive definite,
$$C = -A^{-1}YY'(A + YY')^{-1}$$

Now
$$Y'(A + YY')^{-1}Y = Y'A^{-1}Y - (Y'A^{-1}Y)(Y'(A + YY')^{-1}Y),$$

or
$$Y'(A + YY')^{-1}Y = \frac{Y'A^{-1}Y}{1 + Y'A^{-1}Y}$$

Q.E.D.

Notations

For any p-vector $Y = (Y_1, \ldots, Y_p)'$ and any $p \times p$ matrix $A = (a_{ij})$ we shall write for $i = 1, \ldots, k$ and $k \leq p$

$$Y = (Y_{(1)}, \ldots, Y_{(k)})',$$
$$Y_{[i]} = (Y_{(1)}, \ldots, Y_{(i)})',$$
$$A_{[ij]} = (A_{(i1)}, \ldots, A_{(ij)}),$$
$$A_{[ji]} = (A_{(1i)}, \ldots, A_{(ji)}),$$

$$A = \begin{pmatrix} A_{(11)} & \cdots & A_{(1k)} \\ \vdots & & \vdots \\ A_{(k1)} & \cdots & A_{(kk)} \end{pmatrix}$$

$$A_{[ii]} = \begin{pmatrix} A_{(11)} & \cdots & A_{(1i)} \\ \vdots & & \vdots \\ A_{(i1)} & \cdots & A_{(ii)} \end{pmatrix}$$

where $Y_{(i)}$ are subvectors of Y of dimension $p_i \times 1$, and $A_{(ii)}$ are submatrices of A of dimension $p_i \times p_i$, where the p_i are arbitrary integers including zero such that $\sum_1^k p_i = p$. Let us now define $R_1, \ldots, R_k$ by

$$\sum_{j=1}^{i} R_j = n\bar{X}'_{[i]}(S_{[ii]} + N\bar{X}_{[i]}\bar{X}'_{[i]})^{-1}\bar{X}_{[i]}$$

$$= \frac{N\bar{X}'_{[i]}S_{[ii]}^{-1}\bar{X}_{[i]}}{1 + N\bar{X}'_{[i]}S_{[ii]}^{-1}\bar{X}_{[i]}}, \quad i = 1, \ldots, k.$$

(6.62)

Basic Multivariate Sampling Distributions

Since S is positive definite with probability 1 (we shall assume $N > p$), $S_{[ii]}, i = 1, \ldots, k$, are positive definite and hence $R_i \geq 0$ for $i = 1, \ldots, k$ with probability 1. We are interested here in showing that the joint probability density function of $R_1, \ldots, R_k$ is given by

$$f_{R_1,\ldots,R_k}(r_1,\ldots,r_k) = \Gamma\left(\frac{N}{2}\right)\left[\Gamma\left(\frac{1}{2}\left(N - \sum_1^k p_i\right)\right)\prod_{i=1}^k\left(\Gamma\left(\frac{1}{2}p_i\right)\right)\right]^{-1}$$

$$\times \prod_{i=1}^k (r_i)^{p_i/2-1}\left(1 - \sum_{i=1}^k r_i\right)^{\left(N-\sum_1^k p_i\right)/2 - 1} \quad (6.63)$$

$$\times \exp\left\{-\frac{1}{2}\sum_1^k \delta_i^2 + \frac{1}{2}\sum_{j=1}^k r_j \sum_{i>j} \delta_i^2\right\}$$

$$\times \prod_{i=1}^k \phi\left(\frac{1}{2}(N - \sigma_{i-1}), \frac{1}{2}p_i; \frac{1}{2}r_i\delta_i^2\right)$$

where

$$\sigma_i = \sum_{j=1}^i p_j \quad (\text{with} \quad \sigma_0 = 0)$$

$$\sum_{j=1}^i \delta_j^2 = N\mu'_{[i]}\Sigma_{[ii]}^{-1}\mu_{[i]}, \quad i = 1,\ldots,k$$

and $\phi(a, b; x)$ is the confluent hypergeometric function given by

$$\phi(a, b; x) = 1 + \frac{a}{b}x + \frac{a(a+1)}{b(b+1)}\frac{x^2}{2!} + \frac{a(a+1)(a+2)}{b(b+1)(b+2)}\frac{x^3}{3!} + \cdots. \quad (6.64)$$

For $k = 1$,

$$R_1 = \frac{N\bar{X}'S^{-1}\bar{X}}{1 + N\bar{X}'S^{-1}\bar{X}}, \quad \delta_1^2 = N\mu'\Sigma^{-1}\mu.$$

From (6.59) its probability density function is given by

$$f_{R_1}(r_1) = \frac{\Gamma(\frac{1}{2}N)}{\Gamma(\frac{1}{2}(N-p))\Gamma(\frac{1}{2}p)} r_1^{p/2-1}(1-r_1)^{(N-p)/2-1}$$

$$\times \exp\left\{-\frac{1}{2}\delta_1^2\right\}\phi\left(\frac{1}{2}N, \frac{1}{2}p; \frac{1}{2}r_1\delta_1^2\right) \quad (6.65)$$

which agrees with (6.63). To prove (6.63) in general we first consider the case $k=2$ and then use this result for the case $k=3$. The desired result for the general case will then follow from theses cases. The statistics $R_1, \ldots, R_k$ play an important role in tests of hypotheses concerning means of multivariate distributions with unknown covariance matrices (see Chapter 7) and tests of hypotheses concerning discriminant coefficients of discriminant analysis (see Chapter 9). Let us now prove (6.63) for $k=2, p_1+p_2=p$. Let

$$S = \begin{pmatrix} S_{(11)} & S_{(12)} \\ S_{(21)} & S_{(22)} \end{pmatrix}$$

$$W = S_{(22)} - S_{(21)}S_{(11)}^{-1}S_{(12)}, \qquad U = S_{(21)}S_{(11)}^{-1}, \qquad V = S_{(11)}. \qquad (6.66)$$

Identifying $S_{(22)}$ with $S_{(11)}$, $S_{(21)}$ with $S_{(12)}$, and vice versa in Theorem 6.4.1 we obtain: W is distributed as $W_{p_2}(N-1-p_1, \Sigma_{(22)} - \Sigma_{(21)}\Sigma_{(11)}^{-1}\Sigma_{(12)})$, the conditional distribution of U, given that $S_{(11)} = s_{(11)}$ is normal with mean $\Sigma_{(21)}\Sigma_{(11)}^{-1}$ and covariance matrix $(\Sigma_{(22)} - \Sigma_{(21)}\Sigma_{(11)}^{-1}\Sigma_{(12)}) \otimes s_{(11)}^{-1}$; V is distributed as $W_{p_1}(N-1, \Sigma_{(11)})$ and W is independent of (U, V). Hence the conditional distribution of

$$\sqrt{N} S_{(21)} S_{(11)}^{-1} \bar{X}_{(1)}$$

given that $\bar{X}_{(1)} = \bar{x}_{(1)}, S_{(11)} = s_{(11)}$ is a p_2-variate normal with mean $\sqrt{N}\Sigma_{(21)}\Sigma_{(11)}^{-1}\bar{x}_{(1)}$ and covariance matrix

$$(N\bar{x}_{(1)}' S_{(11)}^{-1} \bar{x}_{(1)})(\Sigma_{(22)} - \Sigma_{(21)}\Sigma_{(11)}^{-1}\Sigma_{(12)}).$$

Now let

$$W_1 = N\bar{X}_{(1)}' S_{(11)}^{-1} \bar{X}_{(1)} = R_1(1-R_1)^{-1},$$

$$W_2 = \frac{N(\bar{X}_{(2)} - S_{(21)}S_{(11)}^{-1}\bar{X}_{(1)})'(S_{(22)} - S_{(21)}S_{(11)}^{-1}S_{(12)})^{-1}(\bar{X}_{(2)} - S_{(21)}S_{(11)}^{-1}\bar{X}_{(1)})}{1+W_1}$$

$$= \{(R_1+R_2)(1-R_1-R_2)^{-1} \qquad (6.67)$$

$$- R_1(1-R_1)^{-1}\}(1+R_1(1-R_1)^{-1})^{-1}$$

$$= R_2(1-R_1-R_2)^{-1}.$$

Basic Multivariate Sampling Distributions

Then

$$N\bar{X}'S^{-1}\bar{X} = N\bar{X}'_{(1)}S_{(11)}^{-1}\bar{X}_{(1)}$$
$$+ N(\bar{X}_{(2)} - S_{(21)}S_{(11)}^{-1}\bar{X}_{(1)})'(S_{(22)} - S_{(21)}S_{(11)}^{-1}S_{(12)})^{-1}$$
$$\times (\bar{X}_{(2)} - S_{(21)}S_{(11)}^{-1}\bar{X}_{(1)})$$
$$= W_1 + W_2(1 + W_1).$$

Similarly, from $N\mu'\Sigma^{-1}\mu = \delta_1^2 + \delta_2^2$, $\delta_1^2 = N\mu'_{(1)}\Sigma_{(11)}^{-1}\mu_{(1)}$, we get

$$\delta_2^2 = N(\mu_{(2)} - \Sigma_{(21)}\Sigma_{(11)}^{-1}\mu_{(1)})'(\Sigma_{(22)} - \Sigma_{(21)}\Sigma_{(11)}^{-1}\Sigma_{(12)})^{-1} \quad (6.68)$$
$$\times (\mu_{(2)} - \Sigma_{(21)}\Sigma_{(11)}^{-1}\mu_{(1)}).$$

Since $\sqrt{N}\bar{X}$ is independent of S and is distributed as a p-variate normal with mean $\sqrt{N}\mu$ and covariance matrix Σ, the conditional distribution of $N\bar{X}_{(2)}$ given that $S_{(11)} = s_{(11)}$ and $\bar{X}_{(1)} = \bar{x}_{(1)}$ is a p_2-variate normal with mean $\sqrt{N}(\mu_{(2)} + \Sigma_{(21)}\Sigma_{(11)}^{-1}(\bar{x}_{(1)} - \mu_{(1)}))$ and covariance matrix $\Sigma_{(22)} - \Sigma_{(21)}\Sigma_{(11)}^{-1}\Sigma_{(12)}$. Furthermore this conditional distribution is independent of the conditional distribution of $\sqrt{N}S_{(21)}S_{(11)}^{-1}\bar{X}_{(1)}$ given that $S_{(11)} = s_{(11)}$ and $\bar{X}_{(1)} = \bar{x}_{(1)}$. Hence the conditional distribution of

$$\sqrt{N}(\bar{X}_{(2)} - S_{(21)}S_{(11)}^{-1}\bar{X}_{(1)})(1 + W_1)^{-1/2}$$

given that $S_{(11)} = s_{(11)}, \bar{X}_{(1)} = \bar{x}_{(1)}$ is a p_2-variate normal with mean $\sqrt{N}(\mu_{(2)} - \Sigma_{(21)}\Sigma_{(11)}^{-1}\mu_{(1)})(1 + w_1)^{-1/2}$ (w_1 is that value of W_1 corresponding to $S_{(11)} = s_{(11)}, \bar{X}_{(1)} = \bar{x}_{(1)}$) and covariance matrix $\Sigma_{(22)} - \Sigma_{(21)}\Sigma_{(11)}^{-1}\Sigma_{(12)}$. Since $S_{(22)} - S_{(21)}S_{(11)}^{-1}S_{(12)}$ is distributed independently of $(S_{(21)}, S_{(11)})$ and $\bar{X}$ as $W_{p_2}(N - 1 - p_1, \Sigma_{(22)} - \Sigma_{(21)}\Sigma_{(11)}^{-1}\Sigma_{(12)})$, by Theorem 6.8.1, the conditional distribution of W_2 given that $S_{(11)} = s_{(11)}, \bar{X}_{(1)} = \bar{x}_{(1)}$, is

$$\chi^2_{p_2}(\delta_2^2(1 + w_1)^{-1})/\chi^2_{N-p_1-p_2} \quad (6.69)$$

where $\chi^2_{p_2}(\delta_2^2(1 + w_1)^{-1})$ and $\chi^2_{N-p_1-p_2}$ are independent. Furthermore by the same theorem W_1 is distributed as

$$\chi^2_{p_1}(\delta_1^2)/\chi^2_{N-p_1}, \quad (6.70)$$

where $\chi^2_{p_1}(\delta_1^2)$ and $\chi^2_{N-p_1}$ are independent. Thus the joint probability density function of (W_1, W_2) is given by

$$f_{W_1,W_2}(w_1, w_2)$$

$$= \exp\left\{-\frac{1}{2}\delta_2^2(1+w_1)^{-1}\right\}$$

$$\times \sum_{\beta=1}^{\infty} \frac{(\frac{1}{2}\delta_2^2(1+w_1)^{-1})^\beta (w_2)^{p_2/2+\beta-1}\Gamma(\frac{1}{2}(N-p_1)+\beta)}{\beta!(1+w_2)^{(N-p_1)/2+\beta}\Gamma(\frac{1}{2}p_2+\beta)\Gamma(\frac{1}{2}(N-p))} \quad (6.71)$$

$$\times \exp\left(-\frac{1}{2}\delta_1^2\right) \sum_{j=0}^{\infty} \frac{(\frac{1}{2}\delta_1^2)^j w_1^{p_1/2+j-1}\Gamma(\frac{1}{2}(N+j))}{j!(1+w_1)^{N/2+j}\Gamma(\frac{1}{2}p_1+j)\Gamma(\frac{1}{2}(N-p_1))}.$$

Now transforming $(W_1, W_2) \to (R_1, R_2)$ as given by (6.67) the joint probability density function of R_1, R_2 is

$$f_{R_1,R_2}(r_1, r_2) = \Gamma\left(\frac{1}{2}N\right)\left[\Gamma\left(\frac{1}{2}(N-p)\right)\Gamma\left(\frac{1}{2}p_1\right)\Gamma\left(\frac{1}{2}p_2\right)\right]^{-1}$$

$$\times (r_1)^{p_1/2-1}(r_2)^{p_2/2-1}(1-r_1-r_2)^{(N-p)/2-1} \quad (6.72)$$

$$\times \exp\left\{-\frac{1}{2}(\delta_1^2+\delta_2^2)+\frac{1}{2}\delta_2^2 r_1\right\} \prod_{i=1}^{2} \phi\left(N-\sigma_{i-1}, \frac{1}{2}p_i; \frac{1}{2}r_i\delta_i^2\right),$$

which agrees with (6.63) for $k = 2$. Let us now consider the case $k = 3$. Let

$$W_3 = (N\bar{X}'S^{-1}\bar{X} - N\bar{X}'_{[2]}S^{-1}_{[22]}\bar{X}_{[2]})/(1 + N\bar{X}'_{[2]}S^{-1}_{[22]}\bar{X}_{[2]}). \quad (6.73)$$

Now $S_{(33)} - S_{(32)}S^{-1}_{[22]}S_{[23]}$ is distributed as

$$W_{p_3}(N-1-p_1-p_2, (\Sigma_{(33)}-\Sigma_{[32]}\Sigma^{-1}_{[22]}\Sigma_{[23]}))$$

and is independent of $S_{[22]}$ and $S_{[32]}$. Also, the conditional distribution of $\sqrt{N}\bar{X}_{(3)}$ given that $S_{[22]} = s_{[22]}$, $\bar{X}_{[2]} = \bar{x}_{[2]}$ is normal with mean $\sqrt{N}(\mu_{(3)} - \Sigma_{[32]}\Sigma^{-1}_{[22]}(\bar{x}_{[2]} - \mu_{[2]}))$ and covariance matrix $\Sigma_{(33)} - \Sigma_{[32]}\Sigma^{-1}_{[22]}\Sigma_{[23]}$ and is independent of the conditional distribution of $\sqrt{N}S_{[32]}S^{-1}_{[22]}\bar{X}_{[2]}$ given that $S_{[22]} = s_{[22]}$ and $\bar{X}_{[2]} = \bar{x}_{[2]}$, which is normal with mean $\sqrt{N}\Sigma_{[32]}\Sigma^{-1}_{[22]}\bar{x}_{[2]}$ and covariance matrix $(N\bar{x}'_{[2]}S^{-1}_{[22]}\bar{x}_{[2]})(\Sigma_{(33)} - \Sigma_{[32]}\Sigma^{-1}_{[22]}\Sigma_{[23]})$. Hence as before the conditional distribution of W_3 given that $S_{[22]} = s_{[22]}$ and $\bar{X}_{[2]} = \bar{x}_{[2]}$ or, equivalently, given that $W_1 = w_1$, $W_2 = w_2$, is given by

$$f_{W_3|W_1,W_2}(w_3|w_1, w_2) = \chi^2_{p_3}(\delta_3^2(1+w_1)(w_1+w_2+w_1w_2)^{-1})/\chi^2_{N-p} \quad (6.74)$$

Basic Multivariate Sampling Distributions

where $\chi^2_{p_3}(\cdot)$ and χ^2_{N-p} are independent. Thus the joint probability density function of W_1, W_2, W_3 is

$$f_{W_1,W_2,W_3}(w_1, w_2, w_3) = \frac{\chi^2_{p_3}(\delta^2_3(1+w_1)(w_1+w_2+w_1w_2)^{-1})}{\chi^2_{N-p}}$$

$$\times \frac{\chi^2_{p_2}(\delta^2_2(1+w_1)^{-1})}{\chi^2_{N-p_1-p_2}} \times \frac{\chi^2_{p_1}(\delta^2_1)}{\chi^2_{N-p}}. \quad (6.75)$$

Now replacing the W_i by R_i we get

$$W_1 = R_1(1-R_1)^{-1}, \quad W_2 = R_2(1-R_1-R_2)^{-1}$$

$$W_3 = ((R_1+R_2+R_3)(1-R_1-R_2-R_3)^{-1}$$

$$- (R_1+R_2)(1-R_1-R_2)^{-1})(1-R_1-R_2)$$

$$= R_3(1-R_1-R_2-R_3)^{-1}.$$

From (6.75) the joint probability density function of R_1, R_2, R_3 is given by

$$f_{R_1,R_2,R_3}(r_1, r_2, r_3) = \Gamma\left(\frac{1}{2}N\right)\left[\Gamma\left(\frac{1}{2}(N-p)\right)\prod_{i=1}^{3}\Gamma\left(\frac{1}{2}p_i\right)\right]^{-1}$$

$$\times \prod_{i=1}^{3}(r_i)^{\frac{p_i}{2}-1}\left(1-\sum_{1}^{3}r_i\right)^{(N-p)/2-1}$$

$$\times \exp\left\{-\frac{1}{2}\sum_{1}^{3}\delta^2_j + \frac{1}{2}\sum_{j=1}^{3}r_j\sum_{i>j}^{3}\delta^2_i\right\} \quad (6.76)$$

$$\times \prod_{i=1}^{3}\phi\left(\frac{1}{2}(N-\sigma_{i-1}), \frac{1}{2}p_i; \frac{1}{2}r_i\delta^2_i\right)$$

which agrees with (6.63) for $k = 3$. Proceeding exactly in this fashion we get (6.63) for general k.

6.9. MULTIPLE AND PARTIAL CORRELATION COEFFICIENTS

Let S be distributed as $W_p(N-1, \Sigma)$ and let

$$S = (S_{ij}) = \begin{pmatrix} S_{11} & S_{(12)} \\ S_{(21)} & S_{(22)} \end{pmatrix}, \quad \Sigma = (\Sigma_{ij}) = \begin{pmatrix} \Sigma_{11} & \Sigma_{(12)} \\ \Sigma_{(21)} & \Sigma_{(22)} \end{pmatrix}$$

We shall first find the distribution of

$$R^2 = \frac{S_{(12)}S_{(22)}^{-1}S_{(21)}}{S_{11}}. \tag{6.77}$$

From this

$$\begin{aligned}
\frac{R^2}{1-R^2} &= \frac{S_{(12)}S_{(22)}^{-1}S_{(21)}}{S_{11} - S_{(12)}S_{(22)}^{-1}S_{(21)}} \\
&= \left(\frac{S_{(12)}S_{(22)}^{-1}S_{(21)}}{\Sigma_{11} - \Sigma_{(12)}\Sigma_{(22)}^{-1}\Sigma_{(21)}}\right) \Big/ \left(\frac{S_{11} - S_{(12)}S_{(22)}^{-1}S_{(21)}}{\Sigma_{11} - \Sigma_{(12)}\Sigma_{(22)}^{-1}\Sigma_{(21)}}\right) = \frac{X}{Y},
\end{aligned} \tag{6.78}$$

where

$$X = \frac{S_{(12)}S_{(22)}^{-1}S_{(21)}}{\Sigma_{11} - \Sigma_{(12)}\Sigma_{(22)}^{-1}\Sigma_{(21)}}, \qquad Y = \frac{S_{(11)} - S_{(12)}S_{(22)}^{-1}S_{(21)}}{(\Sigma_{11} - \Sigma_{(12)}\Sigma_{(22)}^{-1}\Sigma_{(21)})}$$

From Theorem 6.4.1, Y is distributed as central chi-square with $N-p$ degrees of freedom and is independent of $(S_{(12)}, S_{(22)})$ and the conditional distribution of $S_{(12)}S_{(22)}^{-1/2}$ given that $S_{(22)} = s_{(22)}$ is a $(p-1)$-variate normal distribution with mean $\Sigma_{(12)}\Sigma_{(22)}^{-1}s_{(22)}^{1/2}$ and covariance matrix $(\Sigma_{11} - \Sigma_{(12)}\Sigma_{(22)}^{-1}\Sigma_{(21)})I$. Hence the conditional distribution of X given that $S_{(22)} = s_{(22)}$ is noncentral chi-square

$$\chi^2_{p-1}\left(\frac{\Sigma_{(12)}\Sigma_{(22)}^{-1}s_{(22)}\Sigma_{(22)}^{-1}\Sigma_{(21)}}{\Sigma_{11} - \Sigma_{(12)}\Sigma_{(22)}^{-1}\Sigma_{(21)}}\right). \tag{6.79}$$

Since $S_{(22)}$ is distributed as $W_{p-1}(N-1, \Sigma_{(22)})$ (see Theorem 6.4.1) by Exercise 4,

$$\frac{\Sigma_{(12)}\Sigma_{(22)}^{-1}s_{(22)}\Sigma_{(22)}^{-1}\Sigma_{(21)}}{\Sigma_{12}\Sigma_{22}^{-1}\Sigma_{21}} \tag{6.80}$$

is distributed as χ^2_{N-1}. Since

$$\frac{\Sigma_{(12)}\Sigma_{(22)}^{-1}\Sigma_{(21)}}{\Sigma_{11} - \Sigma_{(12)}\Sigma_{(22)}^{-1}\Sigma_{(21)}} = \frac{\rho^2}{1-\rho^2}, \tag{6.81}$$

where

$$\rho^2 = \frac{\Sigma_{(12)}\Sigma_{(22)}^{-1}\Sigma_{(21)}}{\Sigma_{11}}, \tag{6.82}$$

$R^2/(1-R^2)$ is distributed as the ratio (of independent random variables) X/Y, where Y is distributed as χ^2_{N-p} and X is distributed as $\chi^2_{p-1}((\rho^2/(1-\rho^2))\chi^2_{(N-1)})$

Basic Multivariate Sampling Distributions

with random noncentrality parameter

$$(\rho^2/(1-\rho^2))\chi^2_{N-1}.$$

Since, from (6.5), a noncentral chi-square $Z = \chi^2_N(\lambda)$ is distributed as χ^2_{N+2K} where K is a Poisson random variable with parameter $\lambda/2$, i.e., its probability density function is given by

$$f_Z(z) = \sum_{k=0}^{\infty} f_{\chi^2_{N+2k}}(z) p_K(k),$$

where $p_K(k)$ is the Poisson probability mass function with parameter $\lambda/2$, it follows that the conditional distribution of X given that $\chi^2_{N-1} = t$ is χ^2_{p-1+2K}, where the conditional distribution of K given that $\chi^2_{N-1} = t$ is Poisson with parameter $\frac{1}{2}t(\rho^2/(1-\rho^2))$. Let $\lambda/2 = \frac{1}{2}(\rho^2/(1-\rho^2))$. The unconditional probability mass function of K is given by

$$\begin{aligned} p_K(k) &= \int_0^{\infty} \exp\left\{-\frac{1}{2}\lambda t\right\} \frac{(\lambda t/2)^k}{k!} \frac{t^{((N-1)/2)-1} \exp\{-\frac{1}{2}t\}dt}{2^{(N-1)/2}\Gamma(\frac{1}{2}(N-1))} \\ &= \frac{\Gamma(\frac{1}{2}(N-1)+k)}{k!\Gamma(\frac{1}{2}(N-k))}(\rho^2)^k(1-\rho^2)^{(N-1)/2}, \qquad k=0,1,2,\ldots \end{aligned} \qquad (6.83)$$

This implies that the unconditional distribution of K is negative binomial. Hence the probability density function of X is given by

$$f_X(x) = \sum_{k=0}^{\infty} f_{\chi^2_{p-1+2k}}(x) p_K(k), \qquad (6.84)$$

where $f_{\chi^2_{p-1+2k}}(x)$ is the probability density function of χ^2_{p-1+2k} and $p_K(k)$ is given by (6.83). Thus we get the following theorem.

Theorem 6.9.1. *The probability density function of $R^2/(1-R^2)$ is given by the probability density function of the ratio X/Y of two independently distributed random variables X, Y, where X is distributed as a chi-square random variable χ^2_{p-1+2K} with K a negative binomial random variable with probability mass function given in (6.83) and Y is distributed as χ^2_{N-p}.*

It is now left to the reader as an exercise to verify that the probability density function of R^2 is given by

$$f_{R^2}(r^2) = \begin{cases} \dfrac{(1-\rho^2)^{(N-1)/2}(1-r^2)^{(N-p-2)/2}}{\Gamma(\frac{1}{2}(N-1))\Gamma(\frac{1}{2}(N-p))} \\ \quad \times \sum_{j=0}^{\infty} \dfrac{(\rho^2)^j (r^2)^{(p_1)/2+j-1}\Gamma^2(\frac{1}{2}(N-1)+j)}{j!\Gamma(\frac{1}{2}(p-1)+j)} \\ \qquad\qquad\qquad\qquad\qquad\qquad \text{if } r^2 \geq 0, \\ 0, \qquad\qquad\qquad\qquad\qquad\qquad \text{otherwise.} \end{cases} \quad (6.85)$$

This derivation is studied by Fisher (1928). For $p = 2$, the special case studied by Fisher in 1915, the probability density function of R is given by

$$\begin{aligned} f_R(r) &= \frac{2^{N-3}(1-\rho^2)^{(N-1)/2}(1-r^2)^{(N-4)/2}}{(N-3)!\pi} \sum_{j=0}^{\infty} \frac{(2\rho r)^j}{j!}\Gamma^2\left(\frac{1}{2}(N-1)+j\right) \quad (6.86) \\ &= \frac{(1-\rho^2)^{(N-1)/2}(1-r^2)^{(N-4)/2}}{\pi(N-3)!}\left[\frac{d^{n-1}}{dx^{n-1}}\left\{\frac{\cos^{-1}(-x)}{(1-x^2)^{1/2}}\right\}|x=rp\right], \end{aligned}$$

which follows from (6.85) with $p = 2$ and the fact that

$$\Gamma(n)\Gamma\left(n + \frac{1}{2}\right) = \sqrt{\pi}\Gamma(2n)/2^{2n-1}. \quad (6.87)$$

It is well known that as $N \to \infty$, the distribution of $(\sqrt{N}(R-\rho))/(1-\rho^2)$ tends to normal distribution with mean 0 and variance 1.

Let $X = (X_1, \ldots, X_p)'$ be normally distributed with mean μ and covariance matrix $\Sigma = (\sigma_{ij})$ and let $\rho_{ij} = \sigma_{ii}/(\sigma_{ii}\sigma_{jj})^{1/2}$. It is now obvious that the distribution of the sample correlation coefficient R_{ij} between the ith and jth components of X, based on a random sample of size N, is obtained from (6.86) by replacing ρ by ρ_{ij}. Let $X^\alpha = (X_{\alpha 1}, \ldots, X_{\alpha p})'$, $\alpha = 1, \ldots, N$, be a random sample of size N from the distribution of X and let

$$S = \sum_{\alpha=1}^{N}(X^\alpha - \bar{X})(X^\alpha - \bar{X})'$$

be partitioned as

$$S = \begin{pmatrix} S_{(11)} & S_{(12)} \\ S_{(21)} & S_{(22)} \end{pmatrix}$$

where $S_{(11)}$ is a $q \times q$ submatrix of S. We observed in Chapter 5 that the sample partial correlation coefficients $r_{ij \cdot 1, \ldots, q}$ can be computed from $s_{(11)} - s_{(21)}s_{(11)}^{-1}s_{(12)}$ in the same way that the sample simple correlation coefficients r_{ij} are computed from s. Furthermore, we observed that to obtain the distribution of R_{ij} (random

Basic Multivariate Sampling Distributions

variable corresponding to r_{ij}) we needed the fact that S is distributed as $W_p(N-1, \Sigma)$. Since, from Theorem 6.4.1, $S_{(22)} - S_{(21)}S_{(11)}^{-1}S_{(12)}$ is distributed as $W_{p-q}(N-q-1, \Sigma_{(22)} - \Sigma_{(21)}\Sigma_{(11)}^{-1}\Sigma_{(12)})$ and is independent of $(S_{(11)}, S_{(12)})$, it follows that the distibution $R_{ij.1,\ldots,q}$ based on N obserations is the same as that of the simple correlation coefficient R_{ij} based on $N - q$ observations with a corresponding population parameter $\rho_{ij.1,\ldots,q}$.

6.10. DISTRIBUTION OF MULTIPLE PARTIAL CORRELATION COEFFICIENTS

Let X^α, $\alpha = 1, \ldots, N$ be a random sample of size N from $N_p(\mu, \Sigma)$ and let $\bar{X} = \frac{1}{N}\sum_{\alpha=1}^{N} X^\alpha$, $S = \sum_{\alpha=1}^{N}(X^\alpha - \bar{X})(X^\alpha - \bar{X})'$. Assume that $N > p$ so that S is positive definite with probability one. Partition S and Σ as

$$\Sigma = \begin{pmatrix} \Sigma_{11} & \Sigma_{12} & \Sigma_{13} \\ \Sigma_{21} & \Sigma_{22} & \Sigma_{13} \\ \Sigma_{31} & \Sigma_{32} & \Sigma_{33} \end{pmatrix}, \quad S = \begin{pmatrix} S_{11} & S_{12} & S_{13} \\ S_{21} & S_{22} & S_{23} \\ S_{31} & S_{32} & S_{33} \end{pmatrix} \quad (6.88)$$

with Σ_{22}, S_{22} each of dimension $p_1 \times p_1$; Σ_{33}, S_{33} each of dimension $p_2 \times p_2$ where $p_1 + p_2 = p - 1$. Let

$$\rho_1^2 = \Sigma_{12}\Sigma_{22}^{-1}\Sigma_{21}/\Sigma_{11}$$

$$\rho^2 = \rho_1^2 + \rho_2^2 = (\Sigma_{12}\Sigma_{13})\begin{pmatrix} \Sigma_{22} & \Sigma_{23} \\ \Sigma_{32} & \Sigma_{33} \end{pmatrix}^{-1}(\Sigma_{12}\Sigma_{13})'/\Sigma_{11}$$

$$\bar{R}_1 = S_{12}S_{22}^{-1}S_{21}/S_{11}$$

$$R^2 = \bar{R}_1 + \bar{R}_2 = (S_{12}S_{13})\begin{pmatrix} S_{22} & S_{23} \\ S_{32} & S_{33} \end{pmatrix}^{-1}(S_{12}S_{13})'/S_{11}$$

We shall term ρ_1^2, ρ_2^2 as population multiple partial correlation coefficients and $\bar{R}_1, \bar{R}_2$ as sample multiple partial correlation coefficients. The following theorem gives the joint probability density function of $\bar{R}_1, \bar{R}_2$. A more general case has been treated by Giri and Kiefer (1964).

Theorem 6.10.1. *The joint pdf of $\bar{R}_1, \bar{R}_2$ on the set $\{(\bar{r}_1, \bar{r}_2) : \bar{r}_1 + \bar{r}_2 < 1\}$ is given by*

$$f_{\bar{R}_1,\bar{R}_2}(\bar{r}_1, \bar{r}_2) = K(1-\rho^2)^{N/2}(1-\bar{r}_1-\bar{r}_2)^{1/2(N-p-1)}$$

$$\times \prod_{i=1}^{2}(\bar{r}_i)^{\frac{1}{2}p_i-1}\left[1+\sum_{i=1}^{2}\bar{r}_i\left(\frac{(1-\rho^2)}{\gamma_i}-1\right)\right]^{-N/2} \quad (6.89)$$

$$\times \sum_{\beta_1=0}^{\infty}\sum_{\beta_2=0}^{\infty}\prod_{i=1}^{2}\left[\frac{\Gamma(\frac{1}{2}(N+p_i-\sigma_i))\Gamma(\frac{1}{2}+\beta_i)\theta_i^{\beta_i}}{(2\beta_i)!\Gamma(\frac{1}{2}p_i+\beta_i)}\right]$$

where

$$\gamma_i = 1 - \sum_{j=1}^{i}\rho_j^2, \quad \gamma_0 = 1, \quad \sigma_i = \sum_{j=1}^{i}p_j$$

$$\alpha_i^2 = \rho_i^2(1-\rho^2)/\gamma_i\gamma_{i-1}$$

$$\theta_i = \frac{4\bar{r}_i\alpha_i^2}{1+\sum_{i=1}^{2}\bar{r}_i\left[\frac{1-\rho^2}{\gamma_i}-1\right]}$$

and K is the normalizing constant.

Proof. We prove that $(\bar{R}_1, \bar{R}_2)$ is a maximal invariant statistic and in deriving its distribution we can, without any loss of generality assume that

$$\Sigma_{11} = 1, \Sigma_{22} = I_{p_1}, \Sigma_{33} = I_{p_2}, \Sigma_{13} = 0, \Sigma_{23} = 0$$

where I_k is the $k \times k$ identity matrix. Let

$$U_1 = \frac{\bar{R}_1 S_{11}}{1-\rho_1^2}$$

$$S_{[13].2} = \begin{pmatrix} S_{11.2} & S_{13.2} \\ S_{31.2} & S_{33.2} \end{pmatrix}$$

$$= \begin{pmatrix} S_{11} & S_{13} \\ S_{31} & S_{33} \end{pmatrix} - \begin{pmatrix} S_{12} \\ S_{32} \end{pmatrix} S_{22}^{-1}(S_{21} S_{23})$$

$$\Sigma_{[13].2} = \begin{pmatrix} \Sigma_{11.2} & \Sigma_{13.2} \\ \Sigma_{31.2} & \Sigma_{33.2} \end{pmatrix}$$

$$= \begin{pmatrix} \Sigma_{11} & \Sigma_{13} \\ \Sigma_{31} & \Sigma_{33} \end{pmatrix} - \begin{pmatrix} \Sigma_{12} \\ \Sigma_{32} \end{pmatrix} \Sigma_{22}^{-1}(\Sigma_{21} \Sigma_{23}).$$

Basic Multivariate Sampling Distributions

By Theorem 6.4.1 U_1 and $S_{[13].2}$ are independent and $S_{[13].2}$ is distributed as Wishart

$$W_{1+p_2}(N-1+p_2, \Sigma_{[13.2]})$$

and the *pdf* of U_1 is given by

$$f_{U_1}(u_1) = (1-\bar{\rho}_1^2)^{(N-1)/2} \frac{1}{2^{p_1/2}\Gamma(p_1/2)}$$

$$\times \exp\left\{-\frac{1}{2}u_1\right\}(u_1)^{p_1/2-1}$$

$$\times \phi\left(\frac{N-1}{2}, \frac{p_1}{2}; \frac{u_1\bar{\rho}_1^2}{2}\right)$$

where ϕ is the confluent hypergeometric function given in (6.64). Define

$$U_2 = \frac{S_{13.2}S_{33.2}^{-1}S_{31.2}}{\Sigma_{11.2} - \Sigma_{13.2}\Sigma_{33.2}^{-1}\Sigma_{31.2}}$$

$$= \frac{S_{11}\bar{R}_2}{(1-\bar{\rho}_1^2-\bar{\rho}_2^2)}$$

$$U_3 = \frac{S_{11.2} - S_{13.2}S_{33.2}^{-1}S_{31.2}}{\Sigma_{11.2} - \Sigma_{13.2}\Sigma_{33.2}^{-1}\Sigma_{31.2}}$$

$$= \frac{S_{11}(1-\bar{R}_1-\bar{R}_2)}{1-\bar{\rho}_1^2-\bar{\rho}_2^2}$$

Applying Theorem 6.4.1 to $S_{[13].2}$ we conclude that U_2 and U_3 are independent and U_2 is distributed as χ^2_{N-p}. The *pdf* of U_3 is given by

$$f_{U_3}(u_3) = \left(\frac{1-\bar{\rho}_1^2-\bar{\rho}_2^2}{1-\bar{\rho}_1^2}\right)^{(N-p_1-1)/2} \frac{1}{2^{p_2/2}\Gamma(p_2/2)} \qquad (6.90)$$

$$\times \exp\left\{-\frac{1}{2}u_3\right\}(u_3)^{p_2/2-1}$$

$$\times \phi\left(\frac{1}{2}(N-p_1-1), p_2/2; u_3\bar{\rho}_2^2/2(1-\bar{\rho}_1^2)\right)$$

From (6.90) the joint *pdf* of T_2, T_3 where

$$T_2 = \frac{U_1(1-\rho_1^2)}{(1-\rho_1^2-\rho_2^2)(U_2+U_3)} = \frac{\bar{R}_1}{1-\bar{R}_1}$$

$$T_3 = \frac{U_2}{U_2+U_3} \qquad (6.91)$$

$$= \bar{R}_2/(1-\bar{R}_1)$$

is given by $\left(\text{writing } B(m,n) = \dfrac{\Gamma(m)\Gamma(n)}{\Gamma(m+n)}\right)$

$f_{T_2,T_3}(t_2, t_3)$

$$= \sum_{j=0}^{\infty}\sum_{k=0}^{\infty} \frac{\alpha_j \beta_k}{j!k!} \frac{\left[\left(\dfrac{1-\rho_1^2-\rho_2^2}{1-\rho_1^2}\right)^j t_2^{p_1/2+j-1}\left[1+\left(\dfrac{1-\rho_1^2-\rho_2^2}{1-\rho_1^2}\right)t_2\right]^{-(\frac{n}{2}+j+k)}\right]}{B(p_1/2+j, (n-p_1)/2+k)}$$

$$\times \left\{\frac{(t_3)^{p_2/2+k-1}(1-t_3)^{(n-p_1-p_2)/2-1}}{B(p_2/2+k, (n-p_1-p_2)/2)}\right\} \qquad (6.92)$$

where

$$\alpha_j = \Gamma\left(\frac{n}{2}+j\right)(1-\rho_1^2)^{n/2}(\rho_1^2)^j/\Gamma(n/2)$$

$$\beta_k = \Gamma((n-p_1)/2+k)\left(\frac{\rho_2^2}{1-\rho_1^2}\right)^k\left(\frac{1-\rho_1^2-\rho_2^2}{1-\rho_1^2}\right)^{n/2}/\Gamma((n-p_1)/2).$$

From (6.92), using (6.91) we get (6.89). Q.E.D.

6.11. BASIC DISTRIBUTIONS IN MULTIVARIATE COMPLEX NORMAL

Theorem 6.11.1. *Let $Z = (Z_1, \ldots, Z_p)'$ be a p-variate complex normal random vector with mean α and positive definite Hermitian covariance matrix Σ having property (4.19). Then $Z^*\Sigma^{-1}Z$ is distributed as $\chi^2_{2p}(2\alpha^*\Sigma^{-1}\alpha)$.*

Proof. Let $U = CX$, C is a $p \times p$ nonsingular complex matrix such that $C\Sigma C^* = I$. By Theorem 4.2.3 U is distributed as a p-variate complex random vector with mean $\beta = C\alpha$ and covariance I. Write $U = (U_1, \ldots, U_p)'$ with $U_j = X_j + iY_j$, $\beta = (\beta_1, \ldots, \beta_p)'$ with $\beta_j = \beta_{jR} + i\beta_{jC}$. We obtain $X_1 - \beta_{1R}, \ldots, X_p - \beta_{pR}$; $Y_1 - \beta_{1C}, \ldots, Y_p - \beta_{pC}$ are independently and identically distributed real normal random variables with the same mean 0 and

Basic Multivariate Sampling Distributions 249

the same variance $1/2$. Hence $2\sum_{j=1}^{p}(X_j^2 + Y_j^2) = 2U^*U = 2Z^*\Sigma^{-1}Z$ is distributed as $\chi_{2p}^2(\lambda^2)$ where $\lambda^2 = 2\sum_{j=1}^{p}(\beta_{jR}^2 + \beta_{jC}^2) = 2\beta^*\beta = 2\alpha^*\Sigma^{-1}\alpha$.

Q.E.D.

Theorem 6.11.2. Let $Z^j = (Z_{j1}, \ldots, Z_{jp})', j = 1, \ldots, N$ be independently and identically distributed as $CN_p(\alpha, \Sigma)$ and let $\bar{Z} = 1/N \sum_{j=1}^{N} Z^j$, $A = \sum_{j=1}^{N}(Z^j - \bar{Z})(Z^j - \bar{Z})^*$. Then $T_C^2 = N\bar{Z}^*A^{-1}\bar{Z}$ is distributed as the ratio $\chi_{2p}^2(2N\alpha^*\Sigma^{-1}\alpha)/\chi_{2(N-p)}^2$ where $\chi_{2p}^2(2N\alpha^*\Sigma^{-1}\alpha)$ and $\chi_{2(N-p)}^2$ are independent.

Proof. Since Σ is Hermitian positive definite, by Theorem 1.8.4 there exists a complex $p \times p$ nonsingular matrix C such that $C\Sigma C^* = I$. Let $V = (V_1, \ldots, V_p)' = \sqrt{N}C\bar{Z}$, $W = CAC^*$, and $\nu = \sqrt{N}C\alpha$. From (5.71) W is distributed as the complex Wishart $W_c(n, I)$ with $n = N - 1$ degrees of freedom and parameter I. Let Q be an $p \times p$ unitary matrix with first row

$$(V_1(V^*V)^{-1/2}, \ldots, V_p(V^*V)^{-1/2})$$

and the remaining rows are defined arbitrarily. Writing $U = (U_1, \ldots, U_p)' = QV$, $B = QWQ^*$ we obtain

$$T_c^2 = U^*B^{-1}U = (U_1^*U_1^*)/(B_{11} - B_{12}B_{22}^{-1}B_{12}^*)$$

$$= (V^*V)/(B_{11} - B_{12}B_{22}^{-1}B_{12}^*),$$

where B is partitioned as

$$B = \begin{pmatrix} B_{11} & B_{12} \\ B_{21} & B_{22} \end{pmatrix}$$

where B_{22} is $(p-1) \times (p-1)$. From (5.71) taking $\Sigma = I$ the joint pdf of B_{22}, B_{12} and $H = (B_{11} - B_{12}B_{22}^{-1}B_{21})$ is

$$I_o(\det B_{22})^{N-p-1}(\det H)^{N-p-1} \exp\{-tr(H + B_{12}B_{22}^{-1}B_{12}^* + B_{22})\} \quad (6.93)$$

where

$$I_o = \pi^{p(p-1)/2} \prod_{j=1}^{p} \Gamma(N-j).$$

From (6.93), using (5.71) we conclude that H is independent of B_{22} and B_{12}; and H is distributed as complex Wishart with degrees of freedom $N - p$ and parameter unity. From Theorem 5.3.4, the conditional distribution of B given Q, is that of $\sum_{\alpha=2}^{N} V^{\alpha}(V^{\alpha})^*$, where conditionally given Q, $V_{\alpha}, \alpha = 2, \ldots, N$, are independent and each has a p-variate complex normal distribution mean 0 and covariance I. Hence $B_{11} - B_{12}B_{22}^{-1}B_{21}$, given Q, is distributed as $\sum_{\alpha=1}^{N-p} W_\alpha W_\alpha^*$ where $W_\alpha, \alpha = 1, \ldots, N - p$, given Q are independent and each has a single variate complex normal distribution with mean 0 and variance unity. From Theorem 6.11.1 and the fact that the sum of independent chi-squares is a chi-square we conclude that

$$2(B_{11} - B_{12}B_{22}^{-1}B_{21}),$$

conditionally given Q, is distributed as $\chi^2_{2(N-p)}$. Since this distribution does not involve $Q 2(B_{11} - B_{12}B_{22}^{-1}B_{21})$ is unconditionally distributed as $\chi^2_{2(N-p)}$. The quantity $2V^*V$ (using Theorem 6.11.1) is distributed as $\chi^2_{2p}(2N\alpha^*\Sigma^{-1}\alpha)$. Hence from Theorem 5.3.5 we get the Theorem. Q.E.D.

Theorem 6.11.3. *Let A, Σ be similarly partitioned into submatrices as*

$$A = \begin{pmatrix} A_{11} & A_{12} \\ A_{21} & A_{22} \end{pmatrix}, \quad \Sigma = \begin{pmatrix} \Sigma_{11} & \Sigma_{12} \\ \Sigma_{21} & \Sigma_{22} \end{pmatrix}$$

where A_{11} and Σ_{11} are 1×1. Then the *pdf* of

$$R_c^2 = A_{12}A_{22}^{-1}A_{12}^*/A_{11}$$

is given by

$$f_{R_c^2}(r_c^2) = \frac{\Gamma(N-1)}{\Gamma(p-1)\Gamma(N-p)}(1-\rho_c^2)^{N-1}(r_c^2)^{p-2}(1-r_c^2)^{N-p-1}$$

$$\times F((N-1), N-1; p-1, r_c^2\rho_c^2)$$

where F is given in exercise 17b. We refer to Goodman (1963) for the proof. For more relevent results in connection with multivariate distributions we refer to Bartlett (1933), Giri (1965, 1971, 1972, 1973), Giri, Kiefer and Stein (1963), Karlin and Traux (1960), Kabe (1964, 1965), Khatri (1959), Khirsagar (1972), Mahalanobis, Bose and Roy (1937), Olkin and Rubin (1964), Roy and Ganadesikan (1959), Stein (1969), Wijsman (1957), Wishart (1948), and Wishart and Bartlett (1932, 1933).

6.12. BASIC DISTRIBUTIONS IN SYMMETRICAL DISTRIBUTIONS

We discuss here some basic distribution results related to symmetrical distributions.

Basic Multivariate Sampling Distributions

Theorem 6.12.1. Let $X = (X_1, \ldots, X_p)'$ be distributed as $E_p(0, I)$ with $P(X = 0) = 0$ and let $W = (W_1, \ldots, W_p)'$ be distributed as $N_p(0, I)$.

(a) $\dfrac{X}{\|X\|}$ and $\dfrac{W}{\|W\|}$; where $\|X\|^2 = X'X$, $\|W\|^2 = W'W$, are identically distributed.

(b) Let $Z = (Z_1, \ldots, Z_p)' = \dfrac{X}{\|X\|}$ and let $e_i = Z_i^2, i = 1, \ldots, p$. Then $(e_1, \ldots, e_p)$ has the Dirichlet distribution $D(\tfrac{1}{2}, \ldots, \tfrac{1}{2})$ with probability density function

$$f(e_1, \ldots, e_p) = \frac{\Gamma(p/2)}{(\Gamma(\tfrac{1}{2}))^p} \left[\prod_{j=1}^{p-1}(e_j)^{1/2-1}\right]\left(1 - \sum_{j=1}^{p-1} e_j\right)^{1/2-1} \quad (6.94)$$

(c) where $0 \le e_i \le 1$, $\sum_{j=1}^{p} e_j = 1$.

Proof.

(a) Let $U = (U_{ij}) = (U_1, \ldots, U_p)$, $U_i = (U_{i1}, \ldots, U_{ip})', i = 1, \ldots, p$, be a $p \times p$ random matrix such that $U_1, \ldots, U_p$, are independently and identically distributed $N_p(0, I)$ and U is independent of X.

Let $S(U_{\alpha 1}, \ldots, U_{\alpha k})$ denote the subspace spanned by the values of $U_{\alpha 1}, \ldots, U_{\alpha k}$. Under above assumptions on U.

$$P\{U_1, \ldots, U_p, \text{ are linearly dependent}\}$$
$$\le \sum_{i=1}^{p} p\{U_i \in S(U_1, \ldots, U_{i-1}, U_{i+1}, \ldots, U_p\}$$
$$= pP\{U_1 \in S(U_2, \ldots, U_P)\}$$
$$= pE(P\{U_1 \in S(U_2, \ldots, U_p)|U_2 = u_2, \ldots, U_p = u_p\})$$
$$= pE(0) = 0,$$

since the probability that U_1 lies in a space of dimension less than p is zero. Hence U is nonsingular with probability one. Let $\phi = \phi(U)$ be an $p \times p$ orthogonal matrix obtained by applying Gram-Schmidt orthogonalization process on $U_1, \ldots, U_p$, such that $U = \phi T$, where T is a $p \times p$ upper triangular matrix with positive diagonal elements. Obviously $\phi(OU) = O\phi(U)$, for any $p \times p$ orthogonal matrix O. Since U and OU are identically distributed, $\phi(U)$ and $O\phi(U)$ are also identically distributed. Let $Z = X/\|X\|$. Since X and OX have the same distribution, Z and $\phi'Z$ have the same distribution. Hence, for $t \in E^p$,

$$\begin{aligned} E(\exp(it'Z)) &= E(E(\exp\{it'\phi'(U)Z\}|U)) \\ &= E(E(\exp\{it'\phi'(U)Z\}|Z)) \\ &= E(E(\exp\{it'\phi'(U)O'Z\})|Z) \\ &= E(\exp\{it'\phi_1(U)\}), \end{aligned}$$

where O is a $p \times p$ orthogonal matrix such that $O'Z = (1, 0, \ldots, 0)'$ and ϕ_1 is the first row of ϕ. Since the characteristic function determines uniquely the distribution function we conclude that $X/\|X\|$ and $\phi_1(U)$ are identically distributed whatever may be the distribution of X in $E_p(0, I)$ provided that $P(X = 0) = 0$. Now, since $N_p(0, I)$ is also a member of $E_p(0, I)$ we conclude that $X/\|X\|$ and $W/\|W\|$ have the same distribution.

(b) Let $V_i = W_i^2, i = 1, \ldots, p, V = (V_1, \ldots, V_p)'; L = \sum_{i=1}^p V_i$. Then

$$f_V(v) = \frac{1}{(2\pi)^{p/2}} \exp\left\{-\frac{1}{2}\sum_{i=1}^p v_i\right\} \left\{\prod_{i=1}^p (v_i)^{1/2-1}\right\}. \tag{6.95}$$

From (6.95) it follows that the joint probability density function of $e_1, \ldots, e_p$ is given by (6.94).

Q.E.D.

Theorem 6.12.2. *Let $X = (X_1, \ldots, X_p)'$ be distributed as $E_p(0, I)$. The probability density function of $L = X'X$ is given by*

$$f_L(l) = \frac{\pi^{p/2}}{\Gamma(p/2)} l^{p/2} q(l), \qquad l \geq 0, \tag{6.96}$$

where the probability density function of X is $q(x'x)$.

Proof. Let $r = \|X\|$ and let

$$\begin{aligned} X_1 &= r \sin \mathcal{O}_1 \sin \mathcal{O}_2 \cdots \sin \mathcal{O}_{p-2} \sin \mathcal{O}_{p-1} \\ X_2 &= r \sin \mathcal{O}_1 \sin \mathcal{O}_2 \cdots \sin \mathcal{O}_{p-2} \cos \mathcal{O}_{p-1} \end{aligned}$$

$$\vdots$$

$$\begin{aligned} X_{p-1} &= r \sin \mathcal{O}_1 \cos \mathcal{O}_1 \\ X_p &= r \cos \mathcal{O}_1 \end{aligned}$$

with $r > 0, 0 < \mathcal{O}_i \leq \pi, i = 1, \ldots, p-2, 0 < \mathcal{O}_{p-1} \leq 2\pi$.

We first note that

$$X_1^2 + \cdots X_p^2 = r^2.$$

Basic Multivariate Sampling Distributions

The Jacobian of the transformation from $(X_1, \ldots, X_p)$ to $(r, \mathcal{O}_1, \ldots, \mathcal{O}_p)$ is

$$([(\sin \mathcal{O}_1)^{p-2}(\sin \mathcal{O}_2)^{p-3} \cdots (\sin \mathcal{O}_{p-2})]). \tag{6.98}$$

Hence the probability density function of $L, \mathcal{O}_1, \ldots, \mathcal{O}_{p-1}$ is

$$\left(\left[\frac{1}{2}(l)^{p/2-1} q(l)(\sin \mathcal{O}_2)^{p-2}(\sin \mathcal{O}_2)^{p-3} \cdots (\sin \mathcal{O}_{p-2})\right]\right). \tag{6.99}$$

Thus L is independent of $(\mathcal{O}_1, \ldots, \mathcal{O}_p)$ and the probability density function of $\mathcal{O}_i$ is proportional to $(\sin \mathcal{O}_i)^{p-1-i}, i = 1, \ldots, p-1$.

Since the integration of $[(\sin \mathcal{O}_1)^{p-2}(\sin \mathcal{O}_2)^{p-3} \cdots (\sin \mathcal{O}_{p-2})]$ with respect to $\mathcal{O}_1, \ldots, \mathcal{O}_{p-1}$ results $\frac{2\pi^{p/2}}{\Gamma(p/2)}$, the probability density function of L is given by (6.96). Q.E.D.

Example 6.12.1. Let $X = (X_1, \ldots, X_p)'$ be distributed as $E_p(\mu, \Sigma, q)$ and let $q(Z) = (2\pi)^{-1/2} \exp\{-\frac{1}{2}Z\}$. Then $L = (X-\mu)'\Sigma^{-1}(X-\mu)$ has the probability density function

$$f_L(l) = \frac{2^{-1/2p}}{\Gamma(1/2p)} l^{\frac{1}{2}p} \exp\{-\frac{1}{2}l\}$$

which is a central χ_p^2. In other words if X is distributed as $N_p(\mu, \Sigma)$ then L is distributed as χ_p^2.

Theorem 6.12.3. $X = (X_1, \ldots, X_p)'$ is distributed as $E_p(\mu, \Sigma, q)$ with Σ positive definite if and only if X is distributed as $\mu + R\Sigma^{1/2}U$, where $\Sigma^{1/2}$ is the symmetric matrix satisfying $\Sigma^{1/2}\Sigma^{1/2} = \Sigma$ and $L = R^2$ is distributed, independently of U, with pdf given by (6.96) and U is uniformly distributed on $\{U \in R^p \text{ with } U'U = 1\}$.

Proof. Let $\Sigma^{-1/2}(X-\mu) = Y$. Then Y is distributed as $E_p(0, I, q)$. From Theorems 4.12.1 and 4.12.2 we conclude that $Y = RU$ with $Y'Y = R^2 = L$ and U is a function of angular variables $\theta_1, \ldots, \theta_{p-1}$, the variables R and U are independent. Thus the distribution of Y is characterized by the distribution of R and U. For all Y, U is uniformly distributed on $\{U \in R^p \text{ with } U'U = 1\}$. Q.E.D.

Theorem 6.12.4. Let $X = (X_1, \ldots, X_p)' = (X'_{(1)}, X'_{(2)})'$ with $X_{(1)} = (X_1, \ldots, X_k)', k \leq p$ be distributed as $E_p(0, I, q)$ and let

$$R_1^2 = X'_{(1)}X_{(1)}, \qquad R_2^2 = X'_{(2)}X_{(2)}.$$

Then

(a) *the joint probability density function of* (R_1, R_2) *is given by (with $s = p - k$)*

$$f_{R_1 R_2}(r_1, r_2) = \frac{4\pi^{\frac{1}{2}p}}{\Gamma(\frac{1}{2}k)\Gamma(\frac{1}{2}s)} r_1^{k-1} r_2^{s-1} q(r_1^2 + r_2^2), \qquad (6.100)$$

(b) $U = X'_{(1)} X_{(1)} / X'X$ *is distributed as Beta* $(\frac{1}{2}k, \frac{1}{2}s)$ *independently of q.*

Proof.

(a) Transform

$$X_1 = R_1(\sin \theta_1) \cdots (\sin \theta_{k-2})(\sin \theta_{k-1}),$$
$$X_2 = R_1(\sin \theta_1) \cdots (\sin \theta_{k-2})(\cos \theta_{k-1}),$$
$$\vdots$$
$$X_{k-1} = R_1(\sin \theta_1)(\cos \theta_1)$$
$$X_k = R_1 \cos \theta_1$$

with $R_1 > 0, 0 < \theta_i < \pi, i = 1, \ldots, k-2, 0 < \theta_{k-1} < 2\pi$,

$$X_{k+1} = R_2(\sin \phi_1) \cdots (\sin \phi_{r-1})$$
$$X_{k+2} = R_2(\sin \phi_1) \cdots (\sin \phi_{r-2})(\cos \phi_{r-1})$$
$$\vdots$$
$$X_{p-1} = R_2(\sin \phi_1)(\cos \phi_1)$$
$$X_p = R_2 \cos \phi_1,$$

with $R_2 > 0, 0 < \phi_i < \pi, i = 1, \ldots, r-2, 0 < \phi_{r-1} < 2\pi$. Using calculations of Theorem 6.12.1 we get (6.100).

(b) Let $L_1 = R_1^2, L_2 = R_2^2$. From (6.100)

$$f_{L_1, L_2}(l_1, l_2) = \frac{\pi^{\frac{1}{2}p}}{\Gamma(\frac{1}{2}k)\Gamma(\frac{1}{2}s)} l_1^{\frac{1}{2}k-1} l_2^{\frac{1}{2}s-1} q(l_1, l_2). \qquad (6.101)$$

Transform $(L_1, L_2) \to (X = L_1 + L_2, L_1)$. The joint *pdf* of (X, U) is

$$f_{X, L_1}(x, l_1) = \frac{\pi^{\frac{1}{2}p}}{\Gamma(\frac{1}{2}k)\Gamma(\frac{1}{2}s)} l_1^{\frac{1}{2}k-1} (x - l_1)^{\frac{1}{2}s-1} q(x). \qquad (6.102)$$

Basic Multivariate Sampling Distributions

From (6.101) the joint *pdf* of (X, U) is

$$f_{X,U}(x, u) = \frac{\Gamma(\frac{1}{2}p)}{\Gamma(\frac{1}{2}k)\Gamma(\frac{1}{2}s)} u^{\frac{1}{2}k-1}(1-u)^{\frac{1}{2}s-1} x^{\frac{1}{2}p-1} q(x).$$

Hence the *pdf* of U is given by (using Theorem 6.12.1)

$$f_U(u) = \frac{\Gamma(\frac{1}{2}p)}{\Gamma(\frac{1}{2}k)\Gamma(\frac{1}{2}s)} u^{\frac{1}{2}k-1}(1-u)^{\frac{1}{2}s-1}.$$

Q.E.D.

Notation

Let $Y = (Y_1, \ldots, Y_p)'$ have *pdf* $E_p(0, I, q)$. We shall denote the *pdf* of $L = Y'Y$ by

$$g_p(l) = \frac{\pi^{\frac{1}{2}p}}{\Gamma(\frac{1}{2}p)} l^{\frac{1}{2}p-1} \exp\left\{-\frac{1}{2}l\right\} \tag{6.103}$$

which is χ_p^2.

Hence $g_k(l)$ will denote the *pdf* of $L = Y'Y$ where $Y = (Y_1, \ldots, Y_k)'$ and Y is distributed as $E_k(0, I, q)$.

Example 6.12.2. Let $Y = (Y_1, \ldots, Y_p)'$ be distributed as $E_p(0, I, q)$ and

$$q(z) = (2\pi)^{-\frac{1}{2}p} \exp\left\{-\frac{1}{2}z\right\}.$$

Then

$$g_p(l) = \frac{1}{2^{\frac{1}{2}p}\Gamma(\frac{1}{2}p)} l^{\frac{1}{2}p-1} \exp\left\{-\frac{1}{2}l\right\}$$

To prove the next two theorems we need the following lemma which can be proved using Theorem 6.4.1.

Lemma 6.12.1. *Let* $Y = (Y_1, \ldots, Y_p)'$. *If*

$$\sum_{i=1}^{p} Y_i^2 = Q_1 + \cdots + Q_k \tag{6.104}$$

where $Q_1, \ldots, Q_k$ *are nonnegative quadratic forms in* Y *of ranks* $p_1, \ldots, p_k$ *respectively, then a necessary and sufficient condition that there exists an orthogonal transformation* $Y = OZ$ *with* $Z = (Z_1, \ldots, Z_p)'$ *such that*

$$Q_1 = \sum_{i=1}^{p_1} Z_i^2, \quad Q_2 = \sum_{i=p_1+1}^{p_1+p_2} Z_i^2, \ldots, Q_k = \sum_{i=p-p_k+1}^{p} Z_i^2$$

is $p_1 + \cdots + p_k = p$.

Theorem 6.12.5. *Let* $X = (X_1, \ldots, X_p)'$ *be distributed as* $E_p(0, \Sigma, q)$ *and let the fourth moment of the components of X be finite. For any $p \times p$ matrix B of rank $k \leq p$, $X'BX$ is distributed as $g_k(\cdot)$ if and only if*

$$\text{rank}(B) + \text{rank}(\Sigma^{-1} - B) = p.$$

Proof. Suppose that $\text{rank}(B) + \text{rank}(\Sigma^{-1} - B) = p$. Since $\Sigma > 0$, by Theorem 1.5.5 there exists a nonsingular matrix C such that $\Sigma = CC'$. Let X be transformed to $Y = C^{-1}X$. Since

$$X'\Sigma^{-1}X = X'BX + X'(\Sigma^{-1} - B)X \tag{6.105}$$

we get

$$Y'Y = Y'C'BCY + Y'(I - C'BC)Y \tag{6.106}$$

and the *pdf* of Y is $f_Y(y) = q(y'y)$. Using Lemma 6.12.1 we obtain, from (6.106), that

$$Y'(C'BC)Y = \sum_{i=1}^{k} Z_i^2$$

where $Z_{(1)} = (Z_1, \ldots, Z_k)'$ is distributed as $E_k(0, I, q)$. Thus $Y'C'BCY = X'BX$ is distributed as $g_k(\cdot)$.

To prove the necessity part let us assume that $X'BX$ is distributed as $g_k(\cdot)$ which implies that $X'BX = \sum_{i=1}^{k} Z_i^2$ where $(Z_1, \ldots, Z_k)'$ is distributed as $E_k(0, I, q)$. Since C is nonsingular and $\text{rank}(B) = k$, there exists a $p \times p$ orthogonal matrix O such that $O'C'BCO$ is a diagonal matrix D with k nonzero diagonal elements $\lambda_1, \ldots, \lambda_k$ (say). Hence

$$Y'C'BCY = \sum_{i=1}^{k} \lambda_i Z_i^2 \tag{6.107}$$

Basic Multivariate Sampling Distributions

where $OY = Z = (Z_1, \ldots, Z_p)'$. Since $q(z'z)$ is an even function of z we conclude that

$$E(Z_i) = 0 \quad \text{for all} \quad i,$$
$$E(Z_i^2) = a \quad \text{for all} \quad i,$$
$$E(Z_i^2 Z_j^2) = b \quad \text{for all} \quad i \neq j,$$
$$E(Z_i^4) = c \quad \text{for all} \quad i,$$

where a, b, c are positive constants. Since $X'BX$ is distributed as $g_k(\cdot)$ we obtain

$$ka = \left(\sum_{i=1}^{k} \lambda_i\right) a,$$

$$\left(\sum_{i=1}^{k} \lambda_i^2\right) c + \sum_{i \neq j} \lambda_i \lambda_j b \qquad (6.108)$$

$$= kc + bk(k-1)$$
$$= k(c-b) + bk^2.$$

From (6.108) we conclude that

$$\sum_{i=1}^{k} \lambda_i = \sum_{i=1}^{k} \lambda_i^2 = k.$$

This implies that $\lambda_i = 1$ for all i. Hence the equation (6.105) can be written as

$$Z'Z = \sum_{i=1}^{k} Z_i^2 + \sum_{i=k+1}^{p} Z_i^2.$$

Applying Lemma 6.12.1 we get $\text{rank}(B) + \text{rank}(\Sigma^{-1} - B) = p$. Q.E.D.

Theorem 6.12.6. *Let $X = (X_1, \ldots, X_p)'$ be distributed as $E_P(0, \Sigma, q)$ and assume that forth moments of all components of X are finite. Then $X'BX$, with $\text{rank}(B) = k$, is distributed as $g_k(\cdot)$ if and only if $B\Sigma B = B$.*

Proof. Let $\Sigma = CC'$ where C is a $p \times p$ nonsingular matrix and let $Y = C^{-1}X$. Then Y is distributed as $E_p(0, I, q)$ and $X'BX = Y'DY$ with $D = C'BC$ and $\text{rank}(B) = \text{rank}(D)$. Since $B\Sigma B = B$ implies $C'BCC'BC = C'BC$ or $DD = D$, we need to prove that $Y'DY$ has the *pdf* $g_k(\cdot)$ if and only if D is idempotent of rank k.

If D is idempotent of rank k then there exists an $p \times p$ orthogonal matrix θ such that (see Theorem 1.5.8)

$$\theta D \theta' = \begin{pmatrix} I & 0 \\ 0 & 0 \end{pmatrix}$$

where I is the $k \times k$ identity matrix. Write

$$Z = \theta Y = (Z_1, \ldots, Z_p)'.$$

Then Z is distributed as $E_p(0, I, q)$ and

$$Y'DY = Z' \begin{pmatrix} I & 0 \\ 0 & 0 \end{pmatrix} Z = \sum_{i=1}^{k} Z_i^2.$$

Hence $X'BX$ is distributed as $g_k(\cdot)$.

To prove the necessity of the condition let us assume that $Y'DY$ has the *pdf* $g_k(\cdot)$. If the rank of D is m then there exists an $p \times p$ orthogonal matrix θ such that $\theta D \theta'$ is diagonal matrix with m nonzero diagonal elements $\lambda_1, \ldots, \lambda_m$. Assuming without any loss of generality that the first m diagonal elements of $\theta D \theta'$ are nonzero we get, with $Z = \theta Y$, $Y'DY = \sum_{i=1}^{m} \lambda_i Z_i^2$. Proceeding exactly in the same way as in Theorem 6.12.5 we conclude that $m = k$ and $\sum_{i=1}^{k} \lambda_i = \sum_{i=1}^{k} \lambda_i^2$. Hence $\lambda_i = 1, i = 1, \ldots, k$, which implies that D is idempotent of rank k.

Q.E.D.

EXERCISES

1. Show that if a quadratic form is distributed as a noncentral chi-square, the noncentrality parameter is the value of the quadratic form when the variables are replaced by their expected values.
2. Show that the sufficiency condition of Theorem 6.2.3 is also necessary for the independence of the quadratic form $X'AX$ and the linear form BX.
3. Let $X = (X_1, \ldots, X_p)'$ be normally distributed with mean μ and covariance matrix Σ. Show that two quadratic forms $X'AX$ and $X'BX$ are independent if $A\Sigma B = 0$.
4. Let S be distributed as $W_p(n, \Sigma)$. Show that for any nonnull p-vector $l = (l_1, \ldots, l_p)'$ (a) $l'Sl/l'\Sigma l$ is distributed as χ_n^2, (b) $l'\Sigma^{-1}l/l'S^{-1}l$ is distributed as χ_{n-p+1}^2.
5. Let $S = (S_{ij})$ be distributed as $W_p(n, \Sigma)$, $n \geq p$, $\Sigma = (\sigma_{(ij)})$. Show that
$$E(S_{ij}) = n\sigma_{ij}, \quad \text{var}(S_{ij}) = n(\sigma_{ij}^2 + \sigma_{ii}\sigma_{jj}),$$
$$\text{cov}(S_{ij}, S_{kl}) = n(\sigma_{ik}\sigma_{jl} + \sigma_{il}\sigma_{jk}).$$

Basic Multivariate Sampling Distributions

6 Let $S_0, S_1, \ldots, S_k$ be independently distributed Wishart random variables $W_p(n_i, I)$, $i = 1, \ldots, k$, and let $V_j = S_0^{-1/2} S_j S_0^{-1/2}$, $j = 1, \ldots, k$. Show that the joint probability density function of $V_1, \ldots, V_k$ is given by

$$C \prod_{i=1}^{k} (\det V_i)^{(n_i-p-1)/2} \left(\det \left(I + \sum_{1}^{k} V_i \right) \right)^{-\left(\sum_{1}^{k} n_i\right)/2}$$

where C is the normalizing constant.

7 Let $S_0, S_1, \ldots, S_k$ be independently distributed as $W_p(n_i, \Sigma)$, $i = 0, 1, \ldots, k$.
  (a) Let, for $j = 1, \ldots, k$,

$$W_j = \left(\sum_0^k S_j \right)^{-1/2} S_j \left(\sum_0^k S_j \right)^{-1/2}, \qquad V_j = S_0^{-1/2} S_J S_0^{-1/2}$$

$$Z_j = \left(I + \sum_1^k V_j \right)^{-1/2} V_j \left(I + \sum_1^k V_i \right)^{-1/2}.$$

Show that the joint probability density function of $W_1, \ldots, W_k$ is given by (with respect to the Lebesgue measure $\prod_{j=1}^{k} dw_j$)

$f_{W_1,\ldots,W_k}(w_1, \ldots, w_k)$

$$= C \prod_{j=1}^{k} (\det w_j)^{(n_i-p-1)/2} \left(\det \left(I - \sum_{j=1}^{k} w_j \right) \right)^{(n_0-p-1)/2}$$

where C is the normalizing constant. Also verify that the joint probability density function of $Z_1, \ldots, Z_k$ is the same as that of $W_1, \ldots, W_k$.
  (b) Let T_j be a lower triangular nonsingular matrix such that

$$S_1 + \cdots + S_j = T_j T_j', \qquad j = 1, \ldots, k-1,$$

and let

$$W_j = T_j^{-1} S_{j+1} T_j'^{-1}, \qquad j = 1, \ldots, k-1.$$

Show that $W_1, \ldots, W_{k-1}$ are independently distributed.
  (c) Let

$$Y_j = (S_1 + \cdots + S_{j+1})^{-1/2} S_{j+1} (S_1 + \cdots + S_{j+1})^{-1/2},$$

$$j = 1, \ldots, k-1.$$

Show that $Y_1, \ldots, Y_{k-1}$ are stochastically independent.

8 Let S be distributed as $W_p(n, \Sigma)$, $n \geq p$, let $T = (T_{ij})$ be a lower triangular matrix such that $S = TT'$, and let

$$X_{ii}^2 = T_{ii}^2 + \sum_{j=1}^{i-1} T_{ij}^2, \quad i = 1, \ldots, p,$$

$$X_{ii-1} = (T_{ii}^2 - T_{i1}^2 - \cdots - T_{ii-2}^2)^{1/2} \left(\frac{T_{ii-1}/(n-i+1)^{1/2}}{1 + T_{ii-1}^2/(n-i+1)} \right),$$

$$X_{ii-2} = (T_{ii}^2 - T_{i1}^2 - \cdots - T_{ii-3}^2)^{1/2} \left(\frac{T_{ii-2}/(n-i+2)^{1/2}}{1 + T_{ii-1}^2/(n-i+2)} \right),$$

$$\vdots$$

$$X_{i1} = T_{ii} \frac{T_{i1}/(n-1)^{1/2}}{1 + T_{i1}^2/(n-1)}.$$

Obtain the joint probability density function of $X_{ii}^2, i = 1, \ldots, p$, and all $T_{ij}, i \neq j, i < j$. Show that the X_{ii}^2 are distributed as central chi-squares whereas the T_{ij} have Student's t-distributions.

9 Let Y be a $k \times n$ matrix and let D be a $(p-k) \times n$ matrix, $n > p$. Show that

$$\int_{YY'=G, YD'=V} dY = 2^{-k} \left(\prod_{i=1}^{k} C(n-p+i) \right) (\det(DD'))^{-p/2}$$

$$\times (\det(G - V(DD')^{-1}V'))^{(n-p-1)/2},$$

where $C(n)$ is the surface area of a unit n-dimensional sphere.

10 Let S be distributed as $W_p(n, \Sigma)$, $n \geq p$, and let S and Σ be similarly partitioned into

$$S = \begin{pmatrix} S_{(11)} & S_{(12)} \\ S_{(21)} & S_{(22)} \end{pmatrix}, \quad \Sigma = \begin{pmatrix} \Sigma_{(11)} & \Sigma_{(12)} \\ \Sigma_{(21)} & \Sigma_{(22)} \end{pmatrix}$$

Show that if $\Sigma_{(12)} = 0$, then

$$\frac{\det S}{(\det S_{(11)})(\det S_{(22)})}$$

is distributed as a product of independent beta random variables.

11 Let $S_i, i = 1, 2, \ldots, k$, be independently distributed Wishart random variables $W_p(n_i, \Sigma)$, $n_i \geq p$. Show that
  (a) the characteristic roots of $\det(S_1 - \lambda(S_1 + S_2)) = 0$ are independent of $S_1 + S_2$;
  (b) $(\det S_1)/\det(S_1 + S_2)$ is distributed independently of $S_1 + S_2$;

Basic Multivariate Sampling Distributions 261

(c) $(\det S_1)/\det(S_1 + S_2)$, $\det(S_1 + S_2)/\det(S_1 + S_2 + S_3)$, ... are all independently distributed.

12 Let $\bar{X}, S$ be based on a random sample of size N from a p-variate normal population with mean μ and covariance matrix Σ, $N > p$, and let X be an additional random observation from this population. Find the distribution of
(a) $X - \bar{X}$
(b) $(N/(N+1))(X - \bar{X})'S^{-1}(X - \bar{X})$.

13 Show that given the joint probability density function of $R_1, \ldots, R_k$ as in (6.63), the marginal probability density function of $R_1, \ldots, R_j$, $1 \le j \le k$, can be obtained from (6.63) by replacing k by j. Also show that for $k = 2$, $\delta_1^2 = 0$, $\delta_2^2 = 0$, $(1 - R_1 - R_2)/(1 - R_1)$ is distributed as the beta random variable with parameter $(\frac{1}{2}(N - p_1 - p_2), \frac{1}{2}p_2)$.

14 (Square root of Wishart). Let S be distributed as $W_p(n, \Sigma), n \le p$, and let $S = CC'$ where C is a nonsingular matrix of dimension $p \times p$. Show that the probability density function of C with respect to the Lebesgue measure dc in the space of all nonsingular matrices c of dimension $p \times p$ is given by

$$K(\det \Sigma)^{-n/2} \exp\left\{-\frac{1}{2} tr \Sigma^{-1} cc'\right\}(\det(cc'))^{(n-p)/2}.$$

[Hint: Write $C = T\theta$ where T is the unique lower triangular matrix with positive diagonal elements such that $S = TT'$ and θ is a random orthogonal matrix distributed independently of T. The Jacobian of the transformation $C \to (T, \theta)$ is $\prod_{i=1}^{p}(t_{ii})^{p-i}h(\theta)$ where $h(\theta)$ is a function of θ only (see Roy, 1959).]

15 (a) Let G be the set of all $p \times r(r \ge p)$ real matrices g and let $\alpha = (R, 0, \ldots, 0)'$ be a real p-vector, $\beta = (\delta, 0, \ldots, 0)'$ a real r-vector. Show that for $k > 0$ (dg stands for the Lebesgue measure on G)

$$\int_G (\det(gg'))^k \exp\left\{-\frac{1}{2}tr(gg' - 2g\theta\alpha')\right\} dg$$

$$= \exp\left\{-\frac{1}{2}R^2\delta^2\right\}(2\pi)^{pr/2} E(\chi_r^2(R^2\delta^2))^k \prod_{i=1}^{p-1} E(\chi_{r-i}^2)^k.$$

(b) Let $x = (x_1, \ldots, x_p)'$, $y = (y_1, \ldots, y_r)'$, $\delta = (y'y)^{1.2}$, $R = (x'x)^{1/2}$. Show that for $k > 0$

$$\int_G (\det(gg'))^k \exp\left\{-\frac{1}{2}tr(gg' - 2gyx')\right\} dg$$

$$= \exp\left\{-\frac{1}{2}R^2\delta^2\right\}(2\pi)^{pr/2} E(\chi_r^2(R^2\delta^2))^k \prod_{i=1}^{p-1} E(\chi_{r-i}^2)^k.$$

16 Consider Exercise 12. Let Σ be a positive definite matrix of dimension $p \times p$. Show that for $k > 0$,

$$\int_G (\det(gg'))^k \exp\left\{-\frac{1}{2} tr\Sigma(gg' - 2gyx')\right\} dg$$

$$= (\det \Sigma)^{-(2k-p)/2} \exp\left\{-\frac{1}{2}(x'\Sigma x)(y'y)\right\}$$

$$\times \int_G (\det(gg'))^k \exp\left\{-\frac{1}{2} tr(g - zy')(g - zy')'\right\} dg$$

where $z = Cx$ and C is a nonsingular matrix such that $\Sigma = CC'$.

Let B be the unique lower triangular matrix with positive diagonal elements such that $S = BB'$, where S is distributed independently of $\sqrt{N}\bar{X}$ (normal with mean $\sqrt{N}\mu$ and covariance Σ), as $W_p(N-1, \Sigma)$, and let $V = B^{-1}\bar{X}$. Show that the probability density function of V is given by

$$f_V(v) = 2^p C \int_{G_T} \exp\left\{-\frac{1}{2} tr(gg' + N(gv - \rho)(gv - \rho)')\right\}$$

$$\times \prod_{i=1}^p |g_{ii}|^{N-i} \prod_{i \geq j} dg_{ij},$$

where

$$C = N^{p/2} \left[2^{Np/2} \pi^{p(p+1)/4} \prod_{i=1}^p \Gamma((N-i)/2) \right]^{-1}$$

and $g = (g_{ij}) \in G_T$ where G_T is the group of $p \times p$ nonsingular triangular matrices with positive diagonal elements. Use the distribution of V to find the probability density function of $R_1, \ldots, R_p$ as defined in (6.63) with $k = p$.

17 Let ξ^α, $\alpha = 1, \ldots, N(N > p)$, be a random sample of size N from a p-variate complex normal distribution with mean α and complex positive definite Hermitian matrix Σ, and let

$$S = \sum_{\alpha=1}^N (\xi^\alpha - \bar{\xi})(\xi^\alpha - \bar{\xi})^*, \qquad \bar{\xi} = \frac{1}{N}\sum_1^N \xi^\alpha.$$

(a) Show that the probability density function of S is given by

$$f_S(s) = k(\det \Sigma)^{-(N-1)}(\det s)^{N-p-1} \exp\{-tr\Sigma^{-1}s\}$$

where $K^{-1} = \pi^{p(p-1)/2} \prod_{i=1}^p \Gamma(N-i)$.

Basic Multivariate Sampling Distributions

(b) Let $S = (S_{(ij)})$, $\Sigma = (\Sigma_{ij})$ be similarly partitioned into submatrices

$$S = \begin{pmatrix} S_{11} & S_{(12)} \\ S_{(21)} & S_{(22)} \end{pmatrix}, \quad \Sigma = \begin{pmatrix} \Sigma_{11} & \Sigma_{(12)} \\ \Sigma_{(12)} & \Sigma_{(22)} \end{pmatrix},$$

where S_{11} and Σ_{11} are 1×1. Show that the probability density function of

$$R_c^2 = \frac{S_{(12)} S_{(22)}^{-1} S_{(21)}}{S_{11}}$$

is given by

$$f_{R_c^2}(r_c^2) = \frac{\Gamma(N-1)}{\Gamma(p-1)\Gamma(N-p)}(1-\rho_c^2)^{N-1}(r_c^2)^{p-2}(1-r_c^2)^{N-p-1}$$

$$\times F(N-1, N-1; p-1; r_c^2 \rho_c^2),$$

where

$$\rho_c^2 = \frac{\Sigma_{(12)} \Sigma_{(22)}^{-1} \Sigma_{(21)}}{\Sigma_{11}}$$

$$F(a, b; c; x) = 1 + \frac{ab}{c}x + \frac{a(a+1)b(b+1)}{c(c+1)}\frac{x^2}{2!} + \cdots.$$

(c) Let $T = (T_{ij})$ be a complex upper triangular matrix with positive real diagonal elements T_{ii} such that $T^*T = S$. Show that the probability density function of T is given by

$$K(\det \Sigma)^{N-1} \prod_{j=1}^{p}(T_{jj})^{2n-(2j-1)} \exp\{-tr\Sigma^{-1}T^*T\}.$$

(d) Define $R_1, \ldots, R_k$ in terms of S (complex) and $\bar{\xi}$, $\delta_1, \ldots, \delta_k$ in terms of α, and Σ in the same way as in (6.63) and (6.64) for the real case. Show that the joint probability density function of $R_1, \ldots, R_k$ is given by ($k \leq p$)

$$f_{R_1,\ldots,R_k}(r_1, \ldots, r_k)$$

$$= \Gamma(N)\left[\Gamma(N-p)\prod_{i=1}^{k}\Gamma(p_i)\right]^{-1}\left(1 - \sum_{1}^{k} r_i\right)^{N-p-1}\prod_{i=1}^{k} r_i^{p_i-1}$$

$$\times \exp\left\{-\sum_{1}^{k}\delta_j^2 + \sum_{1}^{k} r_j \sum_{l>j}^{k}\delta_i^2\right\}\prod_{i=1}^{k}\phi(N - \sigma_{i-1}, p_i; r_i\delta_i^2)$$

REFERENCES

Anderson, T. W. (1945). The noncentral Wishart distribution and its applications to problems of Multivariate Analysis. Ph.D. thesis, Princeton Univ. Princeton, New Jersey.

Anderson, T. W. (1946). The noncentral Wishart distribution and certain problems of multivariate analysis. *Ann. Math. Statist.* 17:409–431.

Bartlett, M. S. (1933). On the theory of statistical regression. *Proc. R. Soc. Edinburgh* 33:260–283.

Cochran, W. G. (1934). The distribution of quadratic form in a normal system with applications to the analysis of covariance. *Proc. Cambridge Phil. Soc.* 30:178–191.

Constantine, A. G. (1963). Some noncentral distribution problems in multivariate analysis. *Ann. Math. Statist.* 34:1270–1285.

Eaton, M. L. (1972). Multivariate Statistical Analysis. Inst. of Math. Statist., Univ. of Copenhagen, Denmark.

Elfving, G. (1947). A simple method of deducting certain distributions connected with multivariate sampling. *Skandinavisk Aktuarietidskrift* 30:56–74.

Fisher, R. A. (1915). Frequency distribution of the values of the correlation coefficient in samples from an indefinitely large population. *Biometrika* 10:507–521.

Fisher, R. A. (1928). The general sampling distribution of the multiple correlation coefficient. *Proc. R. Soc. A* 121:654–673.

Giri, N. (1965). On the complex analogues of T^2- and R^2-tests. *Ann. Math. Statist.* 36:644–670.

Giri, N. (1971). On the distribution of a multivariate statistic. *Sankhya* A33:207–210.

Giri, N. (1972). On testing problems concerning the mean of a multivariate complex Gaussian distribution. *Ann. Inst. Statist. Math.* 24:245–250.

Giri, N. (1973). An integral—its evaluation and applications. *Sankhya* A35:334–340.

Giri, N. (1974). *Introduction to Probability and Statistics*, Part I, Probability. New York: Dekker.

Giri, N. (1993). *Introduction to Probability and Statistics*, 2nd ed. New York: Marcel Dekker.

Giri, N., Kiefer, J. (1964). Minimax character of R^2 test in the simplest case. *Ann. Math. Statist.* 35:1475–1490.

Giri, N., Kiefer, J., Stein, C. (1963). Minimax character of Hotelling's T^2-test in the simplest case. *Ann. Math. Statist.* 34:1524–1535.

Goodman (1963). Statistical analysis based on a certain multivariate complex Gaussian distribution (An Introduction). *Ann. Math. Statist.* 34:152–177.

Graybill, F. A. (1961). *Introduction to Linear Statistical Model.* New York: McGraw-Hill.

Hogg, R. V., Craig, A. T. (1958). On the decomposition of certain χ^2 variables. *Ann. Math. Statist.* 29:608–610.

Ingham, A. E. (1933). An integral that occurs in statistics. *Proc. Cambridge Phil. Soc.* 29:271–276.

James, A. T. (1955). The noncentral Wishart distribution. *Proc. R. Soc. London A*, 229:364–366.

James, A. T. (1954). Normal multivariate analysis and the orthogonal group. *Ann. Math. Statist.* 25:40–75.

James, A. T. (1964). The distribution of matrix variates and latent roots derived from normal samples. *Ann. Math. Statist.* 35:475–501.

Karlin, S., Traux, D. (1960). Slippage problems. *Ann. Math. Ststist.* 31:296–324.

Kabe, D. G. (1964). Decomposition of Wishart distribution. *Biometrika* 51:267.

Kabe, D. G. (1965). On the noncentral distribution of Rao's U-Statistic. *Ann. Math. Statist.* 17:15.

Khatri, C. G. (1959). On the conditions of the forms of the type $X'AX$ to be distributed independently or to obey Wishart distribution. *Calcutta Statist. Assoc. Bull.* 8:162–168.

Khatri, C. G. (1963). Wishart distribution. *J. Indian Statist. Assoc.* 1:30.

Khirsagar, A. M. (1959). Bartlett decomposition and Wishart distribution. *Ann. Math. Statist.* 30:239–241.

Khirsagar, A. M. (1972). *Multivariate Analysis.* New York: Dekker.

Mahalanobis, P. C., Bose, R. C., Roy, S. N. (1937). Normalization of variates and the use of rectangular coordinates in the theory of sampling distributions. *Sankhya* 3:1–40.

Mauldon, J. G. (1956). Pivotal quantities for Wishart's and related distributions, and a paradox in fiducial theory. *J. Roy. Stat. Soc. B*, 17:79–85.

MacDuffee, C. (1946). *The Theory of Matrices*. New York: Chelsea.

MacLane, S., Birkoff, G. (1967). *Algebra*. New York, Macmillan.

Markus, M., Mine, H. (1967). *Introduction to Linear Algebra*. New York: Macmillan.

Nachbin, L. (1965). *The Haar Integral*. Princeton, NJ: Van Nostrand-Reinhold.

Narain, R. D. (1948). A new approach to sampling distributions of the multivariate normal theory. I. *J. Ind. Soc. Agri. Stat*. 3:175–177.

Ogawa, J. (1949). On the independence of linear and quadratic forms of a random sample from a normal population. *Ann. Inst. Math. Statist*. 1:83–108.

Ogawa, J. (1953). On sampling distributions of classical statistics in multivariate analysis. *Osaka Math. J*. 5:13–52.

Olkin, I., Roy, S. N. (1954). On multivariate distribution theory. *Ann. Math. Statist*. 25:325–339.

Olkin, I., Rubin, H. (1964). Multivariate beta distributions and independence properties of Wishart distribution. *Ann. Math. Statist*. 35:261–269.

Perlis, S. (1952). *Theory of Matrices*. Reading, Massachusetts: Addison Wesley.

Rao, C. R. (1965). *Linear Statistical Inference and its Applications*. New York: Wiley.

Rasch, G. (1948). A functional equation for Wishart distribution. *Ann. Math. Statist*. 19:262–266.

Roy, S. N. (1957). *Some Aspects of Multivariate Analysis*. New York: Wiley.

Roy, S. N., Ganadesikan, R. (1959). Some contributions to ANOVA in one or more dimensions II. *Ann. Math. Statist*. 30:318–340.

Stein, C. (1969). Multivariate Analysis I (Notes recorded by M. L. Eaton). Dept. of Statist., Stanford Univ., California.

Sverdrup, E. (1947). Derivation of the Wishart distribution of the second order sample moments by straightforward integration of a multiple integral. *Skand. Akturaietidskr*. 30:151–166.

Wijsman, R. A. (1957). Random orthogonal transformations and their use in some classical distribution problems in multivariate analysis. *Ann. Math. Statist*. 28:415–423.

Wilks, S. S. (1932). Certain generalizations in the analysis of variance. *Biometrika* 24:471–494.

Wishart, J. (1928). The generalized product moment distribution from a normal multivariate distribution. *Biometrika* 20A:32–52.

Wishart, J. (1948). Proof of the distribution law of the second order moment statistics. *Biometrika* 35:55–57.

Wishart, J., Bartlett, M. S. (1932). The distribution of second order moment statistics in a normal system. *Proc. Cambridge Phil. Soc.* 28:455–459.

Wishart, J., Bartlett, M. S. (1933). The generalized product moment distribution in a normal system. *Proc. Cambridge Phil. Soc.* 29:260–270.

7
Tests of Hypotheses of Mean Vectors

7.0. INTRODUCTION

This chapter deals with testing problems concerning mean vectors of multivariate distributions. Using the same developments of the appropriate test criteria we will also construct the confidence region for a mean vector. It will not be difficult for the reader to construct the confidence regions for the other cases discussed in this chapter. The matrix Σ is rarely known in most practical problems and tests of hypotheses concerning the mean vectors must be based on an appropriate estimate of Σ. However, in cases of long experience with the same experimental variables, we can sometimes assume Σ to be known. In deriving suitable test criteria for different testing problems we shall use mainly the well-known likelihood ratio principle and the approach of invariance as outlined in Chapter 3. The heuristic approach of Roy's union-intersection principle of test construction also leads to suitable test criteria. We shall include it as an exercise. For further material on this the reader is referred to Giri (1965), books on multivariate analysis by Anderson (1984), Eaton (1988), Farrell (1985), Kariya (1985), Kariya and Sinha (1989), Muirhead (1982), Rao (1973), and Roy (1957). Nandi (1965) has shown that the test statistic obtained from Roy's union-intersection principal is consistent if the component tests (univariate) are so, unbiased under certain conditions, and admissible if again the component tests are admissible. We first deal with testing problems concerning means of multivariate normal populations, then we treat the case of multivariate complex normal and that of elliptically symmetric distributions. In Section 7.3.1 we treat the problem of mean vector against one-sided alternatives for the multivariate normal populations.

7.1. TESTS: KNOWN COVARIANCES

Let $X^\alpha = (X_{\alpha 1}, \ldots, X_{\alpha p})'$, $\alpha = 1, \ldots, N$, be a random sample of size N from a p-variate normal population with mean μ and positive definite covariance matrix Σ. We will consider the problem of testing the hypothesis H_0; $\mu = \mu_0$ (specified) and a related problem of finding the confidence region for μ under the assumption that Σ is known. In the univariate case ($p = 1$) we use the fact that the difference between the sample mean and the population mean is normally distributed with mean 0 and known variance and use the existing table of standard normal distributions to determine the significance points or the confidence interval. In the multivariate case such a difference has a p-variate normal distribution with mean 0 and known covariance matrix, and hence we can set up the confidence interval or prescribe the test for each component as in the univariate case.

Such a solution has several drawbacks. First, the choice of confidence limits is somewhat arbitrary. Second, for testing purposes it may lead to a test whose performance may be poor against some alternatives. Finally, and probably most important for $p > 2$ detailed tables for multivariate normal distributions are not available. The procedure suggested below can be computed easily and can be given a general intuitive and theoretical justification. Let $\bar{X} = (1/N)\Sigma_{\alpha=1}^{N} X^\alpha$. By Theorem 6.2.2, under H_0, $N(\bar{X} - \mu_0)'\Sigma^{-1}(\bar{X} - \mu_0)$ has central chi-square distribution with p degrees of freedom and hence the test which rejects $H_0 : \mu = \mu_0$ whenever

$$N(\bar{x} - \mu_0)'\Sigma^{-1}(\bar{x} - \mu_0) \geq \chi^2_{p,\alpha}, \qquad (7.1)$$

where $\chi^2_{p,\alpha}$ is a constant such that $P(\chi^2_p \geq \chi^2_{p,\alpha}) = \alpha$, has the power function which increases monotonically with the noncentrality parameter $N(\mu - \mu_0)'\Sigma^{-1}(\mu - \mu_0)$. Thus the power function of the test given in (7.1) has the minimum value α (level of significance) when $\mu = \mu_0$ and its power is greater than α when $\mu \neq \mu_0$. For a given sample mean $\bar{x}$, consider the inequality

$$N(\bar{x} - \mu)'\Sigma^{-1}(\bar{x} - \mu) \leq \chi^2_{p,\alpha}. \qquad (7.2)$$

The probability is $1 - \alpha$ that the mean of a sample of size N from a p-variate normal distribution with mean μ and known positive definite covariance matrix Σ satisfies (7.2). Thus the set of values of μ satisfying (7.2) gives the confidence region for μ with confidence coefficient $1 - \alpha$, and represents the interior and the

Tests of Hypotheses of Mean Vectors

surface of an ellipsoid with center $\bar{x}$, with shape depending on Σ and size depending on Σ and $\chi^2_{p,\alpha}$.

For the case of two p-dimensional normal populations with mean vectors μ, v but with the same known positive definite covariance matrix Σ we now consider the problem of testing the hypothesis $H_0 : \mu - v = 0$ and the problem of setting a confidence region for $\mu - v$ with confidence coefficient $1 - \alpha$. Let $X^\alpha = (X_{\alpha 1}, \ldots, X_{\alpha p})'$, $\alpha = 1, \ldots, N_1$, be a random sample of size N_1 from the normal distribution with mean μ and covariance matrix Σ, and let $Y^\alpha = (Y_{\alpha 1}, \ldots, Y_{\alpha p})'$, $\alpha = 1, \ldots, N_2$, be a random sample of size N_2 (independent of X^α, $\alpha = 1, \ldots, N_1$) from the other normal distribution with mean v and the same covariance matrix Σ. If

$$\bar{X} = \frac{1}{N_1} \sum_{\alpha=1}^{N_1} X^\alpha, \qquad \bar{Y} = \frac{1}{N_2} \sum_{\alpha=1}^{N_2} Y^\alpha,$$

then by Theorem 6.2.2, under H_0,

$$\frac{N_1 N_2}{N_1 + N_2} (\bar{X} - \bar{Y})' \Sigma^{-1} (\bar{X} - \bar{Y})$$

is distributed as chi-square with p degrees of freedom. Given sample observations x^α, $\alpha = 1, \ldots, N_1$, and y^α, $\alpha = 1, \ldots, N_2$, the test rejects H_0 whenever

$$\frac{N_1 N_2}{N_1 + N_2} (\bar{x} - \bar{y})' \Sigma^{-1} (\bar{x} - \bar{y}) \geq \chi^2_{p,\alpha}, \tag{7.3}$$

has a power function which increases monotonically with the noncentrality parameter

$$\frac{N_1 N_2}{N_1 + N_2} (\mu - v)' \Sigma^{-1} (\mu - v); \tag{7.4}$$

its power is greater than α (the level of significance) whenever $\mu \neq v$, and the power function attains its minimum value α whenever $\mu = v$. Given x^α, $\alpha = 1, \ldots, N_1$, and y^α, $\alpha = 1, \ldots, N_2$, the confidence region of $\mu - v$ with confidence coefficient $1 - \alpha$ is given by the set of values of $\mu - v$ satisfying

$$\frac{N_1 N_2}{N_1 + N_2} (\bar{x} - \bar{y} - (\mu - v))' \Sigma^{-1} (\bar{x} - \bar{y} - (\mu - v)) \leq \chi^2_{p,\alpha}, \tag{7.5}$$

which is an ellipsoid with center $\bar{x} - \bar{y}$ and whose shape depends on Σ. In this context it is worth noting that the quantity

$$(\mu - v)' \Sigma^{-1} (\mu - v) \tag{7.6}$$

is called the *Mahalanobis distance* between two p-variate normal populations with the same positive definite covariance matrix Σ but with different mean vectors. Consider now k p-variate normal populations with the same known covariance matrix Σ but with different mean vectors $\mu_i, i = 1,\ldots,k$. Let $\bar{X}_i$ be the mean of a random sample of size N_i from the ith population and let $\bar{x}_i$ be its sample value. An appropriate test for the hypothesis

$$H_0 : \sum_{i=1}^{k} \beta_i \mu_i = \mu_0, \tag{7.7}$$

where the β_i are known constants and μ_0 is a known p-vector, rejects H_0 whenever

$$C\left(\sum_{1}^{k} \beta_i \bar{x}_i - \mu_0\right)' \Sigma^{-1} \left(\sum_{1}^{k} \beta_i \bar{x}_i - \mu_0\right) \leq \chi^2_{p,\alpha}, \tag{7.8}$$

where the constant C is given by

$$C^{-1} = \sum_{i=1}^{k} \frac{\beta_i^2}{N_i} \tag{7.9}$$

Obviously

$$C\left(\sum_{i=1}^{k} \beta_i \bar{X}_i - \mu_0\right)' \Sigma^{-1} \left(\sum_{i=1}^{k} \beta_i \bar{X}_i - \mu_0\right)$$

is distributed as noncentral chi-square with p degrees of freedom and with noncentrality parameter $C(\mu - \mu_0)'\Sigma^{-1}(\mu - \mu_0)$ where $\Sigma_1^k \beta_i \mu_i = \mu$. Given, $\bar{x}_i, i = 1,\ldots,k$, the $(1 - \alpha)$ 100% confidence region for μ is given by the ellipsoid

$$C\left(\sum_{i=1}^{k} \beta_i \bar{x}_i - \mu\right)' \Sigma^{-1} \left(\sum_{i=1}^{k} \beta_i \bar{x}_i - \mu\right) \leq \chi^2_{p,\alpha} \tag{7.10}$$

with center $\Sigma_{i=1}^{k} \beta_i \bar{x}_i$.

7.2. TESTS: UNKNOWN COVARIANCES

In most practical problems concerning mean vectors the covariance matrices are rarely known and statistical testing of hypotheses about mean vectors has to be carried out assuming that the covariance matrices are unknown. We shall first consider testing problems concerning the mean μ of a p-variate normal population with unknown covariance matrix Σ. Testing problems concerning mean vectors of more than one multivariate normal population with unknown covariance matrices will be treated as applications of these problems.

7.2.1. Hotelling's T^2-Test

Let $x^\alpha = (x_{\alpha 1}, \ldots, x_{\alpha p})'$, $\alpha = 1, \ldots, N$, be a sample of size $N(N > p)$ from a p-variate normal distribution with unknown mean μ and unknown positive definite covariance matrix Σ. On the basis of these observations we are interested in testing the hypothesis $H_0 : \mu = \mu_0$ against the alternatives $H_1 : \mu \neq \mu_0$ where Σ is unknown and μ_0 is specified. In the univariate case ($p = 1$) this is a basic problem in statistics with applications in every branch of applied science, and the well-known Student t-test is its optimum solution. For the general multivariate case we shall show that a multivariate analog of Student's t is an optimum solution. This problem is commonly known as Hotelling's problem since Hotelling (1931) first proposed the extension of Student's t-statistic for the two-sample multivariate problem and derived its distribution under the null hypothesis. We shall now derive the likelihood ratio test of this problem. The likelihood of the observations x^α, $\alpha = 1, \ldots, N$ is given by

$$L(x^1, \ldots, x^N | \mu, \Sigma)$$
$$= (2\pi)^{-Np/2} (\det \Sigma)^{-N/2} \exp\left\{-\frac{1}{2} \operatorname{tr} \Sigma^{-1} \left(\sum_{\alpha=1}^{N} (x^\alpha - \mu)(x^\alpha - \mu)'\right)\right\} \qquad (7.11)$$

Given x^α, $\alpha = 1, \ldots, N$, the likelihood L is a function of μ, Σ, for simplicity written as $L(\mu, \Sigma)$. Let Ω be the parametric space of (μ, Σ) and let ω be the subspace of Ω when $H_0; \mu = \mu_0$ is true. Under ω the likelihood function reduces to

$$(2\pi)^{-Np/2} (\det \Sigma)^{-N/2} \exp\left\{-\frac{1}{2} \operatorname{tr} \Sigma^{-1} \left(\sum_{\alpha=1}^{N} (x^\alpha - \mu_0)(x^\alpha - \mu_0)'\right)\right\}. \qquad (7.12)$$

By Lemma 5.1.1, we obtain from (7.12)

$$\max_\omega L(\mu, \Sigma)) = (2\pi)^{Np/2} \left[\det\left(\frac{1}{N} \sum_{\alpha=1}^{N} (x^\alpha - \mu_0)(x^\alpha - \mu_0)'\right)\right]^{-N/2}$$
$$\times \exp\left\{-\frac{1}{2} Np\right\}. \qquad (7.13)$$

We observed in Chapter 5 that under Ω, $L(\mu, \Sigma)$ is maximum when

$$\mu = \frac{1}{N} \sum_{\alpha=1}^{N} x^\alpha = \bar{x}, \qquad \Sigma = \frac{1}{N} \sum_{\alpha=1}^{N} (x^\alpha - \bar{x})(x^\alpha - \bar{x})' = \frac{S}{N}.$$

Hence

$$\max_{\Omega} L(\mu, \Sigma)) = (2\pi)^{Np/2} \left[\det\left(\frac{1}{N}\sum_{\alpha=1}^{N}(x^{\alpha} - \bar{x})(x^{\alpha} - \bar{x})'\right)\right]^{-N/2}$$
$$\times \exp\left\{-\frac{1}{2}Np\right\} \quad (7.14)$$

From (7.13) and (7.14) the likelihood ratio test criterion for testing $H_0 : \mu = \mu_0$ is given by

$$\lambda = \frac{\max_{\omega} L(\mu, \Sigma)}{\max_{\Omega} L(\mu, \Sigma)} = \left[\frac{\det s}{\det(\sum_{\alpha=1}^{N}(x^{\alpha} - \mu_0)(x^{\alpha} - \mu_0)')}\right]^{N/2}$$

$$= \left[\frac{\det s}{\det(s + N(\bar{x} - \mu_0)(\bar{x} - \mu_0)')}\right]^{N/2} \quad (7.15)$$

$$= (1 + N(\bar{x} - \mu_0)' s^{-1}(\bar{x} - \mu_0))^{-N/2}$$

The right-hand side of (7.15) follows from Exercise 1.12. Since λ is a monotonically decreasing function of $N(\bar{x} - \mu_0)' s^{-1}(\bar{x} - \mu_0)$, the likelihood ratio test of $H_0; \mu = \mu_0$ when Σ is unknown rejects H_0 whenever

$$N(N - 1)(\bar{x} - \mu_0)' s^{-1}(\bar{x} - \mu_0) \geq c, \quad (7.16)$$

where c is a constant depending on the level of significance α of the test.

Note In connection with tests we shall use c as the generic notation for the significance point of the test. From (6.60) the distribution of

$$T^2 = N(N-1)(\bar{X} - \mu_0)' S^{-1}(\bar{X} - \mu_0)$$

is given by

$$f_{T^2}(t^2|\delta^2) = \frac{\exp\{-\frac{1}{2}\delta^2\}}{(N-1)\Gamma(\frac{1}{2}(N-p))}$$
$$\times \sum_{j=0}^{\infty} \frac{(\frac{1}{2}\delta^2)^j (t^2/(N-1))^{p/2+j-1} \Gamma(\frac{1}{2}N + j)}{j!\Gamma(\frac{1}{2}p + j)(1 + t^2/(N-1))^{N/2+j}}, \quad t^2 \geq 0 \quad (7.17)$$

where $\delta^2 = N(\mu - \mu_0)' \Sigma^{-1}(\mu - \mu_0)$. This is often called the distribution of T^2 with $N - 1$ degrees of freedom. Under H_0, $\mu = \mu_0$, $\delta^2 = 0$, and $(T^2/(N-1)) \times ((N-p)/p)$ is distributed as central F with parameter $(p, N - p)$. Thus for any

Tests of Hypotheses of Mean Vectors

given level of significance $\alpha, 0 < \alpha < 1$, the constant c of (7.16) is given by

$$c = \frac{(N-1)p}{N-p} F_{p,N-p,\alpha}, \qquad (7.18)$$

where $F_{p,N-p,\alpha}$ is the $(1-\alpha)$ 100% point of the F-distribution with degrees of freedom $(p, N-p)$. Tang (1938) has tabulated the type II error (1-power) of this test for various values of $\delta^2, p, N-p$ and for $\alpha = 0.05$ and 0.01. Lehmer (1944) has computed values of δ^2 for given values of α and type II error. This table is useful for finding the value δ^2 (or equivalently, the value of N for given μ and Σ) needed to make the probability of accepting H_0 very small whenever H_0 is false. Hsu (1938) and Bose and Roy (1938) have also derived the distribution of T^2 by different methods. Another equivalent test procedure for testing H_0 rejects H_0 whenever

$$r_1 = \frac{N\bar{x}'s^{-1}\bar{x}}{1 + N\bar{x}'s^{-1}\bar{x}} \geq c. \qquad (7.19)$$

From (6.66) the probability density function of R_1 (random variable corresponding to r_1) is

$$f_{R_1}(r_1) = \frac{\Gamma(\frac{1}{2}N)}{\Gamma(\frac{1}{2}p)\Gamma(\frac{1}{2}(N-p))} r_1^{p/2-1}(1-r_1)^{(N-p)/2-1}$$
$$\times \exp\left\{-\frac{1}{2}\delta^2\right\} \phi\left(\frac{1}{2}(N-p), \frac{1}{2}p; \frac{1}{2}r_1\delta\right), 0 < r_1 < 1 \qquad (7.20)$$

Thus under H_0, R_1 has a central beta distribution with parameter $(\frac{1}{2}p, \frac{1}{2}(N-p))$. The significance points for the test based on R_1 are given by Tang (1938). From (7.17) and (7.20) it is obvious that the power of Hotelling's T^2-test or its equivalent depends only on the quantity δ^2 and increases monotonically with δ^2.

7.2.2. Optimum Invariant Properties of the T^2-Test

To examine various optimum properties of the T^2-test, we need to verify that the statistic T^2 is the maximal invariant in the sample space under the group of transformations acting on the sample space which leaves the present testing problem invariant. In effect we will prove a more general result since it will be useful for other testing problems concerning mean vectors considered in this chapter. It, is also convenient to take $\mu_0 = 0$, which we can assume without any loss of generality.

Let

$$\bar{X} = \frac{1}{N}\sum_1^N X^\alpha, \qquad S = \sum_{\alpha=1}^N (X^\alpha - \bar{X})(X^\alpha - \bar{X})'$$

be partitioned as

$$\bar{X} = (\bar{X}_{(1)}, \ldots, \bar{X}_{(k)})', \qquad S = \begin{pmatrix} S_{(11)} & \cdots & S_{(1k)} \\ \vdots & & \vdots \\ S_{(k1)} & \cdots & S_{(kk)} \end{pmatrix} \qquad (7.21)$$

where the $\bar{X}_{(i)}$ are subvectors of $\bar{X}$ of dimension $p_i \times 1$ and the $S_{(ij)}$ are submatrices of S of dimension $p_i \times p_j$ such that $\Sigma_1^k p_i = p$. Let

$$\bar{X}_{[i]} = (\bar{X}_{(1)}, \ldots, \bar{X}_{(i)})', \qquad S_{[ii]} = \begin{pmatrix} S_{(11)} & \cdots & S_{(1i)} \\ \vdots & & \vdots \\ S_{(i1)} & \cdots & S_{(ii)} \end{pmatrix} \qquad (7.22)$$

We shall denote the space of values of $\bar{X}$ by $\mathcal{X}_1$ and the space of values of S by $\mathcal{X}_2$ and write $\mathcal{X} = \mathcal{X}_1 \times \mathcal{X}_2$, the product space of $\mathcal{X}_1, \mathcal{X}_2$. Let G_{BT} be the multiplicative group of nonsingular lower triangular matrices g

$$g = \begin{pmatrix} g_{(11)} & 0 & 0 & \cdots & 0 \\ g_{(21)} & g_{(22)} & 0 & \cdots & 0 \\ \vdots & \vdots & & & \vdots \\ g_{(k1)} & g_{(k2)} & & \cdots & g_{(kk)} \end{pmatrix} \qquad (7.23)$$

of dimension $p \times p$ where the $g_{(ij)}$ are submatrices of g of dimension $p_i \times p_j, j = 1, \ldots, k$; and let G_{BT} operate on $\mathcal{X}$ as

$$(\bar{X}, S) \to (g\bar{X}, gSg'), \qquad g \in G_{BT}.$$

Define $R_1^*, \ldots, R_k^*$ by

$$\sum_{j=1}^i R_j^* = N\bar{X}_{[i]}' S_{[ii]}^{-1} \bar{X}_{[i]}, \qquad i = 1, \ldots, k. \qquad (7.24)$$

Since $N > p$ by assumption, S is positive definite with probability 1 and hence $R_i^* > 0$ for all i with probability 1. It may be observed that if $p_1 = p$, $p_i = 0, i = 2, \ldots, k$, then $R_1^* = T^2/(N-1)$.

Tests of Hypotheses of Mean Vectors

Lemma 7.2.1. *The statistic* $(R_1^*, \ldots, R_k^*)$ *is a maximal invariant under* G_{BT}, *operating as*

$$(\bar{X}, S) \to (g\bar{X}, gSg') \qquad g \in G_{BT}.$$

Proof. We shall prove the lemma for the case $k = 2$, the general case following obviously from this. First let us observe the following.

(a) If $(\bar{X}, S) \to (g\bar{X}, gSg'), g \in G_{BT}$, then

$$(\bar{X}_{(1)}, S_{(11)}) \to (g_{(11)}\bar{X}_{(1)}, g_{(11)}S_{(11)}g'_{(11)}).$$

Thus $(R_1^*, R_1^* + R_2^*)$ is invariant under G_{BT}.

(b) Since

$$N\bar{X}'S^{-1}\bar{X} = N\bar{X}'_{(1)}S_{(11)}^{-1}\bar{X}_{(1)}$$
$$+ N(\bar{X}_{(2)} - S_{(21)}S_{(11)}^{-1}\bar{X}_{(1)})'(S_{(22)} - S_{(21)}S_{(11)}^{-1}S_{(12)})^{-1}$$
$$\times (\bar{X}_{(2)} - S_{(21)}S_{(11)}^{-1}\bar{X}_{(1)}), \qquad (7.25)$$

$$R_2^* = N(\bar{X}_{(2)} - S_{(21)}S_{(11)}^{-1}\bar{X}_{(1)})'(S_{(22)} - S_{(21)}S_{(11)}^{-1}S_{(12)})^{-1}$$
$$\times (\bar{X}_{(2)} - S_{(21)}S_{(11)}^{-1}\bar{X}_{(1)}).$$

(c) For any two p-vectors $X, Y \in E^p, X'X = Y'Y$ if and only if there exists an orthogonal matrix O of dimension $p \times p$ such that $X = OY$.

Let $\bar{X}, \bar{Y} \in \mathcal{X}_1$ and $S, T \in \mathcal{X}_2$ be similarly partitioned and let

$$N\bar{X}'_{(1)}S_{(11)}^{-1}\bar{X}_{(1)} = N\bar{Y}'_{(1)}T_{(11)}^{-1}\bar{Y}_{(1)} \qquad (7.26)$$

$$N\bar{X}'S^{-1}\bar{X} = N\bar{Y}'T^{-1}\bar{Y}. \qquad (7.27)$$

To show that (R_1^*, R_2^*) is a maximal invariant under G_{BT} we must show that there exists a $g_1 \in G_{BT}$ such that

$$\bar{X} = g_1\bar{Y}, \qquad S = g_1Tg'_1.$$

Choose

$$g = \begin{pmatrix} g_{(11)} & 0 \\ g_{(21)} & g_{(22)} \end{pmatrix}$$

with $\quad g_{(11)} = S_{(11)}^{-1/2}, g_{(22)} = (S_{(22)} - S_{(21)}S_{(11)}^{-1}S_{(12)})^{-1/2}, \quad$ and $\quad g_{(21)} = -g_{(22)}S_{(21)}S_{(11)}^{-1}.$

Then
$$gSg' = I \tag{7.28}$$
Similarly, choose $h \in G_{BT}$ such that
$$hTh' = I. \tag{7.29}$$
Since (7.26) implies
$$(g_{(11)}\bar{X}_{(1)})'(g_{(11)}\bar{X}_{(1)}) = (h_{(11)}\bar{Y}_{(1)})'(h_{(11)}\bar{Y}_{(1')}) \tag{7.30}$$
from (c) we conclude that there exists an orthogonal matrix θ_1 of dimension $p_1 \times p_1$ such that
$$g_{(11)}\bar{X}_{(1)} = \theta_1 h_{(11)}\bar{Y}_{(1)}. \tag{7.31}$$
From (7.27), (7.28), and (7.29) we get
$$(g\bar{X})'(g\bar{X}) = \|g_{(11)}\bar{X}_{(1)}\|^2 + \|g_{(21)}\bar{X}_{(1)} + g_{(22)}\bar{X}_{(2)}\|^2$$
$$= (h\bar{Y})'(h\bar{Y}) = \|h_{(11)}\bar{Y}_{(1)}\|^2 + \|h_{(21)}\bar{Y}_{(1)} + h_{(22)}\bar{Y}_{(2)}\|^2$$
where $\|\ \|$ denotes the norm, and hence from (7.30) we obtain
$$\|g_{(21)}\bar{X}_{(1)} + g_{(22)}\bar{X}_{(2)}\|^2 = \|h_{(21)}\bar{Y}_{(1)} + h_{(22)}\bar{Y}_{(2)}\|^2 \tag{7.32}$$
From this we conclude that there exists an orthogonal matrix θ_2 of dimension $p_2 \times p_2$ such that
$$g_{(21)}\bar{X}_{(1)} + g_{(22)}\bar{X}_{(2)} = \theta_2(h_{(21)}\bar{Y}_{(1)} + h_{(22)}\bar{Y}_{(2)}). \tag{7.33}$$
Letting
$$\theta = \begin{pmatrix} \theta_1 & 0 \\ 0 & \theta_2 \end{pmatrix},$$
we get from (7.31) and (7.33)
$$\bar{X} = g^{-1}\theta h\bar{Y}' = g_1\bar{Y},$$
where $g_1 = g^{-1}\theta h \in G_{BT}$, and from
$$gSg' = I = hTh' = \theta hTh'\theta'$$
we get $S = g_1 T g_1'$. Hence $(R_1^*, R_1^* + R_2^*)$ or, equivalently, (R_1^*, R_2^*) is a maximal invariant under G_{BT} on $\mathcal{X}$. The proof for the general case is established by showing that $(R_1^*, R_1^* + R_2^*, \ldots, R_1^* + \cdots + R_k^*)$ is a maximal invariant under G_{BT}. The orthogonal matrix θ needed is a diagonal matrix in the block form.

Q.E.D.

Tests of Hypotheses of Mean Vectors

It may be remarked that the statistic $(R_1, \ldots, R_k)$ defined in Chapter 6 is a one to one transformation of $(R_1^*, \ldots, R_k^*)$, and hence $(R_1, \ldots, R_k)$ is also a maximal invariant under G_{BT}. The induced transformation G_{BT}^T on the parametric space Ω corresponding to G_{BT} on $\mathcal{X}$ is identically equal to G_{BT} and is defined by

$$(\mu, \Sigma) \to (g\mu, g\Sigma g'), \qquad g \in G_{BT} = G_{BT}^T.$$

Thus a corresponding maximal invariant in Ω under G_{BT} is $(\delta_1^2, \ldots, \delta_k^2)$, where

$$\sum_{j=1}^{i} \delta_j^2 = N\mu'_{[i]} \Sigma_{[ii]}^{-1} \mu_{[i]}, \qquad i = 1, \ldots, k. \tag{7.34}$$

The problem of testing the hypothesis $H_0 : \mu = 0$ against the alternatives $H_1 : \mu \neq 0$ on the basis of observations x^α, $\alpha = 1, \ldots, N (N > p)$, remains invariant under the group G of linear transformations g (set of all $p \times p$ nonsingular matrices) which transform each x^α to gx^α. These transformations induce on the space of the sufficient statistic $(\bar{X}, S)$ the transformations

$$(\bar{x}, s) \to (g\bar{x}, gsg').$$

Obviously $G = G_{BT}$ if $k = 1$ and $p_1 = p$. A maximal invariant in the space of $(\bar{X}, S)$ is $(N-1)R_1^* = T^2 = N(N-1)\bar{X}'S^{-1}\bar{X}$. The corresponding maximal invariant in the parametric space Ω under G is $\delta_1^2 = N\mu'\Sigma^{-1}\mu = \delta^2$ (say). Its probability density function is given in (7.17). The following two theorems give the optimum character of the T^2-test among the class of all invariant level α tests for $H_0 : \mu = 0$. To state them we need the following definition of a statistical test.

Definition 7.2.1. *Statistical test.* A statistical test is a function of the random sample X^α, $\alpha = 1, \ldots, N$, which takes values between 0 and 1 inclusive such that $E(\phi(X^1, \ldots, X^N)) = \alpha$, the level of the test when H_0 is true.

In this terminology $\phi(x^1, \ldots, x^N)$ is the probability of rejecting H_0 when $x^1, \ldots, x^N$ are observed.

Theorem 7.2.1. *Let x^α, $\alpha = 1, \ldots, N$, be a sequence of N observations from the p-variate normal distribution with mean μ and unknown positive definite covariance matrix Σ. Among all (statistical) tests $\phi(X^1, \ldots, X^N)$ of level α for testing $H_0 : \mu = 0$ against the alternatives $H_1 : \mu \neq 0$ which are invariant with respect to the group of transformations G transforming $x^\alpha \to gx^\alpha$, $\alpha = 1, \ldots, N$, $g \in G$ Hotelling's T^2-test or its equivalent (7.19) is uniformly most powerful.*

Proof. Let $\phi(X^1, \ldots, X^N)$ be a statistical test which is invariant with respect to G. Since $(\bar{X}, S)$ is sufficient for (μ, Σ), $E(\phi(X^1, \ldots, X^N)|\bar{X} = \bar{x}, S = s)$ is independent of (μ, Σ) and depends only on $(\bar{x}, s)$. Since ϕ is invariant, i.e., $\phi(X^1, \ldots, X^N) = \phi(gX^1, \ldots, gX^N)$, $g \in G$, $E(\phi|\bar{X} = \bar{x}, S = s)$, is invariant under G. Since $E(E(\phi|\bar{X}, S)) = E(\phi)$, $E(\phi|\bar{X}, S)$ and ϕ have the same power function. Thus each test in the larger class of level α tests which are functions of X^α, $\alpha = 1, \ldots, N$, can be replaced by one in the smaller class of tests which are function of $(\bar{X}, S)$ having identical power functions. By Lemma 7.2.1 and Theorem 3.2.2 the invariant test $E(\phi|\bar{X}, S)$ depends on $(\bar{X}, S)$ only through the maximal invariant T^2. Since the distribution of T^2 depends only on $\delta^2 = N\mu'\Sigma^{-1}\mu$, the most powerful level α invariant test of $H_0: \delta^2 = 0$ against the simple alternatives $\delta^2 = \delta_0^2$, where δ_0^2 is specified, rejects H_0 (by the Neyman-Pearson fundamental lemma) whenever

$$\frac{f_{T^2}(t^2|\delta_0^2)}{f_{T^2}(t^2|0)} = \frac{\Gamma(\tfrac{1}{2}p)\exp(-\tfrac{1}{2}\delta_0^2)}{\Gamma(\tfrac{1}{2}N)} \sum_{j=0}^{\infty} \frac{(\tfrac{1}{2}\delta_0^2)^j \Gamma(\tfrac{1}{2}N + j)}{j!\,\Gamma(\tfrac{1}{2}p + j)}$$

$$\times \left(\frac{t^2/(N-1)}{1 + t^2/(N-1)}\right)^j \geq c,$$

(7.35)

where the constant c is chosen such that the test has level α. Since the left-hand side of this inequality is a monotonically increasing function of $t^2/(N - 1 + t^2)$ and hence of t^2, the most powerful level α test of H_0 against the simple alternative $\delta^2 = \delta_0^2 (\delta_0^2 \neq 0)$ rejects H_0 whenever $t^2 \geq c$, where the constant c depends on the level α of the test. Obviously this conclusion holds good for any nonzero value of δ^2 instead of δ_0^2. Hence Hotelling's T^2-test which rejects H_0 whenever $t^2 \geq c$ is uniformly most powerful invariant for testing $H_0: \mu = 0$ against the alternatives $\mu \neq 0$. Q.E.D.

The power function of any invariant test depends only on the maximal invariant in the parametric space. However, in general, the class of tests whose power function depends on the maximal invariant δ^2 contains the class of invariant tests as a subclass. The following theorem proves a stronger optimum property of T^2-test than the one proved in Theorem 7.2.1. Theorem 7.2.2. is due to Semika (1941), although the proof presented here differs from the original proof.

Theorem 7.2.2. *On the basis of the observations x^α, $\alpha = 1, \ldots, N$, from the p-variate normal distribution with mean μ and positive definite covariance matrix Σ, among all tests of $H_0: \mu = 0$ against the alternatives $H_1: \mu \neq 0$ with power functions depending only on $\delta^2 = N\mu'\Sigma^{-1}\mu$, the T^2-test is uniformly most powerful.*

Tests of Hypotheses of Mean Vectors

Proof. In Theorem 7.2.1 we observed that each test in the larger class of tests that are functions of X^α, $\alpha = 1, \ldots, N$, can be replaced by one in the smaller class of tests that are functions of the sufficient statistic $(\bar{X}, S)$, having the identical power function. Let $\phi(\bar{X}, S)$ be a test with power function depending on δ^2. Since δ^2 is a maximal invariant in the parametric space of (μ, Σ) under the transformation $(\mu, \Sigma) \to (g\mu, g\Sigma g')$, $g \in G$, we get

$$E_{\mu,\Sigma} \phi(\bar{X}, S) = E_{g^{-1}\mu, g^{-1}\Sigma g^{-1'}} \phi(\bar{X}, S) = E_{\mu,\Sigma}(g\bar{X}, gSg'). \tag{7.36}$$

Since the distribution of $(\bar{X}, S)$ is boundedly complete (see Chapter 5) and

$$E_{\mu,\Sigma}(\phi(\bar{X}, S) - \phi(g\bar{X}, gSg')) = 0 \tag{7.37}$$

identically in μ, Σ we conclude that

$$\phi(\bar{X}, S) - \phi(g\bar{X}, gSg') = 0$$

almost everywhere (may depend on particular g) in the space of $(\bar{X}, S)$. In other words, ϕ is almost invariant with respect to G (see Definition 3.2.6). As explained in Chapter 3 if the group G is such that there exists a right invariant measure on G, then almost invariance implies invariance. Such a right invariant measure on G is given in Example 3.2.6. Hence if the power of the test $\phi(\bar{X}, S)$ depends only on $\delta^2 = N\mu'\Sigma^{-1}\mu$, then $\phi(\bar{X}, S)$ is almost invariant under G, which for our problem implies that $\phi(\bar{X}, S)$ is invariant with respect to G transforming $(\bar{X}, S) \to (g\bar{X}, gSg')$, $g \in G$. Since by Theorem 7.2.1 the T^2-test is uniformly most powerful among the class of tests which are invariant with respect to G, we conlude the proof of the theorem. Q.E.D.

7.2.3. Admissibility and Minimax Property of T^2

We shall now consider the optimum properties of the T^2-test among the class of all level α tests. In almost all standard hypothesis testing problems in multivariate analysis—in particular, in normal ones—no meaningful nonasymptotic (in the sample size N) optimum properties are known either for the classical tests or for any other tests. The property of being best invariant under a grouo of transformations that leave the problem invariant, which is often possessed by some of these tests, is often unsatisfactory because the Hunt-Stein Theorem (see Chapter 3) is not valid. In particular, for the case of the T^2-test the property of being uniformly most powerful invariant under the full linear group G causes the same difficulty since G does not satisfy the conditions of the Hunt-Stein theorem. The following demonstration is due to Stein as reported by Lehmann (1959, p. 338).

Let $X = (X_1, \ldots, X_p)'$, $Y = (Y_1, \ldots, Y_p)'$ be independently distributed normal p-vectors with the same mean 0 and with positive definite covariance matrices

Σ, $\delta\Sigma$, respectively, where δ is an unknown scalar constant. The problem of testing the hypothesis $H_0 : \delta = 1$ against the alternatives $H_1 : \delta > 1$, remains invariant under the full linear group G transforming $X \to gX$, $Y \to gY$, $g \in G$. Since the full linear group G is transitive (see Chapter 2) over the space of values of (X, Y) with probability 1, the uniformly most powerful level α invariant test under G is the trivial test $\phi(x, y) = \alpha$ which rejects H_0 with constant probability α for all values (x, y) of (X, Y). Thus the maximum power that can be achieved over the alternatives H_1 by any invariant test under G is also α. On the other hand, consider the test which rejects H_0 whenever $x_1^2/y_1^2 \geq c$ for any observed x, y (c depending on α). This test has strictly increasing power function $\beta(\delta)$ whose minimum over the set $\delta \geq \delta_1 > 1$ is $\beta(\delta_1) > \beta(1) = \alpha$.

The admissibility of various classical tests in the univariate and multivariate situations is established by using (1) the Bayes procedure, (2) exponential structure of the parametric space, (3) invariance, and (4) local properties. For a comprehensive presentation of this the reader is referred to Kiefer and Schwartz (1965). The admissibility of the T^2-test was first proved by Stein (1956) using the exponential structure of the parametric space and by showing that no other test of the same level is superior to T^2 when $\delta^2 = N\mu'\Sigma^{-1}\mu$ is large (very far from H_0), and later by Kiefer and Schwartz (1965) using the Bayes procedure. It is the latter method of proof that we reproduce here. A special feature of this proof is that it yields additional information on the behavior of the T^2-test closer to H_0. The technique is to select suitable priors (probability measures or positive constant multiples thereof) Π_1 and Π_0 (say) for the parameters (μ, Σ) under H_1 and for Σ under H_0 so that the T^2-test can be identified as the unique Bayes test which, by standard theory, is then admissible. The T^2-test can be written as

$$X'(YY' + XX')^{-1}X \geq c$$

where $X = \sqrt{N}\bar{X}$, $S = \sum_{\alpha=1}^{N-1} Y^\alpha Y^{\alpha'}$, $Y = (Y^1, \ldots, Y^{N-1})$, Y^αs are independently and identically distributed normal p-vectors with mean 0 and covariance matrix Σ. It may be recalled that if $\theta = (\mu, \Sigma)$ and the Lebesgue density function of $V = (X, Y)$ on a Euclidean set is denoted by $f_V(v|\theta)$, then every Bayes rejection region for the $0 - 1$ loss function is of the form

$$\left\{ v : \left[\int f_V(v|\theta)\Pi_1(d\theta) \middle/ \int f_V(v|\theta)\Pi_o(d\theta) \right] \geq c \right\} \qquad (7.38)$$

for some $c(0 \leq c \leq \infty)$. Since in our case the subset of this set corresponding to equal to c has probability 0 for all θ in the parametric space, our Bayes procedure will be essentially unique and hence admissible.

Let both Π_1 and Π_0 assign all their measure to the θ for which $\Sigma^{-1} = I + \eta\eta'$ for some random p-vector η under both H_0 and H_1, and for which $\mu = 0$ under H_0 and $\mu = \Sigma\eta$ with probability 1 under H_1. Regarding the distribution of η on the

Tests of Hypotheses of Mean Vectors

p-dimension Euclidean space E^p we assume that

$$\text{under } H_1 : \frac{d\Pi_1(\eta)}{d\eta} \propto [\det(I + \eta\eta')]^{-N/2} \exp\left\{\frac{1}{2}\eta'(I + \eta\eta')^{-1}\eta\right\}$$

$$\text{under } H_0 : \frac{(d\Pi_0(\eta)}{d\eta} \propto [\det(I + \eta\eta')]^{-N/2}.$$
(7.39)

That these priors represent bona fide probability measures follows from the fact that if $\eta'(I + \eta\eta')^{-1}\eta$ is bounded by unity and $\det(I + \eta\eta') = 1 + \eta'\eta$ so that

$$\int_{E^p} (1 + \eta'\eta)^{-N/2} d\eta < \infty \qquad (7.40)$$

if and only if $N > p$ (which is our assumption). Since in our case

$$f_V(v|\theta) = f_X(x|\mu, \Sigma) f_Y(y|\Sigma) \qquad (7.41)$$

where

$$f_X(x|\mu, \Sigma) = (2\pi)^{-p/2} (\det \Sigma)^{-1/2} \exp\left\{-\frac{1}{2}\operatorname{tr} \Sigma^{-1}(x - \mu)(x - \mu)'\right\},$$

$$f_Y(y|\Sigma) = (2\pi)^{-(N-1)p/2} (\det \Sigma)^{-(N-1)/2} \exp\left\{-\frac{1}{2}\operatorname{tr} \Sigma^{-1} yy'\right\},$$

it follows from (7.39) that

$$\frac{\int f_V(v|\theta)\Pi_1(d\theta)}{\int f_V(v|\theta)\Pi_0(d\theta)}$$

$$= \int (\det(I + \eta\eta'))^{N/2}$$

$$\times \frac{\exp\{-\frac{1}{2}\operatorname{tr}(I + \eta\eta')(xx' + yy') + \eta x' - \frac{1}{2}(I + \eta\eta')^{-1}\eta\eta'\}\Pi_1(d\eta)}{\int (\det(I + \eta\eta'))^{N/2} \exp\{-\frac{1}{2}\operatorname{tr}(I + \eta\eta')(xx' + yy')\}\Pi_0(d\eta)}$$

$$= \exp\left\{\frac{1}{2}\operatorname{tr}(xx' + yy')^{-1} xx'\right\}$$

$$\times \frac{\int \exp\{-\frac{1}{2}\operatorname{tr}(xx' + yy')^{-1}(\eta - (xx' + yy')^{-1}x)(\eta - (xx' + yy')^{-1}x)'\} d\eta}{\int \exp\{-\frac{1}{2}\operatorname{tr}(xx' + yy')\eta\eta'\} d\eta}$$

$$= \exp\left\{-\frac{1}{2}\operatorname{tr}(xx' + yy')^{-1} xx'\right\}. \qquad (7.42)$$

But $X'(XX' + YY')^{-1} X = c$ has probability 0 for all θ. Hence we conclude the following.

Theorem 7.2.3. *For each $c \geq 0$ the rejection region $X'(XX' + YY')^{-1}X \geq c$ or, equivalently, the T^2-test is admissible for testing $H_0 : \mu = 0$ against $H_1 : \mu \neq 0$.*

We shall now examine the minimax property of the T^2-test for testing $H_0 : \mu = 0$ against the alternatives $N\mu'\Sigma^{-1}\mu > 0$. As shown earlier the full linear group G does not satisfy the conditions of the Hunt-Stein theorem. But the subgroup $G_T(G_{BT}$ with $k = p)$, the multiplicative group of $p \times p$ nonsingular lower triangular matrices which leaves the present problem invariant operating as

$$(\bar{X}, S; \mu, \Sigma) \to (g\bar{X}, gSg'; g\mu, g\Sigma g'), \qquad g \in G_T,$$

satisfies the conditions of the Hunt-Stein theorem (see Kiefer, 1957; or Lehmann, 1959, p. 345). We observed in Chapter 3 that on G_T there exists a right invariant measure. Thus there is a test of level α which is almost invariant under G_T, and hence in the present problem there is such a test which is invariant under G_T and which maximizes among all level α tests the minimum power over H_1. Whereas T^2 was a maximal invariant under G with a single distribution under each of H_0 and H_1 for each δ^2, the maximal invariant under G_T is the p-dimensional statistic $(R_1^*, \ldots, R_p^*)$ as defined in Section 7.2.1 with $k = p, p_1 = \cdots = p_k = 1$, or its equivalent statistic $(R_1, \ldots, R_p)$ as defined in Chapter 6 with $k = p$. The distribution of $R = (R_1, \ldots, R_p)$ has been worked out in Chapter 6. As we have observed there, under $H_0(\delta_1^2 = \cdots = \delta_p^2 = 0)$, R has a single distribution, but under H_1 with δ^2 fixed, it depends continuously on a $(p-1)$-dimensional vector $\Delta = \{(\delta_1^2, \ldots, \delta_p^2) : \delta_i^2 \geq 0, \Sigma_1^p \delta_i^2 = \delta^2\}$ for each fixed δ^2. Thus for $N > p > 1$ there is no uniformly most powerful invariant test under G_T for testing H_0 against $H_1 : N\mu'\Sigma^{-1}\mu > 0$. Let $f_R(r|\Delta), f_R(r|0)$ denote the probability density function of R under H_1 (for fixed δ^2) and H_0, respectively. Because of the compactness of the reduced parametric spaces $\{0\}$ under H_0 and

$$\Gamma = \{(\delta_1^2, \ldots, \delta_p^2) : \delta_i^2 \geq 0, \Sigma_1^p \delta_i^2 = \delta^2\}$$

under H_1 and the continuity of $f_R(r|\Delta)$ in Δ, it follows that (see Wald, 1950) every minimax test for the reduced problem in terms of R is Bayes. In particular, Hotelling's test which rejects H_0 whenever $\Sigma_1^p r_i \geq c$, which has constant power on each contour $N\mu'\Sigma^{-1}\mu = \delta^2$ (fixed) and which is also G_T invariant, maximizes the minimum power over H_1 for each fixed δ^2 if and only if there is a probability density measure λ on Γ such that for some constant K

$$\int_\Gamma \frac{f_R(r|\Delta)}{f_R(r|0)} \lambda(d\Delta) \begin{Bmatrix} > \\ = \\ < \end{Bmatrix} K \qquad (7.43)$$

Tests of Hypotheses of Mean Vectors

according as

$$\sum_{1}^{p} r_i \begin{Bmatrix} > \\ = \\ < \end{Bmatrix} c$$

except possibly for a set of measure 0. Obviously c depends on the level of significance α and the measure λ and the constant K may depend only on c and the specific value of δ^2. From (6.64) with $k = p$, we get

$$\frac{f_R(r|\Delta)}{f_R(r|0)} = \exp\left\{-\frac{1}{2}\delta^2 + \sum_{j=1}^{p} r_j \sum_{i>j} \delta_i^2/2\right\}$$

$$\times \Pi_{i=1}^{p} \phi\left(\frac{1}{2}(N-i+1), \frac{1}{2}; \frac{1}{2} r_i \delta_i^2\right)$$

An examination of the integrand in this expression allows us to replace (7.43) by its equivalent

$$\int_{\Gamma} \frac{f_R(r|\Delta)}{f_R(r|0)} \lambda(d\Delta) = K \quad \text{if} \quad \sum_{i=1}^{p} r_i = c. \quad (7.44)$$

Clearly (7.43) implies (7.44). On the other hand, if there are a λ and a K for which (7.44) is satisfied and if $r^* = (r_1^*, \ldots, r_p^*)'$ is such that $\Sigma_{i=1}^{p} r_i^* = c' > c$, writing $f(r) = f_R(r|\Delta)/f_R(r|0)$ and $r^{**} = cr^*/c'$, we see at once that $f(r^*) = f(c'r^{**}/c) > f(r^{**}) = K$, because of the form of f and the fact that $c'/c > 1$ and $\Sigma_{i=1}^{p} r_i^{**} = c$. This and a similar argument for the case $c' < c$ show that (7.44) implies (7.43). [Of course we do not assert that the left-hand side of (7.44) still depends only on $\Sigma_{i=1}^{p} r_i$ if $\Sigma_{i=1}^{p} r_i \neq c$.]

The computations in the next section are somewhat simplified by the fact that for fixed c and δ^2 we can at this point compute the unique value of K for which (7.44) can possibly be satisfied. Let $\hat{R} = (R_1, \ldots, R_{p-1})'$ and write $f_{\hat{R}}(\hat{r}|\Delta, u)$ for the version of the conditional Lebesgue density of $\hat{R}$ given that $\Sigma_{i=1}^{p} R_i = u$ which is continuous in $\hat{r}$ and u for $r_i > 0$, $\Sigma_{i=1}^{p-1} r_i < u < 1$, and is zero elsewhere. Write $f_U(u|\delta^2)$ for the probability density function of $U = \Sigma_{i=1}^{p} R_i$ which depends on Δ only through δ^2, and is continuous for $0 < u < 1$ and vanishes elsewhere. Then (7.44) can be written as

$$\int f_{\hat{R}}(\hat{r}|\Delta, c)\lambda(d\Delta) = \left[K \frac{f_U(c|0)}{f_U(c|\delta^2)}\right] f_{\hat{R}}(\hat{r}|0, c) \quad (7.45)$$

for $r_i > 0$, $\Sigma_{i=1}^{p-1} r_i < c$. The integral of (7.45), being a probability mixture of probability densities, is itself a probability density in $\hat{r}$, as is $f_{\hat{R}}(\hat{r}|0, c)$. Hence the expression in brackets equals 1. It is well known that, for $0 < c < 1$ (see

Theorem 6.8.1),

$$f_U(c|\delta^2) = \frac{\Gamma(\frac{1}{2}N)\exp\{-\frac{1}{2}\delta^2\}}{\Gamma(\frac{1}{2}p)\Gamma(\frac{1}{2}(N-p))} c^{(p-2)/2}(1-c)^{(N-p-2)/2} \phi\left(\frac{1}{2}N,\frac{1}{2}p;\frac{1}{2}c\delta^2\right).$$

Hence (7.44) becomes

$$\int_\Gamma \exp\left\{\sum_{j=1}^p r_j \sum_{i>j} \frac{1}{2}\delta_i^2\right\} \prod_{i=1}^p \phi\left(\frac{1}{2}(N-i+1), \frac{1}{2}; \frac{1}{2}r_i\delta_i^2\right) \lambda(d\Delta)$$

$$= \phi\left(\frac{1}{2}N, \frac{1}{2}p; \frac{1}{2}c\delta^2\right) \quad \text{if} \quad \sum_{i=1}^p r_i = c.$$

(7.46)

For $p = 2, N = 3$, writing

$$\lambda = c\delta^2, \ \beta_i = \delta_i^2/\delta^2, \ t_i = \lambda r_i/c,$$

$$\Gamma_1 = \left\{(\beta_1, \ldots, \beta_p) : \beta_i \geq 0, \sum_{i=1}^p \beta_i = 1\right\}$$

λ^* for the measure associated with λ on $\Gamma\left[\lambda^*(A) = \lambda(\delta^2 A)\right]$ and noting that $\phi(\frac{3}{2}, \frac{1}{2}; \frac{1}{2}x) = (1+x)\exp\{\frac{1}{2}x\}$, we obtain from (7.46)

$$\int_0^1 [1 + (\gamma - t_2)(1 - \beta_2)]\phi\left(1, \frac{1}{2}; \frac{1}{2}\beta_2 t_2\right) d\lambda^*(\beta_2)$$

$$= \exp\left\{\frac{1}{2}(t_2 - \gamma)\right\} \phi\left(\frac{3}{2}, \frac{1}{2}; \frac{1}{2}\gamma\right).$$

(7.47)

Writing

$$B = \exp\left\{-\frac{1}{2}\gamma\right\} \phi\left(\frac{3}{2}, 1; \frac{1}{2}\gamma\right), \qquad \mu_i = \int_0^1 \beta^i d\lambda^*(\beta),$$

$0 \leq i < \infty$, for the ith moment of λ^* we obtain from (7.47)

$$1 + \lambda - \lambda\mu_1 = B,$$

$$-(2r-1)\mu_{r-1} + (2r+\gamma)\mu_r - \gamma\mu_{r+1} = B\left[\frac{\Gamma(r+\frac{1}{2})}{r!\Gamma(\frac{1}{2})}\right], \qquad r \geq 1.$$

(7.48)

Tests of Hypotheses of Mean Vectors

Giri, et al. (1963) after lengthy calculations showed that there exists an absolutely continuous probability measure λ^* whose derivative $m_\gamma(x)$ is given by

$$m_\gamma(x) = \frac{\exp\{-\frac{1}{2}\gamma x\}}{2\pi x^{1/2}(1-x)^{1/2}} \left\{ \int_0^\infty \exp\left\{-\frac{1}{2}\gamma u\right\} \left[\frac{B}{1+u} - \frac{u^{1/2}}{(1+u)^{3/2}}\right] du \right. \quad (7.49)$$

$$\left. + B \int_0^x \frac{\exp(\frac{1}{2}\gamma u)}{1-u} du \right\},$$

proving that for $p = 2, N = 3$, the T^2-test is minimax for testing H_0 against H_1. Later Salaevskii (1968), using this reduction of the problem, after voluminous computations was able to show that there exists a probability measure λ for general p and N, establishing that the T^2-test is minimax in general.

Giri and Kiefer (1962) developed the theory of local (near the null hypothesis) and asymptotic (far in distance from the null hypothesis) minimax tests for general multivariate problems. This theory serves two purposes. First, there is the obvious point of demonstrating such properties for their own sake, though well-known and valid doubts have been raised as to the extent of meaningfulness of such properties. Second, local and asymptotic minimax properties can give an indication of what to look for in the way of genuine minimax or admissibility properties of certain test procedures, even though the latter do not follow from these properties. We present in the following section the theory of local and asymptotic minimax tests as developed by Giri and Kiefer (1962) and use them to show that the T^2-test possesses both of these properties for every α, N, p. This lends to the conjecture that the T^2-test is minimax for all N, p. For relevant further results in connection with the minimax property of the T^2-test the reader is also referred to Linnik et al. (1966). For a more complete presentation of minimax tests in the multivariate setup the reader is referred to Giri (1975).

7.2.4. Locally and Asymptotically Minimax Tests

Locally Minimax Tests

Let $\mathcal{X}$ be a space with an associated σ-field which, along with the other obvious measurability considerations, we will not mention in what follows. For each point (δ^2, η) in the parametric space $\Omega(\delta^2 \geq 0$ and η may be a vector or matrix) suppose that $f(\cdot; \delta^2, \eta)$ is a probability density function on $\mathcal{X}$ with respect to a σ-finite measure μ. The range of η may depend on δ^2. For fixed $\alpha, 0 < \alpha < 1$, we shall be interested in testing, at level α, the null hypothesis $H_0 : \delta^2 = 0$ against the alternative $H_1 : \delta^2 = \lambda$, where λ is a specified positive value. This is a local theory in the sense that $f(x; \lambda, \eta)$ is close to $f(x; 0, \eta)$ when λ is small. Throughout this presentation, such expressions as $o(1), o(h(\lambda)), \ldots$, are to be interpreted as $\lambda \to 0$.

For each $\alpha, 0 < \alpha < 1$, we shall consider rejection regions of the form $R = \{x : U(x) > C_\alpha\}$, where U is bounded and positive and has a continuous distribution function for each (δ^2, η), equicontinuous in (δ^2, η) for $\delta^2 <$ some δ^2 and where

$$P_{0,\eta}\{R\} = \alpha, \quad P_{\lambda,\eta}\{R\} = \alpha + h(\lambda) + q(\lambda, \eta) \tag{7.50}$$

where $q(\lambda, \eta) = o(h(\lambda))$ uniformly in η, with $h(\lambda) > 0$ for $\lambda > 0$ and $h(\lambda) = o(1)$. We shall also be concerned with probability measures $\xi_{0,\lambda}$ and $\xi_{1,\lambda}$ on the sets $\delta^2 = 0$ and $\delta^2 = \lambda$, respectively, for which

$$\frac{\int f(x; \lambda, \eta)\xi_{1,\lambda}(d\eta)}{\int f(x; \lambda, \eta)\xi_{0,\lambda}(d\eta)} = 1 + h(\lambda)[g(\lambda) + r(\lambda)U(x)] + B(x, \lambda), \tag{7.51}$$

where $0 < C_1 < r(\lambda) < C_2 < \infty$ for λ sufficiently small, and where $g(\lambda) = 0(1)$ and $B(x, \lambda) = o(h(\lambda))$ uniformly in x.

Theorem 7.2.4. *Locally minimax. If R satisfies (7.50) and if for sufficiently small λ there exist $\xi_{0,\lambda}$ and $\xi_{1,\lambda}$ satisfying (7.51), then R is locally minimax of level α for testing $H_0 : \delta^2 = 0$ against $H_1 : \delta^2 = \lambda$ as $\lambda \to 0$; that is,*

$$\lim_{\lambda \to 0} \frac{\inf_\eta P_{\lambda,\eta}\{R\} - \alpha}{\sup_{\phi_\lambda \in Q_\alpha} \inf_\eta P_{\lambda,\eta}\{\phi_\lambda \text{ rejects } H_0\} - \alpha} = 1, \tag{7.52}$$

where Q_α is the class of tests of level α.

Proof. Write

$$\tau_\lambda = 1/\{2 + h(\lambda)[g(\lambda) + C_\alpha r(\lambda)]\}, \tag{7.53}$$

so that

$$(1 - \tau_\lambda)/\tau_\lambda = 1 + h(\lambda)[g(\lambda) + C_\alpha r(\lambda)]. \tag{7.54}$$

A Bayes rejection region relative to a priori distribution $\xi_\lambda = (1 - \tau_\lambda)\xi_{0,\lambda} + \tau_\lambda \xi_{1,\lambda}$ (for $0 - 1$ losses) is, by (7.51) and (7.54),

$$B_\lambda = \left\{x : U(x) + \frac{B(x, \lambda)}{r(\lambda)h(\lambda)} > C_\alpha\right\}. \tag{7.55}$$

Write

$$P_{0,\lambda}^*\{A\} = \int P_{0,\lambda}\{A\}\xi_{0,\lambda}(d\eta), \qquad P_{1,\lambda}^*\{A\} = \int P_{\lambda,\eta}\{A\}\xi_{1,\lambda}(d\eta).$$

Let $V_\lambda = R_\lambda - B_\lambda$ and $W_\lambda = B_\lambda - R$. Using the fact that $\sup_x |B(x, \lambda)/h(\lambda)| = o(1)$ and our continuity assumption on the distribution function of U, we have

$$P_{0,\lambda}^*\{V_\lambda + W_\lambda\} = o(1). \tag{7.56}$$

Tests of Hypotheses of Mean Vectors

Also, for $U_\lambda = V_\lambda$ or W_λ,

$$P^*_{1,\lambda}\{U_\lambda\} = P^*_{0,\lambda}\{U_\lambda\}[1 + O(h(\lambda))]. \tag{7.57}$$

Write $r^*_{1,\lambda}(A) = (1 - \tau_\lambda)P^*_{0,\lambda}\{A\} + \tau_\lambda(1 - P^*_{1,\lambda}\{A\})$. From (7.53) and (7.57), the integrated Bayes risk relative to ξ_λ is then

$$\begin{aligned}
r^*_\lambda(B_\lambda) &= r^*_\lambda(R) + (1 - \tau_\lambda)(P^*_{0,\lambda}\{W_\lambda\} - P^*_{0,\lambda}\{V_\lambda\}) \\
&\quad + \tau_\lambda(P^*_{1,\lambda}\{V_\lambda\} - P^*_{1,\lambda}\{W_\lambda\}) \\
&= r^*_\lambda(R) + (1 - 2\tau_\lambda)(P^*_{0,\lambda}\{W_\lambda\} - P^*_{0,\lambda}\{V_\lambda\}) \\
&\quad + P^*_{0,\lambda}\{V_\lambda + W_\lambda\}O(h(\lambda)) \\
&= r^*_\lambda(R) + o(h(\lambda)).
\end{aligned} \tag{7.58}$$

If (7.52) is false, we could, by (7.53), find a family of tests $\{\phi_\lambda\}$ of level α such that ϕ_λ has power function $\alpha + g(\lambda, \eta)$ on the set $\delta^2 = \lambda$, with

$$\limsup_{\lambda \to 0} \left(\frac{[\inf_\eta g(\lambda, \eta) - h(\lambda)]}{h(\lambda)} \right) > 0.$$

The integrated risk r'_λ of ϕ_λ with respect to ξ_λ would then satisfy

$$\limsup_{\lambda \to 0} \left(\frac{r^*_\lambda(R) - r'_\lambda}{h(\lambda)} \right) > 0,$$

thus contradicting (7.58). Q.E.D.

Asymptotically Minimax Tests

Here we treat the case $\lambda \to \infty$, and expressions such as $o(1)$, $o(H(\lambda))$ are to be interpreted in this light. Suppose that in place of (7.50) R satisfies

$$P_{0,\eta}\{R\} = \alpha, \qquad P_{\lambda,\eta}\{R\} = 1 - \exp\{-H(\lambda)(1 + o(1))\}, \tag{7.59}$$

where $H(\lambda) \to \infty$ with λ and the $o(1)$ term is uniform in η. Suppose, replacing (7.51), that

$$\frac{\int f(x; \lambda, \eta)\xi_{1,\eta}(d\eta)}{\int f(x; 0, \eta)\xi_{0,\lambda}(d\eta)} = \exp\{H(\lambda)[G(\lambda) + R(\lambda)U(x)] + B(x, \lambda)\}, \tag{7.60}$$

where $\sup_x |B(x, \lambda)| = o(H(\lambda))$ and $0 < C_1 < R(\lambda) < C_2 < \infty$. Our only other regularity assumption is that C_α is a point of increase from the left of the distribution of U, when $\delta_2 = 0$, uniformly in η; that is,

$$\inf_\eta P_{0,\eta}\{U \geq C_\alpha - \varepsilon\} > \alpha \tag{7.61}$$

for every $\varepsilon > 0$.

Theorem 7.2.5. *If R satisfies (7.59) and (7.61) and if for sufficiently large λ there exist $\xi_{0,\lambda}$ and $\xi_{1,\lambda}$ satisfying (7.60), then R is asymptotically logarithmically minimax of level α for testing $H_0 : \delta^2 = 0$ against $H_1 : \delta^2 = \lambda$ so that $\lambda \to \infty$; that is,*

$$\lim_{\lambda \to \infty} \frac{\inf_\eta \{-\log[1 - P_{\lambda,\eta}\{R\}]\}}{\sup_{\phi_\lambda \in Q_\alpha} \inf_\eta \{-\log[1 - P_{\lambda,\eta}\{\phi_\lambda \text{ rejects } H_0\}]\}} = 1. \quad (7.62)$$

Proof. Suppose, contrary, to (7.62), that there is an $\varepsilon > 0$ and an unbounded sequence Γ of values λ with corresponding tests ϕ_λ in Q_α for which

$$P_{\lambda,\eta}\{R\} > 1 - \exp\{-H(\lambda)(1 + 5\varepsilon)\} \quad \text{for all} \quad \eta. \quad (7.63)$$

There are two cases: (7.64) and (7.67). If $\lambda \in \Gamma$ and

$$-1 - G(\lambda) \le R(\lambda)C_\alpha + 2\varepsilon, \quad (7.64)$$

consider the a priori distribution given by $\xi_{i,\lambda}$ and by τ_λ satisfying

$$\tau_\lambda/(1 - \tau_\lambda) = \exp\{H(\lambda)(1 + 4\varepsilon)\}. \quad (7.65)$$

The integrated risk of any Bayes procedure B_λ must satisfy

$$\begin{aligned} r_\lambda^*(B_\lambda) \le r_\lambda^*(\phi_\lambda) &\le (1 - \tau_\lambda)\alpha + \tau_\lambda \exp\{-H(\lambda)(1 + 5\varepsilon)\} \\ &= (1 - \tau_\lambda)[\alpha + \exp(-\varepsilon H(\lambda))], \end{aligned} \quad (7.66)$$

by (7.63) and (7.65). But from (7.60) a Bayes critical region is

$$B_\lambda = \left\{ x : \frac{U(x) + B(x, \lambda)}{R(\lambda)H(\lambda)} \ge \frac{-(1 + 4\varepsilon) - G(\lambda)}{R(\lambda)} \right\}.$$

Hence if λ is so large that $\sup_x |B(x, \lambda)/H(\lambda)| < \varepsilon/C_2$, we get from (7.64)

$$B_\lambda \supset \{x : U(x) > C_\alpha - \varepsilon/C_2\} = B_\lambda' \quad \text{say.}$$

The assumption (7.61) implies that

$$P_{0,\eta}\{B_\lambda'\} > \alpha + \varepsilon'$$

with $\varepsilon' > 0$, contradicting (7.66) for large λ. On the other hand, if $\lambda \in \Gamma$ and

$$-1 - G(\lambda) > R(\lambda)C_\alpha + 2\varepsilon, \quad (7.67)$$

let

$$\tau_\lambda/(1 - \tau_\lambda) = \exp\{H(\lambda)(1 + \varepsilon)\}. \quad (7.68)$$

Then by (7.60)

$$B_\lambda = \left\{ x : \frac{U(x) + B(x, \lambda)}{R(\lambda)H(\lambda)} \ge \frac{-(1 + \varepsilon) - G(\lambda)}{R(\lambda)} \right\}.$$

Tests of Hypotheses of Mean Vectors

Hence if $\sup_x |B(x, \lambda)/H(\lambda)R(\lambda)| < \varepsilon/2C_2$, we conclude from (7.67) that $B_\lambda \subset R$, so that, by (7.59) and (7.68),

$$r^*(B_\lambda) > \tau_\lambda \exp\{-H(\lambda)[1 + o(1)]\} \\ = (1 - \tau_\lambda) \exp\{H(\lambda)(\varepsilon - o(1))\}. \tag{7.69}$$

But

$$r^*(B_\lambda) \leq r^*(\phi_\lambda) \leq (1 - \tau_\lambda)\alpha + \tau_\lambda \exp\{-H(\lambda)(1 + 5\varepsilon)\} \\ = (1 - \tau_\lambda)[\alpha + \exp\{-4\varepsilon H(\lambda)\}],$$

which contradicts (7.69) for sufficiently large λ. Q.E.D.

Theorem 7.2.6. *For every $p, N,$ and α, Hotelling's T^2-test is locally minimax for testing $H_0 : \delta^2 = 0$ against $H_1 : \delta^2 = \lambda$ as $\lambda \to 0$.*

Proof. In our search for a locally minimax test as $\lambda \to 0$ we look for a level α test which is almost invariant under G_T and which minimizes among all level α tests the minimum power under H_1 (as discussed in the case of the genuine minimax property of the T^2-test). So we restrict our attention to the space of $R = (R_1, \ldots, R_p)'$, the maximal invariant under G_T in the space of $(\bar{X}, S)$. We now verify the assumption of Theorem 7.2.4 with $x = r$, $\eta_i = \delta_i^2/\delta^2$, $\eta = \eta = (\eta_1, \ldots, \eta_p)'$, and $U(x) = \sum_{i=1}^p r_i$. We can take $h(\lambda) = b\lambda$ with b a positive constant. Of course, $P_{\lambda,\eta}\{R\}$ does not depend on η. From (6.66)

$$\frac{f(r; \lambda, \eta)}{f(r; 0, 0)} = 1 + \frac{\lambda}{2}\left\{-1 + \sum_{j=1}^p r_j\left[\sum_{i>j} \eta_i + (N - j + 1)\eta_j\right]\right\} \tag{7.70}$$
$$+ B(r, \eta, \lambda),$$

where $B(r, \eta, \lambda) = o(\lambda)$ uniformly in r and η. Here the set $\{\delta^2 = 0\}$ is a single point. Also the set $\{\delta^2 = \lambda\}$ is a convex finite-dimensional Euclidian set where in each component η_i is $0(h(\lambda))$. If there exists any $\xi_{1,\lambda}$ satisfying (7.51), the degenerate $\xi'_{1,\lambda}$ which assigns measure 1 to the mean of $\xi_{1,\lambda}$ also satisfies (7.51), and (7.51) is satisfied by letting $\xi_{0,\lambda}$ give measure 1 to the single point $\eta = 0$, whereas $\xi_{1,\lambda}$ gives measure 1 to the single point η^* (say) whose jth coordinate is $(N-j)^{-1}(N-j+1)^{-1}p^{-1}N(N-p)$, so that $\sum_{i>j}\eta_i^* + (N-j+1)\eta_j^* = N/p$ for all j. Applying Theorem 7.2.4 we get the result. Q.E.D.

Theorem 7.2.7. *For every α, p, N, Hotelling's T^2-test is asymptotically (logarithmically) minimax for testing $H_0 : \delta^2 = 0$ against the alternative $H_1 : \delta^2 = \lambda$ as $\lambda \to \infty$.*

Proof. From (6.64) [since $\phi(a, b; x) = \exp(x(1 + o(1)))$ as $x \to \infty$] we get

$$\frac{f(r; \lambda, \eta)}{f(r; 0, \eta)} = \exp\left\{\frac{\lambda}{2}\left[-1 + \sum_{j=1}^{p} r_j \sum_{i \geq j} \eta_i\right](1 + B(r, \eta, \lambda))\right\} \quad (7.71)$$

with $\sup_{r,\eta} |B(r, \eta, \lambda)| = o(1)$ as $\lambda \to \infty$. From this and the smoothness of $f(r; 0, \eta)$ we see (e.g., putting $\eta_p = 1$, the density of U being independent of η) that

$$P_{\lambda,\eta}\{U < C_\alpha\} = \exp\left\{\frac{1}{2}\lambda(C_\alpha - 1)[1 + o(1)]\right\} \quad (7.72)$$

as $\lambda \to \infty$. Thus (7.59) is satisfied with $H(\lambda) = \frac{1}{2}(1 - C_\alpha)$. Next, letting $\xi_{1,\lambda}$ assign measure 1 to the point $\eta_1 = \cdots = \eta_{p-1} = 0$, $\eta_p = 1$, and $\xi_{0,\lambda}$ assign measure 1 to $(0,0)$, we obtain (7.60). Finally (7.61) is trivial. Applying Theorem 7.2.5 we get the result. Q.E.D.

Suppose, for a parameter set $\Omega' = \{(\theta, \eta) : \theta \in \Theta, \eta \in H\}$ with associated distributions, with Θ a Euclidean set, that every test ϕ has a power function $\beta_\phi(\theta, \eta)$ which, for each η is twice continuously differentiable in the components of θ at $\theta = 0$, an interior point of Θ. Let Q_α be the class of locally strictly unbiased level α tests of $H_0 : \theta = 0$ against $H_1 : \theta \neq 0$; our assumption on β_ϕ implies that all tests in Q_α are similar and that $\partial \beta_\phi / \partial \theta_i |_{\theta=0} = 0$ for ϕ in Q_α. Let $\Delta_p(\eta)$ be the determinant of the matrix $B_\phi(\eta)$ of second derivatives of $\beta_\phi(\theta, \eta)$ with respect to the components of θ at $\theta = 0$. We assume the parametrization to be such that $\Delta'_\phi(\eta) > 0$ for all η for at least one ϕ' in Q_α.

A test ϕ^* is said to be of type E if $\phi^* \in Q_\alpha$ and $\Delta_{\phi*}(\eta) = \max_{\phi \in Q_\alpha} \Delta_\phi(\eta)$ for all η. If H is a single point, ϕ^* is said to be of type D.

Write

$$\bar{\Delta}(\eta) = \max_{\phi \in Q_\alpha} \Delta_\phi(\eta).$$

A test ϕ^* will be said to be of type D_A if $\phi \in Q_\alpha$ and

$$\max_\eta [\bar{\Delta}(\eta) - \Delta_{\phi*}(\eta)] = \min_{\phi \in Q_\alpha} \max_\eta [\bar{\Delta}(\eta) - \Delta_\phi(\eta)]$$

and of type D_M if

$$\max_\eta [\bar{\Delta}(\eta)/\Delta_{\phi*}(\eta)] = \min_{\phi \in Q_\alpha} \max_\eta [\bar{\Delta}(\eta)/\Delta_\phi(\eta)].$$

The notion of type D and E regions is due to Isaacson (1951). The D_A and D_M criteria resemble stringency and regret criteria employed elsewhere in statistics. The reader is referred to Giri and Kiefer (1964) for the proof that the T^2-test is not

Tests of Hypotheses of Mean Vectors

of type D among all G_T invariant tests and hence is not of type D_A or D_M or E among all tests.

7.2.5. Applications of the T^2-Test

Confidence Region of Mean Vector

Let $x^\alpha = (x_{\alpha 1}, \ldots, x_{\alpha p})'$, $\alpha = 1, \ldots, N$, be a sample of size N from a p-variate normal distribution with unknown mean μ and unknown positive definite covariance matrix Σ. Let

$$\bar{x} = \frac{1}{N}\sum_1^N x^\alpha, \qquad s = \sum_1^N (x^\alpha - \bar{x})(x^\alpha - \bar{x})'.$$

For the corresponding random sample $X^\alpha, \alpha = 1, \ldots, N$, $N(N-1) \times (\bar{X} - \mu)'S^{-1}(\bar{X} - \mu)$ is distributed as Hotelling's T^2 with $N-1$ degrees of freedom. Let $T_0^2(\alpha)$, for $0 < \alpha < 1$, be such that $P(T^2 \geq T_0^2(\alpha)) = \alpha$. Then the probability of drawing a sample $x^\alpha, \alpha = 1, \ldots, N$, of size N with mean $\bar{x}$ and sample covariance s such that

$$N(N-1)(\bar{x} - \mu)'s^{-1}(\bar{x} - \mu) \leq T_0^2(\alpha)$$

is $1 - \alpha$. Hence given $X^\alpha, \alpha = 1, \ldots, N$, the $100(1-\alpha)\%$ confidence region of μ consists of all p-vectors m satisfying

$$N(N-1)(\bar{x} - m)'s^{-1}(\bar{x} - m) \leq T_0^2(\alpha). \tag{7.73}$$

The boundary of this region is an ellipsoid whose center is at the point $\bar{x}$ and whose size and shape depend on s and α.

7.2.6. Simultaneous Confidence Interval

Let $\beta_1, \ldots, \beta_k$ be a set of parameters and let $I_i, i = 1, \ldots, k$, be the set of confidence intervals for $\beta_i, i = 1, \ldots, k$ satisfying

$$P\{\beta_i \in I_i, i = 1, \ldots, k\} = 1 - \alpha. \tag{7.73a}$$

Then the I_i are called the $(1-\alpha)\%$ confidence intervals of $\beta_1, \ldots, \beta_k$. From (7.73) we obtain simultaneous confidence intervals for linear functions $\ell'\mu$, $\ell \in E^p$ by the use of the following lemma.

Lemma 7.2.2. *Let S be positive definite and symmetric. Then for all $\ell = (\ell_1, \ldots, \ell_p)' \in E^p$,*

$$(\ell'y)^2 \leq (\ell'S\ell)(y'S^{-1}y). \tag{7.73b}$$

where $y = (\bar{X} - \mu)$.

Proof. Put $\gamma = \ell'y(\ell'S\ell)^{-1}$. Since S is positive definite and symmetric we get

$$(y - \gamma S\ell)'S^{-1}(y - \gamma S\ell) \geq 0.$$

Hence

$$y'S^{-1}y - 2\gamma \ell' SS^{-1}y + \gamma^2 \ell' SS^{-1}S\ell = y'S^{-1}y - \frac{(\ell'y)^2}{\ell'S\ell} \geq 0,$$

which implies (7.73b). Q.E.D.

Now using (7.73) we conclude with confidence $(1 - \alpha)\%$ that the mean vector μ satisfies for all $\ell \in E^p$

$$|\ell'\bar{X} - \ell'\mu| \leq \sqrt{(\ell'S\ell)(T_o^2(\alpha))/N(N-1)}.$$

Test for the Equality of Two Mean Vectors

Let $X^\alpha = (X_{\alpha 1}, \ldots, X_{\alpha p})'$, $\alpha = 1, \ldots, N_1$, be a random sample of size N_1 from a p-variate normal population with mean vector μ and positive definite covariance matrix Σ, and let $Y^\alpha = (Y_{\alpha 1}, \ldots, Y_{\alpha p})'$, $\alpha = 1, \ldots, N_2$, be a random sample of size N_2 from another independent normal population with mean ν and positive definite covariance matrix Σ. Let

$$\bar{X} = \frac{1}{N_1} \sum_1^{N_1} X^\alpha, \qquad \bar{Y} = \frac{1}{N_2} \sum_1^{N_2} Y^\alpha,$$

$$S = \sum_1^{N_1} (X^\alpha - \bar{X})(X^\alpha - \bar{X})' + \sum_1^{N_2} (Y^\alpha - \bar{Y})(Y^\alpha - \bar{Y})'.$$

It can be verified that $(\bar{X}, \bar{Y}, S)$ is a complete sufficient statistic for (μ, ν, Σ), $(N_1 N_2/(N_1 + N_2))^{1/2}(\bar{X} - \bar{Y})$ has p-variate normal distribution with mean $(N_1 N_2/(N_1 + N_2))^{1/2}(\mu - \nu)$ and positive definite covariance matrix Σ, and S is distributed as Wishart $W_p(N_1 + N_2 - 2, \Sigma)$ independently of $(\bar{X}, \bar{Y})$. The problem of testing the hypothesis $H_0 : \mu - \nu = 0$ against the alternatives $H_1 : \mu - \nu \neq 0$ remains invariant under the group of affine transformations $X^\alpha \to gX^\alpha + b$, $\alpha = 1, \ldots, N_1$, $Y^\alpha \to gY^\alpha + b$, $\alpha = 1, \ldots, N_2$, where $g \in G$, $b \in E^p$ (Euclidean p-space). The maximal invariant under the group of affine transformations in the space of $(\bar{X}, \bar{Y}, S)$ is given by

$$T^2 = (N_1 + N_2 - 2)(N_1 N_2/(N_1 + N_2))(\bar{X} - \bar{Y})'S^{-1}(\bar{X} - \bar{Y})$$

Tests of Hypotheses of Mean Vectors

and T^2 is distributed as Hotelling's T^2 with $N_1 + N_2 - 2$ degrees of freedom and the noncentrality parameter

$$\delta^2 = (N_1 N_2/(N_1 + N_2))(\mu - \nu)' \Sigma^{-1}(\mu - \nu).$$

An optimum test for this problem is the Hotelling's T^2-test which rejects H_0 for large values of T^2. This test possesses all the properties of the T^2-test discussed above.

Example 7.2.1. Consider Example 5.3.1 and assume that the two p-variate normal populations have the same positive definite covariance matrix Σ (unknown). Let the mean of population I (1971) be μ and that of population II (1972) be ν. We are interested here in testing the hypothesis $H_0 : \mu - \nu = 0$. Here $N_1 = N_2 = 27$:

$$\bar{x} = (84.89, 186.30, 9.74, 13.46, 304.37, 13.63)'$$
$$\bar{y} = (77.14, 167.18, 10.45, 13.10, 361.55, 14.76)'$$

$$\frac{S}{52} = \begin{pmatrix} & 1 & 2 & 3 & 4 & 5 & 6 \\ 1 & 1143.07 & & & & & \\ 2 & 57.40 & 3.84 & & & & \\ 3 & 70.16 & 4.25 & 25.54 & & & \\ 4 & 79.48 & 0.66 & 23.62 & 326.56 & & \\ 5 & 15.28 & 0.77 & 1.18 & 2.40 & 0.30 & \\ 6 & 21.60 & 1.04 & 2.56 & 4.14 & 0.39 & 0.83 \end{pmatrix}$$

The value of

$$t^2 = (N_1 N_2/(N_1 + N_2))(N_1 + N_2 - 2)(\bar{x} - \bar{y})' s^{-1}(\bar{x} - \bar{y}) = 217.55.$$

The 1% significance value of T^2 is 21.21. Thus we reject the hypothesis that the means of the two populations are equal.

Problem of Symmetry and Tests of Significance of Contrasts

Let $x^\alpha = (x_{\alpha 1}, \ldots, x_{\alpha p})'$, $\alpha = 1, \ldots, N$, be a sample of size N from a p-variate normal population with mean $\mu = (\mu_1, \ldots, \mu_p)'$ and covariance matrix Σ. We are interested in testing the hypothesis

$$H_0 : \mu_1 = \cdots \mu_p = \gamma \quad \text{(unknown)}.$$

Let $E = (1, \ldots, 1)'$ be a p-vector with components all equal to unity. A matrix C of dimension $(p - 1) \times p$ is called a contrast matrix if $CE = 0$.

Example 7.2.2. The $(p-1) \times p$ matrix C_1

$$C_1 = \begin{pmatrix} 1 & -1 & 0 & \cdots & 0 & 0 \\ 0 & 1 & -1 & \cdots & 0 & 0 \\ \vdots & \vdots & \vdots & & \vdots & \vdots \\ 0 & 0 & 0 & \cdots & 1 & -1 \end{pmatrix}$$

is a contrast matrix of rank $p-1$. The $(p-1) \times p$ matrix C_2

$$C_2 = \begin{pmatrix} \frac{1}{(1.2)^{1.2}} & \frac{-1}{(1.2)^{1/2}} & 0 & \cdots & 0 \\ \frac{1}{(2.3)^{1/2}} & \frac{1}{(2.3)^{1/2}} & \frac{-2}{(2.3)^{1/2}} & \cdots & 0 \\ \vdots & \vdots & \vdots & & \vdots \\ \frac{1}{((p-1)p)^{1/2}} & \frac{1}{((p-1)p)^{1/2}} & \frac{1}{((p-1)p)^{1/2}} & \cdots & \frac{-(p-1)}{((p-1)p)^{1/2}} \end{pmatrix}$$

is an orthogonal contrast matrix of rank $(p-1)$ and is known as a Helmert matrix.

Obviously from the relation $CE = 0$ we conclude that all rows of C are orthogonal to E and the sum of the elements of any row of C is zero. Furthermore any two contrast matrices C_1, C_2 are related by

$$C_1 = DC_2, \qquad (7.74)$$

where D is a nonsingular matrix of dimension $(p-1) \times (p-1)$. Under H_0, $\mu = \gamma E$ and hence $E(CX^\alpha) = 0$ for any contrast matrix C. Conversely, if $E(CX^\alpha) = 0$ for some contrast matrix C (for each α), we have $C\mu = 0$. But on account of (7.74) $C = DC_1$, where C_1 is defined in Example 7.2.2, and hence $0 = DC_1\mu$, which implies $C_1\mu = 0$, and thus $\mu_1 = \cdots = \mu_p$.

Furthermore, for any contrast matrix C of dimension $(p-1) \times p$ (of rank $p-1$), the matrix $\binom{E}{C}$ is a nonsingular matrix and hence CX^α, $\alpha = 1, \ldots, N$, are independently and identically distributed $(p-1)$-dimensional normal vectors with mean $C\mu$ and positive definite co-variance matrix $C\Sigma C'$. Hence the appropriate test for $H_0 : C\mu = 0$ rejects H_0 if

$$t^2 = N(N-1)(C\bar{x})'(CsC')^{-1}(C\bar{x}) \geq k,$$

where CSC' is distributed independently of $C\bar{X}$ as Wishart $W_{p-1}(N-1, C\Sigma C')$ and the constant k is chosen such that the test has level α. Obviously the statistic T^2 (in this case) is distributed as Hotelling's T^2 based on a random sample CX^α, $\alpha = 1, \ldots, N$, of size N. It may be noted that T^2 does not depend on the particular choice of the contrast matrix C. As for any other contrast matrix C_1 we

Tests of Hypotheses of Mean Vectors

can write $C_1 = DC$ where D is nonsingular and
$$T^2 = (N-1)N(C_1\bar{X})'(C_1 S C_1')^{-1}(C_1\bar{X})$$
$$= (N-1)N(C\bar{X})'(CSC')^{-1}(C\bar{X}).$$

The noncentrality parameter of this distribution is
$$N(C\mu)'(C\Sigma C')^{-1}(C\mu).$$

Example 7.2.3. An interesting application of this was given by Rao (1948) in the case of a four-dimensional normal vector $X = (X_1, X_2, X_3, X_4)'$, where $X_1, \ldots, X_4$ represent the thickness of cork borings on trees in the four directions north, south, east, and west, respectively. The hypothesis in this case is that of equal bark deposit in every direction. The contrast matrix C in this case is

$$C = \begin{pmatrix} 1 & 1 & -1 & -1 \\ 1 & -1 & 0 & 0 \\ 0 & 0 & 1 & -1 \end{pmatrix}$$

For numerical data and the results the reader is referred to Rao (1948).

Example 7.2.4. Randomized block design with correlated observations. Consider a randomized block design with N blocks and p treatments. Let y_{ij} denote the yield of the ith treatment of the jth block and let Y_{ij} be the corresponding random variables. Assume that the Y_{ij} are normally distributed with

$$E(Y_{ij}) = \mu + \mu_i + \beta_j,$$

$$\mathrm{cov}(Y_{ij}, Y_{i'j'}) = \begin{cases} \sigma_{ii'} & \text{if } j = j', \\ 0 & \text{otherwise,} \end{cases}$$

$$\mathrm{var}(Y_{ij}) = \sigma_{ii},$$

$i = 1, \ldots, p; j = 1, \ldots, N$, where μ_i is the ith treatment effect, and β_j is the jth block effect. Such a case arises when, for example the β_j are random variables (random effect model). Write $Y = (Y^1, \ldots, Y^N)$, $Y^\alpha = (Y_{\alpha 1}, \ldots, Y_{\alpha p})'$, $\alpha = 1, \ldots, N$. Y is a $p \times N$ random matrix of elements Y_{ij} and Σ is a $p \times p$ matrix of elements $\sigma_{ii'}$. Then $\mathrm{cov}(Y) = \Sigma \otimes I$ where I is the identity matrix of dimension $N \times N$. The usual hypothesis in this case is $H_0 : \mu_1 = \cdots = \mu_p$. With the contrast matrix C_1 in Example 7.2.2, under H_0,

$$E(C_1 Y) = \begin{pmatrix} \mu_1 - \mu_2 \\ \mu_2 - \mu_3 \\ \vdots \\ \mu_{p-1} - \mu_p \end{pmatrix} E = 0,$$

where E is an N vector with all components equal to unity and $\text{cov}(C_1 Y) = (C_1 \Sigma C_1') \otimes I$. Under the assumption of normality the column vectors of $C_1 Y$ are independently distributed $(p-1)$-variate normal vectors with mean 0 under H_0 and with covariance matrix $C_1 \Sigma C_1'$. The appropriate test statistic for testing H_0 rejects H_0 when

$$t^2 = (N-1)N(C_1\bar{y})'(C_1 s C_1')^{-1}(C_1\bar{y}) \geq c,$$

where c is a constant depending on the level α of the test and $\bar{y} = (1/N)\sum_1^N y^\alpha$, $s = \sum_1^N (y^\alpha - \bar{y})(y^\alpha - \bar{y})'$. It is easy to see that $C_1 S_1 C_1' (N > p)$ is distributed independently of $\bar{Y}$ as $W_{p-1}(N-1, C_1 \Sigma C_1')$. Thus T^2 is distributed as Hotelling's T^2 with the noncentrality parameter

$$\delta^2 = N(\mu_1 - \mu_2, \mu_2 - \mu_3, \ldots, \mu_{p-1} - \mu_p)'(C_1 \Sigma C_1')^{-1}$$
$$\times (\mu_1 - \mu_2, \mu_2 - \mu_3, \ldots, \mu_{p-1} - \mu_p).$$

Paired T^2-Test and the Multivariate Analog of the Behren-Fisher Problem

Let $X^\alpha = (X_{\alpha 1}, \ldots, X_{\alpha p})'$, $\alpha = 1, \ldots, N_1$, be a random sample of size N_1 from a p-variate normal population with mean μ and positive definite covariance matrix Σ_1, and let $Y^\alpha = (Y_{\alpha 1}, \ldots, Y_{\alpha p})'$, $\alpha = 1, \ldots, N_2$, be a random sample of size N_2 from another independent p-variate normal population with mean v and positive definite covariance matrix Σ_2. We are interested here in testing the hypothesis $H_0 : \mu = v$. It is well known that even for $p = 1$ the likelihood ratio test is very complicated and is not suitable for practical use. If $\Sigma_1 = \Sigma_2$, we have shown that the T^2-test is the appropriate solution. However, if $\Sigma_1 \neq \Sigma_2$ but $N_1 = N_2 = N$, a suitable solution is reached by using the following paired device. Define $Z^\alpha = X^\alpha - Y^\alpha$, $\alpha = 1, \ldots, N$. Obviously Z^α, $\alpha = 1, \ldots, N$, constitute a random sample of size N from a p-variate normal distribution with mean $\theta = \mu - v$ and positive definite covariance matrix $\Sigma_1 + \Sigma_2 = \Sigma$ (say). The testing problem reduces to that of testing $H_0 : \theta = 0$ when Σ is unknown. Define

$$\bar{Z} = \frac{1}{N}\sum_1^N Z^\alpha, \quad S = \sum_1^N (Z^\alpha - \bar{Z})(Z^\alpha - \bar{Z})'.$$

Tests of Hypotheses of Mean Vectors

On the basis of sample observations Z^α, $\alpha = 1, \ldots, N$, the likelihood ratio test of H_0 rejects H_0 whenever

$$t^2 = (N-1)N\bar{z}'s^{-1}\bar{z} \geq c,$$

where the constant c depends on the level α of the test, and it possesses all the optimum properties of Hotelling's T^2-test (obviously in the class of tests based only on the differences Z^α, $\alpha = 1, \ldots, N$).

When $\Sigma_1 \neq \Sigma_2$, the multivariate analog of Scheffé' solution (Scheffé, 1943) gives an appropriate solution. This extension is due to Bennett (1951). Assume without any loss of generality that $N_1 < N_2$. Define

$$Z^\alpha = X^\alpha - \left(\frac{N_1}{N_2}\right)^{1/2} Y^\alpha + \frac{1}{(N_1 N_2)^{1/2}} \sum_1^{N_1} Y^\alpha - \frac{1}{N_2} \sum_1^{N_2} Y^\alpha,$$

$$\alpha = 1, \ldots, N_1.$$

It is easy to verify that Z^α, $\alpha = 1, \ldots, N_1$, are independently distributed normal p-vectors with the same mean $\mu - v$ and the same covariance matrix $\Sigma_1 + (N_1/N_2)\Sigma_2$. Let

$$\bar{Z} = \frac{1}{N_1} \sum_1^{N_1} Z^\alpha, \qquad S = \sum_1^{N_1} (Z^\alpha - \bar{Z})(Z^\alpha - \bar{Z})'.$$

Obviously $\bar{Z}$ and S are independent, $\bar{Z}$ has a p-variate normal distribution with mean $\mu - v$ and with positive definite covariance matrix $(\Sigma_1 + (N_1/N_2)\Sigma_2)$, and S is distributed as $W_p(N_1 - 1, \Sigma_1 + (N_1/N_2)\Sigma_2)$. An appropriate solution for testing $H_0: \mu - v = 0$ is given by

$$t^2 = (N_1 - 1)N_1 \bar{z}'s^{-1}\bar{z} \geq c,$$

where c depends on the level α of the test and T^2 has Hotelling's T^2-distribution with $N_1 - 1$ degrees of freedom and the noncentrality parameter $N_1(\mu - v)'(\Sigma_1 + (N_1/N_2)\Sigma_2)^{-1}(\mu - v)$.

7.3. TESTS OF SUBVECTORS OF μ IN MULTIVARIATE NORMAL

Let $X^\alpha = (X_{\alpha 1}, \ldots, X_{\alpha p})'$, $\alpha = 1, \ldots, N$, be a random sample of size N from a p-variate normal distribution with mean μ and positive definite covariance matrix Σ. We shall use the notations of Section 7.2.2 for the presentation of this section. We shall consider the following two testing problems concerning subvectors of μ. The two-sample analogs of these problems are obvious and their appropriate solutions can be easily obtained from the one-sample results presented here.

(a) In the notation of Section 7.2.2, let $k = 2$, $p_1 + p_2 = p$. We are interested here in testing the hypothesis $H_0 : \mu_{(1)} = 0$ when Σ is unknown. Let Ω be the parametric space of (μ, Σ) and $\omega = \{(0, \mu_{(2)}), \Sigma\}$ be the subspace of Ω when H_0 is true. The likelihood of the observations x^α, $\alpha = 1, \ldots, N$, on X^α, $\alpha = 1, \ldots, N$, is

$$L(\mu, \Sigma) = (2\pi)^{-Np/2} (\det \Sigma)^{-N/2}$$

$$\times \exp\left\{-\frac{1}{2} \operatorname{tr} \Sigma^{-1} \sum_{\alpha=1}^{N} (x^\alpha - \mu)(x^\alpha - \mu)'\right\}.$$

Obviously

$$\max_{\Omega} L(\mu, \Sigma) = (2\pi)^{-Np/2} [\det(s/N)]^{-N/2} \exp\left\{-\frac{1}{2} Np\right\}. \quad (7.75)$$

It is also easy to verify that

$$\max_{\omega} L(\mu, \Sigma) = (2\pi)^{-Np/2} \left[\det\left(\frac{s_{(11)} + N\bar{x}_{(1)}\bar{x}_{(1)}'}{N}\right)\right]^{-N/2}$$

$$\times \left[\det\left(\frac{s_{(22)} - s_{(21)} s_{(11)}^{-1} s_{(12)}}{N}\right)\right]^{-N/2} \exp\left\{-\frac{1}{2} Np\right\}. \quad (7.76)$$

The likelihood ratio criterion for testing H_0 is given by

$$\lambda = \frac{\max_\omega L(\mu, \Sigma)}{\max_\Omega L(\mu, \Sigma)} = \left[\frac{\det s_{(11)}}{\det(s_{(11)} + N\bar{x}_{(1)}\bar{x}_{(1)}')}\right]^{-N/2} \quad (7.77)$$

$$= (1 + r_1^*)^{-N/2}.$$

Thus the likelihood ratio test of H_0 rejects H_0 whenever

$$(N - 1) r_1^* > c,$$

where the constant c depends on the level of significance α of the test. In terms of the statistic R_1, this is also equivalent to rejecting H_0 whenever $r_1 \geq c$. From Chapter 6 the probability density function of R_1 is given by

$$f_{R_1}(r_1 | \delta_2^2) = \frac{\Gamma(\frac{1}{2} N)}{\Gamma(\frac{1}{2} p_1) \Gamma(\frac{1}{2}(N - p_1))} r_1^{p_1/2 - 1} (1 - r_1)^{(N - p_1)/2 - 1}$$

$$\times \left\{-\frac{1}{2} \delta_1^2\right\} \phi\left(\frac{1}{2} N, \frac{1}{2} p_1; \frac{1}{2} r_1 \delta_1^2\right)$$

Tests of Hypotheses of Mean Vectors

provided $r_1 \geq 0$ and is zero elsewhere, where $\delta_1^2 = N\mu'_{(1)}\Sigma^{-1}_{(11)}\mu_{(1)}$ and R_1^* is a Hotelling's T^2-statistic based on the random sample $X^\alpha_{(1)} = (X_{\alpha 1}, \ldots, X_{\alpha p_1})'$, $\alpha = 1, \ldots, N$, from a p_1-variate normal distribution with mean $\mu_{(1)}$ and positive definite covariance matrix $\Sigma_{(11)}$.

Let T_1 be the translation group such that $t_1 \in T_1$ translates the last p_2 components of each X^α and let G_{BT} be as defined in Section 7.2.2 with $k = 2$. This problem remains invariant under the affine group (G_{BT}, T_1) transforming

$$X^\alpha = gX^\alpha + t_1, \qquad \alpha = 1, \ldots, N_1, \qquad g \in G_{BT}, \qquad t_1 \in T_1.$$

Note that t_1 can be regarded as a p-vector with its first p_1 components equal to zero. A maximal invariant in the space of $(\bar{X}, S)$ is R_1 and the corresponding maximal invariant in the parametric space Ω is δ_1^2. From the computations in connection with the T^2-test it is now obvious that this test possesses the same optimum properties as those of Hotelling's T^2-test (Theorems 7.2.1, 7.2.2, 7.2.3, and the minimax property).

(b) In the notation of Section 7.2.2 let $k = 3$, $p_1 + p_2 + p_3 = p$. We are interested here in testing the hypothesis $H_0 : \mu_{[2]} = 0$ when μ, Σ are unknown and the parametric space $\Omega = \{(0, \mu_{(2)}, \mu_{(3)}), \Sigma\}$. It may be verified that

$$\max_\Omega L(\mu, \Sigma) = (2\pi)^{-Np/2}[\det(s/N)]^{-N/2}(1 + N\bar{x}'_{(1)} s^{-1}_{(11)} \bar{x}_{(1)})^{-N/2}$$

$$\times \exp\left\{-\frac{1}{2}Np\right\}, \qquad (7.78)$$

$$\max_\omega L(\mu, \Sigma) = (2\pi)^{-Np/2}[\det(s/N)]^{-N/2}(1 + N\bar{x}'_{[2]} s^{-1}_{[22]} \bar{x}_{[2]})^{-N/2}$$

$$\times \exp\left\{-\frac{1}{2}Np\right\}, \qquad (7.79)$$

where ω is the subspace of Ω when H_0 is true. Hence the likelihood ratio criterion λ is

$$\lambda = \frac{\max_\omega L(\mu, \Sigma)}{\max_\Omega L(\mu, \Sigma)} = \left(\frac{1 + N\bar{x}'_{(1)} s^{-1}_{(11)} \bar{x}_{(1)}}{1 + N\bar{x}'_{[2]} s^{-1}_{[22]} \bar{x}_{[2]}}\right)^{-N/2}$$

$$= \left(\frac{1 + r_1^* + r_2^*}{1 + r_1^*}\right)^{N/2} = \left(\frac{1 - r_1}{1 - r_1 - r_2}\right)^{N/2} \qquad (7.80)$$

Hence the likelihood ratio test of $H_0 : \mu_{[2]} = 0$ rejects H_0 whenever $(1 - r_1 - r_2)/(1 - r_1) \geq c$, where c is a constant depending on the level of significance α. From Chapter 6 the joint probability density function of (R_1, R_2) is given by

$$f_{R_1,R_2}(r_1, r_2 | \delta_1^2, \delta_2^2) = \frac{\Gamma(\frac{1}{2}N)}{\Gamma(\frac{1}{2}p_1)\Gamma(\frac{1}{2}(N-p_1))} r_1^{p_1/2-1} r_2^{p_2/2-1}$$
$$\times (1 - r_1 - r_2)^{(N-p_1-p_2)/2-1}$$
$$\times \exp\left\{-\frac{1}{2}(\delta_1^2 + \delta_2^2) + \frac{1}{2}r_1\delta_2^2\right\} \quad (7.81)$$
$$\times \phi\left(\frac{1}{2}N, \frac{1}{2}p_1; \frac{1}{2}r_1\delta_1^2\right)$$
$$\times \phi\left(\frac{1}{2}(N - p_1), \frac{1}{2}p_2; \frac{1}{2}r_2\delta_2^2\right)$$

provided $r_1 \geq 0, r_2 \geq 0$ and

$$\delta_1^2 = N\mu'_{(1)}\Sigma^{-1}_{(11)}\mu_{(1)}, \qquad \delta_1^2 + \delta_2^2 = N\mu'_{[2]}\Sigma^{-1}_{[22]}\mu_{[2]}.$$

Under H_0, $\delta_1^2 = \delta_2^2 = 0$. From (7.81) it follows that under H_0

$$Z = (1 - R_1 - R_2)/(1 - R_1)$$

is distributed as a central beta random variable with parameter

$$\left(\frac{1}{2}(N - p_1 - p_2), \frac{1}{2}p_2\right).$$

Let T_2 be the transformation group which translates the last p_3 components of each X^α, and let G_{BT} be as defined in Section 7.2.2 with $k = 3, p_1 + p_2 + p_3 = p$. This problem remains invariant under the group (G_{BT}, T_2) of affine transformations, transforming

$$X^\alpha \to gX^\alpha + t, \qquad \alpha = 1, \ldots, N,$$

$g \in G_{BT}$ (with $k = 3$), $t \in T_2$ (t can be considered as a p-vector with the first $p_1 + p_2$ components equal to zero). A maximal invariant in the space of $(\bar{X}, S)$ [the induced transformation on $(\bar{X}, S)$ is $(\bar{X}, S) \to (g\bar{X} + t, gSg')$] is (R_1, R_2) [also its equivalent statistic (R_1^*, R_2^*)]. A corresponding maximal invariant in Ω is (δ_1^2, δ_2^2). Under H_0, $\delta_1^2 = \delta_2^2 = 0$ and under the alternatives H_1, $\delta_2^2 > 0, \delta_1^2 = 0$.

From (7.81) it follows that the likelihood ratio test is not uniformly most powerful (optimum) invariant for this problem and that there is no uniformly most powerful invariant test for the problem. However, for fixed p, the likelihood ratio test is nearly optimum as N becomes large (Wald, 1943). Thus, if p is not large, it seems likely that the sample size occurring in practice was usually large enough for this result to be relevant. However, if the dimension p is large, it may

Tests of Hypotheses of Mean Vectors

be that the sample size N must be extremely large for this result to apply. Giri (1961) has shown that the difference of the powers of the likelihood ratio test and the best invariant test is $o(N^{-1})$ when p_1, p are both equal to $O(N)$ and $\delta_2^2 = O(\sqrt{N})$. For the minimax property Giri (1968) has shown that no invariant test under (G_{BT}, T_2) is minimax for testing H_0 against $H_1: \delta_1^2 = \lambda$ for every choice of λ. However Giri (1968) has shown that the test which rejects $H_0: \delta_2^2 = 0$ against the alternatives $H_1: \delta_2^2 = \lambda > 0$ whenever $R_1 + ((n - p_1)/p_2)R_2 \geq c$ where c depends on the level α of the test is locally best invariant and locally minimax as $\lambda \to 0$.

7.3.1. Test of Mean Against One-sided Alternatives

Let $x^\alpha = (x_{\alpha 1}, \ldots, x_{\alpha p})'$, $\alpha = 1, \ldots, N (N > p)$ be a sample of size N from a p-variate normal distribution with mean $\mu = (\mu_1, \ldots, \mu_p)'$ and positive definite covariance matrix Σ. The problem of testing $H_0: \mu = 0$ against alternatives $H_1: \mu \geq 0$ (i.e. $\mu_i \geq 0$ for all $i = 1, \ldots, p$ with at least one strict inequality for some i) has been treated by many authors including Bartholomeu (1961), Chacko (1963), Kodô (1963), Nüesch (1966), Sharack (1967), Pearlman (1969), Eaton (1970), Marden (1982), Kariya and Cohen (1992), Wang and McDermott (1998). This problem has received considerable interest in statistical literature in the context of clinical trials, particularly for the case Σ is known. We refer to Wang and McDermott (1998) and the references therein for the application aspects of the problem. We derive here the likelihood ratio test of H_0 against H_1 when Σ is known. We refer to Pearlman (1969) for the case Σ is unknown. The maximum likelihood estimates of the parameters under H_1 are not very easy to compute. The algorithm of computing these estimates is not simple because of the dependence between the components.

The likelihood of x^α, $\alpha = 1, \ldots, N$ is given by

$$L(\mu, \Sigma) = L(x^1, \ldots, x^N | \mu, \Sigma)$$

$$= (2\pi)^{-Np/2} (\det \Sigma)^{-N/2} \exp\left\{-\frac{1}{2} \operatorname{tr} \Sigma^{-1} \left(\sum_{\alpha=1}^{N} (x^\alpha - \mu)(x^\alpha - \mu)'\right)\right\}$$

where Σ is a known positive definite matrix.

The likelihood ratio test of H_0 against H_1 rejects H_0 whenever

$$\lambda = \max_{H_0} L(\mu, \Sigma) / \max_{H_1} L(\mu, \Sigma) \leq \lambda_0$$

where λ_0 is a constant depending on the size α of the test. Let $N\bar{x} = \sum_{\alpha=1}^{N} x_\alpha$. Then

$$\lambda = \exp\left\{-\frac{N}{2}\bar{x}'\Sigma^{-1}\bar{x}\right\} - \max_{\mu \geq 0} \exp\left\{-\frac{N}{2}(\bar{x} - \mu)'\Sigma^{-1}(\bar{x} - \mu)\right\}. \quad (7.81a)$$

Evaluation of (7.81a) enables us to examine the following statistic

$$\bar{\chi}^2 = N\{\bar{x}'\Sigma^{-1}\bar{x} - \min_{\mu \geq 0}(\bar{x} - \mu)'\Sigma^{-1}(\bar{x} - \mu)\}.$$

To compute $\bar{\chi}^2$ we need the minimum of the quadratic form $(\bar{x} - \mu)'\Sigma^{-1}(\bar{x} - \mu)$ when $\mu \geq 0$ and Σ is known. In general this can be done by quadratic programming (for example see Nüesch (1966)).

A geometric interpretation of the statistic $\bar{\chi}^2$, which will give us an actual picture useful for its computation as well as for the derivation of its distribution, is as follows. Since $\Sigma > 0$, there exists a $p \times p$ nonsingular matrix A such that $A\Sigma A' = I$. Let $A^{-1} = (a^{ij})$ and $Y = (Y_1, \ldots, Y_p)' = A\bar{X}$. Then Y is distributed as $N_p(m, N^{-1}I)$ and

$$\bar{\chi}^2 = N\{Y'Y - \min_{\mu \geq 0}(Y - m)'(Y - m)\} \quad (7.81b)$$

where $m = (m_1, \ldots, m_p)' = AE(\bar{X}) = A\mu$. Hence $\bar{\chi}^2$ is proportional to the difference between the square length of the vector Y in the p-dimensional Euclidean space of m and the distance between a point Y and a closed convex polyhedral cone C defined by the inequalities

$$\mu_i = \sum_{j=1}^{p} a^{ij} m_j \geq 0, \quad i = 1, \ldots, p.$$

If $Y \in C$ then the second term in (7.81b) vanishes and we have

$$\bar{\chi}^2 = NY'Y = N\bar{X}'\Sigma^{-1}\bar{X}.$$

Complication arises if $Y \notin C$. In any case there exists a vector $\hat{\mu} = (\hat{\mu}_1, \ldots, \hat{\mu}_p)'$ such that $\hat{\mu}_i > 0, i = 1, \ldots, p$ and

$$\min_{\mu \geq 0}(\bar{X} - \mu)'\Sigma^{-1}(\bar{X} - \mu) = (\bar{X} - \hat{\mu})'\Sigma^{-1}(\bar{X} - \hat{\mu}).$$

The point $\hat{\mu}$ is the maximum likelihood estimate of μ under H_1. The following theorem gives some insight about $\hat{\mu}$.

Theorem 7.3.1. *The point $\hat{\mu}$ is the maximum likelihood estimate of μ under H_1 if and only if one of the ith components of the two vectors $\hat{\mu}$ and $\{-\Sigma^{-1}(\bar{X} - \hat{\mu})\}$ is zero and the other is non-negative.*

Tests of Hypotheses of Mean Vectors

Proof. For $\mu \in C$ we have

$$(\bar{X} - \mu)'\Sigma^{-1}(\bar{X} - \mu) - (\bar{X} - \hat{\mu})'\Sigma^{-1}(\bar{X} - \hat{\mu})$$
$$= (\mu - \hat{\mu})'\Sigma^{-1}(\mu - \hat{\mu}) - 2(\mu - \hat{\mu})'\Sigma^{-1}(\bar{X} - \hat{\mu})$$
(7.81c)

Since $\Sigma > 0$, the first term in (7.81c) is positive. The second term is the inner product of $-2\Sigma^{-1}(\bar{X} - \hat{\mu})$ and $(\mu - \hat{\mu})$. From the condition of the Theorem it follows that if a component of the first vector is positive then the corresponding components of $\hat{\mu}$ and $(\mu - \hat{\mu})$ are zero and non-negative respectively and all the non-positive components of the first are zero. Thus the second term in (7.81c) is positive and hence the right-hand of (7.81c) is positive. This establishes the sufficiency of the condition.

To prove the necessity part let us first note that $-2\Sigma^{-1}(\bar{X} - \hat{\mu})$ is the vector of derivatives of

$$g(\hat{\mu}) = (\bar{X} - \hat{\mu})'\Sigma^{-1}(\bar{X} - \hat{\mu})$$

with respect to $\hat{\mu}$. So if the condition is violated we can find a p-vector $X = (X_1, \ldots, X_p)'$ (say) whose components are non-negative in the neighbourhood of $\hat{\mu}$ where the quadratic form has smaller value and this contradicts the assumption that $\hat{\mu}$ is the maximum likelihood estimator. Q.E.D.

The actual computation of the maximum likelihood estimator $\hat{\mu}$ when $\bar{X}$ is given may need successive approximations. As Kudô (1963) observed the convergence of this approximation is not fast but one can sometimes judge at a comparatively early stage of the calculation, by observing $\bar{X}$ only, which of the components should be zero and which should be positive. We refer to this paper for an example concerning the computation of $\hat{\mu}$.

Geometrically the maximum likelihood estimate $\hat{\mu}$ is the projection of the vector $\bar{X}$ along regression planes onto the positive orthant of the sample space. If one uses the linear transformation of $\bar{X}$ to a uncorrelated Y, which exists as $\Sigma > 0$, the projection is orthogonal onto a polyhedral half cone, the affine image of the positive orthant. Thus $\hat{\mu}$ is a vector whose components are either positive or zero. This leads to a partition of the sample space into 2^p disjoint regions. Let us denote by χ_k any of the $\binom{p}{k}$ regions of the sample space with exactly k of the $\hat{\mu}_j$'s positive. We assume, without any loss of generality, that the k positive $\hat{\mu}_j$'s are the last k components. We write

$$\chi_k = \{\bar{X} | \hat{\mu}_{(1)} = 0, \hat{\mu}_{(2)} > 0\}$$

where $\hat{\mu}_{(2)}$ contains the last k components of $\hat{\mu}$. Similarly partition $\bar{X} = (\bar{X}'_{(1)}, \bar{X}'_{(2)})'$. Let $\Sigma^{-1} = \Delta$. Partition

$$\Sigma = \begin{pmatrix} \Sigma_{11} & \Sigma_{12} \\ \Sigma_{21} & \Sigma_{22} \end{pmatrix}, \quad \Delta = \begin{pmatrix} \Delta_{11} & \Delta_{12} \\ \Delta_{21} & \Delta_{22} \end{pmatrix}$$

with Σ_{22}, Δ_{22} are both $k \times k$ submatrices. From Theorem 7.3.1 we get

$$\begin{aligned} -\Delta_{11}\bar{X}_{(1)} + \Delta_{12}(\hat{\mu}_{(2)} - \bar{X}_{(2)}) &> 0 \\ -\Delta_{12}\bar{X}_{(1)} + \Delta_{22}(\hat{\mu}_{(2)} - \bar{X}_{(2)}) &= 0 \end{aligned} \tag{7.81d}$$

solving we get

$$\hat{\mu}_{(2)} = \bar{X}_{(2)} + \Delta_{22}^{-1}\Delta_{21}\bar{X}_{(1)} = \bar{X}_{(2)} - \Sigma_{21}\Sigma_{11}^{-1}\bar{X}_{(1)}. \tag{7.81e}$$

Using (7.81c) and (7.81d) we get

$$-(\Delta_{11} - \Delta_{12}\Delta_{22}^{-1}\Delta_{21})\bar{X}_{(1)} = -\Sigma_{11}^{-1}\bar{X}_{(1)} > 0.$$

Hence

$$\chi_k = \{\bar{X} : \{\Sigma_{11}^{-1}\bar{X}_{(1)} < 0\} \cap \{\bar{X}_{(2)} - \Sigma_{21}\Sigma_{11}^{-1}\bar{X}_{(1)} > 0\}\}.$$

Since Theorem 7.3.1 implies

$$\hat{\mu}'\Sigma^{-1}(\bar{X} - \hat{\mu}) = 0,$$

we can write

$$\bar{\chi}^2 = N\hat{\mu}'\Sigma^{-1}\hat{\mu}$$

and the likelihood ratio test of H_0 against H_1 rejects H_0 whenever $N\hat{\mu}'\Sigma^{-1}\hat{\mu} \geq C$ where C is a constant depends on the size α of the test.

Theorem 7.3.2.

$$P(N\hat{\mu}'\Sigma^{-1}\hat{\mu} \geq C) = \sum_{k=1}^{p} w(p, k) P(\chi_k^2 \geq C)$$

where the weights $w(p, k)$ are the probability content of all χ_k's for a fixed k and χ_k^2 is the central chi-square random variable with k degrees of freedom.

Tests of Hypotheses of Mean Vectors

Proof. Let R denotes the rejection region of the likelihood ratio test of H_0 against H_1:

$$P(R|H_0) = \sum_{k=1}^{p} \sum_{\chi_k} P(R \cap \chi_k)$$

$$= \sum_{k=1}^{p} \sum_{\chi_k} P(\chi_k) P(\chi_k) P(R|\chi_k).$$

Since $R \cap \chi_0 = \phi$ (null set), the summation starts from 1. But

$$P(R|\chi_k) = P(N\hat{\mu}\Sigma^{-1}\hat{\mu} \geq C|H_0)$$

$$= P(N\hat{\mu}\Sigma^{-1}\hat{\mu} \geq C | \{\hat{\mu}_{(1)} = 0\} \cap \{\hat{\mu}_{(2)} > 0\}).$$

Since under H_0 $\hat{\mu}_{(2)}$ is a k-variate normal with mean 0 and covariance $\Sigma_{22.1} = \Sigma_{22} - \Sigma_{21}\Sigma_{11}^{-1}\Sigma_{12}$ we get $\hat{\mu}'_{(2)}\Sigma_{22.1}^{-1}\hat{\mu}_{(2)}$ is distributed as χ_k^2.

In addition $\{\hat{\mu}_{(2)} > 0\}$ and $\{N\hat{\mu}'_{(2)}\Sigma_{22.1}^{-1}\hat{\mu}_{(2)} \geq C\}$ are independent. Hence

$$P(R|H_0) = \sum_{k=1}^{p} w(p,k) P(\chi_k^2 \geq C).$$

Q.E.D.

7.4. TESTS OF MEAN VECTOR IN COMPLEX NORMAL

Let $z^1, \ldots, z^N$ be a sample of size N from $CN_p(\beta, \Sigma)$. We consider the problem of testing $H_0 : \beta = 0$ against the alternative $H_1 = \beta^*\Sigma^{-1}\beta = \delta^2 > 0$ on the basis of these observations. This problem is the complex analog of Hotelling's T^2 problem in the real case. The complex analog of other testing problems of mean vectors, considered in Sections 7.2, 7.3 can be analyzed by minor modifications of the results developed for the real cases (see also Goodman (1962)). The likelihood of the observations $z^1, \ldots, z^N$ is

$$L(z^1, \ldots, z^N | \beta, \Sigma) = \pi^{-Np} (\det \Sigma)^{-N}$$

$$\times \exp\left\{-\operatorname{tr} \Sigma^{-1} \left(\sum_{\alpha=1}^{N} (z^\alpha - \beta)(z^\alpha - \beta)^*\right)\right\}$$

$$= \pi^{-Np} (\det \Sigma)^{-N}$$

$$\times \exp\{-\operatorname{tr} \Sigma^{-1} (A + N(\bar{z} - \beta)(\bar{z} - \beta)^*)\}$$

where $A = \sum_{\alpha=1}^{N} (z^\alpha - \bar{z})(z^\alpha - \bar{z})^*$, $\bar{z} = (1/N) \sum_{\alpha=1}^{N} z^\alpha$.
Using Lemma 5.3.3 and Theorem 5.3.4 we obtain

$$\lambda = \frac{\max_{H_0} L(z^1, \ldots, z^N | \beta, \Sigma)}{\max_{\Omega} L(z^1, \ldots, z^N | \beta, \Sigma)}$$

$$= \left[\frac{\det(A)}{\det(A + N\bar{z}\bar{z}^*)}\right]^N$$

$$= (1 + t_c^2)^{-N}$$

where $t_c^2 = N\bar{z}^* A^{-1} \bar{z}$. Thus the likelihood ratio test rejects H_0 whenever

$$t_c^2 \geq k \tag{7.82}$$

where the constant k is chosen such that

$$P(T_c^2 \geq k | H_0) = \alpha.$$

We are using T_c^2 as the random variables with values t_c^2. From Theorem 6.11.2 the distribution of T_c^2 is given by

$$f_{T_c^2}(t_c^2 | \delta^2) = \frac{\exp\{-\delta^2\}}{\Gamma(N-p)} \sum_{j=0}^{\infty} \frac{(\delta^2)^j \Gamma(N+j)(t_c^2)^{p+j-1}}{j! \Gamma(p+j)(1+t_c^2)^{N+j-1}} \tag{7.83}$$

where $\delta^2 = N\beta^* \Sigma^{-1} \beta$. Under $H_0 : \delta^2 = 0$.

The problem of testing $H_0 : \beta = 0$ against $H_1 : \delta^2 > 0$ remains invariant under the full linear group $G_\ell(p)$ of $p \times p$ nonsingular complex matrices g transforming

$$(\bar{z}, A; \beta, \Sigma) \rightarrow (g\bar{z}, gAg^*; g\beta, g\Sigma g^*).$$

From Section 7.2.2 it follows that T_c^2 is a maximal invariant in the space of $(\bar{Z}, A)$ under $G_\ell(p)$. A corresponding maximal invariant in the parametric space of (β, Σ)

Tests of Hypotheses of Mean Vectors

is δ^2. From (7.83)

$$\frac{f_{T_c^2}(t_c^2|\delta^2=\lambda)}{f_{T_c^2}(t_c^2|\delta^2=0)} = \frac{\exp\{-\lambda\}\Gamma(p)}{\Gamma(N)} \sum_{j=0}^{\infty} \frac{\Gamma(N+j)}{\Gamma(p+j)j!} \left[\frac{\lambda t_c^2}{1+t_c^2}\right]^j. \quad (7.84)$$

Since the right-hand side of (7.84) is a monotonically increasing function of $[t_c^2/1+t_c^2]$ and hence of t_c^2 for all λ (fixed) > 0 we prove (as Theorem 7.2.1) the following:

Theorem 7.4.1. *For testing $H_0 : \beta = 0$ against $H_1 : \delta^2 > 0$ the likelihood ratio test which rejects H_0 whenever $t_c^2 \geq k$, where the constant k is chosen to get the size α, is uniformly most powerful invariant with respect to the group of transformations $G_\ell(p)$ of $p \times p$ nonsingular complex matrices.*

7.5. TESTS OF MEANS IN SYMMETRIC DISTRIBUTIONS

Let $X = (X_{ij}) = (X_1, \ldots, X_n)'$, $X_i' = (X_{ij}, \ldots, X_{ip})'$, $i = 1, \ldots, n(n > p)$ be a $n \times p$ random matrix with *pdf*

$$f_X(x) = (\det \Sigma)^{-n/2} q(\text{tr } \Sigma^{-1}(x - e\mu')' \Sigma^{-1}(x - e\mu'))$$

$$= (\det \Sigma)^{-n/2} q\left(\sum_{i=1}^{n}(x_i - \mu)' \Sigma^{-1}(x_i - \mu)\right) \quad (7.85)$$

where $x = (x_{ij})$ is a value of X, $\mu = (\mu_1, \ldots, \mu_p)' \in E^p$, Σ is a $p \times p$ positive definite matrix, $e = (1, \ldots, 1)'$ n-vector and q is a function only of the sum of n quadratic forms $(x_i - \mu)' \Sigma^{-1}(x_i - \mu)$ satisfying

$$\int q(\text{tr } u'u) du = 1.$$

This is a subclass of the family of elliptically symmetric distributions with location parameter $e\mu'$ and scale matrix Σ. We shall assume that $n > p$ so that by Lemma 5.1.2 $S = X'(I - (ee'/n))X = \sum_{i=1}^{n}(X_i - \bar{X})(X_i - \bar{X})'$ is positive definite with probability one, where $\bar{X} = (1/n) \sum_{i=1}^{n} X_i$.

7.5.1. Test of Mean Vector

Likelihood ratio test. We consider the problem of testing $H_0 : \mu = 0$ when Σ is unknown on the basis of an observation x on X. The likelihood of x is

$$L(x|\mu, \Sigma) = (\det \Sigma)^{-n/2} q\left(\sum_{i=1}^{n}(x_i - \mu)'\Sigma^{-1}(x_i - \mu)\right)$$

$$= (\det \Sigma)^{-n/2} q(\operatorname{tr} \Sigma^{-1}(s + n(\bar{x} - \mu)(\bar{x} - \mu)'))$$

where $s = \sum_{i=1}^{n}(x_i - \bar{x})(x_i - \bar{x})'$. Let $\Omega = \{(\mu, \Sigma)\}$ be the parametric space. From Theorem 5.3.6 the maximum likelihood estimators of μ, Σ are

$$\hat{\mu} = \bar{x}, \quad \hat{\Sigma} = \frac{p}{u_q} s$$

where u_q maximize $u^{np/2} q(u)$. The likelihood ratio is given by

$$\lambda = \frac{\max_{H_0} L(x|\mu, \Sigma)}{\max_\Omega L(x|\mu, \Sigma)}$$

$$= \frac{\left[\det\left(\frac{ps}{u_q}\right)\right]^{n/2} q\left(\frac{u_q}{p}\right)}{\left[\det\left(\frac{p(s + n\bar{x}\bar{x}')}{u_q}\right)\right]^{n/2} q\left(\frac{u_q}{p}\right)} \quad (7.86)$$

$$= (1 + n\bar{x}'s^{-1}\bar{x})^{-n/2}$$

The likelihood ratio test of $H_0 : \mu = 0$ is given by

$$\text{reject } H_0 \text{ whenever } n\bar{x}'s^{-1}\bar{x} \geq C \quad (7.87)$$

or its equivalent, given by

$$r_1 = \frac{n\bar{x}'s^{-1}\bar{x}}{1 + n\bar{x}'s^{-1}\bar{x}} \geq \frac{C}{1-C} \quad (7.88)$$

where C is a constant depending on the size α of the test.

To determine C we need the distribution of $T^2 = n\bar{X}'S^{-1}\bar{X}$ under H_0. The following lemma will show that the distribution of T^2 under H_0 does not depend on a particular choice of q in (7.85). Taking, in particular, $X_1, \ldots, X_n$ to be independently and identically distributed $N_p(0, \Sigma)$ we conclude from (6.60) that $n\bar{X}'S^{-1}\bar{X}$ has Hotelling's T^2 distribution with $n-1$ degrees of freedom under H_0.

Tests of Hypotheses of Mean Vectors

Its *pdf* is given by

$$f_{T^2}(t^2) = \frac{\Gamma\left(\frac{n}{2}\right)}{\Gamma\left(\frac{p}{2}\right)\Gamma\left(\frac{n-p}{2}\right)} \frac{(t^2)^{p/2-1}}{(1+t^2)^{n/2}}, \qquad t^2 \geq 0. \tag{7.89}$$

Lemma 7.5.1. *Let* $Y = (Y_1, \ldots, Y_n)'$, $Y_i = (Y_{i1}, \ldots, Y_{ip})'$, $i = 1, \ldots, n$ *be a* $n \times p$ *random matrix with spherically symmetric distribution with pdf given by*

$$f_Y(y) = q\left(\sum_{i=1}^{n} y_i' y_i\right).$$

(a) *Let* $Y = (Y'_{(1)}, Y'_{(2)})'$, *with* $Y_{(1)} = (Y_1, \ldots, Y_k)'$, $Y_2 = (Y_{k+1}, \ldots, Y_n)'$ *and* $n - k \geq p \geq k$. *The distribution of*

$$Y_{(1)}(Y'_{(2)} Y_{(2)})^{-1} Y'_{(1)}$$

does not depend on a particular choice of q in (7.85).

(b) *If* $k = 1$, $Y_{(1)}(Y'_{(2)} Y_{(2)})^{-1} Y'_{(1)}$ *has Hotelling's* T^2 *distribution with* $(n-1)$ *degrees of freedom under* H_0.

(c) *Let* $k = 1$, $Y_1 = (Y_1^{(1)}, Y_1^{(2)})$ *where* $Y_1^{(1)}, Y_1^{(2)}$ *are* $1 \times q$, $1 \times (p-q)$ *sub-vectors of* Y_1 *respectively, and* $A = Y'_{(2)} Y_{(2)} = \begin{pmatrix} A_{11} & A_{12} \\ A_{21} & A_{22} \end{pmatrix}$ *where* A_{11} *is the left-hand cornered* $q \times q$ *submatrix of* A. *Define*

$$T_1^2 = (Y_1^{(1)}) A_{11}^{-1} (Y_1^{(1)})' = \frac{R_1}{1 - R_1}$$

$$T^2 = T_1^2 + T_2^2 = Y_1 A^{-1} Y_1' = \frac{R_1 + R_2}{1 - R_1 - R_2}.$$

The joint distribution of (T_1^2, T_2^2) *and equivalently the joint distribution of* (R_1, R_2) *under* H_0 *does not depend on q.*

Proof.

(a) Since $n \geq p$, $Y'Y$ is positive definite with probability one. Hence there exists a $p \times p$ upper triangular nonsingular matrix B such that $Y'Y = BB'$. Transform Y to U such that $Y = UB$ where U is a $n \times p$ matrix having the property $U'U = I$. Since the Jacobian of the transformation $Y \to U$ is $(\det(B))^n$, the joint *pdf* of U, B is

$$f_{U,B}(u, b) = q(\operatorname{tr}(bb'))(\det(b))^n.$$

Hence the distribution of U with $U'U = I$ is uniform and U is distributed independently of B. Write

$$U = (U'_{(1)}, U'_{(2)})'$$

where $U_{(1)}, U_{(2)}$, are $k \times p$, $(n-k) \times p$ submatrices of U. Since

$$Y = UB$$

we get

$$Y_{(i)} = U_{(i)}B, \quad i = 1, 2.$$

So

$$Y_{(1)}(Y'_{(2)}Y_{(2)})^{-1}Y'_{(1)} = U_{(1)}B(B'U'_{(2)}U_{(2)}B)^{-1}B'U'_{(1)}$$
$$= U_{(1)}(U'_{(2)}U_{(2)})^{-1}U'_{(1)}.$$

Hence the distribution of $Y_{(1)}(Y'_{(2)}Y_{(2)})^{-1}Y'_{(1)}$ does not depend on any particular choice of q.

(b) Here $k = 1$. Since $Y_{(1)}(Y'_{(2)}Y_{(2)})^{-1}Y'_{(1)}$ has a completely specified distribution for all q in (7.85) we assume without any loss of generality that $Y_1, \ldots, Y_n$ are independently distributed $N_p(0, I)$. Hence $Y'_{(2)}Y_{(2)} = \sum_{i=2}^{n} Y_i Y'_i$ is distributed as Wishart $W_p(n-1, I)$ independently of $Y_{(1)}$. Hence by Theorem 6.8.1 $Y_{(1)}(Y'_{(2)}Y_{(2)})^{-1}Y'_{(1)}$ has Hotelling's T^2 distribution with $n-1$ degrees of freedom.

(c) Let $U = (U_1, \ldots, U_n)'$, $U_i = (U_{i1}, \ldots, U_{ip})'$, $i = 1, \ldots, n$; and let U_1 be similarly partitioned as Y_1. Since $Y_{(2)} = U_{(2)}B$ and $Y_1 = U_1 B$ we get

$$Y'_{(2)}Y_{(2)} = B'CB,$$
$$Y_1^{(1)} = U_1^{(1)}B_{11} \tag{7.90}$$

where $C = U'_{(2)}U_{(2)}$ and B is partitioned as

$$B = \begin{pmatrix} B_{11} & B_{12} \\ 0 & B_{22} \end{pmatrix}$$

with B_{11} a $q \times q$ submatrix. Partition C similarly as

$$C = \begin{pmatrix} C_{11} & C_{12} \\ C_{21} & C_{22} \end{pmatrix}.$$

From (7.90) we get

$$A_{11} = B'_{11}C_{11}B_{11}.$$

Hence

$$T_1^2 = U_1^{(1)} B_{11} (B_{11}' C_{11} B_{11})^{-1} B_{11}' U_1^{(1)'}$$
$$= U_1^{(1)} C_{11}^{-1} U_1^{(1)'},$$
$$T_1^2 + T_2^2 = U_1 (U_{(2)}' U_{(2)})^{-1} U_1'.$$

Hence the joint distribution of (T_1^2, T_2^2) under H_0 does not depend on q. Taking $Y_1, \ldots, Y_n$ to be independent and identically distributed $N_p(0, I)$ and using (6.73) we get

$$f_{R_1, R_2}(r_1, r_2) = \frac{\Gamma\left(\frac{n}{2}\right)}{\Gamma\left(\frac{n-p}{2}\right) \Gamma\left(\frac{q}{2}\right) \Gamma\left(\frac{p-q}{2}\right)} r_1^{q/2-1} r_2^{(p-q)/2-1}$$
$$\times (1 - r_1 - r_2)^{(n-p)/2-1} \qquad (7.91)$$
$$\text{if} \quad 0 < r_i < 1, \qquad i = 1, 2.$$

Q.E.D.

Invariance and Ratio of Densities of T^2

Let $G_\ell(p)$ be the multiplicative group of $p \times p$ nonsingular matrices g transforming $(\bar{X}, S; \mu, \Sigma) \to (g\bar{X}, gSg'; g\mu, g\Sigma g')$. From Section 7.2 a maximal invariant under $G_\ell(p)$ in the space of $(\bar{X}, S)$ is $T^2 = n\bar{X}' S^{-1} \bar{X}$. A corresponding maximal invariant in the parametric space is $\delta^2 = n\mu' \Sigma^{-1} \mu$. The distribution of T^2 depends on the parameters only through δ^2.

Using Theorem 3.8.1 we find the ratio of densities of T^2. Since the Jacobian of the transformation

$$g \to hg$$

$g, h \in G_\ell(p)$ is $(\det h)^p$ (see Example 3.2.8) a left invariant Haar measure on $G_\ell(p)$ is

$$d\mu(g) = \frac{dg}{(\det g)^p},$$

and by Theorem 2.4.2 a left invariant measure on the sample space $\mathcal{X}$ is

$$d\lambda(x) = \frac{dx}{(\det s)^{n/2}}.$$

Hence the *pdf* of $X \in \mathcal{X}$ under $H_1 : \mu \neq 0$ with respect to λ is

$$p_2(x) = (\det s)^{n/2} (\det \Sigma^{-1})^{n/2} q(\operatorname{tr} \Sigma^{-1}(s + n(\bar{x} - \mu)(\bar{x} - \mu)').$$

The *pdf* $p_1(x)$ of X under $H_0: \mu = 0$ is the value of $p_2(x)$ with $\mu = 0$. Using Theorem 3.8.1 we obtain

$$\frac{dP(t^2|\delta^2)}{dP(t^2|0)}$$

$$= \frac{\int_{G_\ell(p)} (\det gsg')^{n/2}(\det\Sigma^{-1})^{n/2} q(\operatorname{tr}\Sigma^{-1}(g(s+n\bar{x}\bar{x}')g' - 2ng\bar{x}\mu' + n\mu\mu'))\frac{dg}{(\det g)^p}}{\int_{G_\ell(p)} (\det gsg')^{n/2}(\det\Sigma^{-1})^{n/2} q(\operatorname{tr}\Sigma^{-1}(g(s+n\bar{x}\bar{x}')g'))\frac{dg}{(\det g)^p}}$$

(7.92)

Since Σ is positive definite we can write

$$\Sigma = \Sigma^{1/2}\Sigma^{1/2}$$

where $\Sigma^{1/2}$ is a $p \times p$ symmetric nonsingular matrix. Let

$$\Sigma^{-\frac{1}{2}}\mu = v, \qquad \Sigma^{-\frac{1}{2}}g = h.$$

Then $h \in G_\ell(p)$ and $n\mu'\Sigma^{-1}\mu = nv'v = \delta^2$. The numerator of (7.92) can be written as

$$\int_{G_\ell(p)} (\det hsh')^{n/2} q(\operatorname{tr}(h(s+n\bar{x}\bar{x}')h') - 2nh\bar{x}v' + nv'v)\frac{dh}{(\det h)^p}$$

Since $s + n\bar{x}\bar{x}'$ is positive definite with probability one we can similarly write

$$s + n\bar{x}\bar{x}' = kk$$

where $k = (s + n\bar{x}\bar{x}')^{1/2}$. Let

$$y = k^{-1}\bar{x}, g = hk \qquad \text{with} \qquad g \in G_\ell(p).$$

Then the ratio (7.92) can be written as

$$\frac{\int_{G_\ell(p)} (\det gg')^{(n-p)/2} q(\operatorname{tr}(gg' - 2ngyv' + nvv'))dg}{\int_{G_\ell(p)} (\det gg')^{(n-p)/2} q(\operatorname{tr}(gg'))dg} \tag{7.93}$$

Using the fact (see Theorem 6.8.1) that given any $\alpha = (\alpha_1, \ldots, \alpha_p)'$ there exists an $p \times p$ orthogonal matrix θ such that

$$\theta\alpha = ((\alpha'\alpha)^{1/2}, 0, \ldots, 0)',$$

Tests of Hypotheses of Mean Vectors

and denoting $\theta g = g = (g_{ij})$ for notational convenience. We rewrite (7.93) as

$$\frac{\int_{G_\ell(p)} (\det gg')^{(n-p)/2} q(\text{tr}(gg' - 2g_{11}\sqrt{r}\delta + \delta^2)) dg}{\int_{G_\ell(p)} (\det gg')^{(n-p)/2} q(\text{tr}(gg')) dg} \quad (7.94)$$

where
$$r = \frac{n\bar{x}'s^{-1}\bar{x}}{1 + n\bar{x}'s^{-1}\bar{x}}.$$

Note From (7.89) the *pdf* of $R = \dfrac{n\bar{X}'S^{-1}\bar{X}}{1 + n\bar{X}'S^{-1}\bar{X}}$ under H_0 is given by

$$f_R(r) = \frac{\Gamma\left(\frac{n}{2}\right)}{\Gamma\left(\frac{p}{2}\right)\Gamma\left(\frac{n-p}{2}\right)} (r)^{p/2-1}(1-r)^{(n-p)/2-1}, \, 0 < r < 1. \quad (7.95)$$

In the following theorem we prove the uniformly most powerful invariant property of the T^2-test in $E_p(\mu, \Sigma)$ under the assumption that q is convex and nonincreasing from $[0, \infty)$ to $[0, \infty)$. In the proof of the theorem the nonincreasing property of q is not used. Since $f_X(x)$ is the *pdf* of X, the convexity property of q implies that q is nonincreasing.

Theorem 7.5.1 *Let $X = (X_1, \ldots, X_n)'$, $X_i = (X_{i1}, \ldots, X_{ip})'$, $i = 1, \ldots, n$ be a $n \times p$ random matrix with pdf*

$$f_X(x) = (\det \Sigma)^{-n/2} q(\text{tr } \Sigma^{-1}(\sum_{i=1}^n (x_i - \mu)(x_i - \mu)')$$

where q is convex and nonincreasing from $[0, \infty)$ to $[0, \infty)$. Assume that if $(\mu, \Sigma) \in \Omega$ (parametric space) then $(\mu, c\Sigma) \in \Omega$ for $c > 0$. Among all tests $\phi(x)$ of level α for testing $H_0 : \mu = 0$ against $H_1 : \mu \neq 0$ which are invariant under the group of transformations $G_\ell(p)$ transforming $X_i \to gX_i, i = 1, \ldots, n, g \in G_\ell(p)$, Hotelling's T^2 test (7.87) or its equivalent (7.88) is uniformly most powerful invariant (UMPI).

Proof. For invariant tests, the problem reduces to testing $H_0 : \delta^2 = n\mu'\Sigma^{-1}\mu = 0$ against the alternatives $H_1 : \delta^2 > 0$. Using the Neyman-Pearson lemma and (7.94) the most powerful invariant test of H_0 against the simple aternative $H_1' : \delta^2 = \delta_0^2$ (δ_0^2 specified) is given by

$$\frac{\int_{G_\ell(p)} (\det gg')^{(n-p)/2} q(\text{tr } gg' - 2g_{11}\sqrt{r}\delta_0 + \delta_0^2) dg}{\int_{G_\ell(p)} (\det gg')^{(n-p)/2} q(\text{tr } gg') dg} \geq c \quad (7.96)$$

where the constant c depends on the level α of the test.

Let $H(\sqrt{r})$ denote the numerator of (7.96). Transforming $g \to -g$ we get

$$H(\sqrt{r}) = \int_{G_\ell(p)} (\det gg')^{(n-p)/2} q(\operatorname{tr} gg' + 2g_{11}\sqrt{r}\delta_0 + \delta_0^2) dg$$

$$= H(-\sqrt{r}).$$

Since q is convex, for $\frac{1}{2} \le \alpha \le 1$, we get

$$H(\sqrt{r}) = \alpha H(\sqrt{r}) + (1-\alpha)H(-\sqrt{r}) \ge H(\sqrt{r(2\alpha-1)}).$$

Hence we conclude that $H(\sqrt{r})$ is a nondecreasing function of $\sqrt{r}$. Since this holds for any $\delta_0 > 0$ we conclude that Hotelling's T^2 test or its equivalent as given in (7.88) is UMPI.   Q.E.D.

7.5.2. Tests of Hypotheses of Subvectors of μ

Let $X = (X_1, \ldots, X_n)'$, $X_i = (X_{i1}, \ldots, X_{ip})'$, $i = 1, \ldots, n (n > p)$ be a $n \times p$ random matrix with *pdf* given in (7.85). Using the notations of Section 7.3 we consider the following two testing problems concerning subvectors of μ.

(a) Let $k = 2, p_1 + p_2 = p$. We consider the problem of testing $H_0 : \mu_{(1)} = 0$ when $\mu_{(2)}, \Sigma$ are unknown. Let Ω be the parametric space of (μ, Σ) with Σ positive definite. Obviously $(\mu, c\Sigma) \in \Omega$ with $c > 0$ and let $w = \{(0, \mu'_{(2)})', \Sigma\}$ be the subspace of Ω when H_0 is true. Using (7.75) and (7.76) when $X_1, \ldots, X_n$ are independent $N_p(\mu, \Sigma)$ and Theorem 5.3.6 we get (as in (7.86))

$$\lambda = \frac{\max_w L(\mu, \Sigma)}{\max_\Omega L(\mu, \Sigma)} = \left[\frac{\det s_{(11)}}{\det(s_{(11)} + n\bar{x}_{(1)}\bar{x}'_{(1)})}\right]^{n/2} \quad (7.97)$$

$$= (1 + n\bar{x}'_{(1)} s_{(11)}^{-1} \bar{x}_{(1)})^{-n/2}.$$

Hence the likelihood ratio test of H_0 rejects H_0 whenever

$$t_1^2 = n\bar{x}'_{(1)} s_{(11)}^{-1} \bar{x}_{(1)} \ge c \quad (7.98)$$

where the constant c depends on the level α of the test. From Lemma 7.5.1 the *pdf* of $T_1^2 = n\bar{X}'_{(1)} S_{(11)}^{-1} \bar{X}_{(1)}$ under H_0 is given by

$$f_{T_1^2}(t_1^2) = \frac{\Gamma\left(\frac{n}{2}\right)}{\Gamma\left(\frac{p_1}{2}\right)\Gamma\left(\frac{n-p_1}{2}\right)} \frac{(t_1^2)^{p_1/2-1}}{(1+t^2)^{n/2}}, \quad t_1^2 \ge 0. \quad (7.99)$$

The invariance of this problem has been discussed in Section 7.3. A maximal invariant in the sample space is T_1^2 and a corresponding maximal

Tests of Hypotheses of Mean Vectors

invariant in the parametric space is $\delta_1^2 = n\mu'_{(1)} \Sigma^{-1}_{(11)} \mu_{(1)}$. Under the assumption that q is convex and nonincreasing and proceeding as in Theorem 7.5.1 it is obvious that the likelihood ratio test for this problem is UMPI.

(b) In the notation of Section 7.3 let $k = 2, p_1 + p_2 = p$. We treat here the problem of testing $H_0 : \mu_{(2)} = 0$ against the alternatives $H_1 : \mu_{(2)} \neq 0$, when it is given that $\mu_{(1)} = 0$. Using (7.78), (7.79) and Theorem 5.3.6 the likelihood ratio test of $H_0 : \mu_{(2)} = 0$ rejects H_0 whenever

$$z = \frac{1 - r_1 - r_2}{1 - r_1} \geq c \qquad (7.100)$$

where the constant c depends on the level α of the test. From (7.91) $(1 - R_1 - R_2/1 - R_1)$ is distributed as central beta under H_0 with parameter $(\frac{1}{2}(n - p), \frac{1}{2}p_2)$. The invariance of the problem and other properties are discussed in Section 7.3.

7.5.3. Locally Minimax Tests

We state only the results and refer the readers to relevant references. For testing $H_0 : \mu = 0$ against $H_1 : n\mu'\Sigma^{-1}\mu = \lambda > 0$ for *pdf* given in (7.85) Giri and Sinha (1987) have shown that Hotelling's T^2 test given in (7.87) or its equivalent (7.88) is locally minimax as $\lambda \to 0$.

For testing $H_0 : \mu_{(1)} = 0$ against $H_1 : \delta_1^2 = n\mu'_{(1)} \Sigma^{-1}_{(11)} \mu_{(1)} = \lambda > 0$ (see problem (a) above) Giri and Sinha (1987) have shown that the likelihood ratio test given in (7.97) is locally minimax as $\lambda \to 0$.

For problem (b) above Giri (1987) has shown that for testing $H_0 : \mu_{(2)} = 0$ against the alternatives $H_1 : \delta_2^2 = n\mu'\Sigma^{-1}\mu - \delta_1^2 = \lambda > 0$ the test which rejects H_0 whenever

$$r_1 + \frac{n - p_1}{p_2} r_2 \geq c$$

where c depends on the level α of the test, is locally minimax as $\lambda \to 0$.

EXERCISES

1 Prove (7.48).
2 Prove (7.50) and (7.51).
3 Test the hypothesis H_0 given in (7.7) when Σ is unknown.
4 Let T^2 be distributed as Hotelling's T^2 with $N - 1$ degrees of freedom. Show that $((N - p)/p)(T^2/(N - 1))$ is distributed as a noncentral F with $(p, N - p)$ degrees of freedom.

5. Let $\sqrt{N}\bar{X}$ be distributed as a p-dimensional normal random variable with mean $\sqrt{N}\mu$ and positive definite covariance matrix Σ and let S be distributed, independently of $\bar{X}$, as $W_p(N-1, \Sigma)$. Show that the distribution of $T^2 = N(N-1)\bar{X}'S^{-1}\bar{X}$ remains unchanged if μ is replaced by $(\delta, 0, \ldots, 0)'$ and Σ by I where $\delta^2 = \mu'\Sigma^{-1}\mu$.

6. *(Test of symmetry of biological organs)*. Let X^α, $\alpha = 1, \ldots, N$, be a random sample of size N from a p-variate normal population with mean μ and positive definite covariance matrix Σ. Assume that p is an even integer, $p = 2k$. Let $\mu = (\mu_{(1)}, \mu_{(2)})$, $\mu_{(1)} = (\mu_1, \ldots, \mu_k)'$. On the basis of the observations x^α on X^α, $\alpha = 1, \ldots, N$, find the appropriate T^2-test of $H_0 : \mu_{(1)} = \mu_{(2)}$.

Note: In many anthropological problems $x_1, \ldots, x_k$ represent measurements on characters on the left side and $x_{k+1}, \ldots, x_p$, represent measurements on the same characters on the right side.

7. *(Profile analysis)*. Suppose a battery of p psychological tests is administered to a group and $\mu_1, \ldots, \mu_p$, are their expected scores. The profile of the group is defined as the graph obtained by joining the points $(i, \mu_i), i = 1, \ldots, p$, successively. For two different groups with expected scores $(\mu_1, \ldots, \mu_p)$ and $(\nu_1, \ldots, \nu_p)$, respectively, for the same battery of tests we obtain two different profiles, one obtained from the points (i, μ_i) and the other obtained from the points $(i, \nu_i), i = 1, \ldots, p$. Two profiles are said to be similar if line segments joining the points $(i, \mu_i), (i+1, \mu_{i+1})$ are parallel to the corresponding line segments joining the points $(i, \nu_i), (i+1, \nu_{i+1})$. For two groups of sizes N_1, N_2, respectively, let $x^\alpha = (x_{\alpha 1}, \ldots, x_{\alpha p})'$, $\alpha = 1, \ldots, N_1$, be the scores of N_1 individuals from the first group and let $y^\alpha = (y_{\alpha 1}, \ldots, y_{\alpha p})'$, $\alpha = 1, \ldots, N_2$, be the scores of N_2 individuals from the second group. Assume that they are samples from two independent p-variate normal populations with different mean vectors $\mu = (\mu_1, \ldots, \mu_p)'$, $\nu = (\nu_1, \ldots, \nu_p)'$ and the same covariance matrix Σ. On the basis of these observations test the hypothesis

$$H_0 : \mu_i - \mu_{i+1} = \nu_i - \nu_{i+1}, \qquad i = 1, \ldots, p-1.$$

Hint Let C_1 be the contrast matrix as defined in Example 7.2.2. Hypothesis H_0 is equivalent to testing the hypothesis that $E(C_1 X^\alpha) = E(C_1 Y^\beta)$, $\alpha = 1, \ldots, N_1, \beta = 1, \ldots, N_2$.

8. *(Union-intersection principle)*. Let $X^\alpha = (X_{\alpha 1}, \ldots, X_{\alpha p})'$, $\alpha = 1, \ldots, N$, be a random sample of size N from a p-variate normal distribution with mean μ and positive definite covariance matrix Σ. The hypothesis $H_0 : \mu = 0$ is true if and only if $H_l : l'\mu = 0$ for any nonnull vector $l \in E^p$ is true. Thus H_0 will be rejected if at least one of the hypothesis $H_l, l \in L = E^p - \{0\}$, is rejected and hence $H_0 = \cap_{l \in L} H_l$. Let ω_l denote the rejection region of the hypothesis

H_l. Obviously the rejection region of H_0 is $\omega = \cup_{l \in L} \omega_l$. The sizes of ω_l should be such that ω is a size α rejection region of H_0. This is known as the union-intersection principle of Roy.

It is evident that $H_l : l'\mu = 0$ is the hypothesis about the scalar mean of the random variables $l'X^\alpha, \alpha = 1, \ldots, N$, with variances $l'\Sigma l$, and that the optimum test for this univariate problem is Student's t-test. Show that the union-intersection principle for testing H_0 leads to the T^2-test.

9. Let $X_i^\alpha = (X_{i\alpha 1}, \ldots, X_{i\alpha p})', \alpha = 1, \ldots, N_i, i = 1, \ldots, k$, be a random sample of size N_i from k independent p-variate normal populations with mean vectors $\mu^i = (\mu_{i1}, \ldots, \mu_{ip})'$ and positive definite covariance matrix Σ_i. Let $N_1 = \min_i N_i$. Define for known scalar constants $\beta_1, \ldots, \beta_k$

$$Y^\alpha = \beta_1 X_1^\alpha + \sum_{i=2}^k \beta_i \left(\frac{N_1}{N_i}\right)^{1/2}$$

$$\times \left(X_i^\alpha - \frac{1}{N_1}\sum_{\beta=1}^{N_1} X_i^\beta + \frac{1}{(N_1 N_i)^{1/2}}\sum_{\gamma=1}^{N_i} X_i^\gamma\right), \quad \alpha = 1, \ldots, N_1$$

$$\bar{Y} = \frac{1}{N_1}\sum_{1}^{N_1} Y^\alpha, \quad S = \sum_{\alpha=1}^{N_1}(Y^\alpha - \bar{Y})(Y^\alpha - \bar{Y})'.$$

Consider the problem of testing $H_0 : \sum_1^k \beta_i \mu^i = \mu_0$ (specified). Show that

$$T^2 = N_1(N_1 - 1)(\bar{Y} - \mu_0)'S^{-1}(\bar{Y} - \mu_0)$$

is disributed as Hotelling's T^2 with $N_1 - 1$ degrees of freedom under H_0.

10. [Giri, 1965]. Let $Z = (Z_1, \ldots, Z_p)'$ be a complex p-dimensional Gaussian random variable with mean $\alpha = (\alpha_1, \ldots, \alpha_p)'$ and positive definite Hermitian complex covariance matrix $\Sigma = E(Z - \alpha)(Z - \alpha)^*$, and let $Z^\alpha = (Z_{\alpha 1}, \ldots, Z_{\alpha p})', \alpha = 1, \ldots, N$, be a sample of size N from the distribution of Z. On the basis of these observations find the likelihood ratio test of the following testing problems.
    (a) To test the hypothesis $H_0 : \alpha = 0$, when Σ is unknown.
    (b) To test the hypothesis $H_0 : \alpha_1 = \cdots = \alpha_{p_1} = 0, p_1 < p$, when Σ is unknown.
    (c) To test the hypothesis $H_0 : \alpha_1 = \cdots = \alpha_{p_1+p_2} = 0, p_1 + p_2 < p$, when it is given that $\alpha_1 = \cdots = \alpha_{p_1} = 0$, when Σ is unknown.

11. Let $X^\alpha = (X_{\alpha 1}, \ldots, X_{\alpha p})', \alpha = 1, \ldots, N_1$, be a random sample of size N_1 from a p-dimensional normal distribution with mean $\mu = (\mu_1, \ldots, \mu_p)'$ and positive definite covariance matrix Σ_1 (unknown), and let $Y^\alpha = (Y_{\alpha 1}, \ldots, Y_{\alpha p})', \alpha = 1, \ldots, N_2$, be a random sample of size N_2 from another

independent p-dimensional normal distribution with mean $v = (v_1, \ldots, v_p)'$ and positive definite covariance matrix Σ_2 (unknown). Find the appropriate test of $H_0 : \mu_i - v_i = 0, i = 1, \ldots, p_1 < p$.

12 Prove (7.87) and (7.100).

REFERENCES

Anderson, T. W. (1984). *An Introduction to Multivariate Statistical Analysis*, 2nd ed. New York: Wiley.

Bartholomeu, D. J. (1961). A test of homogeneity of means under restricted alternatives. *J.R. Statist. Soc. B*, 23:237–281.

Bennett, B. M. (1951). Note on a solution of the generalized Behrens-Fisher problem. *Ann. Inst. Statist. Math.* 2:87–90.

Bose, R. C. and Roy, S. N. (1938). The distribution of Studentized D^2-statistic. *Sankhya* 4:19–38.

Chacko, V. J. (1963). Testing homogeneity against ordered alternatives. *Ann. Math. Statist.* 34:945–956.

Eaton, M. L. (1970). A complete class theorem for multidimensional one-sided alternatives. *Ann. Math. Statist.* 41:1884–1888.

Eaton, M. L. (1988). *Multivariate Statistics, A Vectorspace Approach*. New York: Wiley.

Farrell, R. H. (1985). *Multivariate Calculations*. New York: Springer Verlag.

Giri, N. (1961). On the Likelihood Ratio Tests of Some Multivariate Problems. Ph.D. Thesis, Stanford Univ.

Giri, N. (1965). On the complex analogues of T^2- and R^2-tests. *Ann. Math. Statist.* 36:664–670.

Giri, N. (1968). Locally and asymptotically minimax tests of a multivariate problem. *Ann. Math. Statist.* 39:171–178.

Giri, N. (1972). On a testing problem concerning mean of multivariate complex Gaussian distribution. *Ann. Inst. Statist. Math.* 24:245–250.

Giri, N. (1975). *Invariance and statistical minimax tests*, Selecta Statistica Canadiana, Vol. 3. Hindusthan Publ. Corp. India.

Giri, N. (1987). On a locally best invariant and locally minimax test in symmetrical multivariate distributions. Advances in Multivariate Statistical analysis, D. Reidel Pub. co. 63–83.

Giri, N. and Behara, M. (1971). Locally and asymptotically minimax tests of some multivariate decision problems. *Arch. Math.* 4:436–441.

Giri, N. and Kiefer, J. (1962). Minimax property of Hotelling's and certain other multivariate tests (abstract). *Ann. Math. Statist.* 33:1490–1491.

Giri, N. and Kiefer, J. (1964). Locally and asymptotic minimax properties of multivariate tests. *Ann. Math. Statist.* 35:21–35.

Giri, N., Kiefer, J. and Stein, C. (1963). Minimax character of Hotelling's T^2-test in the simplest case. *Ann. Math. Statist.* 34:1524–1535.

Giri, N. and Sinha, B. K. (1987). Robust tests of mean vector in symmetrical multivariate distributions. *Sankhya A*, 49:254–263.

Goodman, N. R. (1962). Statistical analysis based on a certain multivariate complex Gaussian distribution (an introduction). *Ann. Math. Statist.* 33:152–176.

Hotelling, H. (1931). The generalization of Student's ratio. *Ann. Math. Statist.* 2:360–378.

Hsu, P. L. (1938). Notes on Hotelling's generalized T. *Ann. Math. Statist.* 16:231–243.

Isaacson, S. L. (1951). On the theory of unbiased tests of simple statistical hypotheses specifying the values of two or more parameters. *Ann. Math. Statist.* 22:217–234.

Kariya, T. (1981). A robustness property of Hotelling's T^2 test. *Ann. Math. Statist.*, 9:210–213.

Kariya, T. (1985). *Testing in Multivariate General Linear Model*. New York: Kinokuniya.

Kariya, T. and Sinha, B. K. (1989). *Robustness of Statistical Tests*. New York: Academic Press.

Kariya, T. and Cohen, A. (1992). On the invariance structure of the one-sided testing problem for a multivariate normal mean vector. *J.A.S.A.*, 93:380–386.

Kiefer, J. (1957). Invariance, minimax sequential estimation and continuous time processes. *Ann. Math. Statist.* 28:573–601.

Kiefer, J. and Schwartz, R. (1965). Admissible Bayes character of T^2- and R^2- and other fully invariant tests for classical multivariate normal problems. *Ann. Math. Statist.* 36:747–770.

Kshirsagar, A. M. (1972). *Multivariate Analysis*. New York: Dekker.

Kudô, A. (1963). A multivariate analogue of the one-sided test. *Biometrika* 50:113–119.

Lehmann, E. (1959). *Testing Statistical Hypothesis.* New York: Wiley.

Lehmer, E. (1944). Inverse tables of probabilities of errors of the second kind. *Ann. Math. Statist.* 15:388–398.

Linnik, Ju. Vo. (1966). Teor. Verojatn. Ee. Primen. 561, MR 34.

Linnik, Ju. Vo., Pliss, V. A. and Salaevskie, O. V. (1966). Dokl. Akad. Nauk SSSR 168.

Marden, J. I. (1982). Minimal complete class of tests of hypotheses with multivariate one-sided alternatives. *Ann. Statist.* 10:962–970.

Muirhead, R. J. (1982). *Aspects of Multivariate Statistical Theory.* New York: Wiley.

Nandi, H. K. (1965). On some properties of Roy's union-intersection tests. *Calcutta Statist. Assoc. Bull.* 4:9–13.

Nüesch, P. E. (1966). On problem of testing location in multivariate populations for restricted alternatives. *Ann. Math. Statist.* 37:113–119.

Pearlman, M. D. (1969). One-sided testing problems in multivariate analysis. *Ann. Math. Statist.* 40:549–569.

Rao, C. R. (1948). Tests of significance in multivariate analysis. *Biometrika* 35:58–79.

Rao, C. R. (1973). *Linear Statistical Inference and its Applications.* New York: Wiley.

Roy, S. N. (1953). On a heuristic method of test construction and its use in multivariate analysis. *Ann. Math. Statist.* 24:220–238.

Roy, S. N. (1957). *Some Aspects of Multivariate Analysis.* New York: Wiley.

Salaevskii, O. V. (1968). Minimax character of Hotelling's T_2-test. *Sov. Math. Dokl.* 9:733–735.

Scheffé, H. (1943). On solutions of Behrens-Fisher problem based on the t-distribution. *Ann. Math. Statist.* 14:35–44.

Semika, J. B. (1941). An optimum property of two statistical tests. *Biometrika* 32:70–80.

Sharack, G. R. (1967). Testing against ordered alternatives in model I analysis of variance; normal theory and non-parametric. *Ann. Math. Stat.* 38:1740–1753.

Stein, C. (1956). The admissibility of Hotelling's T^2-test. *Ann. Math. Statist.* 27:616–623.

Tang, P. C. (1938). The power functions of the analysis of variance tests with tables and illustration of their uses. *Statist. Res. Mem.* 2:126–157.

Wald, A. (1943). Tests of statistical hypotheses concerning several parameters when the number of observations is large. *Trans. Amer. Math. Soc.* 54:426–482.

Wald, A. (1950). *Statistical Decision functions*. New York: Wiley.

Wang, Y. and McDermott, M. (1998). Conditional likelihood ratio test for a non-negative normal mean vector. *J.A.S.A.* 93:380–386.

8
Tests Concerning Covariance Matrices and Mean Vectors

8.0. INTRODUCTION

In Sections 8.1–8.7 we develop techniques for testing hypotheses concerning covariance matrices, and covariance matrices and mean vectors of several p-variate normal populations. Then we treat the cases of multivariate complex normals and multivariate elliptically symmetric distributions in Sections 8.8 and 8.9 respectively. The tests discussed are invariant tests, and most of the problems and tests considered are generalizations of univariate ones. In Section 8.1 we discuss the problem of testing the hypothesis that the covariance matrix of a p-variate normal population is a given matrix. In Section 8.2 we consider the sphericity test, that is, where the covariance matrix is equal to a given matrix except for an unknown proportionality factor, which has only a trivial corresponding univariate hypothesis. In Section 8.3 we divide the set of p-variates having a joint multivariate normal distribution into k subsets and study the problem of mutual independence of these subsets. We consider, in detail, the special case of two subsets where the first subset has only one component and where the R^2-test is the appropriate test statistic. Sections 8.4 and 8.5 deal with the admissibility and the minimax properties of tests of independence and the R^2-test. Section 8.7 deals with the multivariate general linear hypothesis. In Section 8.5 we study problems of testing hypotheses of equality of covariance matrices and equality of both covariance matrices and mean vectors. The asymptotic distribution of the likelihood ratio test statistics under the null hypothesis is given for each problem. In Section 8.5.2 we treat the problem of multiple correlation with partial information.

Because of space requirements we treat only the problem of testing $\rho^2 = 0$ in p-variate complex normal in Section 8.8. Section 8.9 will deal with several testing problems concerning the scale matrix Σ in $E_p(\mu, \Sigma)$. Sections 8.10 will treat incomplete data.

8.1. HYPOTHESIS: A COVARIANCE MATRIX IS UNKNOWN

Let $X^\alpha = (X_{\alpha 1}, \ldots, X_{\alpha p})'$, $\alpha = 1, \ldots, N$, be a random sample of size $N(N > p)$ from a p-variate normal distribution with unknown mean μ and unknown positive definite covariance matrix Σ. As usual we assume throughout that $N > p$, so that the sample covariance matrix S is positive definite with probability 1. We are interested in testing the null hypothesis $H_0 : \Sigma = \Sigma_0$ against the alternatives $H_1 : \Sigma \neq \Sigma_0$ where Σ_0 is a fixed positive definite matrix. Since Σ_0 is positive definite there exists a nonsingular matrix $g \in G_l(p)$, the full linear group, such that $g\Sigma_0 g' = I$. In particular, we can take $g^{-1} = \Sigma_0^{1/2}$ where $\Sigma_0^{1/2}$ is a symmetric matrix such that $\Sigma_0 = \Sigma_0^{1/2} \Sigma_0^{1/2}$.

Let $Y^\alpha = gX^\alpha$, $\alpha = 1, \ldots, N$, $\nu = g\mu$, and $\Sigma^* = g\Sigma g'$. Then Y^α, $\alpha = 1, \ldots, N$, constitute a random sample of size N from a p-variate normal distribution with unknown mean ν and unknown positive definite covariance matrix Σ^*. The problem is transformed to testing the null hypothesis $H_0 : \Sigma^* = I$ against alternatives that $\Sigma^* \neq I$ on the basis of sample observations y^α on Y^α, $\alpha = 1, \ldots, N$. The parametric space $\Omega = \{(\nu, \Sigma^*)\}$ is the space of ν and Σ^*, and under H_0 it reduces to the subspace $\omega = \{(\nu, I)\}$. Let

$$\bar{x} = \frac{1}{N}\sum_{\alpha=1}^{N} x^\alpha, \qquad \bar{y} = \frac{1}{N}\sum_{\alpha=1}^{N} y^\alpha, \qquad s = \sum_{\alpha=1}^{N}(x^\alpha - \bar{x})(x^\alpha - \bar{x})'$$

$$b = \sum_{\alpha=1}^{N}(y^\alpha - \bar{y})(y^\alpha - \bar{y})'.$$

Obviously $\bar{y} = g\bar{x}$, $b = gsg'$. The likelihood of the observations y_α, $\alpha = 1, \ldots, N$ is

$$L(\nu, \Sigma^*) = (2\pi)^{-Np/2}(\det \Sigma^*)^{-N/2}$$
$$\times \exp(-\tfrac{1}{2}\operatorname{tr} \Sigma^{*-1}\{\Sigma_{\alpha=1}^{N}(y^\alpha - \nu)(y^\alpha - \nu)'\}) \tag{8.1}$$

Covariance Matrices and Mean Vectors

By Lemma 5.1.1,

$$\max_{\Omega} L(\nu, \Sigma^*)(= 2\pi/N)^{-Np/2}(\det(b))^{-N/2}\exp\{-\tfrac{1}{2}Np\} \quad (8.2)$$

Under H_0, $L(\nu, \Sigma^*)$ reduces to

$$L(\nu, I) = (2\pi)^{-Np/2}\exp\{-\tfrac{1}{2}\operatorname{tr}\sum_{\alpha=1}^{N}(y^{\alpha} - \nu)(y^{\alpha} - \nu)'\}, \quad (8.3)$$

so

$$\max_{\omega} L(\nu, \Sigma^*) = (2\pi)^{-Np/2}\exp\{-\tfrac{1}{2}\operatorname{tr} b\} \quad (8.4)$$

Hence the likelihood ratio criterion for testing $H_0 : \Sigma^* = I(\Sigma = \Sigma_0)$ is given by

$$\begin{aligned}\lambda &= \frac{\max_{\omega} L(\nu, \Sigma^*)}{\max_{\Omega} L(\nu, \Sigma^*)} = \left(\frac{e}{N}\right)^{Np/2}(\det(b))^{N/2}\exp\{-\tfrac{1}{2}\operatorname{tr} b\} \\ &= \left(\frac{e}{N}\right)^{Np/2}(\det(\Sigma_0^{-1}s))^{N/2}\exp\{-\tfrac{1}{2}\operatorname{tr}\Sigma_0^{-1}s\}\end{aligned} \quad (8.5)$$

as $g'g = \Sigma_0^{-1}$. Thus we get the following theorem.

Theorem 8.1.1. *The likelihood ratio test of $H_0 : \Sigma = \Sigma_0$ rejects H_0 whenever*

$$\lambda = (e/N)^{Np/2}(\det\Sigma_0^{-1}s)^{N/2}\exp\{-\tfrac{1}{2}\operatorname{tr}\Sigma_0^{-1}s\} \leq C,$$

where the constant C, is chosen in such a way that the test has size α.

To evaluate the constant C, we need the distribution of λ under H_0. Let B be the random matrix corresponding to b; that is, $B = \sum_{\alpha=1}^{N}(Y^{\alpha} - \bar{Y})(Y^{\alpha} - \bar{Y})'$. Then B has a Wishart distribution with parameter I and $N - 1$ degrees of freedom when H_0 is true. The characteristic function $\phi(t)$ of $-2\log\lambda$ under H_0 is given by (using (6.32) and (6.37)).

$$\begin{aligned}\phi(t) &= E(\exp\{-2it\log\lambda\}) = E(\lambda^{-2it}) \\ &= \int K\left(\frac{e}{N}\right)^{-ipNt}(\det b)^{1/2(N-p-2-2iNt)}\exp\left\{-\frac{1}{2}(1-2it)\operatorname{tr} b\right\}db \\ &= \left(\frac{2e}{N}\right)^{-ipNt}(1-2it)^{-\tfrac{1}{2}p(N-1-2iNt)}\prod_{j=1}^{p}\frac{\Gamma(\tfrac{1}{2}(N-j)-iNt)}{\Gamma(\tfrac{1}{2}(N-j))} \\ &= \prod_{j=1}^{p}\phi_j(t),\end{aligned} \quad (8.6)$$

where $\phi_j(t), j = 1, \ldots, p$, is given by

$$\phi_j(t) = \frac{(2e/N)^{-iNt}(1-2it)^{-(N-1-2iNt)/2}\Gamma(\frac{1}{2}(N-j)-iNt)}{\Gamma(\frac{1}{2}(N-j))} \quad (8.7)$$

But as $N \to \infty$, using Stirling's approximation for the gamma function, we obtain

$$\phi_j(t) \sim 2^{-iNt}e^{-iNt}(1-2it)^{(2iNt-N+1)/2}$$

$$\times \frac{\exp\{-[\frac{1}{2}(N-j)-iNt]\}[\frac{1}{2}(N-j-2)-iNt]^{(N-j-1)/2-iNt}}{\exp\{-[\frac{1}{2}(N-j)]\}[\frac{1}{2}(N-j-2)]^{(N-j-1)/2}}$$

$$= (1-2it)^{-j/2}\left(1 - \frac{it(j+2)}{\frac{1}{2}(N-j-2)(1-2it)}\right)^{(N-j)/2-1/2}$$

$$\times \left(1 - \frac{it(j+2)}{itN(1-2it)}\right)^{-iNt}$$

$$\to (1-2it)^{-j/2}.$$

Thus as $N \to \infty$

$$\phi(t) \to \prod_{j=1}^{p}(1-2it)^{-j/2}. \quad (8.8)$$

Since $(1-2it)^{-j/2}$ is the characteristic function of a chi-square random variable χ_j^2 with j degrees of freedom, as $N \to \infty$, $-2\log\lambda$ is distributed as $\sum_{j=1}^{p}\chi_j^2$, where the χ_j^2 are independent whenever H_0 is true. Thus $-2\log\lambda$ is distributed as $\chi_{p(p+1)/2}^2$ when H_0 is true and $N \to \infty$. For small values of $n = N-1$ Nagarsenker and Pillai (1973) have tabulated the upper 1% and 5% points of the null distribution of $-2\log\lambda^*$, where λ^* is the modified likelihood ratio test criterion.

The problem of testing $H_0: \Sigma^* = I$ against the alternatives $H_1: \Sigma^* \neq I$ remains invariant under the affine group $G = (O(p), E^p)$ where $O(p)$ is the multiplicative group of $p \times p$ orthogonal matrices, and E^p is the translation group, operating as

$$Y^\alpha \to gY^\alpha + a, \quad g \in O(p), \quad a \in E^p, \quad \alpha = 1, \ldots, N. \quad (8.9)$$

This induces in the space of the sufficient statistic $(\bar{Y}, B)$ the transformation $(\bar{Y}, B) \to (g\bar{Y} + a, gBg')$.

Covariance Matrices and Mean Vectors

Lemma 8.1.1. *A set of maximal invariants in the space of $(\bar{Y}, B)$ under the affine group G comprises the characteristic roots of B, that is, the roots of*

$$\det(B - \lambda I) = 0. \tag{8.10}$$

Proof. Since $\det(g(B - \lambda I)g') = \det(B - \lambda I)$, the roots of the equation $\det(B - \lambda I) = 0$ are invariant under G. To see that they are maximal invariant suppose that $\det(B - \lambda I) = 0$ and $\det(B^* - \lambda I) = 0$ have the same roots, where B, B^* are two symmetric positive definite matrices; we want to show that there exists a $g \in O(p)$ such that $B^* = gBg'$. Since B, B^* are symmetric positive definite matrices there exist orthogonal matrices $g_1, g_2 \in O(p)$ such that

$$g_1 B g_1' = \Delta, \qquad g_2 B^* g_2' = \Delta,$$

where Δ is a diagonal matrix whose elements are the roots of (8.10). Since $g_1 B g_1' = g_2 B^* g_2'$ we get

$$B^* = g_2' g_1 B g_1' g_2 = gBg'$$

where $g = g_2' g_1 \in O(p)$. Q.E.D.

We shall denote the characteristic roots of B by $R_1, \ldots, R_p$. Similarly the corresponding maximal invariants in the parametric space of (v, Σ^*) under G are $\theta_1, \ldots, \theta_p$, the roots of $\det(\Sigma^* - \lambda I) = 0$. Under H_0 all $\theta_i = 1$, and under H_1 at least one $\theta_i \neq 1$. The likelihood ratio test criterion λ in terms of the maximal invariants $(R_1, \ldots, R_p)$ can be written as

$$\lambda = (e/N)^{Np/2} \prod_{i=1}^{p} (r_i)^{N/2} \exp\{-\tfrac{1}{2} \Sigma_{i-1}^{p} r_i\}. \tag{8.11}$$

The modified likelihood ratio test for testing $H_0 : \Sigma = \Sigma_0$ rejects H_0 when

$$(e/N)^{Np/2} (\det \Sigma_0^{-1} s)^{(N-1)/2} \exp\{-\tfrac{1}{2} \operatorname{tr} \Sigma_0^{-1} s\} \leq C',$$

where the constant C' depends on the size α of the test. Note that the modified likelihood ratio test is obtained from the corresponding likelihood ratio test by replacing the sample size N by $N - 1$. Since e/N is constant, we do not change the constant term in the modified likelihood ratio test for the sake of convenience only.

It is well known that (see, e.g., Lehmann, 1959, p. 165) for $p = 1$ the rejection region of the likelihood ratio test is not unbiased. The same result also holds in this case (Das Gupta, 1969). However, the modified likelihood ratio test is unbiased. The following theorem is due to Sugiura and Nagao (1968).

Theorem 8.1.2. *For testing $H_0 : \Sigma = \Sigma_0$ against the alternatives $\Sigma \neq \Sigma_0$ for unknown μ, the modified likelihood ratio test is unbiased.*

Proof. Let $g \in O(p)$ be such that $g\Sigma_0^{-1/2}\Sigma\Sigma_0^{-1/2}g'$ is a diagonal matrix Γ where $\Sigma_0^{-1/2}$ is the inverse matrix of the symmetric matrix $\Sigma_0^{1/2}$ such that $\Sigma_0^{1/2}\Sigma_0^{1.2} = \Sigma_0$. As indicated earlier we can assume, without any loss of generality, that $\Sigma_0 = I$ and $\Sigma = \Gamma$, the diagonal matrix whose diagonal elements are the characteristic roots of $\Sigma_0^{-1/2}\Sigma\Sigma_0^{-1/2}$. Hence S has a Wishart distribution with parameter Γ when H_1 is true. Let ω be the acceptance region of the modified likelihood ratio test, that is,

$$\omega = \{s | s \text{ is positive definite and } (e/N)^{Np/2}(\det \Sigma_0^{-1}s)^{(N-1)/2}$$
$$\times \exp\{-\tfrac{1}{2}\operatorname{tr} \Sigma_0^{-1}s\} > C'\}.$$

Then the probability of accepting H_0 when H_1 is true is given by [see (6.32)]

$$P\{\omega | H_1\} = \int_\omega C_{n,p}(\det s)^{(N-p-2)/2}(\det \Gamma)^{-(N-1)/2} \exp\{-\tfrac{1}{2}\operatorname{tr} \Gamma^{-1}s\}\, ds \qquad (8.12)$$
$$= \int_{\omega^*} C_{n,p}(\det u)^{(N-p-2)/2} \exp\{-\tfrac{1}{2}\operatorname{tr} u\}\, du,$$

where $u = \Gamma^{-1/2}s\Gamma^{-1/2}$ and ω^* is the set of all positive definite matrices u such that $\Gamma^{1/2}u\Gamma^{1/2}$ belongs to ω. Note that the Jacobian is $\det(\partial u/\partial s) = (\det \Gamma)^{-(p+1)/2}$. Since $\omega^* = \omega$ when H_0 is true and in the region ω

$$(\det u)^{(N-p-2)/2}\exp\{-\tfrac{1}{2}\operatorname{tr} u\} \geq C'(e/N)^{Np/2}(\det u)^{-(p+1)/2}, \qquad (8.13)$$

we get

$$\int_{\omega-\omega\cap\omega^*}(\det u)^{(N-p-2)/2}\exp\{-\tfrac{1}{2}\operatorname{tr} u\}\, du$$

exists. Also

$$\int_{\omega-\omega\cap\omega^*}(\det u)^{(N-p-2)/2}\exp\{-\tfrac{1}{2}\operatorname{tr} u\}\, du$$
$$\geq C'\left(\frac{e}{N}\right)^{Np/2}\int_{\omega-\omega\cap\omega^*}(\det u)^{-(p-2)/2}du, \qquad (8.14)$$

and

$$-\int_{\omega-\omega\cap\omega^*}(\det u)^{(N-p-2)/2}\exp\{-\tfrac{1}{2}\operatorname{tr} u\}\, du$$
$$\geq C'\left(\frac{e}{N}\right)^{Np/2}\int_{\omega-\omega\cap\omega^*}(\det u)^{-(p-2)/2}du. \qquad (8.15)$$

Covariance Matrices and Mean Vectors

Combining (8.14) and (8.15) with the fact that

$$\int_{\omega \cap \omega^*} (\det u)^{-(p+1)/2} du < \infty,$$

we get

$$P\{\omega|H_0\} - P\{\omega|H_1\}$$

$$= C_{n,p} \left\{ \int_{\omega - \omega \cap \omega^*} - \int_{\omega^* - \omega \cap \omega^*} \right\} (\det u)^{(N-p-2)/2} \exp\{-\tfrac{1}{2} \operatorname{tr} u\} \, du$$

$$\geq C_{n,p} C' \left(\frac{e}{N}\right)^{Np/2} \left\{ \int_{\omega - \omega \cap \omega^*} - \int_{\omega^* - \omega \cap \omega^*} \right\} (\det u)^{-(p+1)/2} du \quad (8.16)$$

$$= C_{n,p} C' \left(\frac{e}{N}\right)^{Np/2} \left\{ \int_{\omega} - \int_{\omega^*} \right\} (\det u)^{-(p+1)/2} du = 0.$$

The last inequality follows from the fact (see Example 3.2.8) that $(\det u)^{-(p+1)/2}$ is the invariant measure in the space of the u under the full linear group $G_l(p)$ transforming $u \to gug'$, $g \in G_l(p)$; that is;

$$\int_{\omega^*} (\det u)^{-(p+1)/2} du = \int_{\omega} (\det u)^{-(p+1)/2} du,$$

and hence the result. Q.E.D.

The acceptance region of the likelihood ratio test does not possess this property, and within the acceptance region we have

$$(\det u)^{(N-p-2)/2} \exp\{-\tfrac{1}{2} \operatorname{tr} u\} \geq C(e/N)^{-Np/2} (\det u)^{-(p+2)/2}$$

instead of (8.13). Anderson and Das Gupta (1964a) showed that (this will follow trivially from Theorem 8.5.2) any invariant test for this problem (obviously it depends only $r_1, \ldots, r_p$) with the acceptance region such that if $(r_1, \ldots, r_p)$ is in the region, so also is $(\bar{r}_1, \ldots, \bar{r}_p)$ with $\bar{r}_i \leq r_i$, $i = 1, \ldots, p$, has a power function that is an increasing function of each θ_i where $\theta_1, \ldots, \theta_p$ are the characteristic roots of Σ^*.

Das Gupta (1969) obtained the following results.

Theorem 8.1.3. *The likelihood ratio test for $H_0 : \Sigma = \Sigma_0$ (i) is biased against $H_1 : \Sigma \neq \Sigma_0$, and (ii) has a power function $\beta(\theta)$ that increases as the absolute deviation $|\theta_i - 1|$ increases for each i.*

Proof. As in the Theorem 8.1.2 we take $\Sigma_0 = I$ and $\Sigma = \Gamma$, the diagonal matrix with diagonal elements $(\theta_1, \ldots, \theta_p)$. S has a Wishart distribution with parameter

Γ and $N-1$ degrees of freedom. Let $S = (S_{ij})$. Then

$$(\det s)^{N/2} \exp\{-\tfrac{1}{2}\operatorname{tr} s\} = \left[\frac{\det s}{\prod_{i=1}^{p} S_{ii}}\right]^{N/2} \left[\prod_{i=1}^{p} s_{ii}^{N/2} \exp\{-\tfrac{1}{2}s_{ii}\}\right]. \quad (8.17)$$

From (6.32), since Γ is a diagonal matrix, $S_{ii}/\theta_i, i = 1, \ldots, p$ are independently distributed χ^2_{N-1} random variables and for any $k(>0)[\det S/\prod_{i=1}^{p} S_{ii}]^k$ and the S_{ii} (or any function thereof) are mutually independent. Furthermore, the distribution of $[\det S/\prod_{i=1}^{p} S_{ii}]^{N/2}$ is independent of $\theta_1, \ldots, \theta_p$. From Exercise 5 it follows that there exists a constant θ_p^* such that $1 < \theta_p^* < N/(N-1)$ and

$$P\{S_{pp}^{N/2} \exp(-\tfrac{1}{2}S_{pp}) \geq C | \theta_p = 1\}$$
$$< P\{S_{pp}^{N/2} \exp(-\tfrac{1}{2}S_{pp}) \geq C | \theta_p^* = \theta_p\} \quad (8.18)$$

irrespective of the value of C chosen. Hence if we evaluate the probability with respect to S_{pp}, keeping $S_{11}, \ldots, S_{p-1,p-1}$ and $[\det S/\prod_{i=1}^{p} S_{ii}]^{N/2}$ fixed, we obtain

$$P\{(\det S)^{N/2} \exp\{-\tfrac{1}{2}\operatorname{tr} S\} \geq C | H_1\}$$
$$> P\{(\det S)^{N/2} \exp\{-\tfrac{1}{2}\operatorname{tr} S\} \geq C | H_0\} \quad (8.19)$$

Thus the acceptance region ω of the likelihood ratio test satisfies $P(\omega|H_1) - P(\omega|H_0) > 0$. Hence the likelihood ratio test is biased. From Exercise 5 it follows that if $2r = m$, then $\beta(\theta)$ increases as $|\theta - 1|$ increases. Hence from the fact noted in connection with the proof of (i) we get the proof of (ii). Q.E.D.

Das Gupta and Giri (1973) proved the following theorem. Consider the class of rejection regions $C(r)$ for $r \geq 0$, given by

$$C(r) = \{s | s \text{ is positive definite and}$$
$$(\det \Sigma_0^{-1} s)^{r/2} \exp\{-\tfrac{1}{2}\operatorname{tr} \Sigma_0^{-1} s\} \leq k\}. \quad (8.20)$$

Theorem 8.1.4. *For testing $H_0 : \Sigma = \Sigma_0$ against the alternatives $H_1 : \Sigma \neq \Sigma_0$: (a)$P\{C(r)|H_1\}$ increases monotonically as each θ_i (characteristic root of $\Sigma_0^{-1/2}\Sigma\Sigma_0^{-1/2}$) deviates from $r/(N-1)$ either in the positive or in the negative direction; (b)$C(r)$ for which $(1 - \det(\Sigma_0^{-1/2}\Sigma\Sigma_0^{-1/2}))(r-n) \geq 0$ is unbiased for H_0 against H_1.*

Covariance Matrices and Mean Vectors

The proof follows from Theorems 8.1.2 and 8.1.3 and the fact that $C(r)$ (with $\Sigma_0 = I$) can also be written as

$$(\det s^*)^{(N-1)/2} \exp\{-\tfrac{1}{2}\operatorname{tr} s^*\} \leq k^*,$$

where $s^* = ((N-1)/r)s$.

Using the techniques of Kiefer and Schwartz (1965), Das Gupta and Giri (1973) have observed that the following rejection regions are unique (almost everywhere), Bayes, and hence admissible for this problem whenever $N - 1 > p$:

(i) $(\det \Sigma_0^{-1} s)^{r/2} \exp\{-\tfrac{1}{2}\operatorname{tr} \Sigma_0^{-1} s\} \leq k$, $1 < r, < \infty$,

(ii) $(\det \Sigma_0^{-1} s)^{r/2} \exp\{-\tfrac{1}{2}\operatorname{tr} \Sigma_0^{-1} s\} \geq k$, $-\infty < r < 0$.

For this problem Kiefer and Schwartz (1965) have shown that the test which rejects H_0 whenever

$$(\det \Sigma_0^{-1} s) \leq C, \tag{8.21}$$

where the constant C is chosen such that the test has the level of significance α, is admissible Bayes against the alternatives that $\Sigma_0 - \Sigma$ is negative definite. The value of C can be determined from Theorem 6.6.1. Note that $\det(\Sigma_0^{-1} S) = \det(\Sigma_0^{-1/2} S \Sigma_0^{-1/2})$. They have also shown that for testing $H_0 : \Sigma = \Sigma_0$, the test which rejects H_0 whenever

$$\operatorname{tr} \Sigma_0^{-1} s \geq C_1 \quad \text{or} \quad \leq C_2, \tag{8.22}$$

where C_1, C_2 are constants depending on the level of significance α of the test, is admissible against the alternatives $H_1 : \Sigma \neq \Sigma_0$. This is in the form that is familiar to us when $p = 1$. It is easy to see that $\operatorname{tr}(\Sigma_0^{-1} S)$ has a χ^2 distribution with $(N-1)p$ degrees of freedom when H_0 is true. John (1971) derived the LBI test for this problem. To establish the LBI property of the test based on $\operatorname{tr} \Sigma_0^{-1} S$ we need the following preliminaries. Let $O(p)$ be the group of $p \times p$ orthogonal matrices $O = (O_{ij}) = (O_1, \ldots, O_p)$, where $O_i = (O_{i1}, \ldots, O_{ip})'$, $i = 1, \ldots, p$. An invariant probability measure μ on $O(p)$ is given by the joint distribution of the O_i, where for each i $(O_{i1}^2, \ldots, O_{ip}^2)$ has Dirichlet distribution $D(\tfrac{1}{2}, \ldots, \tfrac{1}{2})$ as given in Theorem 4.3.6. This measure can be constructed from the normal distribution as follows.

Let $U = (U_1, \ldots, U_p)$ with $U_i = (U_{i1}, \ldots, U_{ip})'$, $i = 1, \ldots, p$ be a $p \times p$ random matrix where $U_1, \ldots, U_p$ are independently and identically distributed $N_p(0, I)$. As in Theorem 4.3.6 write $U = OT$ with $O \in O(p)$ and T is a $p \times p$ upper triangular matrix with positive diagonal elements obtained from U by applying the Gram-Schmidt orthogonalization process on $U_1, \ldots, U_p$, and $O = (O_{ij})$ is the $p \times p$ orthogonal matrix such that for each i $(O_{i1}, \ldots, O_{ip}) = O_i$ is distributed as

$$\left(\frac{U_{i1}}{\|U_i\|}, \ldots, \frac{U_{ip}}{\|U_i\|}\right)$$

This implies that

$$(O_{i1}^2, \ldots, O_{ip}^2)$$

has the Dirichlet distribution $D(\frac{1}{2}, \ldots, \frac{1}{2})$. From this it follows that

$$\int_{O(p)} O_{ij} O_{lk} \mu(dO) = \frac{\delta_{il}\delta_{jk}}{p} \quad (8.23)$$

where δ_{ij} is Kronecker's δ.

Let $A = (a_{ij})$, $B = (b_{ij})$ be $p \times p$ matrices. Since

$$\text{tr } AOBO' = \sum_i \sum_j \sum_k \sum_l a_{ij} b_{kl} O_{jl} O_{il}$$

It follows that

$$\int_{O(p)} (\text{tr } AOBO') d\mu(O) = \frac{(\text{tr } A)(\text{tr } B)}{p} \quad (8.24)$$

It is now left to the reader to verify that

$$\int_{O(p)} (\text{tr } AOBO')^2 d\mu(O) = \frac{(\text{tr}(A^2))(\text{tr}(B^2))}{p(p+1)}$$

$$\int_{O(p)} (\text{tr } AO)(\text{tr}(BO) d\mu(O) = \frac{\text{tr}(AB')}{p}. \quad (8.25)$$

Theorem 8.1.5. *For testing $H_0 : \Sigma = \Sigma_0$ (equivalently $\Sigma^* = \Sigma_0^{-1/2} \Sigma \Sigma_0^{-1/2} = I$) against the alternative $H_1 : \Sigma - \Sigma_0$ is positive definite (equivalently $\Sigma^* - I$ is positive definite), the test which rejects H_0 whenever*

$$\text{tr } \Sigma_0^{-1} s \geq C$$

is LBI under the affine group G of transformations (8.9).

Proof. From Lemma 8.1.1, $R_1, \ldots, R_p$ is a minimal invariant under G in the sample space whose distribution depends only on $\theta_1, \ldots, \theta_p$ the corresponding maximal invariant in the parametric space. Let R and θ be diagonal matrices with diagonal elements $R_1, \ldots, R_p$ and $\theta_1, \ldots, \theta_p$, respectively. From (3.20) the ratio

Covariance Matrices and Mean Vectors

of densities of R is given by (with μ the invariant probability measure on $O(p)$)

$$\frac{dP(R|\theta)}{dP(R|I)} = \frac{\int_{O(p)} (\det \theta)^{-n/2} \exp\{-\frac{1}{2}\text{tr } \theta^{-1}ORO'\} d\mu(O)}{\int_{O(p)} \exp\{-\frac{1}{2}\text{tr } R\} d\mu(O)}$$

Using (3.24) we get

$$\frac{dP(R|\theta)}{dP(R|I)} = 1 + \int_{O(p)} \frac{1}{2}(\text{tr}(I - \theta^{-1})ORO' d\mu(O) - \frac{n}{2}\text{tr}(I - \theta^{-1})$$

$$+ \int_{O(p)} [\tfrac{1}{2}(\text{tr}(I - \theta^{-1})ORO')]^2 d\mu(O)$$

$$+ o\left(\sum_{i=1}^{p}(1 - \theta_i^{-1})\right).$$

Using (8.24) and (8.25) we get

$$\frac{dP(R|\theta)}{dP(R|I)} = 1 + \text{tr}(I - \theta^{-1})\left[\frac{\text{tr } R}{2p} - \frac{n}{2}\right] + o\left(\sum_{i=1}^{p}(1 - \theta_i^{-1})\right)$$

when $\sum_{i=1}^{p}(1 - \theta_i^{-1}) \to 0$. Hence from the results presented in Section 3.7 we get the theorem. Q.E.D.

We now derive several admissible tests for this problem. Given two positive definite matrices Σ_{0L} and Σ_{0U} let us consider the problem of testing $H_0 : \Sigma = \Sigma_0$ against $H_1 : \Sigma$ is one of the pair $(\Sigma_{0L}, \Sigma_{0U})$ or else that either $\Sigma - \Sigma_{0L}$ or $\Sigma_{0U} - \Sigma$ is positive definite and find the admissible test of H_0 against H_1 by using the Bayes approach of Kiefer and Schwartz (1965) as discussed in Section 7.2.3. Let the prior Π_0 under H_0 be such that

$$\Pi_0(\Sigma = \Sigma_0) = 1,$$

$$\Pi_0(dv) = \frac{1}{(2\pi)^{p/2}[\det(\Sigma_{0U} - \Sigma_0)]^{1/2}}$$

$$\times \exp\left\{-\tfrac{1}{2}\text{tr}(v - v_0)'(\Sigma_{0U} - \Sigma_0)^{-1}(v - v_0)\right\}$$

where $v = \sqrt{N}\mu$, v_0 is a fixed vector and let the prior Π_1 under H_1 be given by

$$\Pi_1 = c_1 \Pi_{1a} + (1 - c_1)\Pi_{1d}$$

where $0 < c_1 < 1$ and

$$\Pi_{1a}(\Sigma = \Sigma_{oL}) = 1,$$

$$\Pi_{1a}(dv) = \frac{1}{(2\pi)^{p/2}(\det(\Sigma_{0U} - \Sigma_{0L}))^{1/2}}$$
$$\times \exp\{-\tfrac{1}{2}\mathrm{tr}(v - v_0)'(\Sigma_{0U} - \Sigma_{0L})^{-1}(v - v_0)\},$$
$$\Pi_{1d}(\Sigma = \Sigma_0, v = v_0) = 1.$$

Using (7.38) the rejection region of the admissible Bayes test is given by,

$$\frac{\int f(x^1,\ldots,x^N|\Delta)\Pi_1(d\Delta)}{\int f(x^1,\ldots,x^N|\Delta)\Pi_0(d\Delta)} = \left(\left[c_1(\det\Sigma_{0L})^{-(N-1)/2}(\det\Sigma_{0U})^{-1/2}\right.\right.$$
$$\times \exp\left\{-\frac{tr}{2}(\Sigma_{0L}^{-1}s - G)\right\}$$
$$+ (1 - c_1)(\det\Sigma_{0U})^{-N/2}\exp\frac{tr}{2}\left(\Sigma_{0U}^{-1}s - G\right)\right] /$$
$$\left[(\det\Sigma_0)^{-(N-1)/2}(\det\Sigma_{0U})^{-1/2}\right.$$
$$\left.\left.\times \exp\left\{-\frac{tr}{2}(\Sigma_{0U}^{-1}s - G)\right\}\right]\right) \geq c$$

for some $c, 0 \leq c < \infty$ and

$$G = \Sigma_{0U}^{-1}(\sqrt{N}\bar{x} - v_0)(\sqrt{N}\bar{x} - v_0)'.$$

From above the rejection region of the admissible Bayes test can be written as

$$c_2 \exp\{\tfrac{1}{2}\mathrm{tr}(\Sigma_0^{-1} - \Sigma_{0L}^{-1})s\} + c_3 \exp\left\{\frac{tr}{2}(\Sigma_0^{-1} - \Sigma_{0U}^{-1})s\right\} \geq 1 \qquad (8.26)$$

where c_2, c_3 are nonnegative constants (not both zero).

Let us suppose that there are positive constants a_L and a_U such that $\Sigma_{0L} = a_L\Sigma_0$ and $\Sigma_{0U} = a_U\Sigma_0$, then (8.26) reduces to

$$\mathrm{tr}\,\Sigma_0^{-1}s \geq c_4 \qquad \text{or} \qquad \leq c_5 \qquad (8.27)$$

where c_4 and c_5 are arbitrary constants depending on the level α of the test. Hence we get the following theorem.

Theorem 8.1.6. *For testing $H_0 : \Sigma = \Sigma_0$ against $H_1 : \Sigma = a\Sigma_0, a \neq 1$, the test given in (8.27) is admissible.*

Covariance Matrices and Mean Vectors

Remark. If $c_4 = \infty$ or $c_5 = 0$ (8.27) reduces to one-sided tests.

For testing $H_0 : \Sigma = \Sigma_0$ against the alternatives $H_1 : \Sigma - \Sigma_0$ is positive definite, the admissible Bayes character of the rejection region

$$\det \Sigma_0^{-1} s \geq c \qquad (8.28)$$

where c is a constant depending on the level α of the test, can be established using the Bayes approach by letting Π_1 assign all its measure to $\Sigma^{-1} = \Sigma_0^{-1} + \eta\eta'$ (independently of v) under H_1 with η a $p \times 1$ random vector having

$$\Pi_1(d\eta) = (\det(\Sigma_0^{-1} + \eta\eta'))^{-(N-1)/2}$$

and $\Pi_0(\Sigma = \Sigma_0) = 1$.

The local minimax property of the LBI test follows from the fact that the affine group $(O(p), E^p)$ satisfies the conditions of Hunt-Stein theorem (Section 7.2.3).

8.2. THE SPHERICITY TEST

Let $X^\alpha = (X_{\alpha 1}, \ldots, X_{\alpha p})'$, $\alpha = 1, \ldots, N$, be a random sample of size $N(N > p)$ from a p-variate normal population with unknown mean μ and unknown positive definite covariance matrix Σ. We are interested here in testing the null hypothesis $H_0 : \Sigma = \sigma^2 \Sigma_0$ against the alternatives $H_1 : \Sigma \neq \sigma^2 \Sigma_0$ where Σ_0 is a fixed positive definite matrix and σ^2, μ are unknown. Since Σ_0 is positive definite, there exists a $g \in G_l(p)$ such that $g\Sigma_0 g' = I$. Let $Y^\alpha = gX^\alpha$, $\alpha = 1, \ldots, N$; $v = g\mu$, $\Sigma^* = g\Sigma g'$. Then $Y^\alpha, \alpha = 1, \ldots, N$, constitute a random sample of size N from a p-variate normal population with mean v and positive definite covariance matrix Σ^*. The problem is reduced to testing $H_0 : \Sigma^* = \sigma^2 I$ against $H_1 : \Sigma^* \neq \sigma^2 I$ when σ^2, μ are unknown. Since under $H_0 : \Sigma^* = \sigma^2 I$, the ellipsoid $(y - v)' \Sigma^{*-1}(y - v) = \text{const}$ reduces to the sphere $(y - v)'(y - v)/\sigma^2 = \text{const}$, the hypothesis is called the hypothesis of sphericity. Let $\bar{X}, S, \bar{Y}, B$ be as defined as in Section 8.1. The likelihood of the observations y^α on Y^α, $\alpha = 1, \ldots, N$, is given by

$$L(v, \Sigma^*) = (2\pi)^{-Np/2} (\det \Sigma^*)^{-N/2}$$

$$\times \exp\left\{-\frac{1}{2} \operatorname{tr} \Sigma^{*-1} \left(\sum_{\alpha=1}^{N} (y^\alpha - v)(y^\alpha - v)'\right)\right\}. \qquad (8.29)$$

The parametric space $\Omega = \{(v, \Sigma^*)\}$ is the space of v and Σ^*. Under H_0 it reduces to $\omega = \{(v, \sigma^2 I)\}$. From (8.29)

$$L(v, \sigma^2 I) = (2\pi)^{-Np/2}(\sigma^2)^{-Np/2} \times \exp\left\{-\frac{1}{2\sigma^2}\operatorname{tr}\left(\sum_{\alpha=1}^{N}(y^\alpha - v)(y^\alpha - v)'\right)\right\}. \tag{8.30}$$

Hence

$$\max_\omega L(v, \sigma^2 I) = (2\pi)^{-Np/2}\left(\frac{\operatorname{tr} b}{Np}\right)^{-Np/2} \exp\{-\tfrac{1}{2}Np\}, \tag{8.31}$$

since the maximum likelihood estimate of σ^2 is $\operatorname{tr}(b)/Np$. Thus we get

$$\lambda = \frac{\max_\omega L(v, \sigma^2 I)}{\max_\Omega L(v, \Sigma^*)} = \left[\frac{\det b}{((\operatorname{tr} b)/p)^p}\right]^{N/2} = \left[\frac{\det \Sigma_0^{-1} s}{((\operatorname{tr} \Sigma_0^{-1} s)/p)^p}\right]^{N/2}. \tag{8.32}$$

Theorem 8.2.1. *For testing $H_0 : \Sigma = \sigma^2 \Sigma_0$ where σ^2, μ are unknown and Σ_0 is a fixed positive definite matrix, the likelihood ratio test of H_0 rejects H_0 whenever*

$$\lambda = \frac{(\det \Sigma_0^{-1} s)^{N/2}}{((\operatorname{tr} \Sigma_0^{-1} s)/p)^{Np/2}} \leq c, \tag{8.33}$$

where the constant c is chosen such that the test has the required size α.

The corresponding modified likelihood ratio test of this problem is obtained from the likelihood ratio test by replacing N by $N - 1$.

To find the constant c we need the probability density function of λ when H_0 is true. Mauchly (1940) first derived the test criterion and obtained various moments of this criterion under the null hypothesis. Writing $W = \lambda^{2/N}$, Mauchly showed that

$$E(W^k) = p^{kp} \frac{\Gamma(\tfrac{1}{2}p(N-1))}{\Gamma(\tfrac{1}{2}p(N-1) + pk)} \times \prod_{j=1}^{p}\frac{\Gamma(\tfrac{1}{2}(N-j) + k)}{\Gamma(\tfrac{1}{2}(N-j))}, \quad k = 0, 1, \ldots \tag{8.34}$$

For $p = 2$ (8.34) reduces to

$$E(W^k) = \frac{N-2}{N-2+2k} = (N-2)\int_0^1 (z)^{N-3+2k} dz. \tag{8.35}$$

Thus under H_0, W is distributed as Z^2 where the probability density function of Z is

$$f_Z(z) = \begin{cases} (N-2)z^{N-3} & 0 \le z \le 1 \\ 0 & \text{otherwise.} \end{cases} \qquad (8.36)$$

Khatri and Srivastava (1971) obtained the exact distribution of W in terms of zonal polynomials and Meijer's G-function. Consul (1968) obtained the null distribution of W. Nagarsenker and Pillai (1973) tabulated various percentage points of the distribution of W.

From Anderson (1958, Section 8.6) we obtain

$$P\{-(N-1)\rho \log W \le z\} = P\{\chi_f^2 \le z\} + \omega_2[P\{\chi_{f+4}^2 \le z\} \\ - P\{\chi_f^2 \le z\}] + O(1/N^3), \qquad (8.37)$$

where

$$1 - \rho = \frac{2p^2 + p + 2}{6p(N-1)}, \qquad \omega_2 = \frac{(p+2)(p-1)(p-2)(2p^3 + 6p^2 + 3p + 2)}{288p^2(N-1)^2 p^2},$$

$$f = \tfrac{1}{2}(p)(p+1) + 1.$$

Thus for large N

$$P\{-(N-1)\rho \log W \le z\} = P\{\chi_f^2 \le z\}.$$

The problem of testing $H_0 : \Sigma^* = \sigma^2 I$ against the alternatives $H_1 : \Sigma^* = \sigma^2 V$ where V is an unknown positive definite $p \times p$ matrix not equal to I remains invariant under the group $G = R_+ \times E^p \times O(p)$ of affine transformations

$$g = (b, a, O) \qquad (8.38)$$

with $b \in R_+, a \in E^p, O \in O(p)$ transforming

$$Y^\alpha \to bOY^\alpha + a, \alpha = 1, \ldots, N. \qquad (8.39)$$

A set of maximal invariants in the sample space under G is

$$\left(\frac{R_1}{\sum_{i=1}^p R_i}, \ldots, \frac{R_p}{\sum_{i=1}^p R_i} \right)$$

where $R_1, \ldots, R_p$, are the characteristic roots of B. A corresponding maximal invariant in the parametric space is

$$\left(\frac{\theta_1}{\sum_{i=1}^p \theta_i}, \ldots, \frac{\theta_p}{\sum_{i=1}^p \theta_i} \right)$$

where $\theta_1, \ldots, \theta_p$, are the characteristic roots of V. Under H_0, $\theta_i = 1$ for all i and under H_1 at least one $\theta_i \neq 1$. We shall now prove that the modified likelihood ratio test for this problem is unbiased. This was first proved by Glesser (1966), then by Sugiura and Nagao (1968), whose proof we present here.

Theorem 8.2.2. *For testing $H_0 : \Sigma = \sigma^2 \Sigma_0$ against the alternatives $H_1 : \Sigma \neq \sigma^2 \Sigma$, where σ^2 is an unknown positive quantity, μ is unknown, and Σ_0 is a fixed positive definite matrix, the modified likelihood ratio test with the acceptance region*

$$\omega = \left\{ s : s \text{ is positive definite and } \frac{(\det \Sigma_0^{-1} s)^{(N-1)/2}}{((\operatorname{tr} \Sigma_0^{-1} s)/p)^{(N-1)p/2)}} \geq c' \right\} \quad (8.40)$$

is unbiased.

Proof. As in Theorem 8.1.2, considering $g\Sigma_0^{-1/2} s \Sigma_0^{-1/2} g'$ instead of s where $g \in O(p)$ such that $g\Sigma_0^{-1/2} \Sigma \Sigma_0^{-1/2} g' = \Gamma$ we can without any loss of generality assume that $\Sigma_0 = I$ and $\Sigma = \Gamma$, the diagonal matrix whose diagonal elements are the p characteristic roots $\theta_1, \ldots, \theta_p$, of $(\Sigma_0^{-1/2} \Sigma \Sigma_0^{-1/2})$. Thus S has a Wishart distribution with parameter Γ and $N - 1$ degrees of freedom. Hence

$$\begin{aligned} P\{\omega | H_1\} &= C_{n,p} \int_\omega (\det s)^{(N-p-2)/2} (\det \Gamma)^{-(N-1)/2} \exp\{-\tfrac{1}{2} \operatorname{tr} \Gamma^{-1} s\} \, ds \\ &= C_{n,p} \int_{\omega^*} (\det u)^{(N-p-2)/2} \exp\{-\tfrac{1}{2} \operatorname{tr} u\} \, du, \end{aligned} \quad (8.41)$$

where u and ω^* are defined as in Theorem 8.1.2. Transform u to $v_{11} v$ where the symmetric matrix v is given by

$$v = \begin{pmatrix} 1 & v_{12} & \cdots & v_{1p} \\ v_{21} & v_{22} & \cdots & v_{2p} \\ \vdots & \vdots & & \vdots \\ v_{p1} & v_{p2} & \cdots & v_{pp} \end{pmatrix} \quad (8.42)$$

The Jacobian of this transformation is $v_{11}^{p(p+1)/2-1}$. Since the region remains invariant under the transformation $u \to cu$, where c is a positive real number, we

Covariance Matrices and Mean Vectors

get

$$P\{\omega|H_1\} = C_{n,p} \int_{\omega^*} (v_{11})^{(N-1)p/2-1}(\det v)^{(N-p-2)/2}$$
$$\times \exp\{-\tfrac{1}{2}\operatorname{tr}(v_{11}v)\}dv_{11}dv \qquad (8.43)$$
$$= C_{n,p} 2^{(N-1)p/2}\Gamma(\tfrac{1}{2}(N-1)p)\int_{\omega^{**}} (\det v)^{(N-p-2)/2}$$
$$\times (\operatorname{tr}(v))^{-(N-1)p/2}dv$$

where ω^{**} is the set of positive definite matrices v such that $\Gamma^{1/2}v\Gamma^{1/2}$ belongs to ω. Now proceeding as in Theorem 8.1.2

$$P\{\omega|H_0\} - P\{\omega|H_1\} \geq 2^{p(N-1)/2}\Gamma(\tfrac{1}{2}p(N-1))C_{n,p}c'$$
$$\times \left\{\int_\omega - \int_{\omega^{**}}\right\}(\det v)^{-(p+1)/2}dv \qquad (8.44)$$

Transform $v \to x = \theta_1^{-1}\Gamma^{1/2}v\Gamma^{1/2}$ in the second integral of (8.44). Since the Jacobian of this transformation is $(\det \Gamma)^{(p+1)/2}\theta_1^{-p(p+1)/2}$, we get

$$\int_{\omega^{**}}(\det v)^{-(p+1)/2}dv = \int_\omega (\det x)^{-(p+1)/2}dx,$$

and hence the result. Q.E.D.

Kiefer and Schwartz (1965) showed that the likelihood ratio test for this problem is admissible Bayes whenever $N - 1 > p$. Das Gupta (1969) showed that the likelihood ratio test for this problem is also unbiased for testing H_0 against H_1. The proof proceeds in the same way as that of Theorem 8.1.3.

The following theorem gives the LBI test of $H_0 : \Sigma^* = \sigma^2 I$ against $H_1 : \Sigma^* = \sigma^2 V \neq \sigma^2 I$. In terms of θ_i's H_0 reduces to $\theta_i = 1$ for all i and the local alternatives correspond to the absolute value $|\theta_i^{-1} - 1|$ being small but not equal to zero for all i.

Theorem 8.2.3. *For testing H_0 against H_1 the level α test which rejects H_0 whenever*

$$\frac{\operatorname{tr} B^2}{(\operatorname{tr} B)^2} \geq c \qquad (8.45)$$

where c is a constant depending on the level α of the test, is LBI.

Proof. Since the Jacobian of the inverse transformation (8.39) is $(b)^{np}$ and db/b is an invariant measure on R_+, using (3.21) we get (with R, θ diagonal matrices with diagonal elements $R_1, \ldots, R_p$ and $\theta_1, \ldots, \theta_p$ respectively),

$$\frac{dP(R|\sigma^2\theta)}{dP(R|\sigma^2 I)}$$

$$= \frac{\int_{R_+}\int_{0(p)} (\det\theta)^{-n/2} (b^2)^{\frac{1}{2}(np-1)} \exp\left\{-\frac{b^2}{2\sigma^2}\operatorname{tr}\theta^{-1}ORO'\right\} d\mu(O) db}{\int_{R_+}\int_{0(p)} (b^2)^{\frac{1}{2}(np-1)} \exp\left\{-\frac{b^2}{2\sigma^2}\operatorname{tr} R\right\} d\mu(O) db} \quad (8.46)$$

$$= (\det\theta)^{-n/2} \int_{0(p)} (1+F)^{-np/2} d\mu(O)$$

where $1+F = (\operatorname{tr}\theta^{-1}ORO'/\operatorname{tr} R)$ and μ is the invariant probability measure on $O(p)$. Using (3.24) we expand the integrand in (8.46) as

$$1 - \frac{np}{2}F + \frac{np(np+2)}{8}F^2 - \frac{np(np+2)(np+4)}{48}F^3(1+\alpha F)^{-\frac{(np+6)}{2}} \quad (8.47)$$

where $0 < \alpha < 1$. Since $F = (\operatorname{tr}(\theta^{-1} - I)ORO'/\operatorname{tr} R)$ and $\theta^{-1} - I \le (\sum_{i=1}^p |\theta_i^{-1} - 1|)I$ where $\|\cdot\|$ stand for the absolute value symbol, we get $|F| < \sum_{i=1}^p |\theta_i^{-1} - 1|$. From Equations (8.23–8.25) we get

$$\frac{dP(R|\sigma^2\theta)}{dP(R|\sigma^2 I)} = 1 + \frac{3n(np+2)}{8(p+1)}\left\{\frac{\operatorname{tr} B^2}{(\operatorname{tr} B)^2}\right\}(\operatorname{tr}(\theta^{-1} - I)^2) + o(\operatorname{tr}(\theta^{-1} - I)^2). \quad (8.48)$$

Hence the power of any level α invariant test ϕ is

$$\alpha + \frac{n(np+2)}{8(p+1)} E_{H_0}\left(\phi \frac{\operatorname{tr} B^2}{(\operatorname{tr} B)^2} \gamma^2\right) + o(\gamma^2)$$

where $\gamma^2 = \operatorname{tr}(\theta^{-1} - 1)^2 = \operatorname{tr}(V^{-1} - I)^2$. Using (3.26) we get the theorem.

Q.E.D.

The LBI test was first derived by Sugiura (1972). The local minimax property of this LBI test follows from the fact that the group $G = R_+ \times E^p \times O(p)$ satisfies the conditions of the Hunt-Stein Theorem (Section 7.2.3). Following the Kiefer and Schwartz (1965) approach the likelihood ratio test can be shown to be admissible.

8.3. TESTS OF INDEPENDENCE AND THE R^2-TEST

Let $X = (X_1, \ldots, X_p)'$ be a normally distributed p-vector with unknown mean μ and positive definite covariance matrix Σ. Let $X^\alpha = (X_{\alpha 1}, \ldots, X_{\alpha p})'$,

Covariance Matrices and Mean Vectors

$\alpha = 1, \ldots, N$, be a random sample of size $N (N > p)$ from this population. Let

$$\bar{X} = \frac{1}{N} \sum_{\alpha=1}^{N} X^\alpha, \qquad S = \sum_{\alpha=1}^{N} (X^\alpha - \bar{X})(X^\alpha - \bar{X})'. \qquad (8.49)$$

We shall use the notation of Section 7.2.2. Partition $\bar{X}, \mu, S, \Sigma$ as

$$\mu = (\mu'_{(1)}, \ldots, \mu'_{(k)})',$$

$$\bar{X} = (\bar{X}'_{(1)}, \ldots, \bar{X}'_{(k)})',$$

$$X = (X'_{(1)}, \ldots, X'_{(k)})', \qquad \Sigma = \begin{pmatrix} \Sigma_{(11)} & \cdots & \Sigma_{(1k)} \\ \vdots & & \vdots \\ \Sigma_{(k1)} & \cdots & \Sigma_{(kk)} \end{pmatrix}$$

$$S = \begin{pmatrix} S_{(11)} & \cdots & S_{(1k)} \\ \vdots & & \vdots \\ S_{(k1)} & \cdots & S_{(kk)} \end{pmatrix}.$$

We are interested in testing the null hypothesis that the subvectors $X_{(1)}, \ldots, X_{(k)}$ are mutually independent. The null hypothesis can be stated, equivalently, as

$$H_0 : \Sigma_{(ij)} = 0 \quad \text{for all} \quad i \neq j. \qquad (8.50)$$

Note that both the problems considered earlier in this chapter can be transformed into the problem of independence of components of X.

Let Ω be the parametric space of (μ, Σ). Under H_0, Ω is reduced to $\omega = \{(\mu, \Sigma_D)\}$ where Σ_D is a diagonal matrix in the block form with unknown diagonal elements $\Sigma_{(ii)}, i = 1, \ldots, k$. The likelihood of the sample observations x^α on $X^\alpha, \alpha = 1, \ldots, N$ is given by

$$L(\mu, \Sigma) = (2\pi)^{-Np/2} (\det \Sigma)^{-N/2}$$

$$\times \exp\left\{-\tfrac{1}{2} \operatorname{tr} \Sigma^{-1} \left(\sum_{\alpha=1}^{N} (x^\alpha - \mu)(x^\alpha - \mu)' \right) \right\}. \qquad (8.51)$$

Hence

$$\max_{\Omega} L(\mu, \Sigma) = (2\pi)^{-Np/2} [\det(s/N)]^{-N/2} \exp\{-\tfrac{1}{2} Np\}. \qquad (8.52)$$

Under H_0,

$$L(\mu, \Sigma_D) \prod_{i=1}^{k} (2\pi)^{-Np_i/2} (\det \Sigma_{(ii)})^{-N/2}$$

$$\times \exp\left\{-\frac{1}{2}\operatorname{tr} \Sigma_{(ii)}^{-1}\left(\sum_{\alpha=1}^{N}(x_{\alpha(i)} - \mu_{(i)})(x_{\alpha(i)} - \mu_{(i)})'\right)\right\} \quad (8.53)$$

where $x^{\alpha} = (x_{\alpha(1)}, \ldots, x_{\alpha(k)})'$, and $x_{\alpha(i)}$ is $p_i \times 1$. Now

$$\max_{\omega} L(\mu, \Sigma_D) = \prod_{i=1}^{k} \max_{\Sigma_{(ii)},\mu_{(i)}} \left[(2\pi)^{-Np_i/2}(\det \Sigma_{(ii)})^{-N/2}\right.$$

$$\left. \times \exp\left\{-\frac{1}{2}\operatorname{tr} \Sigma_{(ii)}^{-1} \sum_{\alpha=1}^{N}(x_{\alpha(i)} - \mu_{(i)})(x_{\alpha(i)} - \mu_{(i)})'\right\}\right] \quad (8.54)$$

$$= \prod_{i=1}^{k} \{(2\pi)^{-Np_i/2}[\det(s_{(ii)}/N)]^{-N/2} \exp\{-\tfrac{1}{2}Np_i\}\}.$$

From (8.52) and (8.54), the likelihood ratio criterion λ for testing H_0 is given by

$$\lambda = \frac{\max L(\mu, \Sigma_D)}{\max_{\Omega} L(\mu, \Sigma)} = \left[\frac{\det s}{\prod_{i=1}^{k} \det s_{(ii)}}\right]^{N/2} = v^{N/2}, \quad (8.55)$$

where $v = (\det s)/(\prod_{i=1}^{k} \det s_{(ii)})$. Hence we have the following theorem.

Theorem 8.3.1. *For testing $H_0 : \Sigma = \Sigma_D$, the likelihood ratio test rejects H_0 whenever $\lambda \leq c'$ or, equivalently, $v \leq c$, where c' or c is chosen such that the test has level of significance α.*

Let $s = (s_{ij})$. Writing $r_{ij} = s_{ij}/(s_{ii}s_{jj})^{1/2}$, the matrix r of sample correlation coefficients r_{ij} is

$$r = \begin{pmatrix} 1 & r_{12} & \cdots & r_{1p} \\ r_{21} & 1 & \cdots & r_{2p} \\ \vdots & \vdots & & \vdots \\ r_{p1} & r_{p2} & \cdots & 1 \end{pmatrix}. \quad (8.56)$$

Covariance Matrices and Mean Vectors

Obviously $\det s = (\prod_{i=1}^{p} \det s_{ii}) \det r$. Let us now partition r into submatrices $r_{(ij)}$ similar to s as

$$r = \begin{pmatrix} r_{(11)} & \cdots & r_{(1k)} \\ r_{(21)} & \cdots & r_{(2k)} \\ \vdots & & \vdots \\ r_{(k1)} & \cdots & r_{(kk)} \end{pmatrix}. \tag{8.57}$$

Then

$$\det s_{(ii)} = \det(r_{(ii)}) \prod_{j=p_1+\cdots+p_{i-1}+1}^{p_1+\cdots+p_i} s_{jj}. \tag{8.58}$$

Thus

$$v = \frac{\det(r)}{\prod_{i=1}^{k} \det(r_{(ii)})} \tag{8.59}$$

gives a representation of v in terms of sample correlation coefficients.

Let G_{BD} be the group of $p \times p$ nonsingular block diagonal matrices g of the form

$$g = \begin{pmatrix} g_{(11)} & 0 & \cdots & 0 \\ 0 & g_{(22)} & \cdots & 0 \\ \vdots & \vdots & & \vdots \\ 0 & 0 & \vdots & g_{(kk)} \end{pmatrix}, \tag{8.60}$$

where $g_{(ii)}$ is a $p_i \times p_i$ submatrix of g and $\Sigma_1^k p_i = p$. The problem of testing $H_0 : \Sigma = \Sigma_D$ against the alternatives $H_1 : \Sigma \neq \Sigma_D$ remains invariant under the group g of affine transformations (g, a), $g \in G_{BD}$ and $a \in E^p$, transforming each X^α to $gX^\alpha + a$. The corresponding induced group of transformations in the space of $(\bar{X}, S)$ is given by $(\bar{X}, S) \to (g\bar{X} + a, gSg')$. Obviously this implies that

$$\bar{X}_{(i)} \to g_{(ii)}\bar{X}_{(i)} + a_{(i)} \qquad S_{(ii)} \to g_{(ii)}S_{(ii)}g'_{(ii)}, \tag{8.61}$$

and hence

$$\frac{\det s}{\prod_{i=1}^{k} \det s_{(ii)}} = \frac{\det(gsg')}{\prod_{i=1}^{k} \det(g_{(ii)}s_{(ii)}g'_{(ii)})}. \tag{8.62}$$

To determine the likelihood ratio test or the test based on v we need the distribution of V under H_0. Under H_0, S has a Wishart distribution with parameter Σ_D and $N-1$ degrees of freedom; the $X_{(i)}$ are mutually independent; the marginal distribution of $S_{(ii)}$ is Wishart with parameter $\Sigma_{(ii)}$ and $N-1$ degrees of freedom; and $S_{(ii)}$ is distributed independently of $S_{(jj)}(i \neq j)$. Using these facts it can be shown that under H_0,

$$E(V^h) = \frac{\prod_{i=1}^{p}\Gamma(\frac{1}{2}(N-i)+h)\prod_{i=1}^{k}\{\prod_{j=1}^{p_i}\Gamma(\frac{1}{2}(N-j))\}}{\prod_{i=1}^{p}\Gamma(\frac{1}{2}(N-i))\prod_{i=1}^{k}\{\prod_{j=1}^{p_i}\Gamma(\frac{1}{2}(N-j)+h)\}}, \quad (8.63)$$

$$h = 0, 1, \ldots$$

Since $0 \leq V \leq 1$, these moments determine the distribution of V uniquely. Since these moments are independent of Σ_D when H_0 is true, from (8.63) it follows that when H_0 is true V is distributed as $\prod_{i=2}^{k}\{\prod_{j=1}^{p_i} X_{ij}\}$ where the X_{ij} are independently distributed central beta random variables with parameters

$$(\tfrac{1}{2}(N-\delta_{i-1}-j), \tfrac{1}{2}\delta_{i-1}) \quad \text{with} \quad \delta_j = \sum_{i=1}^{j} p_i, \; \delta_0 = 0.$$

If all the p_i are even, $p_i = 2r_i$ (say), then under H_0, V is distributed as $\prod_{i=2}^{k}\{\prod_{j=1}^{r_i} Y_{ij}^2\}$ where the Y_{ij}, are independently distributed central beta random variables with parameters $((N-\delta_{i-1}-2j), \delta_{i-1})$.

Wald and Brookner (1941) have given a method for deriving the distribution when the p_i are odd. For further results on the distribution we refer the reader to Anderson (1958, Section 9.4)

Let

$$f = \tfrac{1}{2}\left[p(p+1) - \sum_{i=1}^{k} p_i(p_i+1)\right],$$

$$\rho = 1 - \frac{2(p^3 - \sum_{i=1}^{k} p_i^3) + 9(p^2 - \sum_{i=1}^{k} p_i^2)}{6N(p^2 - \sum_{i=1}^{k} p_i^2)},$$

$$a = \rho N, \quad \lambda_2 = \frac{p^4 - \sum_{i=1}^{k} p_i^4}{48} - \frac{5(p^2 - \sum_{i=1}^{k} p_i^2)}{96} - \frac{(p^3 - \sum_{i=1}^{k} p_i^3)^2}{72(p^2 - \sum_{i=1}^{k} p_i^2)}.$$

Using Box (1949), we obtain

$$P\{-a \log V \leq z\} = P\{\chi_f^2 \leq z\} + \frac{\lambda_2}{a^2}[P\{\chi_{f+4}^2 \leq z\} - P\{\chi_f^2 \leq z\}] + o(a^{-3}).$$

Thus for large N

$$P\{-a \log V \leq z\} \simeq P\{\chi_f^2 \leq z\}. \quad (8.64)$$

8.3.1. The R^2-Test

If $k = 2, p_1 = 1, p_2 = p - 1$, then the likelihood ratio test criterion λ is given by

$$\lambda = \left(\frac{\det s}{s_{11} \det(s_{(22)})}\right)^{N/2} = \left(\frac{s_{11} - s_{(12)} s_{(22)}^{-1} s_{(21)}}{s_{11}}\right)^{N/2} \quad (8.65)$$

$$= (1 - r^2)^{N/2}$$

where $r^2 = s_{(12)} s_{(22)}^{-1} s_{(21)}/s_{11}$ is the square of the sample multiple correlation coefficient between X_1 and $(X_2, \ldots, X_p)$. The distribution of $R^2 = S_{(12)} S_{(22)}^{-1} S_{(21)}/S_{11}$ is given in (6.86), and depends on $\rho^2 = \Sigma_{(12)} \Sigma_{(22)}^{-1} \Sigma_{(21)} \Sigma_{11}^{-1}$, the square of the population multiple correlation coefficient between X_1 and $(X_2, \ldots, X_p)$. Since $\Sigma_{(22)}$ is positive definite, $\rho^2 = 0$ if and only if $\Sigma_{(12)} = 0$. From (6.86) under H_0, $(N - p)/(p - 1)(R^2/(1 - R^2))$ is distributed as a central $F_{p-1,N-p}$ with $(p - 1, N - p)$ degrees of freedom.

Theorem 8.3.2. *The likelihood ratio test of $H_0 : \rho^2 = 0$ rejects H_0 whenever*

$$\frac{N-p}{p-1} \frac{r^2}{1-r^2} \geq F_{p-1,N-p,\alpha}$$

where $F_{p-1,N-p,\alpha}$ is the upper significance point corresponding to the level of significance α.

Observe that this is also equivalent to rejecting H_0 whenever $r^2 \geq c$, where the constant c depends on the level of significance α of the test.

Example 8.3.1. Consider the data given in Example 5.3.1. Let ρ^2 be the square of the population multiple correlation coefficient between X_6 and $(X_1, \ldots, X_5)$. The square of the sample multiple correlation coefficient r^2 based on 27 observations for each year's data is given by

$$r^2 = 0.85358 \quad \text{for} \quad \text{1971 observations,}$$

$$r^2 = 0.80141 \quad \text{for} \quad \text{1972 observations.}$$

We wish to test the hypothesis at $\alpha = 0.01$ that the wheat yield is independent of the variables plant height at harvesting (X_1), number of effective tillers (X_2), length of ear (X_3), number of fertile spikelets per 10 ears (X_4), and number of grains per 10 ears (X_5). We compare the value of $(21/5)(r^2/(1 - r^2))$ with $F_{5,21,0.01} = 9.53$ for each year's data. Obviously for each year's data $(21/5)(r^2/(1 - r^2)) > 9.53$, which implies that the result is highly significant. Thus the wheat yield is highly dependent on $(X_1, \ldots, X_5)$.

As stated earlier the problem of testing $H_0 : \Sigma_{(12)} = 0$ against $H_1 : \Sigma_{(12)} \neq 0$ remains invariant under the group G of affine transformations $(g, a), g \in G_{BD}$, with $k = 2, p_1 = 1, p_2 = p - 1, a \in E^p$, transforming $(\bar{X}, S; \mu, \Sigma) \to (g\bar{X} + a, gSg'; g\mu + a, g\Sigma g')$. A maximal invariant in the space of $(\bar{X}, S)$ under G is

$$R^2 = \frac{S_{(12)} S_{(22)}^{-1} S_{(21)}}{S_{11}} \qquad (8.66)$$

and the corresponding maximal invariant in the parametric space Ω is

$$\rho^2 = \frac{\Sigma_{(12)} \Sigma_{(22)}^{-1} \Sigma_{(21)}}{\Sigma_{11}} \qquad (8.67)$$

Under H_0, $\rho^2 = 0$ and under H_1, $\rho^2 > 0$. The probabifity density function of R^2 is given in (6.86).

Theorem 8.3.3. *On the basis of observations $x^i = (x_{i1}, \ldots, x_{ip})'$, $i = 1, \ldots, N (N > p)$, from a p-variate normal distribution with unknown mean μ and unknown positive definite covariance matrix Σ, among all tests $\phi(X_1, \ldots, X^N)$ of $H_0 : \Sigma_{(12)} = 0$ against the alternatives $H_1 : \Sigma_{(12)} \neq 0$ which are invariant under the group of affine transformations G, the test which rejects H_0 whenever the square of the sample multiple correlation coefficient $r^2 > C$, where the constant C depends on the level of significance α of the test (or equivalently the likelihood ratio test), is uniformly most powerful.*

Proof. Let $\phi(X^1, \ldots, X^N)$ be an invariant test with respect to the group of affine transformations G. Since $(\bar{X}, S)$ is sufficient for (μ, Σ), $E(\phi(X^1, \ldots, X^N)|\bar{X} = \bar{x}, S = s)$ is independent of (μ, Σ) and depends only on $(\bar{x}, s)$. As ϕ is invariant under G, $E(\phi|\bar{X} = \bar{x}, S = s)$ is invariant under G, and ϕ, $E(\phi|\bar{X}, S)$ have the same power function. Thus each test in the larger class of level α tests which are functions of $X^i, i = 1, \ldots, N$, can be replaced by one in the smaller class of tests which are functions of $(\bar{X}, S)$ having identical power functions. Since R^2 is a maximal invariant in the space of $(\bar{X}, S)$ under G, the invariant test $E(\phi|\bar{X}, S)$ depends on $(\bar{X}, S)$ only through R^2, whose distribution depends on (μ, Σ) only through ρ^2. The most powerful level α invariant test of $H_0 : \rho^2 = 0$ against the simple alternative $\rho^2 = \rho_0^2$, where ρ_0^2 is a fixed positive number, rejects H_0 whenever [from (6.86)]

$$(1 - \rho_0^2)^{(N-1)/2} \sum_{j=0}^{\infty} \frac{(\rho_0^2)^j (r^2)^{j-1} \Gamma^2(\frac{1}{2}(N-1) + j) \Gamma(\frac{1}{2}(p-1))}{j! \Gamma^2(\frac{1}{2}(N-1)) \Gamma(\frac{1}{2}(p-1) + j)} \geq C', \qquad (8.68)$$

Covariance Matrices and Mean Vectors

where the constant C' is so chosen that the test has level α. From (8.68) it is now obvious that R^2-test which rejects H_0 whenever $r^2 \geq C$ is uniformly most powerful among all invariant level α tests for testing $H_0 : \rho^2 = 0$ against the alternatives $H_1 : \rho^2 > 0$. Q.E.D.

Simaika (1941) proved the following stronger optimum property of the R^2-test than the one presented in Theorem 8.3.3.

Theorem 8.3.4. *On the basis of observations $x^i, i = 1, \ldots, N$, from the p-variate normal distribution with unknown mean μ and unknown positive definite covariance matrix Σ, among all tests (level α) of $H_0 : \rho^2 = 0$ against $H_1 : \rho^2 > 0$ with power functions depending only on ρ^2, the R^2-test is uniformly most powerful.*

This theorem can be proved from Theorem 8.3.3 in the same way as Theorem 7.2.2 is proved from Theorem 7.2.1. It may be added that the proof suggested here differs from Simaika's original proof.

8.4. ADMISSIBILITY OF THE TEST OF INDEPENDENCE AND THE R^2-TEST

The development in this section follows the approach of Section 7.2.2. To prove the admissibility of the R^2-test we first prove the admissibility of the likelihood ratio test of independence using the approach of Kiefer and Schwartz (1965) and then give the modifications needed to prove the admissibility of the R^2-test.

Let $V = (Y, X)$, where $X = \sqrt{N}\bar{X}$, $Y = (Y^1, \ldots, Y^{N-1})$ are such that $S = YY' = \sum_{i=1}^{N-1} Y^i Y^{i'}$, and $Y^1, \ldots, Y^{N-1}$ are independently and identically distributed normal p-vectors with mean 0 and covariance matrix Σ, and X is distributed, independently of $Y^1, \ldots, Y^{N-1}$, as p-variate normal with mean $\nu = \sqrt{N}\mu$ and covariance matrix Σ. It may be recalled that if $\theta = (\mu, \Sigma)$ and the Lebesgue density function of V on a Euclidean set is denoted by $f_V(v|\theta)$, then every Bayes rejection region for the $0 - 1$ loss function is of the form

$$\left\{ v : \frac{\int f_V(v|\theta)\pi_0(d\theta)}{\int f_V(v|\theta)\pi_1(d\theta)} \leq C \right\} \tag{8.69}$$

for some constant $C (0 \leq C \leq \infty)$ where π_1 and π_0 are the probability measures (or positive constant multiples thereof) for the parameter θ under H_1 and H_0, respectively. Since in our case the subset of this set corresponding to equality sign C has probability 0 for all θ in the parametric space, our Bayes procedures will be essentially unique and hence admissible.

Write $Y' = (Y'_{(1)}, \ldots, Y'_{(k)})$, where the $Y'_{(i)}$ are submatrices of dimension $(N-1) \times p_i$. Then $S_{(ii)} = Y_{(i)}Y'_{(i)}$ and the likelihood ratio test of independence rejects H_0 whenever

$$\det(yy') / \prod_{i=1}^{k} \det(y_{(i)}y'_{(i)}) \leq C. \tag{8.70}$$

Let π_1 assign all its measure to values of θ for which $\Sigma^{-1} = I + \eta\eta'$ for some random p-vector η and $V = \Sigma\eta Z$ for some random variable Z. Let the conditional (a priori) distribution of V given Σ under H_1 be such that with a priori probability 1, $\Sigma^{-1}V = \eta Z$ where Z is normally distributed with mean 0 and variance $(1 - \eta'(I + \eta\eta')^{-1}\eta)^{-1}$, and let the marginal distribution π_1^* of Σ under H_1 be given by

$$\frac{d\pi^*(\eta)}{d\eta} = [\det(I + \eta\eta')]^{-(N-1)/2}, \tag{8.71}$$

which is integrable on E^p (Euclidean p-space) provided $N - 1 > p$. Let π_0 assign all its measure to values of θ for which $\Sigma = \Sigma_D$ with

$$\Sigma_D^{-1} = \begin{pmatrix} I_{(1)} + \eta_{(1)}\eta'_{(1)} & 0 & 0 & 0 \\ 0 & I_{(2)} + \eta_{(2)}\eta'_{(2)} & 0 & 0 \\ \vdots & \vdots & \vdots & \vdots \\ 0 & 0 & 0 & I_{(k)} + \eta_{(k)}\eta'_{(k)} \end{pmatrix} \tag{8.72}$$

for some random vector $\eta = (\eta'_{(1)}, \ldots, \eta'_{(k)})'$ where the $\eta_{(i)}$ are subvectors of dimension $p_i \times 1$ with $\sum_{i=1}^{k} p_i = p$ and $V = \Sigma_D \eta Z$ for some random variable Z. Let the conditional a priori distribution of V under H_0 given Σ_D be such that with a priori probability 1, $\Sigma_D^{-1}V = \eta Z$ where Z is normally distributed with mean 0 and variance $(1 - \sum_{i=1}^{k}[\eta'_{(i)}(I_{(i)} + \eta_{(i)}\eta'_{(i)})^{-1}\eta_{(i)}])^{-1}$, and let the marginal (a priori) distribution of Σ under H_0 be given by

$$\frac{d\pi_0^*(\eta)}{d\eta} = \prod_{i=1}^{k}[\det(I_{(i)} + \eta_{(i)}\eta'_{(i)})]^{-(N-1)/2}, \tag{8.73}$$

which is integrable on E^p provided $N - 1 > p$. The fact that these a prioris represent bona fide probability measures follows from Exercise 8.4. Since in our

Covariance Matrices and Mean Vectors

case

$$f_V(v|\theta) = f_X(x|v, \Sigma) f_Y(y|\Sigma)$$
$$= (2\pi)^{-Np/2} (\det \Sigma)^{-N/2} \exp\{-\tfrac{1}{2} \operatorname{tr} \Sigma^{-1}(yy' + (x-v)(x-v)')\},$$

$$\int f_V(v|\theta) \pi_1(d\theta) = \int [(2\pi)^{-(Np+1)/2} (\det(I + \eta \eta'))^{N/2}$$
$$\times \exp\{-\tfrac{1}{2} \operatorname{tr}[(I + \eta \eta')(yy' + xx')$$
$$+ \eta x' z - \tfrac{1}{2} (I + \eta \eta')^{-1} \eta \eta' z^2]\} \quad (8.74)$$
$$\times (\det(I + \eta \eta'))^{-(N-1)/2} (1 - \eta'(I + \eta \eta')^{-1} \eta)^{1/2}$$
$$\times \exp\{-\tfrac{1}{2} z^2 (1 - \eta'(I + \eta \eta')^{-1} \eta)\} \, d\eta \, dz$$
$$= A \exp\{-\tfrac{1}{2} xx'\} \int \exp\{-\tfrac{1}{2} \operatorname{tr}(I + \eta \eta')(yy')\} \, d\eta,$$

where A is a constant independent of η. Similarly,

$$\int f_V(v|\theta) \pi_0(d\theta) = A \exp\{-\tfrac{1}{2} xx'\}$$
$$\times \prod_{i=1}^{k} \int \exp\{-\tfrac{1}{2} \operatorname{tr}(I_{(i)} + \eta_{(i)} \eta'_{(i)}) y_{(i)} y'_{(i)}\} \, d\eta_{(i)} \quad (8.75)$$
$$= A \exp\{-\tfrac{1}{2} \operatorname{tr}(yy' + xx')\}$$
$$\times \prod_{i=1}^{k} \int \exp\{-\tfrac{1}{2} \operatorname{tr}(\eta_{(i)} \eta'_{(i)} y_{(i)} y'_{(i)})\} \, d\eta_{(i)}.$$

From (8.74) and (8.75), using the results of Exercise 8.4 we obtain

$$\frac{\int f_V(v|\theta) \pi_0(d\theta)}{\int f_V(v|\theta) \pi_1(d\theta)} = \left[\frac{\det(yy')}{\prod_{i=1}^{k} \det(y_{(i)} y'_{(i)})} \right]^{1/2} = \left[\frac{\det s}{\prod_{i=1}^{k} \det(s_{(ii)})} \right]^{1/2}. \quad (8.76)$$

Hence we get the following theorem.

Theorem 8.4.1. *For testing $H_0 : \Sigma = \Sigma_D$ against the alternatives $H_1 : \Sigma \neq \Sigma_D$ when μ is unknown the likelihood ratio test that rejects H_0 whenever $[\det(s)/\prod_{i=1}^k \det(s_{(ii)})]^{N/2} \leq C$, where the constant C depends on the level of significance α of the test, is admissible Bayes whenever $N - 1 > p$.*

This approach does not handle the case of the minimum sample size ($N - 1 = p$). In the special case $k = 2, p_1 = 1, p_2 = p - 1$, a slightly different trick, used by Lehmann and Stein (1948), will work even when $N - 1 = p$. Let π_1 assign all its measure under H_1 to values of θ for which

$$\Sigma^{-1} = I + \begin{pmatrix} 1 & \eta' \\ \eta & \eta\eta' \end{pmatrix}$$

where η is a $(p - 1) \times 1$ random vector and the marginal (a priori) distribution of Σ under H_1 is

$$\frac{d\pi_1^*(\eta)}{d\eta} = \left[\det\left(I + \begin{pmatrix} 1 & \eta' \\ \eta & \eta\eta' \end{pmatrix}\right)\right]^{-p/2}, \quad (8.77)$$

which is integrable on E^{p-1}, and let the conditional distribution of V given Σ under H_1 remain the same as the general case above.

Let π_0 assign all its measure to Σ^{-1}, which is of the form

$$\Sigma^{-1} = I + \begin{pmatrix} 1-b & 0 \\ 0 & \eta\eta' \end{pmatrix}, \quad (8.78)$$

where η is a $(p - 1) \times 1$ random vector, $0 \leq b \leq 1$, and the marginal (a priori) distribution of Σ under H_0 is

$$\frac{d\pi_0^*(\eta)}{d\eta} = \left[\det\left(I + \begin{pmatrix} 1-b & 0 \\ 0 & \eta\eta' \end{pmatrix}\right)\right]^{-p/2}, \quad (8.79)$$

which is integrable on E^{p-1}, and let the conditional distribution of V given Σ under H_0 remain the same as the general case above.

Consider the particular Bayes test which rejects H_0 whenever

$$\frac{\int f_V(v|\theta)\pi_0(d\theta)}{\int f_V(v|\theta)\pi_1(d\theta)} \leq 1.$$

Carrying out the integration as in the general case with the modified marginal distribution of Σ under H_0, H_1, we obtain the rejection region

$$\exp\{\tfrac{1}{2}by_{(1)}y'_{(1)}\} / \exp\{\tfrac{1}{2}y_{(1)}y'_{(2)}(y_{(2)}y'_{(2)})^{-1}y_{(2)}y'_{(1)}\} \leq 1.$$

Covariance Matrices and Mean Vectors

Taking logarithms of both sides we finally get the rejection region

$$\frac{s_{(12)} s_{(22)}^{-1} s_{(21)}}{s_{(11)}} \geq b,$$

which in the special case is equivalent to (8.76). Thus we have the following theorem.

Theorem 8.4.2. *For testing $H_0 : \rho^2 = 0$ against the alternatives $H_1 : \rho^2 > 0$, the R^2-test (based on the square of the sample multiple correlation coefficient R^2), which rejects H_0 whenever $r^2 \geq C$, the constant C depending on the level α of the test, is admissible Bayes.*

8.5. MINIMAX CHARACTER OF THE R^2-TEST

The solution presented here is due to Giri and Kiefer (1964b) and parallels that of Giri et al. (1963), as discussed in Section 7.2.3 for the corresponding T^2-results, the steps are the same, the detailed calculations in this case being slightly more complicated. The reader is referred back to Section 7.2.3 for the discussion of the Hunt-Stein theorem, its validity under the group of real lower triangular nonsingular matrices, and its failure under the full linear group.

We have already proved that among all tests based on the sufficient statistic $(\bar{X}, S)$, the R^2-test is best invariant for testing $H_0 : \rho^2 = 0$ against the simple alternative $\rho^2 = \rho_0^2 (>0)$ under the group of affine transformations G. For $p > 2$, this does not imply our minimax result because of the failure of the Hunt-Stein theorem.

We consider, without any loss of generality, test functions which depend on the statistic $(\bar{X}, S)$. It can be verified that the group H of translations $(\bar{X}, S; \mu, \Sigma) \to (\bar{X} + a, S; \mu + a; \Sigma)$ leaves the testing problem in question invariant, that H is a normal subgroup in the group G^* generated by H and the group G_T, the multiplicative group of $p \times p$ nonsingular lower triangular matrices whose first column contains only zeros except for the first element, and that G_T and H (and hence G^*) satisfy the Hunt-Stein conditions. Furthermore it is obvious that the action of the tranformations in H is to reduce the problem to that where $\mu = 0$ (known) and $S = \sum_{\alpha=1}^{N} X^\alpha X^{\alpha'}$ is sufficient for Σ, where N has been reduced by unity from what it was originally. Using the standard method of reduction in steps, we can therefore treat the latter formulation, considering $X^1, \ldots, X^N$ to have 0 mean. We assume also that $N \geq p \geq 2$ (note that N is really $N - 1$ when the mean vector is not 0). Furthermore with this formulation, we need only consider test functions which depend on the sufficient statistic $S = \sum_{\alpha=1}^{N} X^\alpha X^{\alpha'}$, the Lebesgue density of which is given in (6.32).

We now consider the group G_T (of nonsingular matrices). A typical element $g \in G_T$ can be written as

$$g = \begin{pmatrix} g_{11} & 0 \\ 0 & g_{(22)} \end{pmatrix}$$

where $g_{(22)}$ is $(p-1) \times (p-1)$ lower triangular. It is easily seen that the group G_T operating as $(S, \Sigma) \to (gSg', g\Sigma g')$ leaves this reduced problem invariant.

We now compute a maximal invariant in the space of S under G_T in the usual fashion. If a test function ϕ (of S) is invariant under G_T, then $\phi(S) = \phi(gSg')$ for all $g \in G_T$ and for all S. Since S is symmetric, writing

$$S = \begin{pmatrix} S_{11} & S_{(12)} \\ S_{(21)} & S_{(22)} \end{pmatrix}$$

we get

$$\phi(S_{11}, S_{(12)}, S_{(22)}) = \phi(g_{11}S_{11}g_{11}, g_{11}S_{(12)}g'_{(22)}, g_{(22)}S_{(22)}g'_{(22)})$$

Since S is symmetric and positive definite with probability 1 for all Σ, there is an F in G_T with positive diagonal elements such that

$$FF' = \begin{pmatrix} S_{11} & 0 \\ 0 & S_{(22)} \end{pmatrix}.$$

Let $g = LF^{-1}$ where L is any diagonal matrix with values ± 1 in any order on the main diagonal. Then ϕ is a function only of $L_{(22)}F^{-1}_{(22)}S_{(21)}L_{11}/F_{11}$, and hence because of the freedom of choice of L, of $|F^{-1}_{(22)}S_{(21)}/F_{11}|$, or equivalently, of the $(p-1)$-vector whose ith component Z_i ($2 \le i \le p$) is the sum of squares of the first i components of $|F^{-1}_{(22)}S_{(21)}/F_{11}|$ (whose components are indexed $2, 3, \ldots, p$). Write $b_{[i]}$ for the $(i-1)$-vector consisting of the first $i-1$ components of the $(p-1)$-vector b and $C_{[i]}$ for the upper left-hand $(i-1) \times (i-1)$ submatrix of a $(p-1) \times (p-1)$ matrix C. Then Z_i can be written as

$$Z_i = \frac{S_{(12)[i]}(F^{-1}_{(22)[i]})'(F^{-1}_{(22)[i]})S'_{(12)[i]}}{S_{11}} = \frac{S_{12[i]}S^{-1}_{(22)[i]}S'_{(12)[i]}}{S_{11}}. \quad (8.80)$$

The vector $Z = (Z_2, \ldots, Z_p)'$ is thus a maximal invariant under G_T if it is invariant under G_T, and it is easily seen to be the latter. Z_i is essentially the squared sample multiple correlation coefficient computed from the first i coordinates of $X^j, j = 1, \ldots, N$. Let us define a $(p-1)$-vector $R = (R_2, \ldots, R_p)'$ by

$$\sum_{j=1}^{i} R_j = Z_i, \quad 2 \le i \le p. \quad (8.81)$$

Covariance Matrices and Mean Vectors

Obviously $R_i = Z_i - Z_{i-1}$, where we define $Z_1 = 0$. It now follows trivially that R is maximal invariant under G_T and $R_i \geq 0$ for each i, $\Sigma_{i=2}^{p} R_i \leq 1$, and of course

$$\sum_{i=2}^{p} R_i = \frac{S_{(12)} S_{(22)}^{-1} S_{(21)}}{S_{11}} = R^2 \tag{8.82}$$

We shall find it more convenient to work with the equivalent statistic R instead of with Z. A corresponding maximal invariant $\Delta = (\delta_2^2, \ldots, \delta_p^2)'$ in the parametric space of Σ under G_T, when H_1 is true, is given by

$$\sum_{j=2}^{i} \delta_j^2 = \frac{\Sigma_{(12)[i]} \Sigma_{(22)[i]}^{-1} \Sigma_{(12)[i]}'}{\Sigma_{11}}, \quad 2 \leq i \leq p. \tag{8.83}$$

It is clear that $\delta_j^2 \geq 0$ and $\Sigma_{j=2}^{p} \delta_j^2 = \rho^2$. The corresponding maximal invariant under H_0 takes on the single value 0. Thus the Lebesgue density function $f_R(r|\Delta)$ depends only on Δ under H_1 and is fixed $f_R(r|0)$ under H_0.

We can assume $\Sigma_{11} = 1$, $\Sigma_{(22)} = I$ [the $(p-1) \times (p-1)$ identity matrix], and $\Sigma_{(21)} = (\delta_2, \ldots, \delta_p)' = \delta^*$ in (6.32), since $f_R(r|\Delta)$ depends only on Δ. With this choice of $\Sigma(\Sigma^*$, say) we can write (6.32) as [also denote it by $f(s_{11}, s_{(12)}, s_{(22)}|\Sigma)$]

$$W_p(N, \Sigma^*) = K(1 - \rho^2)^{-N/2}$$
$$\times \exp\{-\tfrac{1}{2} \text{tr}[(1-\rho^2)^{-1} s_{11} - 2(1-\rho^2)^{-1} \delta^{*'} s_{(21)} \tag{8.84}$$
$$+ (I - \delta^* \delta^{*'})^{-1} s_{(22)}]\} (\det s)^{(N-p-2)/2}.$$

Let B be the unique lower triangular matrix belonging to G_T with positive diagonal elements $B_{ii}(1 \leq i \leq p)$ such that $S_{(22)} = B_{(22)} B_{(22)}'$, $S_{11} = B_{11}^2$, and let $V = B_{(22)}^{-1} S_{(21)}$. One can easily compute the Jacobians

$$\frac{\partial S_{(22)}}{\partial B_{(22)}} = 2^{p-1} \prod_{i=2}^{p} (B_{ii})^{p+1-i}, \quad \frac{\partial S_{(21)}}{\partial V} = \prod_{i=2}^{p} B_{ii}, \quad \frac{\partial S_{11}}{\partial B_{11}} = 2B_{11}, \tag{8.85}$$

so the joint probability density of B_{11}, V, and $B_{(22)}$ is

$$h(b_{11}, v, b_{(22)}|\Sigma^*) = 2^p f(b_{11}^2, v' b_{(22)}', b_{(22)} b_{(22)}'|\Sigma^*) b_{11} \prod_{i=2}^{p} b_{ii}^{p+2-i}. \tag{8.86}$$

Putting $W = (W_2, \ldots, W_p)'$ with $W_i = |V_i|(2 \leq i \leq p)$, and noting that the $(p-1)$-vector W can arise from any of the 2^{p-1} vectors $V = M_{(22)} V$ where $M_{(22)}$ is a $(p-1) \times (p-1)$ diagonal matrix with diagonal entries ± 1, we write $g = bM$, where with $M_{11} = \pm 1$,

$$M = \begin{pmatrix} M_{11} & 0 \\ 0 & M_{(22)} \end{pmatrix}, \tag{8.87}$$

g ranging over all matrices in G_T. We obtain for the density of W, writing $g_{ij}(i \geq j \geq 2)$ for the elements of $g_{(22)}$,

$$f_W(w|\Sigma^*) = 2^p \int f(g_{11}^2, w'g_{(22)}, g_{(22)}g'_{(22)}) \prod_{i=2}^{p} |g_{ii}|^{p+2-i}$$

$$\times |g_{11}| \prod_{i \geq j \geq 2} dg_{ij} dg_{11}$$

$$= (1-\rho^2)^{-N/2} 2^p K \int \exp\{-\tfrac{1}{2}(1-\rho^2)^{-1}$$

$$\times \operatorname{tr}(g_{11}^2 - \delta^* w' g'_{(22)} - \delta^{*'} g_{(22)} w \qquad (8.88)$$

$$+ (1-\rho^2)(I - \delta^* \delta^{*'})^{-1} g_{(22)} g'_{(22)})\}$$

$$\times \prod_{i=2}^{p} |g_{ii}|^{N+1-i} |g_{11}|^{N-p} (1 - w'w/g_{11}^2)^{(N-p-1)/2}$$

$$\times \prod_{i \geq j \geq 2} dg_{ij} dg_{11}.$$

Writing $W = g_{11} U$ and $R_j = U_j^2 (2 \leq j \leq p)$ we obtain from (8.88) that the probability density function of $R = (R_2, \ldots, R_p)'$ is

$$f_R(r|\Delta) = \frac{(1-\rho^2)^{-N/2} 2K}{\sum_{i=2}^{p} r_i^{1/2}}$$

$$\times \int \exp\left\{-\frac{1}{2}(1-\rho^2)^{-1} \operatorname{tr}(g_{11}^2 - 2g_{11} \delta^* g_{(22)} r^*\right.$$

$$\left. + (1-\rho^2)(I - \delta^* \delta^{*'})^{-1} g_{(22)} g'_{(22)})\right\} \qquad (8.89)$$

$$\times \left(1 - \sum_{j=2}^{p} r_j\right)^{N-p-2)/2} |g_{11}|^{N-1} \prod_{i=2}^{p} |g_{ii}|^{N+1-i}$$

$$\times \prod_{i \geq j \geq 2} dg_{ij} dg_{11},$$

where $r^* = (r_2^{1/2}, \ldots, r_p^{1/2})'$. Let $C = (1-\rho^2)^{-1}(I - \delta^* \delta^{*'})$. Since C is positive definite, there exists a lower triangular $(p-1) \times (p-1)$ matrix T with positive diagonal elements $T_{ii} (2 \leq i \leq p)$ such that $TCT' = I$. Writing $h = Tg_{(22)}$, we

Covariance Matrices and Mean Vectors

obtain

$$\frac{\partial h}{\partial g_{(22)}} = \prod_{i=2}^{p} T_{ii}^{i-1}. \tag{8.90}$$

Let us define for $2 \leq i \leq p$,

$$\lambda_i = 1 - \sum_{j=2}^{i} \delta_j^2, \qquad \lambda_1 = 1 \ (\lambda_p = 1 - \rho^2),$$
$$\alpha_i = (\delta_i^2 \lambda_p / \lambda_{i-1} \lambda_i)^{1/2}, \qquad \alpha = (\alpha_2, \ldots, \alpha_p)'. \tag{8.91}$$

A simple calculation yields $(T_{[i]}\delta_{[i]}^*)'(T_{[i]}\delta_{[i]}^*) = \lambda_p(1 - \lambda_i)/\lambda_i$, so that $\alpha = T\delta^*$. Since $C\delta^* = \delta^*$, by direct computation, we obtain

$$\alpha = TC\delta^* = (T^{-1})'\delta^*.$$

From this and the fact that $\det C = (1 - \rho^2)^{2-p}$, we obtain

$$\alpha = TC\delta^* = (T^{-1})'\delta^*.$$

From this and the fact that $\det C = (1 - \rho^2)^{2-p}$, we obtain

$$f_R(r|\Delta) = 2K(1 - \rho^2)^{-N(p-1)/2} \prod_{i=2}^{p} r_i^{-1/2} \left(1 - \sum_{j=2}^{p} r_j\right)^{(N-p-1)/2}$$
$$\times \int \exp\left\{-\frac{1}{2}(1 - \rho^2)^{-1} g_{11}^2\right\} |g_{11}|^{N-1}$$
$$\times \left\{\int \exp\left\{-\frac{1}{2}(1 - \rho^2)^{-1} \sum_{i \geq j \geq 2} [h_{ij}^2 - 2\alpha_i r_j^{1/2} h_{ij} g_{11}]\right\}\right.$$
$$\left.\times \prod_{i=2}^{p} |h_{ii}|^{N+1-i} \prod_{i \geq j \geq 2} dh_{ij}\right\} dg_{11}, \tag{8.92}$$

the integration being from $-\infty$ to ∞ in each variable. For $i > j$ the integration with respect to h_{ij} yields a factor

$$(2\pi)^{1/2}(1 - \rho^2)^{1/2} \exp\{\alpha_i^2 r_j g_{11}^2/2(1 - \rho^2)\}. \tag{8.93}$$

For $i = j$ we obtain a factor

$$(2\pi)^{1/2}(1-\rho^2)^{(N+2-i)/2}\exp[\alpha_i^2 r_i g_{11}^2/2(1-\rho^2)]$$

$$\times E(\chi_1^2(\alpha_i^2 r_i g_{11}^2/(1-\rho^2))^{(N+1-i)/2}$$

$$= [2(1-\rho^2)]^{(N+2-i)/2}\Gamma(\tfrac{1}{2}(N-i+2)) \qquad (8.94)$$

$$\times \phi\left(\frac{1}{2}(N-i+2), \frac{1}{2}; r_i\alpha_i^2 g_{11}^2/2(1-\rho^2)\right)$$

where $\chi_1^2(\beta)$ is a noncentral chi-square with one degree of freedom and noncentrality parameter β, and ϕ is the confluent hypergeometric function. Integrating with respect to g_{11} we obtain, from (8.93–8.94), that the probability density function of R is (for $r \in H = \{r : r_i \geq 0, 2 \leq i \leq p, \Sigma_{i=2}^p r_i < 1\}$)

$$f_R(r|\Delta) = \frac{(1-\rho^2)^{N/2}(1-\Sigma_{i=2}^p r_i)^{(N-p-1)/2}}{(1+\Sigma_{i=2}^p r_i((1-\rho^2)/\lambda_i - 1))^{N/2}\Gamma(\tfrac{1}{2}(N-p+1))\pi^{(p-1)/2}}$$

$$\times \frac{1}{\Sigma_{i=2}^p \{r_i^{1/2}\Gamma(\tfrac{1}{2}(N-i+2))\}} \sum_{\beta_2=0}^{\infty} \cdots \sum_{\beta_p=0}^{\infty} \Gamma\left(\sum_{j=2}^p \beta_j + \frac{1}{2}N\right)$$

$$\times \prod_{i=2}^p \left\{\frac{\Gamma(\tfrac{1}{2}(N-i+2)+\beta_i)}{(2\beta_i)!}\left[\frac{4r_i\alpha_i^2}{1+\Sigma_{j=2}^p r_j((1-\rho^2)/\lambda_j - 1)}\right]^{\beta_i}\right\}.$$

The continuity of $f_R(r|\Delta)$ in Δ over its compact domain $\Gamma = \{(\delta_2^2, \ldots, \delta_p^2) : \delta_i^2 \geq 0, \Sigma_{j=2}^p \delta_j^2 = \delta^2\}$ is evident. As in the case of the T^2-test, we conclude here also that the minimax character of the critical region $\Sigma_{j=2}^p R_j \geq C$ is equivalent to the existence of a probability measure λ satisfying

$$\int_\Gamma \frac{f_R(r|\Delta)}{f_R(r|0)} \lambda(d\Delta) \begin{Bmatrix} > \\ = \\ < \end{Bmatrix} K \qquad (8.96)$$

according to whether $\Sigma_{i=2}^p r_i$ is greater than, equal to, or less than C, except possibly for a set of measure 0. We can replace (8.96) by its equivalent

$$\int_\Gamma \frac{f_R(r|\Delta)}{f_R(r|0)} \lambda(d\Delta) = K \quad \text{if} \quad \sum_{i=2}^p r_i = C. \qquad (8.97)$$

Clearly (8.96) implies (8.97). On the other hand, if there are a λ and a constant K satisfying (8.97) and if $\bar{r} = (\bar{r}_2, \ldots, \bar{r}_p)'$ is such that $\Sigma_{i=2}^p \bar{r}_i = C' > C$, writing

$$f(r) = [f_R(r|\Delta)/f_R(r|0)] \quad \text{and} \quad \bar{\bar{r}} = (C/C')\bar{r},$$

Covariance Matrices and Mean Vectors

we see at once that

$$f(\bar{r}) = f(C'\bar{r}/C) > f(\bar{r}) = K,$$

because of the form of f and the fact that $C'/C > 1$ and $\Sigma_{i=2}^{p} \bar{r}_i = C$ [note that $\lambda_i^{-1}(1-\rho^2) - 1 = -\Sigma_{j>1} \delta_j^2/\lambda_i$ and that $\lambda_i > 0$]. This and a similar argument for the case $C' < C$ show that (8.96) implies (8.97).

Using the same argument as in the case of the T^2-test, we can similarly show that the value of K which satisfies (8.97) is given by

$$K = (1 - \delta^2)^{N/2} F\left(\frac{1}{2}N, \frac{1}{2}N; \frac{1}{2}(p-1); C\delta^2\right), \tag{8.98}$$

where $F(a, b; c; x)$ is the ordinary $({}_2F_1)$ hypergeometric series, given by

$$F(a, b; c; x) = \sum_{\alpha=0}^{\infty} \frac{x^\alpha \Gamma(a+\alpha)\Gamma(b+\alpha)\Gamma(c)}{\alpha! \Gamma(a)\Gamma(b)\Gamma(c+\alpha)} \tag{8.99}$$

Giri and Kiefer (1964b) considered the case $p = 3, N = 3$ (or $N = 4$ if μ is unknown). Proceeding exactly the same way as in the T^2-test they showed that there exists a probability measure λ whose derivative is given by

$$m_z(x) = \frac{(1-zx)^{1/2}}{2\pi x^{1/2}(1-x)^{1/2}} \left\{ B_z \int_0^x \frac{du}{(1-u)(1-zu)^{3/2}} \right.$$

$$+ \int_0^\infty \left[\frac{B_z u^{1/2}}{(1+u)(z+u)^{3/2}} + \frac{1}{[u(1+u)(z+u)]^{1/2}} \right. \tag{8.100}$$

$$\left. \left. -2 \frac{u^{1/2}}{(1+u)^{1/2}(z+u)^{3/2}} \right] du \right\}$$

where $z = C\delta^2$, $B_z = (1-z)^{5/2} F(\frac{3}{2}; \frac{3}{2}; 1; z)$. The reader is referred to the original references for details of the proof of (8.100) and the other results that follow in this section. Taking (8.100) for granted we have proved the following theorem.

Theorem 8.5.1. *For testing $H_0 : \rho^2 = 0$ against the alternatives $H_1 : \rho^2 > 0$, the R^2-test is minimax for the case $p = 3, N = 3$ (or $N = 4$ if u is unknown).*

Let us examine the local minimax property of the R^2-test in the sense of Giri and Kiefer (1964a) as outlined in Chapter 7. We shall be interested in testing at level α the hypothesis $H_0 : \rho^2 = 0$ against the alternatives $H_1 : \rho^2 > \lambda$, as $\lambda \to 0$.
Let

$$\eta_i = \delta_i^2/\delta^2, \qquad \eta = (\eta_2, \ldots, \eta_p)', \qquad \delta^2 > 0.$$

From (8.95), as $\lambda \to 0$,

$$\frac{f_R(r|\lambda, \eta)}{f_R(r|0, \eta)} = 1 + \tfrac{1}{2}N\lambda \left\{ -1 + \sum_{j=2}^{p} r_j \left[\sum_{i>j} \eta_i + (N-j+2)\eta_j \right] \right\} \quad (8.101)$$
$$+ B(r, \eta, \lambda),$$

where $B(r, \eta, \lambda) = o(\lambda)$ uniformly in η and r. As in the case of the T^2-test (Chapter 7) we see that the assumptions of Theorem 7.2.4 are again satisfied with $U = \sum_{i=2}^{p} R_i = R^2$ with $h(\lambda) = b\lambda$, and $\xi_{1,\lambda}$ assigns measure 1 to the point η whose jth coordinate ($2 \le j \le p$) is

$$(N-j+1)^{-1}(N-j+2)^{-1}(p-1)^{-1}N(N-p+1).$$

Hence we have the following theorem.

Theorem 8.5.2. *For every p, N, and α, the rejection region of the R^2-test is locally minimax for testing $H_0 : \rho^2 = 0$ against $H_1 : \rho^2 = \lambda$ as $\lambda \to 0$.*

The asymptotic minimax property of the T^2-test (Chapter 7) is obviously related to the underlying exponential structure which yields it to the Stein (1956) admissibility result. It is interesting to note that the same departure from this structure (in behavior as $\rho^2 \to 1$) which prevents Stein's method from proving the admissibility of the R^2-test, also prevents us from applying the asymptotic (as $\rho^2 \to 1$) minimax theory in the R^2-test.

8.5.1. Independence of Two Subvectors

We now consider the more general case of two subvectors of dimensions p_1, p_2 respectively with $p_i > 1$, $i = 1, 2$ with $p_1 + p_2 = p$. We assume without any loss of generality that $p_1 < p_2$. Partition Σ, S as

$$\Sigma = \begin{pmatrix} \Sigma_{11} & \Sigma_{12} \\ \Sigma_{21} & \Sigma_{22} \end{pmatrix}, \quad S = \begin{pmatrix} S_{11} & S_{12} \\ S_{21} & S_{22} \end{pmatrix}$$

where Σ_{11}, S_{11} are $p_1 \times p_1$ submatrtices. We consider the problem of testing $H_0 : \Sigma_{12} = 0$ against the alternatives $H_1 : \Sigma_{12} \ne 0$. This problem remains invariant under the group of transformation $G = G_{BD} \times E^p$, where G_{BD} is defined in Section 8.3 with $k = 2$, transforming

$$X^\alpha \to gX^\alpha + a, \quad \alpha = 1, \ldots, N \quad (8.102)$$

Covariance Matrices and Mean Vectors

with

$$g = \begin{pmatrix} g_{(11)} & 0 \\ 0 & g_{(22)} \end{pmatrix} \in G_{BD} \quad \text{and} \quad a \in E^p.$$

The corresponding induced transformation on $(\bar{X}, S)$ is given by

$$(\bar{X}, S) \to (g\bar{X} + a, gSg'). \tag{8.103}$$

A maximal invariant in the space of $(\bar{X}, S)$ under G is $R_1, \ldots, R_{p_1}$, the characteristic roots of $S_{11}^{-1} S_{12} S_{22}^{-1} S_{21}$. A corresponding maximal invariant in the parametric space is given by $\theta_1, \ldots, \theta_{p_1}$, the characteristic roots of $\Sigma_{11}^{-1} \Sigma_{12} \Sigma_{22}^{-1} \Sigma_{21}$. Denote by R, θ, the diagonal matrices with elements $R_1, \ldots, R_{p_1}$ and $\theta_1, \ldots, \theta_{p_1}$ respectively. For invariant tests this problem reduces to testing $H_0: \theta = 0$ against alternatives $H_1: \theta \neq 0$. Several invariant tests are often used for this problem. They are:

i. Roy's test: it rejects H_0 whenever the largest characteristic roots of $S_{11}^{-1} S_{12} S_{22}^{-1} S_{21}$ is greater than a constant depending on the level α of the test;
ii. Lawley-Hotelling's test: it rejects H_0 whenever tr $S_{11}^{-1} S_{12}(S_{22} + S_{21} S_{11}^{-1} S_{12})^{-1} S_{21}$ is greater than a constant depending on the level α of the test;
iii. Pillai's test: it rejects H_0 whenever tr $S_{11}^{-1} S_{12} S_{22}^{-1} S_{21}$ is greater than a constant depending on the level α of the test;
iv. The likelihood ratio test: it rejects H_0 whenever $\det(I - S_{11}^{-1} S_{12} S_{22}^{-1} S_{21})$ is greater than a constant depending on the level α of the test.

Since under the transformation G, Σ_{ij} is transformed to $g_{(ii)} \Sigma_{ij} g'_{(jj)}$, $i, j = 1, 2$, and $g_{(11)}, g_{(22)}$ are nonsingular matrices, we can without any loss of generality assume that

$$\Sigma = \begin{pmatrix} I_{p_1} & \Gamma \\ \Gamma' & I_{p_2} \end{pmatrix}, \quad \Gamma = (\theta, 0).$$

This implies that

$$\Sigma^{-1} = \begin{pmatrix} (I - \Gamma\Gamma')^{-1} & -(I - \Gamma\Gamma')^{-1}\Gamma \\ -(I - \Gamma'\Gamma)^{-1}\Gamma' & (I - \Gamma'\Gamma)^{-1} \end{pmatrix}$$

$$= \begin{pmatrix} (I - \theta\theta')^{-1} & -[(I - \theta\theta')^{-1}\theta, 0] \\ -[(I - \theta\theta')^{-1}\theta, 0]' & \begin{bmatrix} (I - \theta\theta')^{-1} & 0 \\ 0 & I \end{bmatrix} \end{pmatrix} \tag{8.104}$$

Since the Jacobian of the inverse transformation given in (8.103) is $(\det g)^N = (\det g_{(11)})^N (\det g_{(22)})^N$ and the invariant measure on G_{BD} (with $k = 2$) is

$dg/[(\det g_{(11)})^{p_1}(\det g_{(22)})^{p_2}]$, we write the ratio of densities of R (using (3.21)) as

$$\frac{dP(R|\theta)}{dP(R|0)} = \frac{\int [\det(I - \theta\theta')]^{-(N-1)/2} \exp\{-\tfrac{1}{2}\operatorname{tr} \Sigma^{-1} gsg'\}\mu(dg)}{\int \exp\{-\tfrac{1}{2}\operatorname{tr} gsg'\}\mu(dg)} \quad (8.105)$$

where the measure μ is given by

$$\mu(dg) = (\det g_{(11)})^{N-p_1-1}(\det g_{(22)})^{N-p_2-1} dg_{(11)} dg_{(22)}$$

Let $h_{(11)} = g_{(11)} s_{11}^{-1/2}$, $h_{(22)} = g_{(22)} s_{22}^{-1/2}$ and $w = s_{22}^{-1/2} s_{21} s_{11}^{-1/2}$. Then

$$h = \begin{pmatrix} h_{(11)} & 0 \\ 0 & h_{(22)} \end{pmatrix} \in G_{BD}.$$

Hence

$$\operatorname{tr} \Sigma^{-1} gsg' = \operatorname{tr}(I - \theta\theta')^{-1} g_{(11)} s_{11} g'_{(22)}$$

$$- 2\operatorname{tr}[(I - \theta\theta')^{-1}\theta, 0] g_{(22)} s_{22} g'_{(22)}$$

$$+ \operatorname{tr}\begin{bmatrix} (I - \theta\theta')^{-1} & 0 \\ 0 & I \end{bmatrix} g_{(22)} s_{21} g'_{(11)}$$

$$= \operatorname{tr}(I - \theta\theta')^{-1} h_{(11)} h'_{(11)}$$

$$- 2\operatorname{tr}[(I - \theta\theta')^{-1}\theta, 0] h_{(22)} w h'_{(11)}$$

$$+ \operatorname{tr}\begin{bmatrix} (I - \theta\theta')^{-1} & 0 \\ 0 & I \end{bmatrix} h_{(22)} h'_{(22)}.$$

Let $h_{(22)} = (h'_{12}, h'_{22})'$ where h_{12} is $p_1 \times p_2$. Then

$$\operatorname{tr} \Sigma^{-1} gsg' - \operatorname{tr} gsg' = v(h, \theta)(1 + o(\delta^2))$$

where

$$v(h, \theta) = \operatorname{tr} \theta\theta' h_{(11)} h'_{(11)} - 2\operatorname{tr} \theta h_{12} w h'_{(11)} + \operatorname{tr} \theta\theta' h_{12} h'_{12},$$

$$\delta^2 = \sum_{i=1}^{p_1} \theta_i^2,$$

Now

$$(I - \theta\theta') = I + \theta\theta' + o(\delta^2),$$

$$(I - \theta\theta')^{-1}\theta = \theta + o(\delta^2).$$

Covariance Matrices and Mean Vectors

Letting

$$\nu(dh) = \exp\left\{-\frac{\text{tr}}{2}(h_{(11)}h'_{(11)} + h_{(22)}h'_{(22)})\right\} \\ \times (\det h_{(11)})^{N-p_1-1}(\det h_{(22)})^{N-p_2-1}dg_{(11)}dg_{(22)} \tag{8.106}$$

and using (3.24) we rewrite (8.105) as

$$\begin{aligned}\frac{dP(R|\theta)}{dP(R|I)} &= \left(1 + \frac{(N-1)}{2}\text{tr }\theta\theta'\right)\left[1 + \frac{1}{2D}\int \nu(h,\theta)\nu(dh)\right. \\ &\quad + \frac{1}{8D}\int [\nu(h,\theta)]^2\nu(dh) \\ &\quad + \frac{1}{48D}\int [\nu(h,\theta)]^3 \exp\{-\tfrac{1}{2}(\text{tr}(hh') \\ &\quad \left.+ (1-\alpha)\nu(h,\theta))\}\mu(dh)\right]\end{aligned} \tag{8.107}$$

where

$$D = \int \exp\{-\tfrac{1}{2}\text{tr}(h_{(11)}h'_{(11)} + h_{(22)}h'_{(22)})\}\mu(dh). \tag{8.108}$$

It may be verified that (see Kariya and Sinha (1989))

$$\begin{aligned}&\int [\text{tr }\theta\theta'(h_{(11)}h'_{(11)} + h_{12}h'_{12})]^k[\text{tr}(\theta h_{12}wh'_{(11)})]^{2j+1}\nu(dh) = 0, \\ &\qquad k = 1, 2, \quad j = 0, 1, 2, \\ &\int [\text{tr }\theta\theta'(h_{(11)}h'_{(11)} + h_{12}h'_{12})]\nu(dh) = K_1\text{tr }\theta\theta', \\ &\int [\text{tr }\theta\theta'(h_{(11)}h'_{(11)} + h_{12}h'_{12})]^2\nu(dh) = o(\delta^2), \\ &\int [\text{tr }\theta\theta'(h_{(11)}h'_{(11)} + h_{12}h'_{12})][\text{tr}(\theta h_{12}wh'_{(11)})]\nu(dh) = o(\delta^2), \\ &\int [\text{tr }\theta h_{12}wh'_{(11)}]^2\nu(dh) = K_2\text{tr}(\theta\theta')\text{tr}(s_{11}^{-1}s_{12}s_{22}^{-1}s_{21}),\end{aligned} \tag{8.109}$$

where K_1, K_2 are constants and $K_2 > 0$. Using (8.109) and (8.105) we obtain

$$\frac{dP(R|\theta)}{dP(R|0)} = 1 + \left[K_1 + \frac{N-1}{2}\right]\text{tr } \theta\theta' \qquad (8.110)$$

$$+ \tfrac{1}{2}K_2 D^{-1}(\text{tr } \theta\theta')(\text{tr } s_{11}^{-1}s_{12}s_{22}^{-1}s_{21}) + o(\delta^2).$$

From (8.110) the power function of any invariant test ϕ of level α is given by

$$\alpha[((N-1)/2 + K_1)\text{tr}(\theta\theta')] + \frac{K_2}{2D}\text{tr } \theta\theta' E_{H_0}(\phi \text{tr } S_{11}^{-1}S_{12}S_{22}^{-1}S_{21}) + o(\delta^2) \quad (8.111)$$

which is maximized by taking ϕ to be unity whenever $\text{tr } s_{11}^{-1}s_{12}s_{22}^{-1}s_{21}$ is greater than a constant depending on the level α of the test. So we get the following theorem.

Theorem 8.5.3. *For testing $H_0 : \Sigma_{12} = 0$ against the alternatives $H_1 : \Sigma_{12} \neq 0$, the level α test which rejects H_0 whenever*

$$\text{tr } s_{11}^{-1}s_{12}s_{22}^{-1}s_{21} \geq c \qquad (8.112)$$

is LBI when $\delta^2 \to 0$.

The LBI property of Pillai's test was first proved by Schwartz (1967). The following theorem establishes the admissible Bayes character of the likelihood ratio test for testing the independence of several subvectors as treated in Section 8.3.

Theorem 8.5.4. *Let H_0 be given by (8.50). The likelihood ratio test of H_0 is admissible Bayes if $N - 1 > p$.*

Proof. Let $\theta = (\mu, \Sigma)$ and $f(x^1, \ldots, x^N | \theta)$ be the joint *pdf* of $X^\alpha, \alpha = 1, \ldots, N$. Let Π_1 (the a priori under H_1) assign all its measure to Σ of the form $\Sigma = (I + \eta\eta')^{-1}$ and $\mu = (I + \eta\eta')^{-1}\eta z$ where η is a $p \times 1$ random vector with *pdf* proportional to

$$(\det(I + \eta\eta'))^{-\frac{1}{2}(N-1)} = (1 + \eta'\eta)^{-\frac{1}{2}(N-1)} \qquad (8.113)$$

and the conditional distribution of Z given η is normal with mean 0 and variance $(1 + \eta'\eta)/N$. Under H_0 the prior Π_0 assigns all its measure to $(\mu_{(i)}, \Sigma_{(ii)})$ of the form

$$\Sigma_{(ii)} = (I_{p_i} + \eta_{(i)}\eta'_{(i)})^{-1},$$

$$\mu_{(i)} = (I_{p_i} + \eta_{(i)}\eta'_{(i)})^{-1}\eta_{(i)}Z_i, \qquad i = 1, \ldots, k$$

Covariance Matrices and Mean Vectors

where $\eta_{(i)}$ is a $p_i \times 1$ random vector with pdf proportional to

$$(\det(I_{p_i} + \eta_{(i)}\eta'_{(i)}))^{-\frac{1}{2}(N-1)} = (1 + \eta'_{(i)}\eta_{(i)})^{-\frac{1}{2}(N-1)}$$

with $\eta = (\eta'_{(1)}, \ldots, \eta'_{(k)})'$ and the conditional pdf of Z_i given $\eta_{(i)}$ is normal with mean 0 and variance $N^{-1}(1 + \eta'_{(i)}\eta_{(i)})$ and $(\eta_{(1)}, Z_1), \ldots, (\eta_{(k)}, Z_k)$ are mutually independent. Using (7.38) the rejection region of the admissible Bayes test with respect to priors Π_1, and Π_0 is given by

$$\frac{\int f(x^1, \ldots, x^N | \theta) \Pi_1(d\theta)}{\int f(x^1, \ldots, x^N | \theta) \Pi_0(d\theta)} \geq c \qquad (8.114)$$

for some $c, 0 \leq c < \infty$. Now with K a normalizing constant, we can write the numerator of the left-hand side of (8.114) as

$$K \int \left[(1 + \eta'\eta)^{N/2} \exp\left\{ -\frac{1}{2} \sum_{\alpha=1}^{N} (x^\alpha - (I + \eta\eta')^{-1}\eta z)' \right. \right.$$

$$\left. \times (I + \eta\eta')(x^\alpha - (I + \eta\eta')^{-1}\eta z) \right\} \qquad (8.115)$$

$$\left. \times (1 + \eta'\eta)^{-N/2} \exp\left\{ -\frac{\frac{1}{2}Nz^2}{1 + \eta'\eta} \right\} \right] d\eta dz.$$

Using Lemma 6.8.1 we get

$$\sum_{\alpha=1}^{N} (x^\alpha - (I + \eta'\eta)^{-1}\eta z)'(I + \eta\eta')(x^\alpha - (I + \eta'\eta)^{-1}\eta z) + \frac{Nz^2}{1 + \eta'\eta}$$

$$= \sum_{\alpha=1}^{N} x^\alpha x^{\alpha'} - 2Nz\eta'\bar{x} + Nz^2 + \sum_{\alpha=1}^{N} x^{\alpha'} \eta\eta' x^\alpha$$

$$= \text{tr}(s + N\bar{x}\bar{x}') + \eta'(s + N\bar{x}\bar{x}')\eta + N(z - \bar{x}'\eta)^2.$$

Since for h real

$$\int_{E_p} (1 + \eta'\eta)^{-\frac{1}{2}h} d\eta < \infty \qquad \text{if and only if} \qquad h > p,$$

using Lemma 6.8.1 the value of the integral in (8.115) is given by

$$\text{constant } (\det s)^{-1/2} \exp\left\{ -\frac{1}{2} \text{tr}(s + \eta\bar{x}\bar{x}') \right\}.$$

Similarly the denominator of (8.114) is obtained as

$$\text{constant} \left[\prod_{i=1}^{k} (\det s_{(ii)})^{-1/2} \right] \exp\left\{ -\frac{1}{2} \text{tr}(s + \eta \bar{x}\bar{x}') \right\}.$$

Hence the left-hand side of (8.114) is proportional to

$$\left(\frac{\prod_{i=1}^{k} \det s_{(ii)}}{\det s} \right)^{1/2}.$$

Q.E.D.

8.5.2. Test of Multiple Correlation with Partial Information

Let $X = (X_1, \ldots, X_p)'$ be normally distributed p-dimensional random vector with mean μ and positive definite covariance matrix Σ and let X^α, $\alpha = 1, \ldots, N$ ($N > p$) be a random sample of size N from this distribution. Partition $X = (X_1, X'_{(1)}, X'_{(2)})'$ where X_1 is one-dimensional, $X_{(1)}$ is p_1-dimensional and $X_{(2)}$ is p_2-dimensional and $1 + p_1 + p_2 = p$. Let ρ_1^2 and ρ^2 denote the multiple correlation coefficients of X_1 with $X_{(1)}$ and with $(X'_{(1)}, X'_{(2)})'$ respectively. Denote by $\rho_2^2 = \rho^2 - \rho_1^2$. We consider the following testing problems:

a. to test $H_{10} : \rho^2 = 0$ against the alternatives $H_{1\lambda} : \rho_2^2 = 0, \rho_1^2 = \lambda > 0$;
b. to test $H_{20} : \rho^2 = 0$ against the alternatives $H_{2\lambda} : \rho_1^2 = 0, \rho_2^2 = \lambda > 0$.

Let $N\bar{X} = \sum_{\alpha=1}^{N} X^\alpha$, $S = \sum_{\alpha=1}^{N} (X^\alpha - \bar{X})(X^\alpha - \bar{X})'$, $b_{[i]}$ denote the i-vector consisting of the first i components of a vector b and $C_{[i]}$ denote the $i \times i$ upper-left submatrix of a matrix C. Partition S and Σ as

$$S = \begin{pmatrix} S_{11} & S_{(12)} & S_{(13)} \\ S_{(21)} & S_{(22)} & S_{(23)} \\ S_{(31)} & S_{(32)} & S_{(33)} \end{pmatrix}, \quad \Sigma = \begin{pmatrix} \Sigma_{11} & \Sigma_{(12)} & \Sigma_{(13)} \\ \Sigma_{(21)} & \Sigma_{(22)} & \Sigma_{(23)} \\ \Sigma_{(31)} & \Sigma_{(32)} & \Sigma_{(33)} \end{pmatrix}$$

where $S_{(22)}, \Sigma_{(22)}$ are each of dimension $p_1 \times p_1$; $S_{(33)}, \Sigma_{(33)}$ are each of dimension $p_2 \times p_2$. Then

$$\rho_1^2 = \Sigma_{(12)} \Sigma_{(22)}^{-1} \Sigma_{(21)} / \Sigma_{11},$$

$$\rho^2 = \rho_1^2 + \rho_2^2 = (\Sigma_{(12)} \Sigma_{(13)}) \begin{pmatrix} \Sigma_{(22)} & \Sigma_{(23)} \\ \Sigma_{(32)} & \Sigma_{(33)} \end{pmatrix}^{-1} (\Sigma_{(22)} \Sigma_{(13)})' / \Sigma_{11}. \quad (8.115a)$$

Covariance Matrices and Mean Vectors

Let

$$\bar{R}_1 = S_{(12)} S_{(22)}^{-1} S_{(21)} / S_{11},$$

$$\bar{R}_1 + \bar{R}_2 = (S_{(12)} S_{(13)}) \begin{pmatrix} S_{(22)} & S_{(23)} \\ S_{(32)} & S_{(33)} \end{pmatrix}^{-1} (S_{(12)} S_{(13)})' / S_{11}. \qquad (8.115b)$$

The transformation group transforming

$$(\bar{X}, S, \mu, \Sigma) \to (\bar{X} + b, S, \mu + b, \Sigma)$$

$b \in R^p$ leaves the present problem invariant and this, along with the full linear group G of $p \times p$ nonsingular matrices g,

$$g = \begin{pmatrix} g_{11} & 0 & 0 \\ 0 & g_{(22)} & g_{(23)} \\ 0 & g_{(32)} & g_{(33)} \end{pmatrix}$$

with $g_{11} : 1 \times 1$, $g_{(22)} : p_1 \times p_1$, $g_{(22)} : p_2 \times p_2$, generates a group of transformations which leaves the present problem invariant. The action of these transformations is to reduce the problem to that where $\mu = 0$ and $S = \sum_{\alpha=1}^{N} X^\alpha X^{\alpha'}$ is sufficient for Σ, where N has been reduced by one from what it was originally. We treat the latter formulation considering X^α, $\alpha = 1, \ldots, N$ ($N \geq p \geq 2$) to have a zero mean and consider only the group G of transformations g operating as

$$(S, \Sigma) \to (gSg', g\Sigma g')$$

for the invariance of the problem. A maximal invariant in the sample space under G is $(\bar{R}_1, \bar{R}_2)$ as defined in (8.115b). Since $S > 0$ with probability one, $\bar{R}_1 > 0$, $\bar{R}_2 > 0$ and $\bar{R}_1 + \bar{R}_2 = R^2$, the squared sample multiple correlation coefficient between the first and the remaining $p - 1$ components of the random vector X. A corresponding maximal invariant in the parametric space under G is (ρ_1^2, ρ_2^2). From Giri (1979) the joint probability density function of $(\bar{R}_1, \bar{R}_2)$ is given by

$$f_\Delta(\bar{r}_1, \bar{r}_2) = K(1 - \rho^2)^{-N/2} (1 - \bar{r}_1 - \bar{r}_2)^{\frac{1}{2}(N-p-1)} \prod_{i=1}^{2} (\bar{r}_i)^{\frac{1}{2}p_i - 1}$$

$$\times \left[1 + \sum_{i=1}^{2} \bar{r}_i \left(\frac{1 - \rho^2}{\gamma_i} - 1 \right) \right]^{-N/2} \qquad (8.115c)$$

$$\times \sum_{\beta_1=0}^{\infty} \sum_{\beta_2=0}^{\infty} \prod_{i=1}^{2} \frac{\Gamma(\frac{1}{2}(N + p_i - \sigma_i) + \beta_i) \Gamma(\beta_i + \frac{1}{2})}{(2\beta_i)! \Gamma(\frac{1}{2} p_i + \beta_i)} (\theta_i)^{\beta_i}$$

where

$$\gamma_i = 1 - \sum_{j=1}^{i} \rho_j^2, \quad \text{with} \quad \gamma_0 = 1,$$

$$\sigma_i = \sum_{j=1}^{i} p_j, \quad \alpha_i^2 = \rho_i^2(1-\rho^2)/\gamma_i \gamma_{i-1},$$

$$\theta_i = 4\bar{r}_i \alpha_i^2 \left(1 + \sum_{i=1}^{2} \bar{r}_i \left[\frac{1-\rho^2}{\gamma_i} - 1\right]\right)^{-1}$$

and K is the normalizing constant.

By straightforward computations the likelihood ratio test of H_{10} when parameter space $\Omega = \{(\mu, \Sigma) : \Sigma_{(13)} = 0\}$ rejects H_{10} whenever

$$\bar{r}_1 \geq C \tag{8.115d}$$

where the constant C depends on the size α of the test and under H_{10} $\bar{R}_1$ has a central beta distribution with parameter $(\frac{1}{2}p_1, \frac{1}{2}(N - p_1))$. The likelihood ratio test of H_{20} when $\Omega = \{(\mu, \Sigma : \Sigma_{(12)} = 0\}$ rejects H_{20} whenever

$$z = \frac{1 - \bar{r}_1 - \bar{r}_2}{1 - \bar{r}_1} \leq C \tag{8.115e}$$

where the constant C depends on the size α of the test and under H_{20} the corresponding random variable Z is distributed independently of $\bar{R}_1$ as central beta with parameter $(\frac{1}{2}(N - p_1 - p_2), \frac{1}{2}p_2)$.

Theorem 8.5.5. *For testing H_{10} against $H_{1\lambda}$ the likelihood ratio test given in (8.115d) is UMP invariant.*

Proof. Under H_{10} $\gamma_i = 1, i = 0, 1, 2$. Hence $\alpha_i^2 = 0, \theta_i = 0, i = 1, 2$. Under $H_{1\lambda}$, $\rho_1^2 = \lambda, \rho_2^2 = 0, \gamma_0 = 1, \gamma_1 = 1 - \lambda \alpha_1^2 = 1 - \lambda, \alpha_2^2 = 0, \theta_1 = 4\bar{r}_1 \lambda$ and $\theta_2 = 0$. Thus

$$\frac{f_{H_{1\lambda}}(\bar{r}_1, \bar{r}_2)}{f_{H_{10}}(\bar{r}_1, \bar{r}_2)} = K(1-\lambda)^{-N/2} \sum_{i=0}^{\infty} \frac{\Gamma(\frac{1}{2}N + i)(4\bar{r}_1 \lambda)^i}{(2i)!}.$$

Now using Neyman-Pearson Lemma we get the theorem. Q.E.D.

Theorem 8.5.6. *The likelihood ratio test of H_{20} against H_{21} is UMP invariant among all test $\phi(\bar{R}_1, \bar{R}_2)$ based on $(\bar{R}_1, \bar{R}_2)$ satisfying*

$$E_{H_{20}}(\phi(\bar{R}_1, \bar{R}_2)|\bar{R}_1 = \bar{r}_1) = \alpha.$$

Proof. Under $H_{2\lambda}$, $\rho_1^2 = 0$, $\rho_2^2 = \lambda$, $\gamma_0 = 1$, $\gamma_1 = 1$, $\gamma_2 = 1 - \lambda$, $\alpha_1^2 = 0$, $\alpha_2^2 = \lambda$, $\theta_1 = 0$, $\theta_2 = 4\bar{r}_2\lambda(1 - \bar{r}_1\lambda)^{-1}$. Hence

$$f_{H_{2\lambda}}(\bar{r}_2|\bar{r}_1)/f_{H_{20}}(\bar{r}_2|\bar{r}_1) = f_{H_{2\lambda}}(\bar{r}_1, \bar{r}_2)/f_{H_{20}}(\bar{r}_1, \bar{r}_2)$$

$$= K(1 - \lambda)^{-N/2}(1 - \bar{r}_1\lambda)^{-N/2}$$

$$\times \sum_{i=0}^{\infty} \frac{\Gamma(\frac{1}{2}(N - p_1) + i)}{(2i)!} \left(\frac{4\bar{r}_2\lambda}{1 - \bar{r}_1\lambda}\right)^i.$$

Hence $f_{H_{2\lambda}}(\bar{r}_2|\bar{r}_1)$ has a monotone likelihood ratio in $\bar{r}_2 = (1 - z)(1 - \bar{r}_1)$. Now using Lehmann (1939) we get the theorem. Q.E.D.

8.6. MULTIVARIATE GENERAL LINEAR HYPOTHESIS

In this section we generalize the univariate general linear hypothesis and analysis of variance with fixed effect model to vector variates. The algebra is essentially the same as that of the univariate case. Unlike the univariate general linear hypothesis, there is more latitude in the choice of the test criteria in the multivariate case, although the distributions of different test criteria are quite involved. The reader is referred to Giri (1993) for a treatment of the univariate general linear hypothesis, which is very appropriate for following the developments here, to Roy (1953, 1957) for the union-intersection approach for obtaining a suitable test criterion which is also appropriate for this problem, and to Constantine (1963) for some connected distribution results. We shall first state and solve the problem in the most general form and then give the formulation of the multivariate general linear hypothesis in terms of multiple regression. The latter formulation is useful for analyzing multivariate design models.

Let $X^\alpha = (X_{\alpha 1}, \ldots, X_{\alpha p})'$, $\alpha = 1, \ldots, N$, be N independently distributed p-variate normal vectors with mean $E(X^\alpha) = \mu^\alpha = (\mu_{\alpha 1}, \ldots, \mu_{\alpha p})'$ and a common positive definite covariance matrix Σ. A multivariate linear hypothesis is defined in terms of two linear subspaces π_Ω, π_ω of dimensions $s(< N)$, $s - r(0 \le s - r < s)$, respectively. It is assumed throughout that all vectors $(\mu_{1i}, \ldots, \mu_{Ni})'$, $i = 1, \ldots, p$, lie in π_Ω, and it is desired to test the null hypothesis H_0 that they lie in π_ω. We shall also assume that $N - s \ge p$ so that we have enough degrees of freedom to estimate Σ.

Example 8.6.1. Let $N = N_1 + N_2$ and let $X^\alpha, \alpha = 1, \ldots, N_1$, be a random sample of size N_1 from a p-variate normal population with mean $\mu^1 = (\mu_{11}, \ldots, \mu_{1p})'$ and covariance matrix Σ (unknown). Let $X^\alpha, \alpha = N_1 + 1, \ldots, N$, be a random sample of size N_2 from another p-variate normal population with mean $\mu^2 = (\mu_{21}, \ldots, \mu_{2p})'$ and the same covariance matrix Σ. We are interested in testing the null hypothesis $H_0 : \mu^1 = \mu^2$. Here $s = 2$ and $s - r = 1$. Let

$$X = \begin{pmatrix} X_{11} & \cdots & X_{1p} \\ X_{21} & \cdots & X_{2p} \\ \vdots & & \vdots \\ X_{N1} & \cdots & X_{Np} \end{pmatrix} = \begin{pmatrix} X^{1'} \\ X^{2'} \\ \vdots \\ X^{N'} \end{pmatrix}, \Sigma = (\sigma_{ij}). \tag{8.116}$$

This problem can be reduced to a canonical form by applying to each of the N vectors $(X_{1i}, \ldots, X_{Ni})', i = 1, \ldots, p$ an orthogonal transformation which transforms X to $Y = OX$ where O is an $N \times N$ orthogonal matrix

$$O = \begin{pmatrix} O_{11} & \cdots & O_{1N} \\ O_{21} & \cdots & O_{2N} \\ \vdots & & \vdots \\ O_{N1} & \cdots & O_{NN} \end{pmatrix} = \begin{pmatrix} O_1 \\ \vdots \\ O_N \end{pmatrix} \tag{8.117}$$

such that its first s row vectors $O_1, \ldots, O_s$, span π_Ω with $O_{r+1}, \ldots, O_s$ spanning π_ω. Write

$$Y = \begin{pmatrix} Y^{1'} \\ Y^{2'} \\ \vdots \\ Y^{N'} \end{pmatrix}, \quad Y^\alpha = (Y_{\alpha 1}, \ldots, Y_{\alpha p})', \quad \alpha = 1, \ldots, N. \tag{8.118}$$

Thus $E(Y^\alpha) = 0$ for $\alpha = s + 1, \ldots, N$ if and only if all $(\mu_{1i}, \ldots, \mu_{Ni})' \in \pi_\Omega, i = 1, \ldots, p$; and $E(Y^\alpha) = 0, \alpha = 1, \ldots, r, s + 1, \ldots, N$, if and only if all $(\mu_{1i}, \ldots, \mu_{Ni})' \in \pi_\omega, i = 1, \ldots, p$. Now the covariance of $Y_{\alpha i} = \Sigma_{\lambda=1}^N O_{\alpha\lambda} X_{\lambda i}$, $Y_{\beta j} = \Sigma_{\delta=1}^N O_{\beta\delta} X_{\delta j}$ is

$$\text{cov}(Y_{\alpha i}, Y_{\beta j}) = \sum_{\lambda=1}^N \sum_{\delta=1}^N O_{\alpha\lambda} O_{\beta\delta} \text{cov}(X_{\lambda i} X_{\delta j}) = \sigma_{ij} \sum_{\lambda=1}^N O_{\alpha\lambda} O_{\beta\lambda}$$

$$= \begin{cases} \sigma_{ij} & \text{when } \alpha = \beta \\ 0 & \text{when } \alpha \neq \beta, \end{cases}$$

since $\text{cov}(X_{\lambda i}, X_{\delta j}) = \sigma_{ij}$ when $\lambda = \delta$, $\text{cov}(X_{\lambda i}, X_{\delta j}) = 0$ when $\lambda \neq \delta$. Thus the row vectors of Y are independent normal p-vectors with the same covariance

Covariance Matrices and Mean Vectors

matrix Σ and under π_Ω,

$$E(Y^\alpha) = \begin{cases} \mathbf{v}^\alpha \text{ (say)}, & \alpha = 1, \ldots, s, \\ 0, & \alpha = s+1, \ldots, N \end{cases}$$

and under π_ω,

$$E(Y^\alpha) = \begin{cases} \mathbf{v}^\alpha, & \alpha = r+1, \ldots, s, \\ 0, & \alpha = 1, \ldots, r, s+1, \ldots, N \end{cases}$$

Hence in the canonical form we have the following problem: Y^α, $\alpha = 1, \ldots, N$, are independently distributed normal p-vectors with the same positive definite covariance matrix Σ (unknown) and the means $E(Y^\alpha) = 0$, $\alpha = s+1, \ldots, N$. It is desired to test the null hypothesis $H_0 : E(Y^\alpha) = 0$, $\alpha = 1, \ldots, r$.

The likelihood of the observations y^α on Y^α, $\alpha = 1, \ldots, N$, is given by

$$L(\mathbf{v}^1, \ldots, \mathbf{v}^s, \Sigma) = (2\pi)^{-Np/2} (\det \Sigma^{-1})^{N/2}$$

$$\times \exp\left\{-\tfrac{1}{2} \operatorname{tr} \Sigma^{-1} \left[\sum_{\alpha=1}^{s} (y^\alpha - \mathbf{v}^\alpha)(y^\alpha - \mathbf{v}^\alpha)' \right.\right. \quad (8.119)$$

$$\left.\left. + \sum_{\alpha=s+1}^{N} y^\alpha y^{\alpha'} \right]\right\}.$$

Using Lemma 5.1.1 we obtain

$$\max_{\pi_\Omega} L(\mathbf{v}^1, \ldots, \mathbf{v}^s, \Sigma) = (2\pi/N)^{-Np/2}$$

$$\times [\det(\Sigma_{\alpha=s+1}^{N} y^\alpha y^{\alpha'})]^{-N/2} \exp\{-\tfrac{1}{2} Np\}. \quad (8.120)$$

Under H_0, L is reduced to

$$L(\mathbf{v}^{r+1}, \ldots, \mathbf{v}^s, \Sigma) = (2\pi)^{-Np/2} (\det \Sigma^{-1})^{N/2}$$

$$\times \exp\left\{-\tfrac{1}{2} \Sigma^{-1} \left[\sum_{\alpha=1}^{r} y^\alpha y^{\alpha'}\right.\right.$$

$$\left.\left. + \sum_{\alpha=r+1}^{s} (y^\alpha - \mathbf{v}^\alpha)(y^\alpha - \mathbf{v}^\alpha)' + \sum_{\alpha=s+1}^{N} y^\alpha y^{\alpha'} \right]\right\}$$

and

$$\max_{\pi_\omega} L(\mathbf{v}^{r+1}, \ldots, \mathbf{v}^s, \Sigma) = \left(\frac{2\pi}{N}\right)^{-Np/2}$$
$$\times \left[\det\left(\sum_{\alpha=1}^{r} y^\alpha y^{\alpha'} + \sum_{\alpha=s+1}^{N} y^\alpha y^{\alpha'}\right)\right]^{-N/2} \exp\{-\tfrac{1}{2}Np\}. \quad (8.122)$$

Hence the likelihood ratio test of H_0 rejects H_0 whenever

$$\lambda = \left[\frac{\det b}{\det(a+b)}\right]^{N/2} \leq c \quad (8.123)$$

or equivalently,

$$u = \frac{\det b}{\det(a+b)} \leq c', \quad (8.124)$$

where c, c' are constants chosen in such a way that the corresponding test has size α and $a = \sum_{\alpha=1}^{r} y^\alpha y^{\alpha'}$, $b = \sum_{\alpha=s+1}^{N} y^\alpha y^{\alpha'}$. This result is due to Hsu (1941) and Wilks (1932). From Sections 6.3 and 6.5 we conclude that the corresponding random variables $A = \sum_{\alpha=1}^{r} Y^\alpha Y^{\alpha'}$, $B = \sum_{\alpha=s+1}^{N} Y^\alpha Y^{\alpha'}$ are independently distributed Wishart matrices of dimension $p \times p$, and B has a central Wishart distribution with parameter Σ and $N - s$ degrees of freedom. Under H_0, A is distributed as central Wishart with parameter Σ and r degrees of freedom whereas under H_1 it is distributed as noncentral Wishart.

In application to specific problems it is not straightforward to carry out the reduction to the canonical form just given explicitly. The test statistic u can be expressed in terms of the original random variables X. Let $(\hat{\mu}_{1i}, \ldots, \hat{\mu}_{Ni})'$ and $(\hat{\hat{\mu}}_{1i}, \ldots, \hat{\hat{\mu}}_{Ni})'$ be the projections of the vector $(X_{1i}, \ldots, X_{Ni})'$ on π_Ω and π_ω, respectively. Then $\sum_{\alpha=1}^{N}(X_{\alpha i} - \hat{\mu}_{\alpha i})(X_{\alpha i} - \hat{\mu}_{\alpha i})$ is the inner product of two vectors, each of which is the difference of the given vector $(X_{1i}, \ldots, X_{Ni})'$ and its projection on π_Ω, and it remains unchanged under the orthogonal transformation of the coordinate system in which the variables are expressed. Now $O(X_{1i}, \ldots, X_{Ni})'$ can be interpreted as expressing $(X_{1i}, \ldots, X_{Ni})'$ in a new coordinate system with the first s coordinate axes lying in π_Ω. Hence the projection on π_Ω of the transformed vector $(Y_{1i}, \ldots, Y_{Ni})'$ is $(Y_{1i}, \ldots, Y_{si}, 0, \ldots, 0)'$ so that the difference between the vector and its projection is $(0, \ldots, 0, Y_{s+1,i}, \ldots, Y_{Ni})$. The (i, j)th element of $\sum_{\alpha=s+1}^{N} Y^\alpha Y^{\alpha'}$ is therefore given by

$$\sum_{\alpha=s+1}^{N} Y_{\alpha i} Y_{\alpha j} = \sum_{\alpha=1}^{N}(X_{\alpha i} - \hat{\mu}_{\alpha i})(X_{\alpha j} - \hat{\mu}_{\alpha j}). \quad (8.125)$$

Covariance Matrices and Mean Vectors

Similarly, for the transformed vector $(Y_{1i}, \ldots, Y_{Ni})'$ the difference between its projections on π_Ω and π_ω is therefore $(Y_{1i}, \ldots, Y_{ri}, 0, \ldots, 0)'$. Thus $\sum_{\alpha=1}^{r} Y_{\alpha i} Y_{\alpha j}$ is equal to the inner product (for the ith and the jth vectors) of the difference of these projections. Comparing this with the expression of the same inner product in the original coordinate system, we obtain

$$\sum_{\alpha=1}^{r} Y_{\alpha i} Y_{\alpha j} = \sum_{\alpha=1}^{N} (\hat{\mu}_{\alpha i} - \hat{\hat{\mu}}_{\alpha i})(\hat{\mu}_{\alpha j} - \hat{\hat{\mu}}_{\alpha j}) \tag{8.126}$$

In terms of the variable Y the problem of testing H_0 against $H_1 : \Lambda = (\mathbf{v}^1, \ldots, \mathbf{v}^r)' \neq 0$ remains invariant under the following three groups of transformations.

1. The group of translations T which translates $Y^\alpha \to Y^\alpha + d^\alpha$, $\alpha = r + 1, \ldots, s$, and $d^\alpha = (d_{\alpha 1}, \ldots, d_{\alpha p})' \in T$. The maximal invariant under T in the space of Y is $(Y^1, \ldots, Y^r, Y^{s+1}, \ldots, Y^N)$.
2. Let Z be an $r \times p$ matrix such that $Z' = (Y^1, \ldots, Y^r)$, and let W be the $(N - s) \times p$ matrix such that $W' = (Y^{s+1}, \ldots, Y^N)$. The group of $r \times r$ orthogonal transformations $O(r)$ operating in the space of Z as $Z \to OZ$, $O \in O(r)$, and the group of $(N - s) \times (N - s)$ orthogonal transformations $O(N - s)$ operating in the space of W as $W \to OW$, $O \in O(N - s)$, affect neither the independence nor the covariance matrix of the row vectors of Z and W.

Lemma 8.6.1. $Z'Z = \sum_{\alpha=1}^{r} Y^\alpha Y^{\alpha'}$ is a maximal invariant under $O(r)$ in the space of Z.

Proof. Since $(OZ)'(OZ) = Z'Z$, the matrix $Z'Z$ will be a maximal invariant if we show that for any two elements Z^*, Z in the same space, $Z^{*'}Z^* = Z'Z$ implies the existence of an orthogonal matrix $O \in O(r)$ such that $Z^* = OZ$.

Consider first the case $r = p$. Without any loss of generality we can assume that the p columns of Z are linearly independent (the exceptional set of Z's for which this does not hold has probability measure 0). Now $Z^{*'}Z^* = Z'Z$ implies that $O = Z^*Z^{-1}$ is an orthogonal matrix and that $Z^* = OZ$. Consider now the case $r > p$. Without any loss of generality we can assume that the columns of Z are linearly independent. Since for any two p-dimensional subspaces of the r-space there exists an orthogonal transformation transforming one to the other, we assume that after a suitable orthogonal transformation the p column vectors of Z and Z^* lie in the same subspace and the problem is reduced to the case $r = p$. If $r < p$, the first r column vectors of Z can be assumed to be linearly independent. Write $Z = (Z_1, Z_2)$, where Z_1, Z_2 are submatrices of dimensions $r \times r$ and

$r \times (p-r)$, respectively, and similarly for Z^*. Since $Z^{*\prime}Z^* = Z'Z$, we obtain

$$Z_1^{*\prime}Z_1^* = Z_1'Z_1, \qquad Z_1^{*\prime}Z_2^* = Z_1'Z_2 \quad \text{and} \quad Z_2^{*\prime}Z_2^* = Z_2'Z_2. \qquad (8.127)$$

Now by the previous argument $Z_1^{*\prime}Z_1^* = Z_1'Z_1$ implies that there exists an orthogonal matrix $B = (Z_1^{*\prime})^{-1}Z_1'$ such that $Z_1^* = BZ_1$. Also $Z_1^{*\prime}Z_2^* = Z_1'Z_2$ implies that $Z_2^* = BZ_2$. Obviously $Z_2^{*\prime}Z_2^* = Z_2'Z_2$ with $Z_2^* = BZ_2$. Q.E.D.

Similarly a maximal invariant in the space of W under $O(N-s)$ is $W'W = \Sigma_{\alpha=s+1}^{N} Y^\alpha Y^{\alpha\prime}$.

The problem remains invariant under the full linear group $G_l(p)$ (multiplicative group of $p \times p$ nonsingular matrices) of transformation g transforming Z to gZ, W to gW. The corresponding induced transformation in the space of (A, B) is given by $(A, B) \to (gAg', gBg')$. By Exercise 7 the roots of $\det(A - \lambda B) = 0$ (the characteristic roots of AB^{-1}) are maximal invariant in the space of (A, B) under $G_l(p)$. Let $R_1, \ldots, R_p$ denote the roots of $\det(A - \lambda B) = 0$. A corresponding maximal invariant in the parametric space is $(\theta_1, \ldots, \theta_p)$, the characteristic roots of $\Lambda\Lambda'\Sigma^{-1}$ where $\Lambda = E(Z')$. The test statistic U in (8.124) can be written as

$$\det(B(A+B)^{-1}) = \prod_{i=1}^{p}(1+R_i)^{-1}. \qquad (8.128)$$

Anderson (1958) called this statistic $U_{p,r,N-s}$. Some other invariant tests are also proposed for this problem. They are as follows. In all cases the constant c will depend on the level of significance α of the test.

1. Wilks' criterion (Wilks, 1932; Hsu, 1940):

Reject H_0 whenever $\det a(b+a)^{-1} \le c$.

For large N, $W = -[N - s - \frac{1}{2}(p-r+1)] \log U_{p,r,N-s}$ has a limiting χ^2_{pr} distribution with pr degrees of freedom (Box (1949)). Let

$$P(\chi^2_{pr} \le \chi^2_{pr}(\alpha)) = \alpha$$
$$P(U \le u_{p,r,N-s}(\alpha)|H_0) = \alpha. \qquad (8.129)$$

Define

$$C_{p,r,N-s}(\alpha) = \frac{-(N - s - \frac{1}{2}(p-r+1)) \log u_{p,r,N-s}(\alpha)}{\chi^2_{pr}(\alpha)}. \qquad (8.130)$$

Covariance Matrices and Mean Vectors

To test H_0 one computes the chisquare adjustment $C_{p,r,N-s}$ and rejects H_0 at level α if

$$W \geq -[N - s - \tfrac{1}{2}(p - r + 1)] \log U_{p,r,N-s}$$
$$= C_{p,r,N-s}(\alpha)\chi^2_{pr}(\alpha) \quad (8.131)$$

Tables of values of $C_{p,r,N-s}(\alpha)$ have been prepared by Schatzoff (1966), Pillai and Gupta (1969) and Lee (1971) for different values p, r, $N - s$ and α. Tables of Schatzoff, Pillai and Gupta are given in Appendix A.

2. Lawley's V (Lawley, 1938) and Hotelling's T_0^2 (Hotelling, 1951) criterion. Reject H_0 whenever

$$\vartheta = \operatorname{tr} ab^{-1} = \frac{T_0^2}{N - p} \geq C. \quad (8.132)$$

Percentage points of the null distribution of T_0^2 are given in Pillai and Sampson (1959), Davis (1970, 1980) and Hughes and Saw (1972). Asymptotic distribution of T_0^2 in the non-null case has been studied by Siotani (1957, 1971), Ito (1960), Fujikoshi (1970) and Muirhead (1972). In the null case $N\operatorname{tr} AB^{-1}$ is approximately χ^2_{pr} (Morrow (1948)) when $N \to \infty$.

3. The largest and the smallest root criteria of Roy (Roy, 1957)

$$\text{Reject } H_0 \text{ whenever } \max_i r_i \geq C \quad (8.133)$$

$$\text{Reject } H_0 \text{ whenever } \min_i r_i \geq C \quad (8.134)$$

Percentage points of the distribution of $\max_i R_i$ are given in Heck (1960), Pillai and Bantegui (1959) and Pillai (1964, 1965, 1967). Khatri (1972) has obtained the exact distribution of $\max_i R_i$ as a finite series of Laguerre polynomals in a special non-null case. We refer to Krishnaiah (1978) for references and results in this context.

4. Pillai's statistic (Pillai, 1955):

$$\text{Reject } H_0 \text{ whenever } \operatorname{tr} a(a + b)^{-1} \geq C. \quad (8.135)$$

Pillai (1960) obtained 1% and 5% signifance points of $\operatorname{tr} A(A + B)^{-1}$ for $p = 2, \ldots, 8$. Mijares (1964) extended the tables to $p = 50$. Asymptotic expansions of the distribution of $(N - p)\operatorname{tr} A(A + B)^{-1}$ in the non-null case have been obtained by Fujikoshi (1970) and Lee (1971).

These test statistics are functions of $R_1, \ldots, R_p$. Among these invariant tests, test 4 has received much less attention than the others. These tests 1–4, of course, reduce to Hotelling's T^2-test when $r = 1$, and if $r > 1$ and $\min(p, r) > 1$, there does not exist a uniformly most powerful invariant test. All these tests reduce to the univariate F-test when $p = 1$ and to the two-tailed t-test when $p = r = 1$. In theory, we would be able to derive the distribution of these statistics from the joint distribution of $R_1, \ldots, R_p$. Since for any $g \in G_l(p)$, $\det(gAg' - \lambda g B g') = \det(gg') \det(A - \lambda B)$, choosing g such that $g\Sigma g' = I$ we conclude that to find the joint distribution of $(R_1, \ldots, R_p)$ under H_0, we can without any loss of generality assume that $\Sigma = I$. In other words, the joint distribution is independent of Σ under H_0 and under $H_1 : \Lambda \neq 0$ this distribution depends only on $\theta_1, \ldots, \theta_p$.

8.6.1. Distribution of $(R_1, \ldots, R_p)$ under H_0

From Section 6.3, B and A are independently distributed (Wishart matrices) as $W_p(\Sigma, N - s)$ and $W_p(\Sigma, r)$, respectively, provided $N - s \geq p, r \geq p$. Let $N - s = n_2, r = n_1$, let $R_1, \ldots, R_p$, be the characteristic roots of AB^{-1}, and let $R_1 > R_2 > \cdots > R_p > 0$ denote the ordered characteristic roots of AB^{-1} (the probability of two roots being equal is 0). Rather than finding the distribution of $(R_1, \ldots, R_p)$ directly, we will find it convenient to first find the joint distribution of $V_1, \ldots, V_p$ such that $V_i = R_i/(1 + R_i), i = 1, \ldots, p$. Obviously $V_1, \ldots, V_p$ are the characteristic roots of $A(A + B)^{-1}$, that is, the roots of $\det(A - \lambda(A + B)) = 0$. Let V be a diagonal matrix with diagonal elements $V_1, \ldots, V_p$ and let $C = A + B$. We can write

$$C = WW', \quad A = WVW' \tag{8.136}$$

where $W = (W_{ij})$ is a nonsingular matrix of dimension $p \times p$. To determine W uniquely we require here that $W_{i1} \geq 0, i = 1, \ldots, p$ (the probability of $W_{i1} = 0$ is 0). Writing J for Jacobian, the Jacobian of the transformation $(A, B) \to (W, V)$ is equal to

$$J[(A, B) \to (W, V)] = J[(A, B) \to (A, C)] \times J[(A, C) \to (W, V)]. \tag{8.137}$$

It is easily seen that $J[(A, B) \to (A, C)] = 1$. By exercise 8 [see also Olkin (1952)] the Jacobian of the transformation $(A, C) \to (W, V)$ is

$$2^p (\det W)^{p+2} \sum_{i<j}(V_i - V_j). \tag{8.138}$$

As indicated earlier we can take $\Sigma = I$, and hence the joint probability density function of A, B is (by Section 6.3)

$$C_{n_1,p} C_{n_2,p} (\det a)^{(n_1 - p - 1)/2} (\det b)^{(n_2 - p - 1)/2} \exp\{-\tfrac{1}{2} \operatorname{tr}(a + b)\} \tag{8.139}$$

Covariance Matrices and Mean Vectors

where $C_{n,p}$ is given by (6.32). From (8.137–8.139) the joint probability density function of (W, V) is

$$f_{W,V}(w, v) = C_{n_1 p} C_{n_2 p} \prod_{i=1}^{p} [v_i^{(n_1-p-1)/2}(1 - v_i)^{(n_2-p-1)/2}] \qquad (8.140)$$
$$\times \prod_{i<j} (v_i - v_j)(\det(ww'))^{(n_1+n_2-p)/2} \exp\{-\tfrac{1}{2} \operatorname{tr} ww'\}$$

Now integrating out w in (8.140), we obtain the probability density function of V as

$$f_V(v) = K C_{n_1,p} C_{n_2,p} \prod_{i=1}^{p} v_i^{(n_1-p-1)/2}(1 - v_i)^{(n_2-p-1)/2} \prod_{i<j}(v_i - v_j), \qquad (8.141)$$

where

$$K = (2\pi)^{p^2/2} \int \frac{1}{(2\pi)^{p^2/2}} 2^p [\det(ww')]^{(n_1+n_2-p)/2} \exp\{-\tfrac{1}{2} \operatorname{tr} ww'\} dw \qquad (8.142)$$
$$= (2\pi)^{p^2/2} E[\det(WW')]^{(n_1+n_2-p)/2}$$

and $W = (W_{ij})$, the W_{ij} are independently distributed normal random variables with mean 0 and variance 1. Thus the $p \times p$ matrix $S = WW'$ is distributed as $W_p(I, p)$ and its probability density function [by (6.32)] is

$$f_S(s) = C_{p,p}(\det s)^{-\tfrac{1}{2}} \exp\{-\tfrac{1}{2} \operatorname{tr} s\}. \qquad (8.143)$$

Hence

$$E(\det(WW'))^{(n_1+n_2-p)/2} = C_{p,p} \int (\det s)^{(n_1+n_2-p-1)/2} \exp\{-\tfrac{1}{2} \operatorname{tr} s\} ds \qquad (8.144)$$
$$= \frac{C_{p,p}}{C_{n_1+n_2,p}}.$$

Thus

$$K = \frac{(2\pi)^{p^2/2} C_{p,p}}{C_{n_1+n_2,p}} \qquad (8.145)$$

Since $dV_i = (1 + R_i^2)^{-1} dR_i$, from (8.141) the probability density of R, a diagonal matrix with diagonal elements $R_1, \ldots, R_p$, is

$$f_R(r) = C \prod_{i=1}^{p} r_i^{(n_1-p-1)/2}(1 + r_i)^{-(n_1+n_2)/2} \prod_{i<j}(r_i - r_j) \qquad (8.146)$$

where

$$C = \frac{\pi^{p/2} \prod_{i=1}^{p}(\frac{1}{2}(n_1 + n_2 - i + 1))}{\prod_{i=1}^{p} \Gamma(\frac{1}{2}(n_1 - i + 1))\Gamma(\frac{1}{2}(n_2 - i + 1))\Gamma(\frac{1}{2}(p + 1 - i))}. \quad (8.147)$$

Let us now consider the distribution of the characteristic roots of A where A is distributed as $W_p(I, n_1)$. Since B is distributed as $W_p(I, n_2)$, $B/n_2 \to I$ almost surely as $n_2 \to \infty$. Thus the roots of the equation $\det(A - \lambda(B/n_2)) = 0$ converge almost surely to the roots of $\det(A - \lambda I) = 0$. Let $\lambda_1 > \lambda_2 > \cdots > \lambda_p > 0$ be the ordered characteristic roots of A. To find the joint distribution of the λ_i, it is sufficient to find the limit as $n_2 \to \infty$ of the probability density function of the roots of $\det(A - \lambda(B/n_2)) = 0$. From (8.147), the probability density function of the roots $(\lambda_1, \ldots, \lambda_p)$ of $\det(A - \lambda(B/n_2)) = 0$ is given by

$$C(n_2)^{-n_1 p/2} \prod_{i=1}^{p} \lambda_i^{(n_1 - p - 1)/2} \left(1 + \frac{\lambda_i}{n_2}\right)^{-(n_1 + n_2)/2} \prod_{i<j}(\lambda_i - \lambda_j). \quad (8.148)$$

Since

$$Lt_{n_2 \to \infty} \prod_{i=1}^{p}\left(1 + \frac{\lambda_i}{n_2}\right)^{-(n_1 + n_2)/2} = \exp\left\{-\frac{1}{2}\sum_{i=1}^{p} \lambda_i\right\},$$

$$Lt_{n_2 \to \infty} \frac{\Gamma(\frac{1}{2}(n_1 + n_2 - 1))}{(n_2)^{n_1/2} \Gamma(\frac{1}{2}(n_2 - j))} = 2^{-n_1/2},$$

we get

$$Lt_{n_2 \to \infty} C(n_2)^{-n_1 p/2}$$

$$= \pi^{-p/2} \left[2^{n_1 p/2} \prod_{i=1}^{p} \Gamma(\tfrac{1}{2}(n_1 - i + 1))\Gamma(\tfrac{1}{2}(p + 1 - i))\right]^{-1} \quad (8.149)$$

$$= C' \quad \text{(say)}$$

Thus the probability density function of the ordered characteristic roots $\lambda_1, \ldots, \lambda_p$, of A is (with λ a diagonal matrix with diagonal elements $\lambda_1, \ldots, \lambda_p$)

$$f_\lambda(\lambda) = C' \prod_{i=1}^{p} \lambda_i^{(n_1 - p - 1)/2} \exp\left\{-\frac{1}{2} \prod_{i=1}^{p} \lambda_i\right\} \prod_{i<j}(\lambda_i - \lambda_j). \quad (8.150)$$

8.6.2. Multivariate Regression Model

We now discuss a different formulation of the multivariate general linear hypothesis which is very appropriate for the analysis of design models. Let

Covariance Matrices and Mean Vectors

$X^\alpha = (X_{\alpha 1}, \ldots, X_{\alpha p})'$, $\alpha = 1, \ldots, N$, be independently distributed normal p-vectors with means

$$E(X^\alpha) = \beta z^\alpha, \qquad \alpha = 1, \ldots, N, \tag{8.151}$$

where $z^\alpha = (z_{\alpha 1}, \ldots, z_{\alpha s})'$, $\alpha = 1, \ldots, N$ are known vectors and $\beta = (\beta_{ij})$ is a $p \times s$ matrix of unknown elements β_{ij}. As in the general formulation we shall assume that $N - s \geq p$, and that the rank of the $s \times N$ matrix $Z = (z^1, \ldots, z^N)$ is s. Let $\beta = (\beta_1, \beta_2)$, where β_1, β_2 are submatrices of dimensions $p \times r$ and $p \times (s - r)$, respectively. We are interested in testing the null hypothesis

$$H_0 : \beta_1 = \beta_1^0 \quad \text{(a fixed matrix)}$$

where β_2 and Σ are unknown. Here the dimension of π_Ω is $s\,s$ and that of π_ω is $s - r$. The likelihood of the sample observations x^α on X^α, $\alpha = 1, \ldots, N$, is given by

$$L(\beta, \Sigma) = (2\pi)^{-Np/2} (\det \Sigma^{-1})^{N/2}$$

$$\times \exp\left\{-\tfrac{1}{2} \operatorname{tr} \Sigma^{-1} \left(\sum_{\alpha=1}^{N} (x^\alpha - \beta z^\alpha)(x^\alpha - \beta z^\alpha)' \right) \right\} \tag{8.152}$$

Let

$$A = ZZ' = \sum_{\alpha=1}^{N} z^\alpha z^{\alpha\prime}, \qquad C = xZ' = \sum_{\alpha=1}^{N} x^\alpha z^{\alpha\prime}, \qquad x = (x^1, \ldots, x^N).$$

Using Section 1.7, the maximum likelihood estimate $\hat{\beta}$ of β is given by

$$\hat{\beta} A = C. \tag{8.153}$$

Since the rank of Z is s, A is nonsingular and the unique maximum likelihood estimate of β is given by

$$\hat{\beta} = CA^{-1}. \tag{8.154}$$

Now using Lemma 5.1.1, the maximum likelihood estimate of Σ under π_Ω is

$$\hat{\Sigma} = \frac{1}{N} \sum_{\alpha=1}^{N} (x^\alpha - \hat{\beta} z^\alpha)(x^\alpha - \hat{\beta} z^\alpha)' = \frac{1}{N} \left(\sum_{\alpha=1}^{N} x^\alpha x^{\alpha\prime} - \hat{\beta} A \hat{\beta}' \right). \tag{8.155}$$

Thus

$$\max_{\beta,\Sigma} L(\beta, \Sigma) = (2\pi)^{-Np/2} \left\{ \det\left[\sum_{\alpha=1}^{N} \frac{(x^\alpha - \hat{\beta} z^\alpha)(x^\alpha - \hat{\beta} z^\alpha)'}{N} \right] \right\}^{-N/2}$$

$$\times \exp\left\{-\frac{1}{2} Np\right\}. \quad (8.156)$$

To find the maximum of the likelihood function under H_0, let

$$z^\alpha = \begin{pmatrix} z_{(1)}^\alpha \\ z_{(2)}^\alpha \end{pmatrix}, \qquad y^\alpha = x^\alpha - \beta_1^0 z_{(1)}^\alpha, \qquad \alpha = 1, \ldots, N,$$

where $z_{(1)}^\alpha = (z_{\alpha 1}, \ldots, z_{\alpha r})'$. Now $Y^\alpha = X^\alpha - \beta_1^0 z_{(1)}^\alpha$, $\alpha = 1, \ldots, N$, are independently normally distributed with mean $\beta_2 z_{(2)}^\alpha$ and the same covariance matrix Σ. Let $C = (C_1, C_2)$ with C_1 a $p \times r$ submatrix and

$$Z = \begin{pmatrix} Z_1 \\ Z_2 \end{pmatrix}, \qquad A = \begin{pmatrix} A_{(11)} & A_{(12)} \\ A_{(21)} & A_{(22)} \end{pmatrix}, \qquad \hat{\beta} = (\hat{\beta}_1, \hat{\beta}_2)$$

where Z_1 is $r \times N$, $A_{(11)}$ is $r \times r$, and $\hat{\beta}_1$ is $p \times r$. Under H_0, the likelihood function can be written as

$$L(\beta_2, \Sigma) = (2\pi)^{-Np/2} (\det \Sigma^{-1})^{N/2}$$

$$\times \exp\left\{ -\frac{1}{2} \operatorname{tr} \Sigma^{-1} \left[\sum_{\alpha=1}^{N} (y^\alpha - \beta_2 z_{(2)}^\alpha)(y^\alpha - \beta_2 z_{(2)}^\alpha)' \right] \right\}. \quad (8.157)$$

Proceeding exactly in the same way as above we obtain the maximum likelihood estimates of β_2 and Σ under H_0 as

$$\hat{\hat{\beta}}_2 = \sum_{\alpha=1}^{N} y^\alpha z_{(2)}^{\alpha'} \left(\sum_{\alpha=1}^{N} z_{(2)}^\alpha z_{(2)}^{\alpha'} \right)^{-1} = (C_2 - \beta_1^0 A_{(12)}) A_{(22)}^{-1},$$

$$\hat{\hat{\Sigma}} = \frac{1}{N} \sum_{\alpha=1}^{N} (y^\alpha - \hat{\hat{\beta}}_2 z_{(2)}^\alpha)(y^\alpha - \hat{\hat{\beta}}_2 z_{(2)}^\alpha)' \quad (8.158)$$

$$= \frac{1}{N} \sum_{\alpha=1}^{N} (x^\alpha - \beta_1^0 z_{(1)}^\alpha - \hat{\hat{\beta}}_2 z_{(2)}^\alpha)(x^\alpha - \beta_1^0 z_{(1)}^\alpha - \hat{\hat{\beta}}_2 z_{(2)}^\alpha)'.$$

Lemma 8.6.2.

$$N\hat{\hat{\Sigma}} = N\hat{\Sigma} + (\hat{\beta}_1 - \beta_1^0)(A_{(11)} - A_{(12)} A_{(22)}^{-1} A_{(21)})(\hat{\beta}_1 - \beta_1^0)'.$$

Covariance Matrices and Mean Vectors

Proof. Since
$$C = \hat{\beta}A,$$
$$C_2 = (\hat{\beta}_1, \hat{\beta}_2)\begin{pmatrix} A_{(12)} \\ A_{(22)} \end{pmatrix} = \hat{\beta}_1 A_{(12)} + \hat{\beta}_2 A_{(22)}. \tag{8.159}$$

Thus
$$\hat{\beta}_2 = C_2 A_{(22)}^{-1}) - \hat{\beta}_1 A_{(12)} A_{(22)}^{-1}, \hat{\hat{\beta}}_2 - \hat{\beta}_2 = (\hat{\beta}_1 - \beta_1^0) A_{(12)} A_{(22)}^{-1}.$$

Now under H_0
$$\begin{aligned} X - \beta Z &= A - \hat{\beta}Z + (\hat{\beta}_2 - \beta_2)Z_2 + (\hat{\beta}_1 - \beta_1^0)Z_1 \\ &= (X - \hat{\beta}Z) + (\hat{\hat{\beta}}_2 - \beta_2)Z_2 - (\hat{\hat{\beta}}_2 - \hat{\beta}_2)Z_2 + (\hat{\beta}_1 - \beta_1^0)Z_1 \quad (8.160) \\ &= (X - \hat{\beta}Z) + (\hat{\hat{\beta}}_2 - \beta_2)Z_2 + (\hat{\beta}_1 - \beta_1^0)(Z_1 - A_{(12)} A_{(22)}^{-1} Z_2). \end{aligned}$$

Now
$$(Z_1 - A_{(12)} A_{(22)}^{-1} Z_2)Z_2' = A_{(12)} - A_{(12)} = 0.$$

Since
$$(X - \hat{\beta}Z)Z' = XZ' - XZ'(ZZ')^{-1}ZZ' = XZ' - XZ' = 0,$$

which implies that
$$(X - \hat{\beta}Z)Z_i = 0, \quad i = 1, 2,$$

we obtain
$$\begin{aligned} (X &- \beta Z)(X - \beta Z)' \\ &= (X - \hat{\beta}Z)(X - \hat{\beta}Z)' + (\hat{\hat{\beta}}_2 - \beta_2)A_{(22)}(\hat{\hat{\beta}}_2 - \beta_2)' \quad (8.161) \\ &+ (\hat{\beta} - \beta_1^0)(A_{(11)} - A_{(12)} A_{(22)}^{-1} A_{21})(\hat{\beta}_1 - \beta_1^0)'. \end{aligned}$$

Subtracting $(\hat{\hat{\beta}}_2 - \beta_2)Z_2$ from both sides of (8.160), we obtain
$$(X - \beta_1^0 Z_1 - \hat{\hat{\beta}}_2 Z_2) = (X - \hat{\beta}Z) + (\hat{\beta}_1 - \beta_1^0)(Z_1 - A_{(12)} A_{(22)}^{-1} Z_2).$$

Thus
$$\begin{aligned} N\hat{\hat{\Sigma}} &= (X - \hat{\beta}Z)(X - \hat{\beta}Z)' \\ &+ (\hat{\beta}_1 - \beta_1^0)(A_{(11)} - A_{(12)} A_{(22)}^{-1} A_{(21)})(\hat{\beta}_1 - \beta_1^0)'. \end{aligned}$$

Q.E.D.

Using this lemma and (8.156–8.157), we conclude that the likelihood ratio test of $H_0 : \beta_1 = \beta_1^0$ when β_2 and Σ are unkown rejects H_0 whenever

$$u = \frac{\det[\sum_{\alpha=1}^{N}(x^\alpha - \hat{\beta} z^\alpha)(x^\alpha - \hat{\beta} z^\alpha)']}{\det[\sum_{\alpha=1}^{N}(x^\alpha - \hat{\beta} z^\alpha)(x^\alpha - \hat{\beta} z^\alpha)' + (\hat{\beta}_1 - \beta_1^0)(A_{(11)} - A_{(12)}A_{(22)}^{-1}A_{(21)})(\hat{\beta}_1 - \beta_1^0)']}$$

$$\leq C, \tag{8.162}$$

where the constant C depends on the level of significance α of the test. We shall now show that the statistic U is distributed as the statistic U in (8.124). Wilks (1932) first derived the likelihood ratio test criterion for the special case of testing the equality of mean vectors of several populations. Wilks (1934) and Bartlett (1934) extended its use to regression coefficients.

In what follows we do not distinguish between an estimate and the corresponding estimator. For simplicity we shall use the same notation for both. For the maximum likelihood estimator $\hat{\beta}$

$$E(\hat{\beta}) = E\left(\sum_{\alpha=1}^{N} X^\alpha z^{\alpha'} A^{-1}\right) = \beta A A^{-1} = \beta, \tag{8.163}$$

and the covariance between the ith row vector $\hat{\beta}_i$ and the jth row vector $\hat{\beta}_j$ of $\hat{\beta}$ is given by

$$E(\hat{\beta}_i - \beta_i)(\hat{\beta}_j - \beta_j)'$$

$$= A^{-1} E\left\{\sum_{\alpha=1}^{N}(X_{\alpha i} - E(X_{\alpha i}))z^\alpha \left(\sum_{\lambda=1}^{N}(X_{\lambda i} - E(X_{\lambda i}))\right)z^{\lambda'}\right\} A^{-1}$$

$$= A^{-1} \sum_{\alpha=1}^{N} \sigma_{ij} z^\alpha z^{\alpha'} A^{-1} = \sigma_{ij} A^{-1}.$$

Obviously, thus, the row vectors $(\hat{\beta}_1, \ldots, \hat{\beta}_p)$ are normally distributed with mean $(\beta_1, \ldots, \beta_p)$ and covariance matrix

$$\begin{pmatrix} \sigma_{11} A^{-1} & \cdots & \sigma_{1p} A^{-1} \\ \sigma_{21} A^{-1} & \cdots & \sigma_{2p} A^{-1} \\ \vdots & & \vdots \\ \sigma_{p1} A^{-1} & \cdots & \sigma_{pp} A^{-1} \end{pmatrix} = \Sigma \otimes A^{-1}. \tag{8.164}$$

Theorem 8.6.1. $N\hat{\Sigma} = \sum_{\alpha=1}^{N} X^\alpha X^{\alpha'} - \hat{\beta} A \hat{\beta}'$ is distributed independently of $\hat{\beta}$ as $W_p(W - s, \Sigma)$.

Covariance Matrices and Mean Vectors

Proof. Let F be an $s \times s$ nonsingular matrix such that $FAF' = I$. Let $E_2 = FZ$. Then $E_2 E_2' = FZZ'F' = I$. This implies that the s rows of E_2 are orthogonal and are of unit length. Thus it is possible to find an $(N - s) \times N$ matrix E_1 such that

$$E = \begin{pmatrix} E_1 \\ E_2 \end{pmatrix}$$

is an $N \times N$ orthogonal matrix. Let $Y = (Y^1, \ldots, Y^N) = XE'$. Then the columns of Y are independently distributed (normal vectors) with the same covariance matrix Σ and

$$E(Y) = \beta Z E' = \beta F^{-1} E_2(E_1', E_2') = (0, \beta F^{-1}). \tag{8.155}$$

Since

$$XX' = \sum_{\alpha=1}^{N} X^\alpha X^{\alpha'} = YY' = \sum_{\alpha=1}^{N} Y^\alpha Y^{\alpha'}$$

$$\hat{\beta} A \hat{\beta}' = (XZ'A^{-1})A(XZ'A^{-1})' = YEE_2'(F^{-1})'A^{-1}F^{-1}E_2 E' Y' \tag{8.166}$$

$$= Y \begin{pmatrix} 0 \\ I \end{pmatrix} (0, I) Y' = \sum_{\alpha=N-s+1}^{N} Y^\alpha Y^{\alpha'},$$

we get

$$N\hat{\Sigma} = \sum_{\alpha=1}^{N-2} Y^\alpha Y^{\alpha'}, \tag{8.167}$$

where Y^α, $\alpha = 1, \ldots, N - s$, are independently distributed normal p-vectors with means 0 and the same covariance matrix Σ. From (8.166) and (8.167) $N\hat{\Sigma}$ is distributed as $W_p(N - s, \Sigma)$ independently of $\hat{\beta}$. Q.E.D.

Theorem 8.6.2. *Under H_0, $(\hat{\beta} - \beta_1^0)(A_{(11)} - A_{(12)} A_{(22)}^{-1})(\hat{\beta} - \beta_1^0)'$ is distributed as $W_p(\Sigma, r)$ (independently of $N\hat{\Sigma}$).*

Proof. From (8.164) the covariance of the ith and the jth rows of the estimator $\hat{\beta}_1$ is $\sigma_{ij}(A_{(11)} - A_{(12)} A_{(22)}^{-1} A_{(21)})^{-1}$. Let E be an $r \times r$ nonsingular matrix such that

$$E(A_{(11)} - A_{(12)} A_{(22)}^{-1} A_{(21)}) E' = I, \tag{8.168}$$

and let

$$\hat{\beta}_1 - \beta_1^0 = YE = (Y^1, \ldots, Y^r)E. \tag{8.162}$$

Then

$$(\hat{\beta}_1 - \beta_1^0)(A_{(11)} - A_{(12)}A_{(22)}^{-1}A_{(21)})(\hat{\beta}_1 - \beta_1^0)' = \sum_{\alpha=1}^{r} Y^\alpha Y^{\alpha'}. \qquad (8.170)$$

Obviously under H_0 [$E(\cdot)$ denotes the expectation symbol]

$$E(Y) = E[(\hat{\beta}_1 - \beta_1^0)E^{-1}] = 0, \qquad (8.171)$$

since $E(\hat{\beta}_1) = \beta_1^0$. Let the ith and the jth row of Y be Y_i and Y_j, respectively, and let the ith and the jth row of $\hat{\beta}_1$ be $\hat{\beta}_{i1}$ and $\hat{\beta}_{j1}$, respectively. Then

$$\begin{aligned}E(Y_i'Y_j) &= E((E^{-1})'(\hat{\beta}_{i1} - \beta_{i1}^0)'(\hat{\beta}_{j1} - \beta_{j1}^0)E^{-1}) \\ &= \sigma_{ij}[E(A_{(11)} - A_{(12)}A_{(22)}^{-1}A_{(21)})E']^{-1} = \sigma_{ij}I.\end{aligned}$$

Thus $\sum_{\alpha=1}^{r} Y^\alpha Y^{\alpha'}$ is distributed as $W_p(\Sigma, r)$ when H_0 is true.   Q.E.D.

Hence the statistics U as given in (8.124) and (8.162) have identical distributions.

8.6.3. The Distribution of U under H_0

Anderson (1958) called the statistic U, $U_{p,r,N-s}$. Computing various moments of U under H_0 we can show that

$$E(U^k) = \prod_{i=1}^{p} E(X_i^k), \qquad k = 0, 1, \ldots, \qquad (8.172)$$

where $X_1, \ldots, X_p$ are independently distributed central beta random variables with parameter $(\frac{1}{2}(N - s - i + 1), \frac{1}{2}r)$, $i = 1, \ldots, p$. Since U lies between 0 and 1, these moments determine the distribution of U (under H_0) uniquely. Thus, under H_0, U is distributed as

$$U = \prod_{i=1}^{p} X_i. \qquad (8.173)$$

Furthermore, under H_0, $U_{p,r,N-s}$ and $U_{r,p,N-p-s+r}$ have the same distribution. From (8.172) it is easy to see that

(i) $$\frac{1 - U}{U}\left(\frac{N - s}{r}\right) \qquad (8.174)$$

has central F-distribution with degrees of freedom $(r, N-s)$ when $p=1$.

(ii) $$\frac{1-\sqrt{U}}{\sqrt{U}}\left(\frac{N-s-1}{r}\right) \qquad (8.175)$$

has central F-distribution with degrees of freedom $(2r, 2(N-s-1))$ when $p=2$. Box (1949) gave an asymptotic expansion for the distribution of a monotone function of the likelihood ratio statistic $\lambda[=(U_{p,r,N-s})^{N/2}]$ when H_0 is true. The expansion converges extremely rapidly, and therefore the level of significance derived from it will be quite adequate even for moderate values of N. For large N, the Box result is equivalent to the large sample result of Wilks (1938); that is, under H_0, $-2\log\lambda$ is distributed as central χ^2_{pr} with pr degrees of freedom as $N \to \infty$. The Box approximation (with $p \leq r$) is, under H_0,

$$P\{-r\log U_{p,r,N-s} \leq z\} = P\{\chi^2_{pr} \leq z\} + (\gamma/r^2)[P\{\chi^2_{pr+4} \leq z\} \\ - P\{\chi^2_{pr} \leq z\}] + O(N^{-4}), \qquad (8.176)$$

where $\gamma = pr(p^2+r^2-5)/48$. If just the first term is used, the total error of approximation is $O(N^{-2})$; if both terms are used, the error is $O(N^{-4})$. If $r < p$, we use the result that under H_0, $U_{p,r,N-s}$ is distributed as $U_{r,p,N-s-p+r}$.

For the likelihood ratio criterion, exact tables are available only for $p \leq 4$. The Lawley-Hotelling test criterion cannot be used for small samples sizes and appropriate p, since only a result asymptotic in sample size is available (see Anderson, 1958, p. 224; Pillai, 1954). Morrow (1948) has shown that, under H_0, $N\,\text{tr}(AB^{-1})$ has central χ^2_{pr} when $N \to \infty$. The largest and the smallest root criteria of Roy can be used in the general case, although percentage point tables are available only for the restricted values of the parameters. Appropriate tables are given by Foster and Rees (1957), Foster (1957, 1958), Heck (1960), and Pillai (1960). Different criteria for this problem have been compared on the basis of their power functions, in some detail, by Smith et al. (1962) and Gabriel (1969).

8.6.4. Optimum Properties of Tests of General Linear Hypotheses

Using the argument that follows Stein's proof of admissibililty of Hotelling's T^2-test (a generalization of a result of Birnbaum, 1955) Schwartz (1964a) has shown that for testing $H_0 : \Lambda = 0$ against $H_1 : \Lambda \neq 0$, the test (Pillai, 1955) that rejects H_0 whenever $\text{tr}\,a(a+b)^{-1} \geq c$, where the constant c depends on the level of significance α of the test, is admissible. He also obtained the following results:

i. For testing $H_0 : \Lambda = 0$ against the alternatives tr $\Lambda\Lambda'\Sigma^{-1} = \delta$, Pillai's test is locally minimax in the sense of Giri and Kiefer (1964a) as $\delta \to 0$.
ii. Among all invariant level α tests of H_0 which depend only on $R_1, \ldots, R_p$ and which therefore have power functions of the form $\alpha + c\text{tr}(\Lambda\Lambda'\Sigma^{-1}) + o(\Lambda\Lambda'\Sigma^{-1})$, Pillai's test minimizes the value of c.

Ghosh (1964), using Stein's approach, has shown that the Lawley-Hotelling trace test, which rejects H_0 whenever $\text{tr}(ab^{-1}) \geq c$, and Roy's test based on $\max_i(r_i)$ are admissible for testing $H_0 : \Lambda = 0$ against $H_1 : \Lambda \neq 0$. Thus as a consequence of the following result of Anderson et al. (1964), they are unbiased for this problem.

Anderson et al. (1964) gave sufficient conditions on invariant tests (depending only on $R_1, \ldots, R_p$) for the power functions to be monotonically increasing functions of each θ_i, $i = 1, \ldots, p$. Further, they have shown that the likelihood ratio test, the Lawley and Hotelling trace test, and Roy's maximum characteristic root test satisfy these conditions. The monotonicity of the power function of Roy's test has been demonstrated by Roy and Mikhail (1961) using a geometric argument.

Kiefer and Schwartz (1965) have shown, using the Bayes approach, that Pillai's test is admissible Bayes for this problem. The proof proceeds in the same way as that of the admissibility of the T^2- and R^2-tests. The interested reader may consult the original reference for details. This test is fully invariant, similar, and as a consequence of the result given in the preceding paragraph, unbiased. Using the same approach, these authors have also proved the admissibility of the likelihood ratio test under the restriction that $N - s \geq p + r - 1$, although the admissibility of the likelihood ratio test can be proved without this added restriction (see Schwartz, 1964b). Sihna and Giri (1975) proved the Bayes character (and, hence, admissibility) of the likelihood ratio test whenever $N - s > p$. Narain (1950) has shown that the likelihood ratio test is unbiased. We refer to Nandi (1963) for a related admissibility result and to John (1971) for an optimality result.

The unbiasedness property of the likelihood ratio test, Lawley-Hotelling's trace test, Roy's maximum root test and Pillai's trace test has been proved in Anderson, Das Gupta and Mudholkar (1964) and Pearlman and Olkin (1980). A number of numerical comparisons of power functions of these four tests have been made by Schatzoff (1966), Mikhail (1965), Pillai and Jayachandran (1967), Fujikoshi (1970), and Lee (1971) among others. If $\theta_i's$ are not equal the Lawley-Hotelling trace test is better than the likelihood ratio test (LRT) and the LRT is better than Pillai's trace test. If $\theta_i's$ are not very different then the reverse is true. Roy's maximum root test has the largest power among these four tests if the alternative is one-dimensional, i.e. $\theta_2 = \cdots = \theta_r = 0$. However if the alternative is not one-dimensional then it is inferior.

Covariance Matrices and Mean Vectors

8.6.5. Multivariate One-Way, Two-Way Classifications

Most of univariate results in connection with design of experiments can be extended to the multivariate case. We consider here one-way and two-way classifications as examples.

One-Way Classification

Suppose we have r p-variate normal populations with the same positive definite covariance matrix Σ but with different mean vectors $\mu_i, i = 1, \ldots, r$. We are interested here in testing the null hypothesis

$$H_0 : \mu_1 = \cdots = \mu_r.$$

Let $x_{ij} = (x_{ij1}, \ldots, x_{ijp})', j = 1, \ldots, N_i (N_i > p), i = 1, \ldots, r$, be a sample of size N_i from the ith p-variate normal population with mean μ_i and covariance matrix Σ. Define

$$N = \sum_{i=1}^{r} N_i, \qquad N_i x_{i.} = \sum_{j=1}^{N_i} x_{ij}, \qquad N x_{..} = \sum_{i=1}^{r} N_i x_{i.},$$

$$S_i = \sum_{j=1}^{N_i} (x_{ij} - x_{i.})(x_{ij} - x_{i.})', \qquad (8.177)$$

$$s = \sum_{i=1}^{r} \sum_{j=1}^{N_i} (x_{ij} - x_{..})(x_{ij} - x_{..})'.$$

A straightforward calculation shows that the likelihood ratio test of H_0 rejects H_0 whenever

$$\lambda = \left[\frac{\det\left(\sum_{i=1}^{r} s_i\right)}{\det s} \right]^{N/2} \geq c, \qquad (8.178)$$

where c depends on the level of significance of the test.
Since

$$\sum_{i=1}^{r} \sum_{j=1}^{N_i} (x_{ij} - x_{..})(x_{ij} - x_{..})'$$

$$= \sum_{i=1}^{r} \sum_{j=1}^{N_i} (x_{ij} - x_{i.})(x_{ij} - x_{i.})' + \sum_{i=1}^{r} N_i (x_{i.} - x_{..})(x_{i.} - x_{..})',$$

we obtain

$$\lambda = \left[\frac{\det b}{\det(a+b)}\right]^{N/2} \tag{8.179}$$

where

$$a = \sum_{i=1}^{r} N_i(x_{i.} - x_{..})(x_{i.} - x_{..})', \qquad b = \sum_{i=1}^{r} S_i. \tag{8.180}$$

Under H_0, the corresponding random matrices A, B are independently distributed as $W_p(\Sigma, N-r)$, $W_p(\Sigma, r-1)$, respectively. Thus under H_0,

$$U = \frac{\det B}{\det(A+B)} \tag{1.181}$$

is distributed as $U_{p,r-1,N-r}$, and we have discussed its distribution in the context of the general linear hypothesis.

Two-Way Classification

Suppose we have a set of independently normally distributed p-dimensional random vectors $X_{ij} = (X_{ij1}, \ldots, X_{ijp})'$, $i = 1, \ldots, r; j = 1, \ldots, c$, with $E(X_{ij}) = \mu + \alpha_i + \beta_j$, and the same covariance matrix Σ, where

$$\mu = (\mu_1, \ldots, \mu_p)', \qquad \alpha_i = (\alpha_{i1}, \ldots, \alpha_{ip})', \qquad \beta_j = (\beta_{j1}, \ldots, \beta_{jp})',$$

$$\sum_{1}^{r} \alpha_i = 0, \qquad \sum_{j=1}^{c} \beta_j = 0.$$

$$\tag{8.182}$$

We are interested in testing the null hypothesis

$$H_0 : \beta_j = 0 \quad \text{for all} \quad j.$$

In the univariate case, the problem can be treated as a problem of regression by assigning Z suitable values. The same algebra can be used without any difficulty in the multivariate case to reduce the problem to the multiple regression problem. Define

$$X_{..} = \frac{1}{rc}\sum_{i=1}^{r}\sum_{j=1}^{c} X_{ij}, \qquad X_{i.} = \frac{1}{c}\sum_{j=1}^{c} X_{ij}, \qquad X_{.j} = \frac{1}{r}\sum_{i=1}^{r} X_{ij}, \tag{8.183}$$

The statistic U, analogous to the multiple regression model, is

$$U = \frac{\det B}{\det(A+B)} \tag{8.184}$$

Covariance Matrices and Mean Vectors

where

$$B = \sum_{i=1}^{r}\sum_{j=1}^{c}(X_{ij} - X_{i.} - X_{.j} + X_{..})(X_{ij} - X_{i.} - X_{.j} + X_{..})'$$

$$A = r\sum_{j=1}^{C}(X_{.j} - X_{..})(X_{.j} - X_{..})'.$$

(8.185)

Under H_0, U has the distribution $U_{p,r,N-s}$ with $r = c - 1$, $N - s = (r - 1)(c - 1)$. In order for B to be positive definite we need to have $p \le (r - 1)(c - 1)$.

Example 8.6.2. Let us analyze the data in Table 8.1 pertaining to 12 double crosses of barley which were raised during 1971–1972 in Hissar, India. The column indices run over different crosses of barley and the row indices run over four different locations. The observation vector has two components, the first being the height of the barley plant in centimeters and the second the average ear weight in grams. Here

$$b = \begin{pmatrix} 774437.429 & 131452.592 \\ 131452.592 & 22903.067 \end{pmatrix}, \quad a = \begin{pmatrix} 772958.191 & 131499.077 \\ 131499.077 & 22418.604 \end{pmatrix},$$

$\det b / \det(a + b) = 0.4632$. Now

$$\frac{1 - (0.4632)^{1/2}}{(0.4632)^{1/2}}\left(\frac{32}{11}\right) = 1.37$$

is to be compared with $F_{22,64}$ at a 5% level of significance. Thus our data show there is no difference between crosses.

8.7. EQUALITY OF SEVERAL COVARIANCE MATRICES

Let $X_{ij} = (X_{ij1}, \ldots, X_{ijp})'$, $j = 1, \ldots, N_i$, be a random sample of size N_i from a p-variate normal distribution with unknown mean vectors $\mu_i = (\mu_{i1}, \ldots, \mu_{ip})'$ and positive definite covariance matrices Σ_i, $i = 1, \ldots, k$.

We shall consider the problem of testing the null hypothesis

$$H_0 : \Sigma_1 = \cdots = \Sigma_k = \Sigma \quad (say)$$

Table 8.1. Double Crosses of Barley

Location	1	2	3	4	5	6	7	8	9	10	11	12
1	126.60	133.04	113.90	121.52	123.26	133.96	125.42	128.06	137.24	130.50	127.96	129.24
	18.03	23.08	28.56	18.06	20.54	18.78	20.27	27.94	26.74	18.42	20.82	20.75
2	129.26	126.26	115.82	125.10	123.96	127.58	133.74	133.82	140.06	119.36	121.26	130.78
	18.87	22.23	27.70	18.66	19.30	17.42	19.58	26.42	25.85	17.15	20.68	22.46
3	138.76	128.54	107.28	132.56	112.86	118.42	137.08	127.96	129.64	128.04	116.06	137.12
	18.21	24.85	28.16	16.81	20.72	18.40	20.87	25.18	25.90	18.92	22.19	23.46
4	121.40	122.48	118.32	127.64	121.26	133.72	129.56	127.92	134.22	121.98	127.08	132.28
	18.19	24.82	29.32	17.80	19.20	17.09	19.79	25.42	26.35	16.74	23.01	21.92

Covariance Matrices and Mean Vectors

when μ_i, $i = 1, \ldots, k$, are unknown. Let $\Sigma_{i=1}^k N_i = N$ and let

$$S_i = \sum_{j=1}^{N_i} (X_{ij} - X_{i.})(X_{ij} - X_{i.})',$$

$$S = \sum_{i=1}^k S_i, \quad X_{i.} = \frac{1}{N_i} \sum_{j=1}^{N_i} X_{ij}.$$

The parametric space Ω is the space $\{\mu_1, \ldots, \mu_k, \Sigma_1, \ldots, \Sigma_k\}$, which reduces to the subspace $\omega = \{(\mu_1, \ldots, \mu_k, \Sigma\}$ under H_0. The likelihood of the observations x_{ij} on X_{ij} is

$$L(\Omega) = (2\pi)^{-Np/2} \prod_{i=1}^k (\det \Sigma_i)^{-N_i/2}$$

$$\times \exp\left\{-\frac{1}{2} \operatorname{tr} \sum_{i=1}^k \Sigma_i^{-1} \left(\sum_{j=1}^{N_i} (x_{ij} - \mu_i)(x_{ij} - \mu_i)'\right)\right\}.$$

Using Lemma 5.1.1, a straightforward calculation will yield

$$\max_\Omega L(\Omega) = (2\pi)^{-Np/2} \prod_{i=1}^k [\det(s_i/N_i)]^{-N_i/2} \exp\{-\tfrac{1}{2} Np\}. \tag{8.186}$$

When H_0 is true the likelihood function reduces to

$$L(\omega) = (2\pi)^{-Np/2} (\det \Sigma)^{-N/2}$$

$$\times \exp\{-\tfrac{1}{2} \operatorname{tr} \Sigma^{-1} (\Sigma_{i=1}^k \Sigma_{j=1}^{N_i} (x_{ij} - \mu_i)(x_{ij} - \mu_i)')\},$$

and

$$\max_\omega L(\omega) = (2\pi)^{-Np/2} [\det(s/N)]^{-N/2} \exp\{-\tfrac{1}{2} Np\}. \tag{8.187}$$

Thus the likelihood ratio test of H_0 rejects H_0 whenever

$$\lambda = \frac{\max_\omega L(\omega)}{\max_\Omega L(\Omega)} = \frac{N^{pN/2} \prod_{i=1}^k (\det s_i)^{N_i/2}}{(\det s)^{N/2} \prod_{i=1}^k N_i^{pN_i/2}} \leq c, \tag{8.188}$$

where the constant c is chosen so that the test has the required size α.

From Section 6.3 it follows that the S_i are independently distributed $p \times p$ Wishart random matrices with parameters Σ_i and degrees of freedom $N_i - 1 = n_i$ (say). Bartlett, in the univariate case, suggested modifying λ by replacing N_i by n_i and N by $\Sigma_{i=1}^k n_i = n$ (say).

In the case of two populations ($k = 2, p = 1$) the likelihood ratio test reduces to the F-test, and Bartlett in this case gave an intuitive argument for replacing N_i by n_i. He argued that if N_1 (say) is small, s_1 is given too much weight in λ and other effects may be missed. The modified likelihood ratio test in the general case rejects H_0 whenever

$$\lambda' = \frac{n^{np/2} \prod_{i=1}^{k}(\det s_i)^{n_i/2}}{(\det s)^{N/2} \prod_{i=1}^{k} n_i^{pn_i/2}} \leq c', \tag{8.189}$$

where c' is determined so that the test has the required size α. For $p = 1$ the modified likelihood ratio test is based on the F-distribution, but for $p > 1$ the distribution is more complicated.

Define

$$a = 1 - \left(\sum_{i=1}^{k}\frac{1}{n_i} - \frac{1}{n}\right)\frac{2p^2 + 3p - 1}{6(p+1)(k-1)}$$

$$b = \frac{p(p+1)}{48a^3}\left[(p-1)(p+2)\left(\sum_{i=1}^{k}\frac{1}{n_i^2} - \frac{1}{n^2}\right) - 6(k-1)(1-a)^2\right] \tag{8.190}$$

$$f = \tfrac{1}{2}p(p+1)(k-1).$$

It was shown by Box (1949) that a close approximation to the distribution of $\log \lambda$ under H_0 is given by

$$P\{-2a \log \lambda \leq z\} = P\{\chi_f^2 \leq z\} + b[P\{\chi_{f+4}^2 \leq z\} \\ - P\{\chi_f^2 \leq z\}] + O((N-k)^{-3}). \tag{8.191}$$

From this it follows that in large samples under H_0

$$P\{-2a \log \lambda \leq z\} \simeq P\{\chi_f^2 \leq z\}.$$

Giri (1972) has shown that if $\Sigma_1, \ldots, \Sigma_k$ are such that they can be diagonalized by the same orthogonal matrix [a necessary and sufficient condition for this to be true is that $\Sigma_i \Sigma_j = \Sigma_j \Sigma_i$ for all (i,j)], then the test with rejection region

$$\prod_{i=1}^{k}(\det s_i)^{a_i}/(\det s)^b) \leq \text{const}, \tag{8.192}$$

where $b = \sum_{i=1}^{k} a_i = cn$, c being a positive constant, is unbiased for testing H_0 against the alternatives $\det \Sigma_1 \geq \det \Sigma_i$ when $0 < a_i \leq cn_i$ for all i, and against the alternatives $\det \Sigma_1 \leq \det \Sigma_i$ when $a_i > cn_i$ for all i. A special case of this additional restriction, which arises in the analysis of variance components, is the alternatives $H'_1 : \Sigma_1 = l_2\Sigma_2 = \cdots = l_k\Sigma_k$ where the l_i are unknown scalar constants.

Federer (1951) has pointed out that this type of model is also meaningful in certain genetic problems. From the preceding it follows trivially that for testing H_0 against H'_1, the test given in (8.192) is unbiased if $l_i \geq 1$ when $0 < a_i \leq cn_i$ for all i and if $l_i \leq 1$ when $a_i > cn_i$ for all i.

Kiefer and Schwartz (1965) have shown that if $0 < a_i \leq n_i - p$ for all i and b (not necessarily equal to $\Sigma_{i=1}^{k} a_i) \leq n - p$, then the test given in (8.192) is admissible Bayes and similar for testing H_0 against the alternatives that not all Σ_i are equal. It is also similar and fully invariant if $\Sigma_{i=1}^{k} a_i = b$. Such a test can be obtained from the simplest choice of $a_i = 1$ with $b = k$, provided that $n_i > p$ for all i. The likelihood ratio test (respectively the modified likelihood ratio test) can be obtained in this way by setting $a_i = c_1(n_i + 1)$ (respectively $a_i = c_1 n_i$) and $b = \Sigma_{i=1}^{k} a_i$ where $c_1 < 1$.

Some satisfactory solutions to this problem (which cannot be obtained otherwise) can be obtained in the special case $k = 2$. Khatri and Srivastava (1971) have derived the exact nonnull distribution of the modified likelihood ratio test in this case in terms of the H-function. The problem of testing $H_0 : \Sigma_1 = \Sigma_2$ against $H_1 : \Sigma_1 \neq \Sigma_2$ remains invariant under the group of affine transformations $G = (G_l(p), T)$, where $G_l(p)$ is the full linear group of $p \times p$ real nonsingular matrices and T is the group of translations, transforming

$$X_{ij} \to gX_{ij} + b_i, \quad j = 1, \ldots, N_i, \quad i = 1, 2, \quad (8.193)$$

$g \in G_l(p)$, $b_i = (b_{i1}, \ldots, b_{ip})' \in T$. The induced transformation in the space of the sufficient statistic $(X_{1.}, S_1; X_{2.}, S_2)$ is given by

$$(X_{1.}, S_1; X_{2.}, S_2) \to (gX_{1.} + b_1, gS_1g'; gX_{2.} + b_2, gS_2g') \quad (8.194)$$

and the corresponding induced transformation in the parametric space Ω is given by

$$(\mu_1, \Sigma_1; \mu_2, \Sigma_2) \to (g\mu_1 + b_1, g\Sigma_1 g'; g\mu_2 + b_2, g\Sigma_2 g'). \quad (8.195)$$

Theorem 8.7.1. *A maximal invariant in the space of sufficient statistic $(X_{1.}, S_1; X_{2.}, S_2)$ under the group G of transformations (8.194) is $(R_1, \ldots, R_p)$, the characteristic roots of $S_1 S_2^{-1}$.*

Proof. Let R be the diagonal matrix with diagonal elements $R_1, \ldots, R_p$. Since, for $g \in G_l(p)$,

$$(gS_1 g')(gS_2 g')^{-1} = gS_1 S_2^{-1} g^{-1} \quad (8.196)$$

and $S_1 S_2^{-1}$, $gS_1 S_2^{-1} g^{-1}$ have the same characteristic roots, $(R_1, \ldots, R_p)$ is invariant under G. To show that it is a maximal invariant in the space of $(X_{1.}, S_1; X_{2.}, S_2)$ suppose that for any two elements $(Y_{1.}, A_1; Y_{2.}, A_2)$,

$(X_1, S_1; X_2, S_2)$ in this space $S_1 S_2^{-1}, A_1 A_2^{-1}$ have the same characteristic roots $R_1, \ldots, R_p$. By Theorem 1.5.10 there exists g_1, g_2 belonging to $G_l(p)$ such that

$$g_1 S_1 g_1' = R \quad \text{and} \quad g_1 S_2 g_1' = I,$$
$$g_2 A_1 g_2' = R \quad \text{and} \quad g_2 A_2 g_2' = I.$$

Hence, with $g = g_2^{-1} g_1$, we get

$$A_1 = g_2^{-1} R g_2'^{-1} = g_2^{-1} g_1 S_1 g_1' g_2'^{-1} = g S_1 g',$$
$$A_2 = g_2^{-1} g_2'^{-1} = g S_2 g'.$$

Writing

$$b_1 = -g X_1. + Y_1.,$$
$$b_2 = -g X_2. + Y_2.$$

we get $Y_1. = g X_1. + b_1$, $Y_2. = g X_2. + b_2.$ Q.E.D.

A corresponding set of maximal invariants in the parametric space Ω is $(\theta_1, \ldots, \theta_p)$, the characteristic roots of $\Sigma_1 \Sigma_2^{-1}$. In terms of these parameters the null hypothesis can be stated as

$$H_0: \theta_1 = \cdots = \theta_p = 1. \tag{8.197}$$

Several invariant tests have been proposed for this problem:

1. a test based on $\det(S_1 S_2^{-1})$;
2. a test based on $\operatorname{tr}(S_1 S_2^{-1})$;
3. Roy's test based on the largest and the smallest characteristic roots of $S_1 S_2^{-1}$ (Roy, 1953);
4. a test based on $\det[(S_1 + S_2) S_2^{-1}]$ (Kiefer and Schwartz, 1965).

We shall now prove some interesting properties of these tests.

Consider two independent random matrices U_1 of dimension $p \times n_1$ and U_2 of dimension $p \times n_2$, such that the column vectors of U_1 are independently and normally distributed with mean 0 and covariance matrix Σ_1 and the column vectors of U_2 are independently and normally distributed with mean vector 0 and covariance matrix Σ_2. Then

$$S_1 = U_1 U_1', \qquad S_2 = U_2 U_2'.$$

Theorem 8.7.2. *Let ω be a set in the space of $(R_1, \ldots, R_p)$, the characteristic roots of $(U_1 U_1')(U_2 U_2')^{-1}$ such that when a point $(r_1, \ldots, r_p) \in \omega$, so is every*

Covariance Matrices and Mean Vectors

point $(\bar{r}_1, \ldots, \bar{r}_p)$ *for which* $\bar{r}_i \leq r_i, i = 1, \ldots, p$. *Then the probability of the set* ω *depends on* Σ_1 *and* Σ_2 *only through* $(\theta_1, \ldots, \theta_p)$ *and is a monotonically decreasing function of each* θ_i.

Proof. Since Σ_1, Σ_2 are positive definite, there exists a $g \in G_l(p)$ such that $\Sigma_1 = g\theta g', \Sigma_2 = gg'$ where θ is a diagonal matrix with diagonal elements $\theta_1, \ldots, \theta_p$. Write $V_1 = g^{-1} U_1, V_2 = g^{-1} U_2$. It follows that the column vectors of V_1 are independently normally distributed with mean 0 and covariance matrix θ, the column vectors of V_2 are independently normally distributed with mean 0 and covariance matrix I, and $(U_1 U_1')(U_2 U_1')^{-1}$ and $(V_1 V_1')(V_2 V_2')^{-1}$ have the same characteristic roots. Let

$$Q(u_1, u_2) = \{(u_1, u_2) : (r_1, \ldots, r_p) \in \omega\} \quad (8.198)$$

and let $f_{U_i}(u_i|\Sigma_i)$ be the probability density function of $U_i, i = 1, 2$. Then

$$\int_{Q(u_1,u_2)} f_{U_1}(u_1|\Sigma_1) f_{U_2}(u_2|\Sigma_2) du_1 du_2$$
$$= \int_{Q(v_1,v_2)} f_{V_1}(v_1|\theta) f_{V_2}(v_2|I) dv_1 dv_2 = P\{\omega|\theta\} \text{ (say)}. \quad (8.199)$$

Consider $V_2 = v_2$ fixed and let $(v_2 v_2')^{-1} = TT'$ where T is a $p \times p$ nonsingular matrix. The probability density function of $W = TV_1$ is $f_W(w|T\theta T')$. Obviously $(V_1 V_1')(v_2 v_2')^{-1}$ and WW' have the same characteristic roots. Then for $V_2 = v_2$ we have

$$\int_{R(v_1)} f_{V_1}(v_1|\theta) dv_1 = \int_{R(w)} f_W(w|T\theta T') dw \quad (8.200)$$

where $R(v_1) = \{v_1 :$ characteristic roots of $(v_1 v_1')(v_2 v_2')^{-1}$ belong to $\omega\}$. Let θ^* be a diagonal matrix such that $\theta^* - \theta$ is positive semidefinite. It now follows from Exercise 8.11 that (denoting Chi as the ith characteristic root)

$$\text{Ch}i(T\theta^* T') = \text{Ch}i(\theta^{*1/2} T'T \theta^{*1/2}) \geq \text{Ch}i(\theta^{1/2} T'T \theta^{1/2}) = \text{Ch}i(T\theta T'). \quad (8.201)$$

From Exercise 8.12 and from (8.200) we get for $V_2 = v_2$ (fixed)

$$\int_{R(v_1)} f_{V_1}(v_1|\theta) dv_1 \geq \int_{R(v_1)} f_{V_1}(v_1|\theta^*) dv_1. \quad (8.202)$$

Multiplying both sides of (8.202) by $f_{V_2}(v_2|I)$ and integrating with respect to v_2 we obtain $P(\omega|\theta) \geq P(\omega|\theta^*)$ whenever $\theta^* - \theta$ is positive semidefinite. Q.E.D.

From this theorem it now follows that:

Corollary 8.7.1. *If an invariant test with respect to G has an acceptance region ω' such that if $(r_1, \ldots, r_p) \in \omega'$, so is $(\bar{r}_1, \ldots, r_p)$ for $\bar{r}_i \leq r_i, i = 1, \ldots, p$, then the power function of the test is a monotonically increasing function of each θ_i.*

Corollary 8.7.2. *The cumulative distribution function of $R_{i_1}, \ldots, R_{i_k}$ where $i_1, \ldots, i_k$ is a subset of $(1, \ldots, p)$ is a monotonically decreasing function of each θ_i.*

Corollary 8.7.3. *If $g(r_1, \ldots, r_p)$ is monotonically increasing in each of its arguments, a test with acceptance region $g(r_1, \ldots, r_p) \leq$ const has a monotonically increasing power function in each θ_i.*

In particular, Corollary 8.7.3 includes tests with acceptance regions

$$\sum_{i=1}^{k} d_i T_i \leq \text{const} \tag{8.203}$$

where $d_i \geq 0$ and T_i is the sum of all different products of $r_1, \ldots, r_p$ taken i at a time. Special cases of these regions are

1. $\prod_{i=1}^{p} r_i = \det(s_1 s_2^{-1}) \leq$ const
2. $\prod_{i=1}^{p} r_i = \text{tr}(s_1 s_2^{-1}) \leq$ const.

In addition it can be verified that it also includes tests with acceptance region $\sum_{i,j=1}^{p} a_{ij} \omega_{ij} \leq$ const with $a_{ij} \geq 0$ and $\omega_{ij} = T_i/T_j (i > j)$. Roy's tests based on the largest and the smallest characteristic roots with acceptance regions $\max_i r_i \leq$ const and $\min_i r_i \leq$ const, respectively, are also special cases of Corollary 8.7.3.

Sugiura and Nagao (1968) proved the following property of the modified likelihood ratio test.

Theorem 8.7.3. *For testing $H_0 : \Sigma_1 = \Sigma_2$ against the alternatives $H_1 : \Sigma_1 \neq \Sigma_2$ the modified likelihood ratio test with acceptance region*

$$\omega = \left\{ (s_1, s_2) : \prod_{i=1}^{2} (\det s_i(s_1 + s_2)^{-1}))^{n_i/2} \geq c' \right\}, \tag{8.204}$$

where the constant c' is chosen such that the test has size α, is unbiased.

Covariance Matrices and Mean Vectors

Proof. As observed earlier, we can take $\Sigma_2 = I$ and $\Sigma_1 = \theta$, the diagonal matrix with diagonal elements $\theta_1, \ldots, \theta_p$. Now

$$P\{\omega|H_1\} = c_{n_1,p} c_{n_2,p} \int_{(s_1,s_2)\in\omega} (\det s_1)^{(n_1-p-1)/2} (\det s_2)^{(n_2-p-1)/2}$$

$$\times (\det \theta)^{n_1/2} \exp\{-\tfrac{1}{2}\operatorname{tr}(\theta^{-1} s_1 + s_2)\} ds_1 ds_2$$

$$= c_{n_1,p} c_{n_2,p} \int_{(I,u_2)\in\omega} (\det u_1)^{(n-p-1)/2} (\det u_2)^{(n_2-p-1)/2} (\det \theta)^{-n_1/2}$$

$$\times \exp\{-\tfrac{1}{2}\operatorname{tr}(\theta^{-1} + u_2) u_1\} du_1 du_2$$

$$= b \int_{(I,u_2)\in\omega} (\det u_2)^{(n_2-p-1)/2} (\det \theta)^{(n_1/2)} (\det(\theta^{-1} + u_2))^{-n/2} du_2,$$

where $S_1 = U_1$, $S_2 = U_1^{1/2} U_2 U_1^{1/2}$, with $U_1^{1/2}$ a symmetric matrix such that $U_1 = U_1^{1/2} U_1^{1/2}$ and $b = c_{n_1,p} c_{n_2,p}/c_{n,p}$. The Jacobian of the transformation $(s_1, s_2) \to (u_1, u_2)$ is given by

$$\det\left[\frac{\partial(s_1, s_2)}{\partial(u_1, u_2)}\right] = (\det u_1)^{(p+1)/2}.$$

Write $V = \theta^{1/2} U_2 \theta^{1/2}$. Let ω^* be the set of all $p \times p$ positive definite matrices v such that $(I, \theta^{-1/2} v \theta^{-1/2}) \in \omega$, and let $\bar\omega$ be the set of all $p \times p$ positive definite symmetric matrices v such that $(I, v) \in \omega$. Then

$$P\{\omega|H_0\} - P\{\omega|H_1\}$$

$$= b\left\{\int_{\bar\omega} - \int_{\omega^*}\right\} (\det v)^{(n_2-p-1)/2} (\det(I+v))^{-n/2} dv$$

$$= b\left\{\int_{\bar\omega - \bar\omega \cap \omega^*} - \int_{\omega^* - \bar\omega \cap \omega^*}\right\} (\det v)^{n_2/2}$$

$$\times (\det(I+v))^{-n/2} (\det v)^{-(p+1)/2} dv$$

$$\geq bc'\left\{\int_{\bar\omega - \bar\omega \cap \omega^*} - \int_{\omega^* - \bar\omega \cap \omega^*}\right\} (\det v)^{-(p+1)/2} dv$$

$$= bc'\left\{\int_{\bar\omega} - \int_{\omega^*}\right\} (\det v)^{-(p+1)/2} dv = 0$$

since

$$\int_{\bar{\omega}} (\det v)^{(n_2-p-1)/2}(\det(I+v))^{-n/2} dv < \infty, \qquad (8.205)$$

and for any subset of ω' of $\bar{\omega}$

$$\int_{\omega'} (\det v)^{(n_2-p-1)/2}(\det(I+v))^{-n/2} dv \geq c' \int_{\omega'} (\det v)^{-(p+1)/2} dv < \infty. \qquad (8.206)$$

Hence the theorem. Q.E.D.

Subsequently Das Gupta and Girl (1973) considered the following class of rejection regions for testing $H_0 : \Sigma_1 = \Sigma_2$:

$$c(a, b) = \left\{ (s_1, s_2) : \frac{[\det(s_1 s_2^{-1})]^a}{[\det(s_1 s_2^{-1} + I)]^b} \leq k \right\}, \qquad (8.207)$$

where k is a constant depending on the size α of the rejection regions. For the likelihood ratio test of this problem $a = N_1, b = N_1 + N_2$, and for the modified likelihood ratio test $a = n_1 (= N_1 - 1)$ and $b = n_1 + n_2 (= N_1 + N_2 - 2)$. Das Gupta (1969) has shown that the likelihood ratio test is unbiased for testing $H_0 : \Sigma_1 = \Sigma_2$ against $H_1 : \Sigma_1 \neq \Sigma_2$ if and only if $N_1 = N_2$ (it follows trivially from Exercise 5b). In what follows we shall assume that $0 < a < b$, in which case the rejection regions $c(a, b)$ are admissible.

Theorem 8.7.4.

(a) The rejection region $c(a, n_1 + n_2)$ is unbiased for testing $\Sigma_1 = \Sigma_2$ against the alternatives $\Sigma_1 \neq \Sigma_2$ for which $(\det \Sigma_1 - \det \Sigma_2)(n_1 - a) \geq 0$.
(b) The rejection region $C(a, b)$ is biased for testing $\Sigma_1 = \Sigma_2$ against the alternatives $\Sigma_1 \neq \Sigma_2$, for which the characteristic roots of $\Sigma_1 \Sigma_2^{-1}$ lie in the interval with endpoints d and 1, where $d = a(n_1 + n_2)/bn_1$.

Proof. Note that

$$\frac{[\det(s_1 s_2^{-1})]^a}{(\det s_1 s_2^{-1} + I)^n} = [\det(s_1 s_2^{-1})]^{a-n_1} \frac{\prod_{i=1}^{2}(\det s_i)^{n_i}}{(\det(s_1 + s_2))^n}. \qquad (8.208)$$

Covariance Matrices and Mean Vectors

Proceeding exactly in the same way as in Theorem 8.7.3 ($\bar{C}$ being the complement of C) we can get

$$P\{\bar{C}(a,n)|\theta = I\} - P\{\bar{C}(a,n)|\theta\}$$

$$\geq A(p, n_1, n_2, k)\{1 - (\det \theta)^{(a-n_1)/2}\} \int_{\bar{C}(a,n)} (\det v)^{(a-n_1-p-1)/2} dv$$

$$\geq 0$$

where A is a constant. To prove part (b), consider a family of regions given by

$$R(a, b) = \{y : y^a(1+y)^{-b} \geq k, y \geq 0\}.$$

These regions are either intervals or complements of intervals. When $0 < a < b$, $R(a, b)$ is a finite interval not including zero (excluding the trivial extreme case). Consider a random variable Y such that $Y/\sigma(\sigma > 0)$ is distributed as the ratio of independent χ_N^2, $\chi_{N_2}^2$ random variables. Let $\beta(\delta) = P\{Y \in R(a, b)\}$. It can be shown by differentiation that $\beta(\delta) > \beta(1)$ if δ lies in the open interval with endpoints $d, 1$. Define a random variable Z by

$$\frac{(\det S_1)^a (\det S_2)^{b-a}}{[\det(S_1 + S_2)]^b} = \left[\frac{(S_{11}^{(1)})^a (S_{11}^{(2)})^b}{(S_{11}^{(1)} + S_{11}^{(2)})^b}\right] Z \tag{8.209}$$

where $S_k = (S_{ij}^{(k)})$, $k = 1, 2$, and suppose that $\theta_2 = \cdots = \theta_p = 1$. Then the distribution of Z is independent of θ_1 and is independent of the first factor in the right-hand side of (8.209). From Exercise 5b the power of the rejection regions $C(a, b)$ is less than its size if θ_1 lies strictly between d and 1. Q.E.D.

Let θ be the diagonal matrix with diagonal elements $\theta_1, \ldots, \theta_p$. From Theorem 3.2.3 the distribution of R depends only on θ.

Theorem 8.7.5. *Let $f_R(r|\theta)$ be the joint pdf of $R_1, \ldots, R_p$ and let $\sigma = \text{tr}(\theta - I) = \text{tr}(\Sigma_1\Sigma_2^{-1} - I)$. For testing $H_0 : \theta = I$ against $H_1 : \sigma > 0$, the test which rejects H_0 whenever*

$$\text{tr } s_2(s_1 + s_2)^{-1} \leq c$$

where c is a constant depending on the level of significance α of the test, is LBI when $\sigma \to 0$.

Proof. From Example 3.2.6 the Jacobian of the transformation $g \to hg$; $g, h \in G_l(p)$; is $(\det(hh'))^{p/2}$. Hence a left invariant Haar measure on

$G_l(p)$ is

$$d\mu(g) = \frac{dg}{(det(gg'))^{p/2}} \tag{8.210}$$

where $dg = \prod_{ij} dg_{ij}$, $g = (g_{ij})$. Using (3.20) we get

$$\frac{p_R(r|\theta)}{p_R(r|I)}$$

$$= \frac{\int_{G_l(p)} [\det \Sigma_1]^{-(N_1-1)/2} [\det \Sigma_2]^{-(N_2-1)/2}}{\int_{G_l(p)} [\det \Sigma_1]^{-(N-2)/2} \exp\left\{-\frac{1}{2}\text{tr}\Sigma_1^{-1}(gs_1g' + gs_2g')\right\} dg}$$

$$\tag{8.211}$$

where $N = N_1 + N_2$. Using Theorem 1.5.5 we get

$$\beta = \int_{G_l(p)} [\det \Sigma_1]^{-(N-2)/2} \exp\{-\tfrac{1}{2}\text{tr }\Sigma_1^{-1}(gs_1g' + gs_2g')\} [\det gg']^{\frac{N-p-2}{2}} dg$$

$$= \int_{G_l(p)} \exp\{-\tfrac{1}{2}\text{tr}(hs_1h' + hs_2h')\} [\det hh']^{\frac{N-p-2}{2}} dh \tag{8.212}$$

$$= \int_{G_l(p)} \exp\{-\tfrac{1}{2}\text{tr}(grg' + gg')\} [\det gg']^{\frac{N-p-2}{2}} dg$$

where $\Sigma_1^{-1} = g_1'g_1$, $g_1 \in G_l$, $g_1g = h$ and $hg_2 = g$, $g_2 \in G_l(p)$ such that $g_2^{-1}s_2g_2' = I$ and $g_2^{-1}s_1g_2'^{-1} = r$. Applying similar transformations to the numerator of (8.211) we get

$$\frac{p_R(r|\theta)}{p_R(r|I)} = \beta^{-1} \int_{G_l(p)} [\det \theta]^{(N_2-1)/2}$$

$$\times \exp\{-\tfrac{1}{2}\text{tr}(grg' + \theta gg')\} [\det gg']^{\frac{N-p-2}{2}} dg \tag{8.213}$$

Writing $\det \theta = \det(I + \Delta)$ with $\Delta = \theta - I$ we get

$$[\det \theta]^{\frac{1}{2}(N_2-1)} = 1 + \tfrac{1}{2}(N_2 - 1)\sigma + o(\sigma) \tag{8.214}$$

Covariance Matrices and Mean Vectors

as $\sigma = \operatorname{tr} \Delta \to o$. Using (3.24) we get from (8.213)

$$\frac{p_R(r|\theta)}{p_R(r|I)} = 1 + \tfrac{1}{2}(N_2 - 1)\sigma - \tfrac{1}{2}\beta^{-1} \int_{G_l(p)} [\operatorname{tr} \Delta gg']$$
$$\times \exp\{-\tfrac{1}{2}g(r+I)g'\}[\det gg']^{\frac{N-p-2}{2}} dg + o(\sigma) \tag{8.215}$$

Since $\exp\{-\tfrac{1}{2}\operatorname{tr} gg'\}[\det gg']^{\frac{1}{2}(N-p-2)}$ is invariant under the change of sign of g to $-g$ we get

$$\int_{G_l(p)} g_{ij} \exp\{-\tfrac{1}{2}\operatorname{tr} gg'\}[\det gg']^{\frac{1}{2}(N-p-2)} dg = 0, \quad \text{for all} \quad i = j,$$
$$\int g_{ij}g_{i'j'} \exp\{-\tfrac{1}{2}\operatorname{tr} gg'\}[\det gg']^{\frac{1}{2}(N-p-2)} dg = \begin{cases} K, & \text{if } i=i', j=j' \\ 0 & \text{otherwise,} \end{cases} \tag{8.216}$$

where K is a positive constant. The integral in (8.215) is equal to (using (8.216))

$$\int_{G_l(p)} [\operatorname{tr} \Delta g(I+r)^{-1}g'] \exp\{-\tfrac{1}{2}\operatorname{tr} gg'\}[\det gg']^{\frac{1}{2}(N-p-2)} dg$$
$$= \Sigma_{i=1}^p (\theta_i - 1) \Sigma_{j=1}^p (1+r_j)^{-1} \int_{G_l(p)} g_{ij}^2 \exp\{-\tfrac{1}{2}\operatorname{tr} gg'\}[\det gg']^{\frac{1}{2}(N-p-2)} dg$$
$$= K \Sigma_{i=1}^p (\theta_i - 1) \Sigma_{j=1}^p (1+r_j)^{-1}$$
$$= K \operatorname{tr}(\theta - I) \operatorname{tr} s_2(s_1 + s_2)^{-1}$$

Hence from (8.216) we get

$$\frac{p_R(r|\theta)}{p_R(r|I)} = 1 + \tfrac{1}{2}\sigma[(N_2 - 1) - \beta^{-1} K \operatorname{tr} s_2(s_1 + s_2)^{-1}] + o(\sigma). \tag{8.217}$$

Now using the result of Section 3.9 we get the theorem. Q.E.D.

We now prove the locally minimax property of the LBI test. Since $G_l(p)$ does not satisfy the condition of the Hunt-Stein theorem (Section 7.2.3) we replace $G_l(p)$ by $G_T(p)$, the group of $p \times p$ nonsingular lower triangular matrices, for which the theorem holds. As pointed out in Section 3.8 the explicit evaluation of the maximal invariant under $G_T(p)$ is not essential and the ratio of densities of a maximal invariant under $G_T(p)$ is

$$\frac{\int_{G_T(p)} [\det \Sigma_1]^{-\frac{(N_1-1)}{2}} [\det \Sigma_2]^{-\frac{(N_2-1)}{2}} \times \exp\left\{-\frac{1}{2}\text{tr}(\Sigma_1^{-1} g s_1 g' + \Sigma_2^{-1} g s_2 g')\right\} \prod_1^p (g_{ii}^2)^{\frac{N-2-i}{2}} dg}{\int_{G_T(p)} [\det \Sigma_1]^{-\frac{(N-2)}{2}} \exp\left\{-\frac{1}{2}\text{tr}\Sigma_1^{-1}(g s_1 g' + g s_2 g')\right\} \prod_1^p (g_{ii}^2)^{\frac{N-2-i}{2}} dg} \quad (8.218)$$

where $dg = \prod_{i \leq j} dg_{ij}$, $g = (g_{ij}) \in G_T(p)$. Note that (Example 3.2.9) a left invariant Harr measure on $G_T(p)$ is

$$\frac{dg}{\prod_1^p (g_{ii}^2)^{i/2}}.$$

In what follows we write for a $p \times p$ nonsingular symmetric matrix A, $A^{1/2}$ as a $p \times p$ lower triangular nonsingular matrix such that $A^{1/2}(A^{1/2})' = A$, $(A^{1/2})^{-1} = A^{-1/2}$. Let

$$\Gamma = (\Gamma_{ij}) = \Sigma_2^{-1/2} \Sigma_1 (\Sigma_2^{-1/2})',$$

$$\Gamma - I = \phi, \quad V = S_2^{-1/2} S_1 S_2^{-1/2}, \quad (8.219)$$

$$D = \int_{G_T(p)} \exp\{-\tfrac{1}{2}\text{tr } g(v+I)g'\} \prod_1^p (g_{ii}^2)^{\frac{N-i-2}{2}} dg.$$

Using (3.24) and (8.219) the ratio (8.218) can be written as

$$\bar{R} = D^{-1} \int_{G_T(p)} (\det \Gamma)^{\frac{1}{2}(N_2-1)} \exp\{-\tfrac{1}{2}\text{tr}(g(v+\Gamma)g')\} \prod_1^p (g_{ii}^2)^{\frac{N-i-2}{2}} dg$$

$$= 1 + \frac{\sigma}{2}(N_2 - 1) - \frac{D^{-1}}{2} \int_{G_T(p)} \text{tr}(\phi g g') \exp\{-\tfrac{1}{2}\text{tr } g(v+I)g'\}$$

$$\times \prod_1^p (g_{ii}^2)^{\frac{N-i-2}{2}} dg + o(\sigma) \quad (8.220)$$

$$= 1 + \frac{\sigma}{2}(N_2 - 1) - \frac{D^{-1}}{2} \int_{G_T} \text{tr}(\phi g (I+v)^{-1} g') \exp\{-\tfrac{1}{2}\text{tr } g g'\}$$

$$\times \prod_1^p (g_{ii}^2)^{\frac{N-i-2}{2}} dg + o(\sigma).$$

Covariance Matrices and Mean Vectors

To prove the locally minimax result we first prove the following lemma whose proof is straightforward and hence is omitted.

Lemma 8.7.1. *Write $G_T(p) = G_T$. Let $g = (g_{ij}) \in G_T$. Then*

(a) $\int_{G_T} g_{ij} \exp\{-\frac{1}{2} \operatorname{tr} gg'\} \prod_1^p (g_{ii}^2)^{N-i-2/2} dg = 0,$

(b) $D^{-1} \int_{G_T} g_{ij} g_{i'j'} \exp\{-\frac{1}{2} \operatorname{tr} gg'\} \prod_1^p (g_{ii}^2)^{N-i-2/2}$

$dg = \begin{cases} 1, & \text{if } (i,j) = (i',j')(i \neq j) \\ 0 & \text{if } (i,j) \neq (i',j'), \end{cases}$

(c) $D^{-1} \int_{G_T} g_{ii}^2 \exp\{-\frac{1}{2} \operatorname{tr} gg'\} \prod_1^p (g_{ii}^2)^{N-i-2/2} dg = (N-i-1),$

(d) $D^{-1} \int_{G_T} g(I+V)^{-1} g' \exp\{-\frac{1}{2} \operatorname{tr} gg'\} \prod_1^p (g_{ii}^2)^{N-i-2/2} dg = H$ where $H = (h_{ij})$ is a diagonal $p \times p$ matrix with diagonal elements

$$d_{ii} = (N-i-1)W_{ii} + \Sigma_{j<i} W_{ij}$$

with $W = (W_{ij}) = (I+V)^{-1}$,

(e) let $\phi = (\phi_{ij}) = \Gamma - I$,

$$D^{-1} \int_{G_T} \operatorname{tr}(\phi gg') \exp\{-\frac{1}{2} g(V+I)g')\} \prod_1^p (g_{ii}^2)^{N-i-2/2} dg$$

$$= \Sigma_1^p W_{ii}[(N-i-1)\phi_{ii} + \Sigma_{j>i} \phi_{jj}] \tag{8.221}$$

$$= \sigma \Sigma_1^p W_{ii}[(N-i-1)\eta_{ii} + \Sigma_{j>i} \eta_{jj}],$$

where $\eta_{jj} = \phi_{jj}/\sigma$.

Theorem 8.7.6. *For testing $H_0 : \Gamma = I$ against $H_1 : \sigma > 0$, the LBI test is locally minimax as $\sigma \to 0$.*

Proof. From (8.220), using Lemma 8.7.1 we get

$$\int \bar{R}\xi(d\eta) = 1 + \frac{\sigma}{2}(N_2 - 1 - \Sigma_1^p w_{ii}[(N-i-1)\eta_{ii}^0 + \Sigma_{j>i}\eta_{jj}^0]) + o(\sigma) \tag{8.222}$$

where $\eta = (\eta_{11}, \ldots, \eta_{pp})$ and ξ assigns all measure to the single point η^0 (say) whose j-th coordinate is $\eta_{jj}^0 = (N-2-j)^{-1}(N-1-j)^{-1}(N-2)(N-2-p)$ so that

$$\Sigma_{j>i} \eta_{jj}^0 + \eta_{ii}^0 (N-1-i) = \frac{N-2}{p}. \tag{8.223}$$

From (8.222) we get

$$\int \bar{R}\xi(d\eta) = 1 + \frac{\sigma}{2}\left[N_2 - 1 + \frac{N-2}{p}\operatorname{tr} s_2(s_1+s_2)^{-1}\right] + o(\sigma) \quad (8.224)$$

where the term $o(\sigma)$ is uniform in w, η. From Theorem 8.7.5 it follows that the power function of the level α LBI test is of the form $\alpha + h(\sigma) + o(\sigma)$ where $h(\sigma) = b\sigma$ with b a positive constant. From Theorem 7.2.4 we prove the result. Q.E.D.

For further relevant results of the test we refer the reader to Brown (1939) and Mikhail (1962).

Example 8.7.1. Consider Example 5.3.1. Assume that the data pertaining to 1971, 1972 constitute two independent samples from two six-variate normal populations with mean vectors, μ_1, μ_2 and positive definite covariance matrices Σ_1, Σ_2, respectively. We are interested in testing $H_0 : \Sigma_1 = \Sigma_2$ when μ_1, μ_2 are unknown. Here $N_1 = N_2 = 27$. From (8.190),

$$-2a \log \lambda = 49.7890, \quad b = 0.0158, \quad f = 21,$$

since asymptotically

$$P\{-2a \log \lambda \leq z\} = P\{\chi_f^2 \leq z\} = 1 - \alpha, \quad (8.225)$$

for

$$\alpha = 0.05, \quad z = 32.7; \quad \alpha = 0.01, \quad z = 38.9.$$

Hence we reject the null hypothesis H_0. Since the hypothesis is rejected our method of solution of Example 7.2.1 is not appropriate. It is necessary to test the equality of mean vectors when the covariance matrices are unequal, using the Behrens-Fisher approach.

8.7.1. Test of Equality of Several Multivariate Normal Distributions

Consider the problem as formulated in the beginning of Section 8.5. We are interested in testing the null hypothesis

$$H_0 : \Sigma_1 = \cdots = \Sigma_k, \quad \mu_1 = \cdots = \mu_k.$$

In Section 8.6 we tested the hypothesis $\mu_1 = \cdots = \mu_k$, given that $\Sigma_1 = \cdots = \Sigma_k$, and in this section we tested the hypothesis $\Sigma_1 = \cdots = \Sigma_k$. Let λ_1 be the likelihood ratio test criterion for testing the null hypothesis $\mu_1 = \cdots = \mu_k$ given that $\Sigma_1 = \cdots = \Sigma_k$ and let λ_2 be the likelihood ratio test criterion for testing the null hypothesis $\Sigma_1 = \cdots = \Sigma_k$ when $\mu_1, \ldots, \mu_k$ are unknown. It is easy to

Covariance Matrices and Mean Vectors

conclude that the likelihood ratio test criterion λ for testing H_0 is given by

$$\lambda = \lambda_1 \lambda_2 = \frac{N^{pN/2} \prod_{i=1}^{k} (\det s_i)^{N_i/2}}{(\det b)^{N/2} \prod_{i=1}^{k} N_p^{pN_i/2}}$$

where

$$b = \sum_{i=1}^{k} \sum_{\alpha=1}^{N_i} N_i (x_{ij} - x_{..})(x_{ij} - x_{..})' = \sum_{i=1}^{k} s_i + \sum_{i=1}^{k} N_i (x_{i.} - x_{..})(x_{i.} - x_{..})',$$

and the likelihood ratio test rejects H_0 whenever

$$\lambda \leq C,$$

where C depends on the level of significance α. The modified likelihood ratio test of H_0 rejects H_0 whenever

$$w = \frac{n^{pn/2} \prod_{i=1}^{k} (\det s_i)^{n_i/2}}{(\det b)^{n/2} \prod_{i=1}^{k} n_i^{pn_i/2}} \leq C'$$

where C' depends on level α. To determine C' we need to find the probability density function of W under H_0. Using Box (1949), the distribution of W under H_0 is given by

$$P\{-2\rho \log W \leq z\} = P\{\chi_f^2 \leq z\} + \omega_2 [P\{\chi_{f+4}^2 \leq z\}] + o(N^{-3}),$$

where

$$f = \tfrac{1}{2}(k-1)p(p+1),$$

$$1 - \rho = \left(\sum_{i=1}^{k} \frac{1}{n_i} - \frac{1}{n} \right) \left(\frac{2p^2 + 3p - 1}{6(k-1)(p+3)} \right) + \frac{p-k+2}{n(p+3)},$$

$$w_2 = \frac{p}{288\rho^2} \Bigg[6 \left(\sum_{i=1}^{k} \frac{1}{n_i^2} - \frac{1}{n^2} \right)(p+1)(p+2)(p-1)$$

$$- \sum_{i=1}^{k} \left(\frac{1}{n_i} - \frac{1}{n} \right)^2 \frac{(2p^2 + 3p - 1)^2}{(k-1)(p+3)} - 12 \left(\sum_{i=1}^{k} \frac{1}{n_i} - \frac{1}{n} \right)$$

$$\times \frac{(2p^2 + 3p - 1)(p-k+2)}{n(p+3)} - 36 \frac{(k-1)(p-k+2)^2}{n^2(p+3)}$$

$$- \frac{12(k-1)}{n^2}(-2k^2 + 7k + 3pk - 2p^2 - 6p - 4) \Bigg].$$

Thus under H_0 in large samples

$$P\{-2\rho \log W \leq z\} = P\{P\chi_f^2 \leq z\}.$$

8.8. COMPLEX ANALOG OF R^2-TEST

Let Z be a p-variate complex Gaussian random vector with $\alpha = E(Z)$ and Hermitian positive definite complex covariance matrix Σ. Partition Σ as

$$\Sigma = \begin{pmatrix} \Sigma_{11} & \Sigma_{12} \\ \Sigma_{12}^* & \Sigma_{22} \end{pmatrix} \quad (8.226)$$

where Σ_{22} is the $(p-1) \times (p-1)$ lower right-hand submatrix of Σ. Let $\rho_c^2 = \Sigma_{12}\Sigma_{22}^{-1}\Sigma_{12}^*/\Sigma_{11}$. Consider the problem of testing $H_0 : \Sigma_{12} = 0$ against $H_1 : \rho_c^2 > 0$ on the basis of z^α, $\alpha = 1, \ldots, N(N > p)$ observations from $CN_p(\alpha, \Sigma)$. The likelihood of $z^1, \ldots, sz^N$ is

$$L(z^1, \ldots, z^N) = \pi^{-Np}(\det \Sigma)^{-N} \exp\{-\text{tr } \Sigma^{-1}(A + N(\bar{z} - \alpha)(\bar{z} - \alpha)^*)\}, \quad (8.227)$$

where $A = \sum_1^N (z^\alpha - \bar{z})(z^\alpha - \bar{z})^*$, $N\bar{z} = \sum_1^N z^\alpha$. Let A be partitioned similarly as Σ. Using Theorem 5.3.4 we get

$$\max_\Omega L(z^1, \ldots, z^N) = \pi^{-Np} \left(\det\left(\frac{A}{N}\right)\right)^{-N} \exp\{-Np\},$$

$$\max_{H_0} L(z^1, \ldots, z^N) = \pi^{-Np} \left(\det\left(\frac{A_{11}}{N}\right)\right)^{-N} (\det(\frac{A_{22}}{N}))^{-N} \exp\{-Np\}.$$

Hence

$$\lambda = \frac{\max_{H_0} L(z^1, \ldots, z^N)}{\max_\Omega L(z^1, \ldots, z^N)} = \left(\frac{A_{11} - A_{12}A_{22}^{-1}A_{12}^*}{A_{11}}\right)^N = (1 - R_c^2)^N \quad (8.228)$$

where $R_c^2 = A_{12}A_{22}^{-1}A_{22}^*/A_{11}$. From (8.228) it follows that the likelihood ratio test of $H_0 : \Sigma_{12} = 0$ rejects H_0 whenever

$$R_c^2 \geq k \quad (8.229)$$

where the constant k depends on the level α of the test.

The problem of testing H_0 against $H_1 : \rho_c^2 > 0$ is invariant under transformations

$$(\bar{z}, A; \alpha, \Sigma) \to (\bar{z} + b, A; a + b, \Sigma)$$

where b is any arbitrary complex p-column vector. The action of these transformations is to reduce the problem to that where $\alpha = 0$ and $A = \sum_1^N z^\alpha(z^\alpha)^*$

Covariance Matrices and Mean Vectors

is sufficient for Σ. In this formulation N has been reduced by 1 from what it was originally. We consider this latter formulation where $Z^\alpha, \alpha = 1, \ldots, N$ are independently and identically distributed $CN_p(0, \Sigma)$ to test H_0 against H_1. Let G be the group of $p \times p$ nonsingular complex matrices g whose first row and first column contain all zeroes except for the first element. The group G operating as $(A; \Sigma) \to (gAg^*; g\Sigma g^*), g \in G$ leaves this testing problem invariant and a maximal invariant under G is R_c^2. The distribution of R_c^2 is given in Theorem 6.11.3. From this it is easy to conclude that the ratio of the pdf of R_c^2 under H_1 to that of R_c^2 under H_0 is an increasing function of R_c^2 for a given ρ_c^2. Hence we prove the following theorem.

Theorem 8.8.1. *The likelihood ratio test is uniformly most powerful invariant for testing* $H_0 : \Sigma_{12} = 0$ *against* $H_1 : \rho_c^2 > 0$.

In concluding this section we give some developments regarding the complex multivariate general linear hypothesis which is defined for the complex multivariate normal distributions in the same way as that for the multivariate normal distributions. The distribution of statistics based on characteristic roots of complex Wishart matrices is also helpful in multiple time series analysis (see Hannan, 1970). The joint noncentral distributions of the characteristic roots of complex Wishart matrices associated with the complex multivariate general linear hypothesis model were given explicitly by James (1964) in terms of zonal polynomials, whereas Khatri (1964a) expressed them in the form of integrals. In the case of central complex Wishart matrices and random matrices connected with the complex multivariate general linear hypotheses, the distribution of extreme characteristic roots were derived by Pillai and Young (1970) and Pillai and Jouris (1972). The noncentral distributions of the individual characteristic roots of the matrices associated with the complex multivariate general hypothesis and that of traces are given by Khatri (1964b, 1970). Khatri and Bhavsar (1990) have obtained the asymptotic confidence bounds on location parameters for linear growth curve model for multivariate complex Gaussian random variables.

8.9. TESTS OF SCALE MATRICES IN $E_p(\mu, \Sigma)$

The presentation in this section is not a complete one. We include only a selected few problems which are appropriate for our purpose. We refer to Kariya and Sinha (1989) for more results.

8.9.1. The Sphericity Test

Let $X = (X_1, \ldots, X_N)'$, $X_i' = (X_{i1}, \ldots, X_{ip})$ be a $N \times p (N > p)$ random matrix with *pdf*

$$f_X(x) = (\det \Sigma)^{-N/2} q\left(\sum_{i=1}^{N}(x_i - \mu)' \Sigma^{-1}(x_i - \mu)\right) \qquad (8.230)$$

where q is a function on $[0, \infty)$ of the sum of quadratic forms, $\mu = (\mu_1, \ldots, \mu_p)' \in E_p$ and Σ is positive definite. We are interested in testing $H_0 : \Sigma = \Sigma_0$ against the alternatives $H_1 : \Sigma \neq \Sigma_0$ with Σ_0 a fixed positive definite matrix and μ, Σ are unknown. Transform $X \to Y = (Y_1, \ldots, Y_N)'$, $Y_i = \Sigma_0^{-1/2} X_i$, $i = 1, \ldots, N$. Let $\Sigma^* = \Sigma_0^{-1/2} \Sigma (\Sigma_0^{-1/2})'$, $\nu = \Sigma_0^{-1/2} \mu$. The *pdf* of Y is

$$\begin{aligned}f_Y(y) &= (\det \Sigma^*)^{-N/2} q(\Sigma_{i=1}^N (y_i - \nu)'(\Sigma^*)^{-1}(y_i - \nu)) \\ &= (\det \Sigma^*)^{-N/2} q(\mathrm{tr}(\Sigma^*)^{-1}(A + N(\bar{y} - \nu)(\bar{y} - \nu)')\end{aligned} \qquad (8.231)$$

where

$$\begin{aligned}A &= \sum_{i=1}^{N}(y_i - \bar{y})(y_i - \bar{y})' = \Sigma_0^{-1/2} s \Sigma_0^{-1/2}, \quad s = \sum_{i=1}^{N}(x_i - \bar{x})(x_i - \bar{x})', \\ N\bar{y} &= \sum_{i=1}^{N} y_i = N \Sigma_0^{-1/2} \bar{x}, \quad N\bar{x} = \sum_{i=1}^{N} x_i.\end{aligned} \qquad (8.232)$$

On the basis of Y the problem is transformed to testing $H_0 : \Sigma^* = I$. Under the alternatives $\Sigma^* \neq I$. It is invariant under the affine group $G = (O(p), E^p)$ of transformations where $O(p)$ is the multiplicative group of $p \times p$ orthogonal matrices and E^p is the translation group operating as

$$Y_i \to gY_i + b, g \in O(p), b \in E^p, i = 1, \ldots, N.$$

The induced transformation in the space of sufficient statistic $(\bar{Y}, A)$ is given by

$$(\bar{y}, A) \to (g\bar{y} + b, gAg').$$

From Lemma 8.1.1 a maximal invariant in the space of $(\bar{Y}, A)$ under G is $R_1, \ldots, R_p$, the characteristic roots of A. A corresponding maximal invariant in the parametric space is $\theta_1, \ldots, \theta_p$, the characteristic roots of Σ^*. In what follows in this section we write R, θ as diagonal matrices with diagonal elements $R_1, \ldots, R_p$, and $\theta_1, \ldots, \theta_p$ respectively. From (3.20) the probability ratio of the maximal invariant R is given by ($O \in O(p)$)

$$\frac{dP(R|\theta)}{dP(R|I)} = q(\mathrm{tr}\, R) \int_{O(p)} (\det \theta)^{-N/2} q(\mathrm{tr}\, \theta^{-1} O R O') d\mu(O) \qquad (8.233)$$

Covariance Matrices and Mean Vectors

where $d\mu(O)$ is the invariant probability measure on $O(p)$. But

$$(\det \theta^{-1})^{N/2} = 1 + \frac{N}{2}\text{tr}(\theta^{-1} - I) + o(\text{tr}(\theta^{-1} - I))$$

and (by (3.24))

$$q(\text{tr } \theta^{-1}ORO') = q(\text{tr } R) + [\text{tr}[(\theta^{-1} - I)ORO']]q^{(1)}(\text{tr } R)$$
$$+ \tfrac{1}{2}[\text{tr}[(\theta^{-1} - I)ORO']]^2 q^{(2)}(z) \quad (8.234)$$

where

$$z = \text{tr } R + \alpha\text{tr}(\theta^{-1} + I)ORO'$$
$$\geq \text{tr } R(1 + \delta)$$

with $0 \leq \alpha \leq 1$, $\delta = \text{tr}(\theta^{-1} - I)$, $q^{(i)}(x) = (d^i q(x)/dx^i)$. From (3.20), (8.23 and 8.24) the probability ratio in (8.233) is evaluated as (assuming $q^{(2)}(x) \geq 0$ for all x)

$$1 + \frac{N}{2}\delta + \int_{O(p)} \frac{q^{(1)}(\text{tr } R)}{q(\text{tr } R)}[\text{tr}(\theta^{-1} - I)ORO']d\mu(O) + o(\delta)$$
$$= 1 + \delta\left[\frac{q^{(1)}(\text{tr } R)}{q(\text{tr } R)}\text{tr}(R) + \frac{N}{2}\right] + o(\delta). \quad (8.235)$$

Using (8.233), the power function of any invariant level α test $\phi(R)$ of $H_0 : \Sigma^* = I$ against $H_1 : \Sigma^* - I$ is positive definite, can be written as

$$\alpha + \delta E_{H_0}\left\{\phi(R)\left[\text{tr}(R)\left(\frac{q^{(1)}(\text{tr } R)}{q(\text{tr } R)}\right) + \frac{N}{2}\right]\right\} + o(\delta) \quad (8.236)$$

If $xq^{(1)}(x)/q(x)$ is a decreasing function of x, the second term in (8.236) is maximized, by $\phi(R) = 1$ whenever tr $R >$ constant and $\phi(R) = 0$ otherwise. Thus we get the following theorem.

Theorem 8.9.1. *If $q^{(2)}(x) \geq 0$ for all x and $xq^{(1)}(x)/q(x)$ is a decreasing function of x, the test which rejects H_0 whenever tr $\Sigma_0^{-1}s \geq C$, the constant C depends on the level α of the test is LBI for testing H_0 against $H_1 : \Sigma^* - I$ is positive definite when $\delta = \text{tr}(\Sigma^{*-1} - I) \to 0$.*

From Section 8.1 it follows that the LBI test is locally minimax as $\delta \to 0$.

8.9.2. The Sphericity Test

In the notations of Section 8.9.1 consider the problem of testing $H_0 : \Sigma = \sigma^2\Sigma_0$ against the alternatives $H_1 : \Sigma \neq \sigma^2\Sigma_0$ on the basis of X with *pdf* (8.227) with $\sigma^2 > 0$ unknown. In terms of $Y_1, \ldots, Y_N$ with $Y_i = \Sigma_0^{-1/2}X_i$ this problem is reduced to testing $H_0 : \Sigma^* = \Sigma_0^{-1/2}\Sigma(\Sigma_0^{-1/2})' = \sigma^2 I$ against $H_1 : \Sigma^* = \sigma^2 V \neq \sigma^2 I$. From Theorem 5.3.6 and (8.231) we get

$$\lambda = \frac{\max_{H_0} f_Y(y)}{\max_\Omega f_Y(y)} = \frac{(\det A)^{N/2}}{\left(\dfrac{\operatorname{tr} A}{p}\right)^{Np/2}} \tag{8.237}$$

Thus the likelihood ratio test rejects H_0 whenever

$$\left[\frac{\operatorname{tr} A}{\left(\dfrac{\operatorname{tr} A}{p}\right)^{p/2}}\right]^{N/2} \leq C \tag{8.238}$$

where the constant C depends on the level α of the test. This problem is invariant under the affine group of transformation $G = R_+ \times E^p \times O(p)$ (see (8.39)) transforming

$$Y_i \to bOY_i + a, \, i = 1, \ldots, N \tag{8.239}$$

with $b \in R_+, a \in E^p$ and $O \in O(p)$. From (3.27) and Theorem 8.2.3 the probability ratio of a maximal invariant under G is given by

$$\bar{R} = \frac{\int_{R_+}\int_{O(p)} (\det \theta)^{-N/2}(b^2)^{\frac{1}{2}(Np-1)} q(b^2 \operatorname{tr}(\theta^{-1}ORO'))d\mu(O)db}{\int_{R_+}\int_{O(p)} (b^2)^{\frac{1}{2}(Np-1)} q(b^2 \operatorname{tr} R)d\mu(O)db}$$

$$= (\det \theta)^{-N/2} \int_{O(p)} (\operatorname{tr} \theta^{-1}ORO')^{-Np/2} d\mu(O) \tag{8.240}$$

$$= (\det \theta)^{-N/2} \int_{O(p)} (1 + F)^{-Np/2} d\mu(O)$$

where $F = \dfrac{\operatorname{tr}(\theta^{-1} - I)ORO'}{\operatorname{tr} R}$. Using (3.24) we now expand $(1 + F)^{-Np/2}$ as

$$1 - \frac{Np}{2}F + \frac{Np(Np+2)}{8}F^2 - \frac{Np(Np+2)(Np+4)}{48}F^3(1 + \alpha F)^{-\frac{(Np+6)}{2}}$$

for $0 < \alpha < 1$. Since

$$\theta^{-1} - I \leq (\Sigma_{i=1}^p (\theta_i^{-1} - 1))I$$

Covariance Matrices and Mean Vectors

we get the absolute value $|F| < \Sigma_1^p |\theta_i^{-1} - 1|$. From (8.23–8.24) we can write

$$\bar{R} = 1 + \frac{3N(Np+2)}{8(p+1)}[\operatorname{tr}(\theta^{-1} - I)^2]\left[\frac{\operatorname{tr} A^2}{(\operatorname{tr} A)^2}\right] + o(\operatorname{tr}(\theta^{-1} - I)^2). \tag{8.241}$$

Hence the power function of an invariant test $\phi(R)$ of level α can be written as

$$\alpha + \frac{3N(Np+2)\delta^2}{8(p+1)} E_{H_0}\left[\phi \frac{\operatorname{tr} A^2}{(\operatorname{tr} A)^2}\right] + o(\delta^2) \tag{8.242}$$

where $\delta^2 = \operatorname{tr}(\theta^{-1} - I)^2 = \operatorname{tr}(V^{-1} - I)^2$. Hence we get the following theorem.

Theorem 8.9.2. *For testing $H_0 : \Sigma^* = \sigma^2 I$ against $H_1 : \Sigma^* = \sigma^2 V \neq \sigma^2 I$ the test which rejects H_0 for large values of $\operatorname{tr} A^2/(\operatorname{tr} A)^2$ is LBI when $\delta^2 \to 0$ for all q.*

It may be noted from (8.230) that the distributions of $\operatorname{tr} A^2/(\operatorname{tr} A)^2$ under H_0 and under H_1 does not depend on a particular choice of q and hence they are the same as under $N_p(\mu, \Sigma)$ (see Section 8.2). As concluded in Section 8.2 the LBI test is also locally minimax.

8.9.3. Tests of $\Sigma_{12} = 0$

Let $X = (X_1, \ldots, X_N)'$, $X_i' = (X_{i1}, \ldots, X_{ip})$ be a $N \times p$ random matrix with *pdf* given (8.230). Let S, Σ be partitioned as

$$S = \begin{pmatrix} S_{11} & S_{12} \\ S_{21} & S_{22} \end{pmatrix}, \quad \Sigma = \begin{pmatrix} \Sigma_{11} & \Sigma_{12} \\ \Sigma_{21} & \Sigma_{22} \end{pmatrix}$$

where S_{11}, Σ_{11} are 1×1 and $S = \sum_{i=1}^N (X_i - \bar{X})(X_i - \bar{X})'$, $N\bar{X} = \sum_{i=1}^N X_i$. We are interested here to test the null hypothesis $H_0 : \Sigma_{12} = 0$. The multivariate normal analog of this problem has been treated in Section 8.3 and the invariance of this problem has been treated in Section 8.3.1. This problem remains invariant under the group G of affine transformations

$$(g, a), g \in G_{BD}, g = \begin{pmatrix} g_{(11)} & 0 \\ 0 & g_{(22)} \end{pmatrix}, g_{(11)}$$

is 1×1 and g is nonsingular, $a \in E^p$ transforming

$$(\bar{X}, S; \mu, \Sigma) \to (g\bar{X} + a, gSg'; g\mu + a, g\Sigma g'). \tag{8.243}$$

A maximal invariant in the space of $(\bar{X}, S)$ is

$$R^2 = S_{11}^{-1} S_{12} S_{22}^{-1} S_{21}, \tag{8.244}$$

and a corresponding maximal invariant in the parametric space is $\rho^2 = \Sigma_{11}^{-1}\Sigma_{12}\Sigma_{22}^{-1}\Sigma_{21}$. From Theorem 5.3.6 and (8.230)

$$\frac{\max_{H_0} f_X(x)}{\max_{\Omega} f_X(x)} = \left[\frac{\det s}{(\det s_{22})s_{11}}\right]^{N/2} = (1 - R^2)^{N/2}. \qquad (8.245)$$

Hence the likelihood ratio test rejects H_0 whenever $r^2 \leq C$, the constant C depends on the level of significance α of the test. The distribution of R^2 under H_0 is the same as that of R^2 in the multivariate normal case (6.86) with $\rho_0^2 = 0$. If q in (8.230) is convex this test is uniformly most powerful invariant for testing H_0 against $H_1 : \rho^2 > 0$. The proof is similar to that of Theorem 8.3.4. We refer to Giri (1988) and Kariya and Sinha (1989) for details and other relevent results.

8.10. TESTS WITH MISSING DATA

Until now we have discussed statistical inference when the same set of measurements is taken on each observed individual event. In practice, however, it is not always realized due to the inherent nature of the population samples (with skeletal materials all observations can not be taken on each speciman) the nature of the phase sampling employed (different subsets of measurements are taken on different individual events for economical reasons) and etc.

Example 8.10.1. An Air Force Flight Dynamics Laboratory is conducting experiments on pilot performance by changing keyboards in the cockpit of the aircrafts on different flights. The aim of the experiments is to investigate which of the keyboards is significantly better than the others. In these experiments, the scores of the pilot, based on different variables such as pitch steering error, bank steering error and so on, are measured. Situations arise when a particular pilot may be able to participate in experiments involving only some of the keyboards. Also due to the malfunction of the measuring device the scores of certain pilots on some keyboards may not be recorded or there may be unexpected environmental condition like turbulence when some of the experiments are conducted. In order to compare the keyboards the data conducted under similar environmental conditions has to be used. Data collected under abnormal environmental conditions should be discarded when comparing the keyboards. Situations, thus, arise when some data are missing.

Example 8.10.2. In sample surveys when two types of questionnaires, one partial and the other complete one, are distributed, the data contain additional observations. It may also occur when we combine data from different agricultural field experiments.

Covariance Matrices and Mean Vectors

In statistical literature missing data are referred to additional data. We will not discuss further about missing/additional data. Instead we refer the readers to Kariya, Krishnaiah and Rao (1983) for overviews of the subject and exhaustive bibliography. We treat here two testing problems one for the mean vector and the other for the covariance matrix to see the complications associated with this type of problems.

Problem of Mean Vector

Consider the problem of testing $H_0 : \mu = 0$ against $H_1 : \mu \neq 0$ on the basis of a random sample $(X'_\alpha, Y'_\alpha), \alpha = 1, \ldots, N (N \geq p + q + 1)$ on $Z = (X', Y')'$ with $X : p \times 1, Y : q \times 1$ which is distributed as $N_{p+q}(\mu, \Sigma)$. Write $\mu = (\mu'_1, \mu'_2)'$ with $\mu_1 : p \times 1, \mu_2 : q \times 1$, and

$$\Sigma = \begin{pmatrix} \Sigma_{11} & \Sigma_{12} \\ \Sigma_{21} & \Sigma_{22} \end{pmatrix}$$

with $\Sigma_{11} : p \times p, \Sigma_{22} : q \times q$ and an independent sample $U_1, \ldots, U_M$ of size $M > p + 1$ on U which is distributed as $N_p(\mu_1, \Sigma_{11})$. The reduced set-up of the problem in the canonical form is the following.

Let, $\xi, \xi_1, \ldots, \xi_n (n + 1 = N)$ be independent $(p + q)$-dimensional normal vectors with means $E(\xi) = \delta = (\delta'_1, \delta'_2)'; E(\xi_i) = 0, i = 1, \ldots, n$ and common nonsingular covariance matrix

$$\Sigma = \begin{pmatrix} \Sigma_{11} & \Sigma_{12} \\ \Sigma_{21} & \Sigma_{22} \end{pmatrix}$$

where $\delta_1 : p \times 1, \delta_2 : q \times 1$. Let $V, W_1, \ldots, W_m (m + 1 = M)$ be independent normal p-vectors with $E(V) = c\delta_1, E(W_i) = 0, i = 1, \ldots, m$ and a common nonsigular covariance matrix Σ_{11}. Obviously $\xi = \sqrt{N}\bar{Z}, \delta_1 = \sqrt{N}\mu_1$, $\delta_2 = \sqrt{N}\mu_2, c = \sqrt{(M/N)}, \sum_{i=1}^M W_i W'_i = \sum_{i=1}^M (U_i - \bar{U})(U_i - \bar{U})' = S_{UU}$ (say) where $\bar{U} = M^{-1} \sum_{i=1}^M U_i$. Write $\xi_i = (\xi'_{i1}, \xi'_{i2})', i = 1, \ldots, n$ and

$$\sum_{i=1}^n \xi_{i1} \xi'_{i1} = \sum_{\alpha=1}^N (X_\alpha - \bar{X})(X_\alpha - \bar{X})' = S_{XX},$$

$$\sum_{i=1}^n \xi_{i1} \xi'_{i2} = \sum_{\alpha=1}^N (X_\alpha - \bar{X})(Y_\alpha - \bar{Y})' = S_{XY}, \quad (8.246)$$

$$\sum_{i=1}^n \xi_{i2} \xi'_{i2} = \sum_{\alpha=1}^N (Y_\alpha - \bar{Y})(Y_\alpha - \bar{Y})' = S_{YY},$$

so that

$$\sum_{i=1}^{n} \xi_i \xi_i' = \begin{pmatrix} S_{XX} & S_{XY} \\ S_{YX} & S_{YY} \end{pmatrix}.$$

For testing $H_0 : \delta = 0$ against $H_1 : \delta \neq 0$ Hotelling's T^2-test based on $(\xi, \xi_1, \ldots, \xi_n)$ rejects H_0 whenever

$$\xi' \left(\xi\xi' + \sum_{i=1}^{n} \xi_i \xi_i' \right)^{-1} \xi \geq C \tag{8.247}$$

where C is a positive constant such that the level is α.

Eaton and Kariya (1975) derived the likelihood ratio test with additional data and proved that there does not exist a locally most powerful invariant test for this problem. Bhargava (1975) derived the likelihood ratio test for a more general form of missing data and discussed its null distribution. The likelihood ratio test is not very practical because of its very complicated distribution even under H_0. We will show that the Hotelling's T^2 test given in (8.247) is admissible in the presence of additional information given by $V, W_1, \ldots, W_m$.

The joint distribution of ξ's, W's and V is given by

$$(2\pi)^{-\frac{1}{2}((M+N)p+Mq)} |\Sigma|^{-\frac{N}{2}} |\Sigma_{11}|^{-\frac{M}{2}}$$

$$\times \exp\left\{ -\frac{1}{2}(\delta'\Sigma^{-1}\delta + c^2 \delta_1' \Sigma_{11}^{-1} \delta_1) \right\} \tag{8.248}$$

$$\times \exp\left\{ -\frac{1}{2} \text{tr} \left[\Sigma^{-1} \left(\xi\xi' + \sum_{i=1}^{n} \xi_i \xi_i' \right) + \Sigma_{11}^{-1} \left(vv' + \sum_{i=1}^{m} w_i w_i' \right) \right] \right\}.$$

Following Stein (1956) we conclude from (8.248) that it is an exponential family (χ, μ, θ, P) where χ is the $(\frac{1}{2}(p+q)(p+q+1) + \frac{1}{2}p(p+1) + 2p+q)$-dimensional space of (s, s^*, ξ, v) where $s = \sum_{i=1}^{n} \xi_i \xi_i'$ is a $(p+q) \times (p+q)$ matrix, $s^* = \sum_{i=1}^{m} w_i w_i'$ is a $p \times p$ matrix and ξ, v are $p+q$- and q-dimensional vectors respectively. The measure μ is given by

$$\mu(A) = \nu(f^{-1}(A))$$

where the function f: the original sample space $\to \chi$, and is defined by

$$f(\xi, \xi_1, \ldots, \xi_n, v, w_1, \ldots, w_m)$$

$$= \left(\xi\xi' + \sum_{i=1}^{n} \xi_i \xi_i', vv' + \sum_{i=1}^{m} w_i w_i', \xi, v \right). \tag{8.249}$$

Covariance Matrices and Mean Vectors

The adjoint space χ' has the element

$$(\Gamma, \Gamma^*, \eta, \eta^*)$$

with Γ a $(p+q) \times (p+q)$ symmetric positive definite matrix, Γ^* a $p \times p$ symmetric positive definite matrix and η, η^* $(p+q)$- and p-dimensional vectors respectively and is defined by

$$(\Gamma, \gamma^*, \eta, \eta')(s, s^*, \xi, v) = -\frac{1}{2}\text{tr}(\Gamma s + \Gamma^* s^* + \eta' \xi + c\eta^{*'} v). \tag{8.250}$$

The elements $\Gamma, \Gamma^*, \eta, \eta^*$ make the parameter space Θ. The correspondence between this designation of the parameter point and the one in terms of $(\Sigma, \Sigma_{11}, \delta, \delta_1)$ is given by

$$\Gamma = \Sigma^{-1}, \qquad \Gamma^* = \Sigma_{11}^{-1}, \qquad \eta = \Sigma^{-1}\delta, \qquad \eta^* = \Sigma_{11}^{-1}\delta_1. \tag{8.251}$$

The element $(\Gamma, \Gamma^*, 0, 0)$ constitute Θ_0. Finally (as in Stein (1956)) the function P is given by

$$P_{\Gamma,\Gamma^*,\eta,\eta^*}(A) = \frac{1}{\psi(\Gamma, \Gamma^*, \eta, \eta^*)} \int_A \exp[(\Gamma, \Gamma^*, \eta, \eta^*)(s, s^*, \xi, v)] d\mu(s, s^*, \xi, v)$$

with

$$\psi(\Gamma, \Gamma^*, \eta, \eta^*) = \int_\chi \exp[(\Gamma, \Gamma^*, \eta, \eta^*)(s, s^*, \xi, v)] d\mu(s, s^*\xi, v).$$

Now writing in terms of the elements of χ the acceptance region of Hotelling's T^2 test based on $\xi, \xi_1, \ldots, \xi_n$ is given by

$$\{(s, s^*, \xi, v) : \xi' s^{-1} \xi \leq k\} \tag{8.252}$$

where the constant k depends on the size α of the test.

Theorem 8.10.1. *For testing $H_0 : \delta = 0$ against $H_1 : \delta \neq 0$, Hotelling's T^2 test with the acceptance region (8.252) is admissible.*

Proof. The set (2.852) is equivalent to the set A which is the intersection of half spaces of the form

$$\left\{(s, s^*, \xi, v) : \eta'\xi - \frac{1}{2}\text{tr } \eta\eta' s \leq \frac{k}{2}\right\} \tag{(8.253)}$$

and

$$\left\{(s, s^*, \xi, v) : -\frac{1}{2}\text{tr } \eta\eta' s \leq 0\right\}. \tag{(8.254)}$$

These two sets can also be written as

$$(\eta\eta', 0, \eta, 0)(s, s^*, \xi, v) \leq \frac{k}{2}, \tag{8.255}$$

$$(\eta\eta', 0, 0, 0)(s, s^*, \xi, v) \leq 0, \tag{8.256}$$

Thus if $\zeta = (\Gamma, \Gamma^*, \eta, \eta^*)$ is any point in χ' for which

$$\{(s, s^*, \xi, v) : (\Gamma, \Gamma^*, \eta, \eta^*)(s, s^*, \xi, v) > k\} \cap A = \phi \quad \text{(null set)},$$

then it follows (see Stein (1956)) that ζ must be a limit of positive linear combination of elements of the type (8.255) or (8.256). This yields, in particular, that Γ is a positive definite matrix, Γ^* is a null matrix and η^* is a null vector. Choose a parameter point θ_1, $\theta_1 = (\Gamma^{(1)}, \Gamma^*, \eta^{(1)}, \eta^{*(1)})$ with $\eta^{*(1)} \neq 0$. Then it follows that for any $\lambda > 0$, $\Gamma^{(1)} + \lambda\Gamma$ is positive definite and for safficiently large λ, $\eta^{(1)} + \lambda\eta$ is different from a null vector. Hence

$$\theta_1 + \lambda\zeta = (\Gamma^{(1)} + \lambda\Gamma, \Gamma^{*(1)}, \eta^{(1)} + \lambda\eta, \eta^{*(1)}) \in \Theta - \Theta_0.$$

The proof is completed by applying Stein's theorem (1955). Q.E.D.

Note This admissible test ignores additional observations. Sinha, Clement and Giri (1985) obtained other admissible tests depending also on additional data by using the Bayes approach of Kiefer and Schwartz (1963) and compared their power functions.

Problem of Covariance Matrix

Let $X^\alpha = (X_{\alpha 1}, \ldots, X_{\alpha p})'$, $\alpha = 1, \ldots, n$ be independently identically distributed p-variate normal random variable with mean 0 and positive definite covariance matrix Σ. For each α, X^α is partitioned as

$$X^\alpha = (X_1^{\alpha'}, X_2^{\alpha'}, X_3^{\alpha'})'$$

where X_i^α is a subvector of dimension p_i with $p_1 = 1, p_1 + p_2 + p_3 = p$.

In addition consider vector Y_i^α, $i = 1, 2$, $\alpha = 1, \ldots, n_i$ that are independent and distributed as X_i^1. We treat here the problem of testing independence of X_1^1 and X_3^1 ($\Sigma_{13} = 0$) knowing that X_1^1, X_2^1 are independent ($\Sigma_{12} = 0$). Write

$$X = \sum_{\alpha=1}^{n} X^\alpha X^{\alpha'},$$

$$W_i = \sum_{\alpha=1}^{n_i} Y_i^\alpha Y_i^{\alpha'}, \quad i = 1, 2. \tag{8.257}$$

Covariance Matrices and Mean Vectors

Partition

$$S = \begin{pmatrix} S_{11} & S_{12} & S_{13} \\ S_{21} & S_{22} & S_{23} \\ S_{31} & S_{32} & S_{33} \end{pmatrix}, \quad \Sigma = \begin{pmatrix} \Sigma_{11} & \Sigma_{12} & \Sigma_{13} \\ \Sigma_{21} & \Sigma_{22} & \Sigma_{23} \\ \Sigma_{31} & \Sigma_{32} & \Sigma_{33} \end{pmatrix} \quad (8.258)$$

with S_{ii}, Σ_{ii} are both $p_i \times p_i$ submatrices. The matrices S, W_1, W_2 are independent Wishart matrices and they form a sufficient statistic. The pair (W_1, W_2) is what we call additional information. We will assume Σ to be of the form (8.258) with $\Sigma_{12} = 0$ and want to test $H_0 : \Sigma_{13} = 0$ against $H_1 : \Sigma_{13} \neq 0$ on the basis of (S, W_1, W_2). This problem was treated by Perron (1991). The likelihood ratio test for testing H_0 against H_1 rejects H_0 whenever

$$R = \left(\frac{n-p_2}{p_3}\right) \frac{S_{13.2} S_{33.2}^{-1} S_{31.2}}{S_{11.2}} \quad (8.259)$$

is large where

$$S_{ij \cdot k} = S_{ij} - S_{ik} S_{kk}^{-1} S_{kj}.$$

This test statistic does not take into account the additional data. When no additional data is available the locally best invariant test (Giri (1979)) rejects H_0 whenever

$$\phi_1 = (R-1)\left(\frac{n}{n-p_2}\right) \frac{S_{11.2}}{S_{11}} \quad (8.260)$$

is large. We find here the locally best invariant test of H_0 against H_1 when additional data is available which uses (W_1, W_2). Eaton and Kariya (1983) have shown that when $p_1 \geq 1$, $p_2 = 0$ and W_3 is Wishart $W_{p_3}(n_3, \Sigma_{33})$ the locally best invariant test of H_0 against H_1 rejects H_0 whenever

$$\phi_2 = \frac{(n+n_2)(n+n_3)}{p_1 p_3} \operatorname{tr}\{(S_{11} + W_1)^{-1} S_{13}(S_{33} + W_3)^{-1} S_{31}\} \\ - \sum_{i=1,3} \frac{n+n_i}{p_i} \operatorname{tr}(S_{ii} + W_i)^{-1} S_{ii} \quad (8.261)$$

is large.

Let G be the group of transformations given by

$$G = \left\{ g = \begin{pmatrix} g_{11} & 0 & 0 \\ 0 & g_{22} & 0 \\ 0 & g_{32} & g_{33} \end{pmatrix}, g \in G_\ell(p), g_{ii} \in G_\ell(p_i), i = 1, 2, 3 \right\}.$$

Corresponding to $g \in G$ the transformation on the sufficient statistic (S, W_1, W_2) and Σ are given by

$$(S, W_1, W_2) \to \tilde{g}(S, W_1, W_2) = (gSg', g_{11}^2 W_1, g_{22} W_2 g_{22}'),$$
$$\Sigma \to g\Sigma g' \qquad (8.262)$$

where $\tilde{g}$ is the induced transformation on the sufficient statistic corresponding to $g \in G$. A maximal invariant in the parameter space is given by

$$\rho = \Sigma_{11}^{-1} \Sigma_{13} \Sigma_{33.2}^{-1} \Sigma_{31}. \qquad (8.263)$$

Since the power function of any invariant test is constant on each orbit of the parametric space of Σ, there is no loss of generality in working on a class of its representatives instead of working on the original parametric space. Let

$$A(\delta) = \begin{pmatrix} 1 & 0 & \delta D' \\ 0 & I & 0 \\ \delta D & 0 & I \end{pmatrix} \qquad (8.264)$$

with $D = (0, 0, \ldots, 0, 1)'$. The set $\{A(\delta), \delta \in [0, 1)\}$ consists of a class of representatives for the orbits of the parameter space and $\rho(A(\delta)) = \delta^2$. We will show in Theorem 8.10.2 below that the locally best invariant test, of $H_0 : \delta = 0$ against $H_1 : \delta = \lambda$ as $\lambda \to 0$, rejects H_0 whenever

$$\phi_3 = \frac{(n - p_2)}{(n + n_1)} (1 - R) \frac{S_{11.2}}{(S_{11} + W_1)} \qquad (8.265)$$

is large.

Note that the statistic ϕ_3 is the product of two factors. The first factor is equivalent to the likelihood ratio test statistic and it essentially measures the multiple correlation between X_1 and X_3 after removing the effect of X_2, where

$$X = \begin{pmatrix} X^{1'} \\ \vdots \\ X^{n'} \end{pmatrix} = (X_1, X_2, X_3)$$

with X_i $n \times p_i$ submatrices. The second factor is the ratio of two estimates of Σ_{11}. The additional data is used to get an improved estimator in the denominator. Giri's test (Giri, 1979) has $n_1 = 0$, $W_1 = 0$. The second factor provides a measure of orthogonality with X_1 and the columns of X_2. The fact that this test is locally most powerful suggests that as X_1 becomes more nearly orthogonal to the

Covariance Matrices and Mean Vectors

columns of X_2, the first factor becomes more effective in detecting near-zero correlation. The test ϕ_3 does not involve Y_2. In this context we note that Giri's test uses X_2 only through the projection matrix $X_2(X_2'X_2)^{-1}X_2'$ which contains no information on Σ_{22}. It is not surprising that additional information on Σ_{22} is ignored.

Using Theorem 3.9.1 the ratio $\bar{R}$ of the probability densities of the maximal invariant (under G) under H_1 to that under H_0 is given by

$$\bar{R} = \frac{r(\delta, s, w_1, w_2)}{r(o, s, w_1, w_2)} \tag{8.266}$$

where

$$r(\delta, s, w_1, w_2) = \int f(\tilde{g}(s, w_1, w_2) | A(\delta)) \lambda(dg),$$

$\lambda(dg)$ is a left invariant measure on G and $f(\cdot | \Sigma)$ is the joint probability density function of (S, W_1, W_2) with respect to an invariant measure μ when the parameter is Σ. The measures λ and μ are unique up to a multiplicative constant.

Let

$$\lambda(dg) = |g_{33}g_{33}'|^{-\frac{p_2}{2}} \prod_{i=1}^{2} \lambda_{p_i}(dg_{ii}) \tag{8.267}$$

and

$$\mu(d(s, w_1, w_2)) = \mu_p(ds)\mu_{p_1}(dw_1)\mu_{p_2}(dw_2) \tag{8.268}$$

where $\lambda_q(dh) = |hh'|^{-\frac{q}{2}}dh$ is a left-invariant measure on the space of all $q \times q$ matrices h and $\mu_q(dw) = |w|^{-\frac{(q+1)}{2}} \prod_{i,j} dw_{ij}$ is an invariant measure on the space of all $q \times q$ positive definite matrices $w = (w_{ij})$. The joint probability density of (s, w_1, w_2) with respect to the measure μ is given by

$$f(s, w_1, w_2 | \Sigma) = K|\Sigma^{-1}s|^{\frac{1}{2}n} \prod_{i=1}^{2} |\Sigma_{ii}^{-1} w_i|^{\frac{1}{2}n_i}$$
$$\times \exp\left\{-\frac{1}{2}\left(\operatorname{tr} \Sigma^{-1} s + \sum_{i=1}^{2} \operatorname{tr} \Sigma_{ii}^{-1} w_i\right)\right\} \tag{8.269}$$

where K is the normalizing constant independent of Σ.

Theorem 8.10.2.

(a) $\bar{R} = (1-\delta^2)^{\frac{1}{2}n}(1 + \dfrac{n+n_1}{2} + \dfrac{n-p_2}{2}(R-1)\dfrac{n+n_1}{n-p_2}\dfrac{s_{11.2}}{s_{11}+w_1} + o(\delta^3))$.

(b) The locally best invariant test of H_0 against $H_1 : \delta^2 = \lambda$ as $\lambda \to 0$ rejects H_0 whenever ϕ_3 is large.

Proof.

(a) From (8.269)

$$f(\bar{g}(s, w_1, w_2)|A(\delta))$$

$$= K(1-\delta^2)^{\frac{n}{2}}|s|^{\frac{n}{2}}|w_1|^{\frac{n_1}{2}}|w_2|^{\frac{n_2}{2}}$$

$$\times (g_{11}^2)^{\frac{1}{2}(n_1+n_2)}|g_{22}g'_{22}|^{\frac{1}{2}(n+n_2)}|g_{33}g'_{33}|^{\frac{1}{2}n}$$

$$\times \exp\bigg(-\frac{1}{2}[\{s_{11}+(1-\delta^2)w_1\}g_{11}^2 + \text{tr}\{g_{22}(s_{22}+w_2)g'_{22}\}$$

$$+ \text{tr } g_{33}s_{33}g'_{33} + \text{tr } g_{32}s_{22}g'_{32} + 2\text{tr } g_{32}s_{23}g'_{33}$$

$$+ 2\delta g_{11}D'g_{32}s_{21} + 2\delta g_{11}D'g_{32}s_{21} + 2\delta g_{11}D'g_{33}s_{31}]\bigg).$$

Let

$$h_{32} = (g_{32} + g_{33}s_{32}s_{22}^{-1} + \delta D g_{11}s_{12}s_{22}^{-1})s_{22}^{\frac{1}{2}}$$

$$h_{ii} = g_{ii}v_i,$$

$$v_1 = (s_{11}+w_1)^{\frac{1}{2}},$$

$$v_2 = (s_{22}+w_2)^{\frac{1}{2}},$$

$$v_3 = s_{33.2}^{\frac{1}{2}},$$

$$\lambda(dg) = |s_{22}|^{-\frac{1}{2}p_3}\lambda(dh).$$

Covariance Matrices and Mean Vectors

Then, with $z = s_{11}^{-1/2} w_1$ and $r_1 = s_{12} s_{22}^{-1} s_{21} s_{11}^{-1}$, we get

$r(\delta, s, w_1, w_2)$

$= K(1-\delta^2)^{n/2} |s|^{n/2} |s_{22}|^{-\frac{1}{2}p_3} v_1^{-(n+n_1)} |v_2|^{-(n+n_2)}$

$\times |v_3|^{-n} w_1^{\frac{1}{2}n_1} |w_2|^{\frac{1}{2}n_2} \int_G (h_{11}^2)^{\frac{1}{2}(n+n_1)} |h_{22} h'_{22}|^{\frac{1}{2}(n+n_2)}$

$\times |h_{33} h'_{33}|^{\frac{1}{2}(n-p_2)} \exp[-\frac{1}{2} \{h_{11}^2 + \operatorname{tr} h_{22} h'_{22} + \operatorname{tr} h_{33} h'_{33}$

$+ \operatorname{tr} h_{32} h'_{32} + 2\delta D' h_{33} v_3^{-1} s_{31.2} v_1^{-1} h_{11} - 2\delta^2 h_{11}^2 s_{11} v_1^{-2} (r_1 + z)\}]$

$\lambda(dh)$.

Integrating over h_{22}, h_{32} and expanding the exponential close to zero we get

$r(\delta, s, w_1, w_2)$

$= (1-\delta^2)^{n/2} b(s, w_1, w_2)$

$\times \Bigg(\int_{G_\ell(p_1)} \int_{G_\ell(p_3)} [1 - D' h_{33} v_3^{-1} s_{31.2} v_1^{-1} h_{11} \delta$

$\frac{1}{2} \{(D' h_{33} v_3^{-1} s_{31.2})^2 + s_{11}(r_1 + z)\} \delta^2 v_1^{-2} h_{11}^2]$

$\times (h_{11}^2)^{\frac{1}{2}(n+n_1)} |h_{33} h'_{33}|^{\frac{1}{2}(n-p_3)}$

$\times \exp[-\frac{1}{2} \{h_{11}^2 + \operatorname{tr} h_{33} h'_{33}\}] \lambda_{p_1}(dh_{11}) \lambda_{p_3}(dh_{33}) \Bigg)$

$+ Q(\delta, s, w_1, w_2)$,

where $Q = o(\delta^3)$ uniformly in s, w_1, w_2 and b is a function depending only on s, w_1, w_2.

We now decompose $G_\ell(p) = G_T(p) \times O(p)$ where $G_T(p)$ is the group of $p \times p$ nonsingular lower triangular matrices and $O(p)$ is the group of $p \times p$ orthogonal matrices. Write $g = to$, $g \in G_\ell(p), t \in G_T(p), o \in O(p)$. According to this decomposition we can write $\lambda_p(dg) = \tau_p(dt) \times \nu_p(do)$ where τ_p is a left-invariant measure on $G_T(p)$ and ν_p is a left invariant probability measure on $O(p)$. Using James (1954) we get, for A a $p \times p$

matrix,

$$\int_{O(p)} \text{tr}(AO) v_p(dO) = 0,$$

$$\int_{O(p)} \text{tr}^2(AO) v_p(dO) = \text{tr}(AA')/p.$$

Hence

$$r(\delta, s, w_1, w_2)$$
$$= (1-\delta^2)^{n/2} b(s, w_1, w_2)$$
$$\times \left[\int_{G_T(p_1)} \int_{G_T(p_3)} \left\{ 1 + \tfrac{1}{2} \left(\frac{D' T_{33} T'_{33} D r_2}{p_3} + z + r_1 \right) \right. \right.$$
$$\times (1+z)^{-1} T_{11}^2 \delta^2 \} T_{11}^{n+n_1} |T_{33}|^{n-p_2}$$
$$\times \exp[-\tfrac{1}{2} \{ T_{11}^2 + \text{tr}(T_{33} T'_{33}) \} \tau_{p_1}(dT_1) \tau_{p_3}(dT_{33}) + o(\delta^3)].$$

Using the Bartlett decomposition of a Wishart matrix (Giri, 1996) we obtain

$$\frac{r(\delta, s, w_1, w_2)}{r(0, s, w_1, w_2)}$$

$$= (1-\delta^2) \left\{ 1 + \tfrac{1}{2} E \left(\frac{D' U_1 D r_2}{p_3} + z + r_1 \right) \right.$$
$$\left. U_2 (1+z)^{-1} \delta^2 + o(\delta^3) \right\}$$
$$= (1+\delta^2)^{n/2} \left\{ 1 + \frac{n+n_1}{2(1+z)} \left(\frac{n-p_2}{p_3} r_2 + z + r_1 \right) \delta^2 + o(\delta^3) \right\}$$
$$= (1+\delta^2)^{n/2} \left(1 + \frac{n+n_1}{2} + \frac{n-p_2}{2}(r-1) \right.$$
$$\left. + \frac{n+n_1}{n-p_2} \frac{s_{11.2}}{(s_{11}+w_1)} + o(\delta^3) \right),$$

where U_1 is $W_{p_1}(I, n-p_2)$, U_2 is $\chi^2_{n_1+n_2}$ with $R_2 = S_{13.2} S_{33.2}^{-1} S_{31.2} S_{11}^{-1}$, $R = ((n-p_2)/p_3)(R_2/(1-R_1))$.

Part (b) follows from part (a). Q.E.D.

Covariance Matrices and Mean Vectors

EXERCISES

1. Prove (8.3).
2. Prove (8.17).
3. Prove (8.18), (8.21b) and (8.21c).
4. Show that if η and z are $(p \times m)$ matrices and t is a $p \times p$ positive definite matrix, then (E^{mp}, Euclidean space of dimension mp)
   (a) $\int_{E_{mp}} \exp\{-\frac{1}{2}\text{tr}(t\eta\eta' - 2z\eta')\}d\eta = C(\det t)^{-m/2} \exp\{-\frac{1}{2}\text{tr}(t^{-1}zz')\}$ where C is a constant.
   (b) Show that
   $$\int_{E^{mp}} [\det(I + \eta\eta')]^{-h/2} d\eta < \infty$$
   if and only if $h > m + p - 1$.
5. (a) Let X be a random variable such that $X/\theta (\theta > o)$ has a central chi-square distribution with m degrees of freedom. Then show that for $r > 0$
   $$\beta(\theta) = P\{X^r \exp\{-\tfrac{1}{2}X\} \geq C\}$$
   satisfies
   $$\frac{d\beta(\theta)}{d\theta} \begin{Bmatrix} > \\ = \\ < \end{Bmatrix} 0 \quad \text{according as} \quad \theta \begin{Bmatrix} < \\ = \\ > \end{Bmatrix} \frac{2r}{m}.$$
   (b) Let Y be a random variable such that $\delta Y (\delta > 0)$ is distributed as a central F-distribution with $(n_1 - 1, n_2 - 1)$ degrees of freedom, and let
   $$\beta(\delta) = P\left\{\frac{Y^{n_1}}{(1+Y)^{n_1+n_2}} \geq k | \delta\right\}.$$
   Assuming that $n_1 < n_2$, show that there exists a constant $\lambda (< 1)$ independent of k such that
   $$\beta(\delta) > \beta(1)$$
   for all δ lying between λ and 1.
6. Prove Theorem 8.3.4.
7. (a) Let A, B be defined as in Section 8.6. Show that the roots of $\det(A - \lambda B) = 0$ comprise a maximal invariant in the space of (A, B) under $G_l(p)$ transforming $(A, B) \to (gAg', gBg'), g \in G_l(p)$.
   (b) Show that if $r + (N - s) > p$, the $p \times p$ matrix $A + B$ is positive definite with probability 1.

(c) Show that the roots of $\det(A - \lambda B)) = 0$ also comprise a maximal invariant in the space of (A, B) under $G_l(p)$.

8 Show that for the transformation given in (8.137) the Jacobian of the transformation $(A, C) \to (W, V)$ is

$$2^p (\det W)^{p+2} \prod_{i<j} (V_i - V_j).$$

9 (*Two-way classifications with K observations per cell*). Let $X_{ijk} = (X_{ijk1}, \ldots, X_{ijkp})'$, $i = 1, \ldots, I; j = 1, \ldots, J; k = 1, \ldots, K$, be independently normally distributed with

$$E(X_{ijk}) = \mu + \alpha_i + \beta_j + \lambda_{ij}, \qquad \mathrm{cov}(X_{ijk}) = \Sigma$$

where $\mu = (\mu_1, \ldots, \mu_p)'$, $\alpha_i = (\alpha_{i1}, \ldots, \alpha_{ip})'$, $\beta_j = (\beta_{j1}, \ldots, \beta_{jp})'$, $\lambda_{ij} = (\lambda_{ij1}, \ldots, \lambda_{ijp})'$, Σ is positive definite, and

$$\sum_{i=1}^{I} \alpha_i = \sum_{j=1}^{J} \beta_j = \sum_{i=1}^{I} \lambda_{ij} = \sum_{j=1}^{J} \lambda_{ij} = 0.$$

Assume that $p \leq IJ(K-1)$.

(a) Show that the likelihood ratio test of $H_0 : \alpha_i = 0$ for all i rejects H_0 whenever

$$u = \det b / \det(a + b) \leq C,$$

where C is a constant depending on the level of significance, and

$$b = \sum_{i=1}^{I} \sum_{j=1}^{J} \sum_{k=1}^{K} (x_{ijk} - x_{ij.})(x_{ijk} - x_{ij.})'$$

$$a = JK \sum_{i=1}^{I} (x_{i..} - x_{...})(x_{i..} - x_{...})'$$

$x_{ij.} = \frac{1}{K}\sum_{k=1}^{K} x_{ijk}$, $x_{i..} = \frac{1}{JK}\sum_{j,k} x_{ijk}$, $x_{...} = \frac{1}{IJK}\sum_{i,j,k} x_{ijk}$, and so forth.

(b) Find the distribution of the corresponding test statistic U under H_0.

(c) Test the hypothesis $H_0 : \beta_1 = \cdots = \beta_J = 0$.

10 Let X_j denote the change in the number of people residing in Montréal, Canada from the year j to the year $j+1$, who would prefer to live in integrated neighborhoods, $j = 1, 2, 3$. Suppose $X = (X_1, \ldots, X_3)'$ with $E(X) = z\beta$ where

$$z = \begin{pmatrix} 1 & 0 \\ 0 & 1 \\ 1 & 1 \end{pmatrix}, \qquad \beta = \begin{pmatrix} \beta_1 \\ \beta_2 \end{pmatrix}$$

Covariance Matrices and Mean Vectors

of unknown quantities β_1, β_2 and

$$\text{cov } X = \sigma^2 \begin{pmatrix} 1 & \rho & \rho \\ \rho & 1 & \rho \\ \rho & \rho & 1 \end{pmatrix}, \qquad -\tfrac{1}{2} < \rho < 1.$$

Let $X = (-4, 6, 2)'$.
(a) Estimate β.
(b) If $\rho = 0$, estimate β and σ^2.

11 Let A be a positive definite matrix of dimension $p \times p$ and D, D^* be two diagonal matrices of dimension $p \times p$ such that $D^* - D$ is positive semidefinite and D is positive definite. Then show that the ith characteristic root satisfies

$$\chi(ADA') \geq \chi(AD^*A') \qquad \text{for } i = 1, \ldots, p. \tag{8.270}$$

12 Anderson and Das Gupta (1964a). Let X be a $p \times n (n \geq p)$ random matrix having probability density function

$$f_X(x|\Sigma) = (2\pi)^{-np/2} (\det \Sigma)^{-n/2} \exp\{-\tfrac{1}{2} \operatorname{tr} \Sigma^{-1} xx'\}, \tag{8.271}$$

where Σ is a symmetric positive definite matrix.
(a) Show that the distribution of the characteristic roots of XX' is the same as the distribution of the characteristic roots of $(\Delta Y)(\Delta Y)'$ where Y is a $p \times n$ random matrix having the probability density function f with $\Sigma = I$, and Δ is a diagonal matrix with diagonal elements $\theta_1, \ldots, \theta_p$, the characteristic roots of Σ.
(b) Let $C_1 \geq C_2 \geq \cdots \geq C_p$ be the characteristic roots of XX' and let ω be a set in the space of $(C_1, \ldots, C_p)$ such that when a point $(C_1, \ldots, C_p)$ is in ω, so is every point $(\bar{C}_1, \ldots, \bar{C}_p)$ for which $\bar{C}_i \leq C_i (i = 1, \ldots, p)$. Then show that the probability of the set ω depends on Σ only through $\theta_1, \ldots, \theta_p$ and is a monotonically decreasing function of each θ_i.

13 Analyze the data in Table 8.2 pertaining to 10 double crosses of barley which were raised in Hissar, India during 1972. Column indices run over different crosses of barley; the row indices run over four different locations. The observation vector has four components $(x_1, \ldots, x_4)$,

x_1 plant height in centimeters,
x_2 average number of grains per ear,
x_3 average yield in grams per plant,
x_4 average ear weight in grams.

14 Let $X^\alpha = (X_{\alpha 1}, \ldots, X_{\alpha p})', \alpha = 1, \ldots, N$, be independently normally distributed with mean μ and positive definite covariance matrix Σ. On the basis

Table 8.2. Double Crosses of Barley

Replication	1	2	3	4	5	6	7	8	9	10
1	136.24	121.62	135.52	116.14	115.76	132.58	118.00	124.66	127.82	123.46
	48.2	52.6	64.6	59.8	56.4	49.6	54.4	50.6	48.6	59.8
	72.46	90.26	117.26	123.46	118.94	97.56	109.78	96.22	114.32	84.96
	15.82	18.27	19.94	21.89	22.61	22.79	20.64	20.65	27.09	20.13
2	128.82	138.04	133.62	119.66	119.76	147.56	125.46	121.88	126.72	129.64
	54.2	47.8	56.4	63.6	46.8	52.6	56.8	52.8	47.8	47.4
	82.98	76.76	108.12	135.62	97.06	111.72	107.56	110.38	115.26	91.78
	16.08	17.21	20.10	22.76	21.76	24.18	19.70	22.43	27.84	20.67
3	128.74	125.18	142.00	124.78	132.02	141.76	111.32	115.10	127.76	123.12
	46.6	44.6	56.4	61.6	51.4	48.2	49.8	55.4	48.4	50.2
	69.16	75.12	107.26	135.98	96.44	95.86	92.9	111.62	110.28	89.52
	15.86	17.97	19.78	22.97	20.78	21.70	18.72	21.80	27.57	19.21
4	124.62	134.32	123.06	125.86	121.34	141.26	120.68	126.26	122.222	125.04
	53.2	46.2	54.8	63.2	50.8	40.4	58.2	46.8	39.8	46.2
	80.24	75.02	101.30	140.28	106.48	86.58	112.62	91.78	100.36	85.92
	16.11	19.33	19.04	23.15	21.64	22.43	20.78	20.70	26.69	20.85

of observations x^α on X^α find the likelihood ratio test of $H_0 : \Sigma = \Sigma_0, \mu = \mu_0$ where Σ_0 is a fixed positive definite matrix and μ_0 is also fixed. Show that the likelihood ratio test is unbiased for testing H_0 against the alternatives $H_1 : \Sigma \neq \Sigma_0, \mu \neq \mu_0$.

15 Let $X_{ij} = (X_{ij1}, \ldots, X_{ijp})', j = 1, \ldots, N_i$ be a random sample of size N_i from a p-variate normal population with mean μ_i and positive definite covariance matrix $\Sigma_i, i = 1, \ldots, k$. On the basis of observations on the X_{ij}, find the likelihood ratio test of $H_0 : \Sigma_i = \sigma^2 \Sigma_{i0}, i = 1, \ldots, k$, when the Σ_{i0} are fixed positive definite matrices and $\mu_1, \ldots, \mu_k, \sigma^2$ are unknown. Show that both the likelihood ratio test and the modified likelihood ratio test are unbiased for testing H_0 against $H_1 : \Sigma_i \neq \sigma^2 \Sigma_{i0}, i = 1, \ldots, k$.

16 Prove (8.162).

REFERENCES

Anderson, T. W. (1955). The integral of a symmetric unimodal function over a symmetric convex set and some probability inequalities. *Proc. Am. Math. Soc.* 6:170–176.

Anderson, T. W. (1958). *An Introduction to Multivariate Analysis*. New York: Wiley.

Anderson, T. W. and Das Gupta, S. (1964a). A monotonicity property of power functions of some tests of the equality of two covariance matrices. *Ann. Math. Statist.* 35:1059–1063.

Anderson, T. W. and Das Gupta, S. (1964b). Monotonicity of power functions of some tests of independence between two sets of variates. *Ann. Math. Statist.* 35:206–208.

Anderson, T. W., Das Gupta, S. and Mudolkar, G. S. (1964). Monotonicity of power functions of some tests of multivariate general linear hypothesis. *Ann. Math. Statist.* 35:200–205.

Bartlett, M. S. (1934). The vector representation of a sample. *Proc. Cambridge Philo. Soc.* 30, Edinburgh 30:327–340.

Bhargava, R. P. (1975). Some one-sample hypothesis testing problems when there is monotone sample from a multivariate normal populations. *Ann. Inst. Statist. Math.* 27:327–340.

Birnbaum, A. (1955). Characterization of complete class of tests of some multiparametric hypotheses with application to likelihood ratio tests. *Ann. Math. Statist.* 26:21–36.

Box, G. E. P. (1949). A general distribution theory for a class of likelihood ratio criteria. *Biometrika* 36:317–346.

Brown, G. W. (1939). On the power of L_1-test for the equality of variances. *Ann. Math. Statist.* 10:119–128.

Constantine, A. G. (1963). Some noncentral distribution problems of multivariate linear hypotheses. *Ann. Math. Statist.* 34:1270–1285.

Consul, P. C. (1968). The exact distribution of likelihood ratio criteria for different hypotheses. In: Krishnaiah, P. R., ed. *Multivariate Analysis* Vol. II. New York: Academic Press.

Das Gupta, S. (1969). Properties of power functions of some tests concerning dispersion matrices. *Ann. Math. Statist.* 40:697–702.

Das Gupta, S. and Giri, N. (1973). Properties of tests concerning covariance matrices of normal distributions. *Ann. Statist.* 1:1222–1224.

Davis, A. W. (1980). Further of Hotelling's generalized T_0^2. *Communications in Statistics*, B9:321–336.

Davis, A. W. (1970). Exact distribution of Hotelling's generalized T_0^2. *Biometrika* 57:187–191.

Davis, A. W. and Field, J. B. F. (1971). Tables of some multivariate test criteria, Technical Report No. 32. Division of Math. Stat., C. S. I. R. O., Canberra, Australia.

Eaton, M. L. and Kariya, T. (1975). Tests on means with additional information, Technical Report No. 243. School of Statistics, Univ. of Minnesota.

Eaton, M. L. and Kariya, T. (1983). Multivariate test with incomplete data. *Ann. Statist.* 11:653–665.

Federer, W. T. (1951). Testing proportionality of covariance matrices. *Ann. Math. Statist.* 22:102–106.

Foster, R. G. (1957). Upper percentage point of the generalized beta distribution II. *Biometrika* 44:441–453.

Foster, R. G. (1958). Upper percentage point of the generalized beta distribution III. *Biometrika* 45:492–503.

Foster, R. G. and Rees, D. D. (1957). Upper percentage point of the generalized beta distribution I. *Biometrika* 44:237–247.

Fujikoshi, Y. (1970). Asymptotic expansions of the distributions of test statistic in multivariate analysis. *J. Sci. Hiroshima Univ. Ser. A-1*, 34:73–144.

Gabriel, K. R. (1969). Comparison of some methods of simultaneous inference in MANOVA. In: Krishnaiah, P. R., ed *Multivariate Analysis* Vol. II. New York: Academic Press, pp. 67–86.

Ghosh, M. N. (1964). On the admissibility of some tests of Manova. *Ann. Math. Statist.* 35:789–794.

Giri, N. (1968). On tests of the equality of two covariance matrices. *Ann. Math. Statist.* 39:275–277.

Giri, N. (1972). On a class of unbiased tests for the equality of K covariance matrices. In: Kabe, D. G. and Gupta, H., eds. *Multivariate Statistical Inference*. Amsterdam: North-Holland Publ., pp. 57–62.

Giri, N. (1979). Locally minimax tests of multiple correlations. *Canad. J. Statist.* 7:53–60.

Giri, N. (1988). Robust tests of Independence. *Can. J. Stat.* 16:419–428.

Giri, N. (1993). *Introduction to Probability and Statistics*, 2nd Edition. New York: Dekker.

Giri, N. and Kiefer, J. (1964a). Local and asymptotic minimax property of multivariate tests. *Ann. Math. Statist.* 35:21–35.

Giri, N., Kiefer, J. and Stein, C. (1963). Minimax character of Hotelling's T^2-test in the simplest case. *Ann. Math. Statist.* 35:1524–1535.

Glesser, L. J. (1966). A note on sphericity test. *Ann. Math. Statist.* 37:464–467.

Hannan, E. J. (1970). *Multiple Time Series*. New York: Wiley.

Heck, D. L. (1960). Charts of some upper percentage points of the distribution of the largest characteristic root. *Ann. Math. Statist.* 31:625–642.

Hotelling, H. (1951). A generalized T-test and measure of multivariate dipersion. *Proc. Barkeley Symp. Prob. Statist.*, 2nd:23–41.

Hsu, P. L. (1940). On generalized analysis of variance (I). *Biometrika* 31: 221–237.

Hsu, P. L. (1941). Analysis of variance from the power function standpoint. *Biometrika* 32:62–69.

Hughes, D. T. and Saw, J. G. (1972). Approximating the percentage points of Hotelling's T_0^2 statistic. *Biometrika* 59:224–226.

Ito, K. (1960). Asymptotic formulae for the distribution of Hotelling's generalized T_0^2 statistic II. *Ann. Math. Statist.* 31:1148–1153.

James, A. T. (1954). Normal multivariate analysis and the orthogonal group. *Ann. Math. Statist.* 25:40–75.

James, A. T. (1964). Distribution of matrix variates and latent roots derived from normal samples. *Ann. Math. Statist.* 35:475–501.

John, S. (1971). Some optimal multivariate tests. *Biometrika* 58:123–127.

Kariya, T., Krishnaiah, P. R. and Rao, C. R. (1983). Inferance on parameters of multivariate normal populations when some data is missing. In: *Development in Statistics* Vol. 4, Academic Press, pp. 137–183.

Kariya, T. and Sinha, B. (1989). *Robustness of Statistical Tests*. New York: Academic Press, Inc.

Khatri, C. G. (1964a). Distribution of the generalized multiple correlation matrix in the dual case. *Ann. Math. Statist.* 35:1801–1805.

Khatri, C. G. (1964b). Distribution of the largest or the smallest characteristic root under null hypothesis concerning complex multivariate normal populations. *Ann. Math. Statist.* 35:1807–1810.

Khatri, C. G. (1970). On the moments of traces of two matrices in three situations for complex multivariate normal populations. *Sankhya* 32:65–80.

Khatri, C. G. (1972). On the exact finite series distribution of the smallest or the largest of matrices in three dimensions. *Jour. Mult. Analysis* 2:201–207.

Khatri, C. G. and Bhavsar, C. D. (1990). Some asymptotic inference problems connected with elliptical distributions. *Jour. Mult. Analysis* 35:66–85.

Khatri, C. G. and Srivastava, M. S. (1971). On exact non-null distributions of likelihood ratio criteria for sphericity test and equality of two covariance matrices. *Sankhya* 201–206.

Kiefer, J. and Schwartz, R. (1965). Admissible Bayes character of T^2- and R^2- and other fully invariant tests for classical normal problems. *Ann. Math. Statist.* 36:747–760.

Krishnaiah, P. R. (1978). Some recent developments on real multivariate distribution. In: Krishnaiah, P. R. ed. *Developments in Statistics* Vol. 1. New York: Academic Press, pp. 135–169.

Lawley, D. N. (1938). A generalization of Fisher's Z-test. *Biometrika* 30:180–187.

Lee, Y. S. (1971). Asymptotic formulae for the distribution of a multivariate test statistic: power comparisons of certain multivariate tests. *Biometrika* 58:647–651.

Lehmann, E. L. (1959). *Testing Statistical Hypotheses*. New York: Wiley.

Lehmann, E. L. and Stein, C. (1948). Most powerful tests of composite hypotheses, I. *Ann. Math. Statist.* 19:495–516.

Mauchly, J. W. (1940). Significance test of sphericity of a normal n-variate distribution. *Ann. Math. Statist.* 11:204–207.

Mijares, T. A. (1964). Percentage points of the sum $V_1^{(s)}$ of s roots ($s = 1 - 150$). The statistical center, University of Philipines, Manila.

Mijares, T. A. (1964). Percentage points of the sum $V_1^{(s)}$ of s roots ($s = 1 - 150$). The statistical center, University of Philipines, Manila.

Mikhail, W. F. (1962). On a property of a test for the equality of two normal dispersion matrices against one sided alternatives. *Ann. Math. Statist.* 33:1463–1465.

Mikhail, N. N. (1965). A compariaon of tests of the Wilks-Lawley hypothesis in multivariate analysis. *Biometrika* 52:149–156.

Morrow, D. J. (1948). On the distribution of the sums of the characteristic roots of a determinantal equation (Abstract). *Bull. Am. Math. Soc.* 54:75.

Muirhead, R. J. (1972). The asymptotic noncentral distribution of Hotelling's generalized T_0^2. *Ann. Math. Statist.* 43:1671–1677.

Nagarsenker, B. N. and Pillai, K. C. S. (1973). The distribution of sphericity test criterion. *Jour. Mult. Analysis* 3:226–235.

Nandi, H. K. (1963). Admissibility of a class of tests. *Calcutta Statist. Assoc. Bull.* 15:13–18.

Narain, R. D. (1950). On the completely unbiased character of tests of independence in multivariate normal system. *Ann. Math. Statist.* 21:293–298.

Olkin, I. (1952). Note on the jacobians of certain matrix tranformations useful in multivariate analysis. *Biometrika* 40:43–46.

Pillai, K. C. S. (1954). On some Distribution Problems in Multivariate Analysis. Inst. of Statist., Univ. of North Carolina, Chapel Hill, North Carolina.

Pillal, K. C. S. (1955). Some new test criteria in multivariate analysis. *Ann. Math. Statist.* 26:117–121.

Pillai, K. C. S. (1960). Statistical Tables for Tests of Multivariate Hypotheses. Manila Statist. Center, Univ. of Philippines.

Pillai, K. C. S. (1964). On the moments of elementary symmetric functions of the roots of two matrices. *Ann. Math. Statist.* 35:1704–1712.

Pillai, K. C. S. (1965). On the distribution of the largest characteristic roots of a matrix in multivariate analysis. *Biometrika* 52:405–414.

Pillai, K. C. S. (1967). Upper percentage points of the largest root of a matrix in multivariate analysis. *Biometrika* 54:189–194.

Pillai, K. C. S. and Bantegui, C. G. (1959). On the distribution of the largest of six roots of a matrix in multivariate analysis. *Biometrika* 46:237–240.

Pillai, K. C. S. and Gupta, A. K. (1969). On the exact distribution of Wilk's criterion. *Biometrika* 56:109–118.

Pillai, K. C. S. and Jayachandran, K. (1967). Power comparisons of tests of two multivariate hypotheses based on four criteria. *Biometrika* 54:195–210.

Pillai, K. C. S. and Jouris, G. M. (1972). An approximation to the distribution of the largest root of a matrix in the complex Gaussian case. *Ann. Inst. Statist. Math.* 24:517–525.

Pillai, K. C. S. and Sampson, P. (1959). On the Hotelling's generalization of T^2. *Biometrika* 46:160–168.

Pillal, K. C. S. and Young, D. L. (1970). An approximation to the distribution of the largest root of a matrix in the complex Gaussian case. *Ann. Inst. Statist. Math.* 22:89–96.

Roy, S. N. (1953). On a heuristic method of test construction and its use in multivariate analysis. *Ann. Math. Statist.* 24:220–238.

Roy, S. N. (1957). *Some Aspects of Multivariate Analysis*. New York: Wiley.

Roy, S. N. and Mikhail, W. F. (1961). On the monotonic character of the power functions of two multivariates tests. *Ann. Math. Statist.* 32:1145–1151.

Schatzoff, M. (1966). Exact distributions of Wilk's likelihood ratio criterion. *Biometrika* 53:347–358.

Schwartz, R. (1964a). Properties of tests in Manova (Abstract). *Ann. Math. Statist.* 35:939.

Schwartz, R. (1964b). Admissible invariant tests in Manova (Abstract). *Ann. Math. Statist.* 35:1398.

Schwartz, R. (1967). Admissible tests in multivariate analysis of variance. *Ann. Math. Statist.* 38:698–710.

Simaika, J. B. (1941). An optimim property of two statistical tests. *Biometrika* 32:70–80.

Sinha, B. and Giri, N. (1975). On the optimality and non-optimality of some multivariate normal test procedures (to be published).

Sinha, B. K., Clement, B. and Giri, N. (1985). Tests for means with additional information. *Commun. Stat. Theor. Math.* 14:1427–1451.

Siotani, M. (1957). Note on the utilization of the generalized student ratio in the analysis of variance or dispersion. *Ann. Inst. Stat. Math.* 9:157–171.

Siotani, M. (1971). An asymptotic expansion of the non-null distribution of Hotelling's generalized T_0^2-statistic. *Ann. Math. Statist.* 42:560–571.

Smith, H., Gnanadeshikhan, R., and Huges, J. B. (1962). Multivariate analysis of variance. *Biometrika* 18:22–41.

Stein, C. (1956). The admissibility of Hotelling's T^2-test. *Ann. Math. Statist.* 27:616–623.

Sugiura, N. (1972). Locally best invariant test for sphericity and the limiting distributions. *Ann. Math. Statist.* 43:1312–1316.

Sugiura, N. and Nagao, H. (1968). Unbiasedness of some test criteria for the equality of one or two covariance matrices. *Ann. Math. Statist.* 39:1689–1692.

Wald, A. and Brookner, R. J. (1941). On the distribution of Wilks statistic for testing the independence of several groups of variates. *Ann. Math. Statist.* 12:137–152.

Wilks, S. S. (1932). Certain generalizations of analysis of variance. *Biometrika* 24:471–494.

Wilks, S. S. (1934). Moment generating operators for determinant product moments in samples from normal system. *Ann. Math. Statist.* 35:312–340.

Wilks, S. S. (1938). The large sample distribution of likelihood ratio for testing composite hypotheses. *Ann. Math. Statist.* 9:101–112.

9
Discriminant Analysis

9.0. INTRODUCTION

The basic idea of discriminant analysis consists of assigning an individual or a group of individuals to one of several known or unknown distinct populations, on the basis of observations on several characters of the individual or the group and a sample of observations on these characters from the populations if these are unknown. In scientific literature, discriminant analysis has many synonyms, such as classification, pattern recognition, character recognition, identification, prediction, and selection, depending on the type of scientific area in which it is used. The origin of discriminant analysis is fairly old, and its development reflects the same broad phases as that of general statistical inference, namely, a Pearsonian phase followed by Fisherian, Neyman-Pearsonian, and Waldian phases.

Hodges (1950) prepared an exhaustive list of case studies of discriminant analysis, published in various scientific literatures. In the early work, the problem of discrimination was not precisely formulated and was often viewed as the problem of testing the equality of two or more distributions. Various test statistics which measured in some sense the divergence between two populations were proposed. It was Pearson (see Tildesley, 1921) who first proposed one such statistic and called it the coefficient of racial likeness. Later Pearson (1926) published a considerable amount of theoretical results on this coefficient of racial likeness and proposed the following form for it:

$$\frac{N_1 N_2}{N_1 + N_2}(\bar{x} - \bar{y})' s^{-1} (\bar{x} - \bar{y})$$

on the basis of sample observations $x^\alpha = (x_{\alpha 1}, \ldots, x_{\alpha p})'$, $\alpha = 1, \ldots, N_1$, from the first distribution, and $y^\alpha = (y_{\alpha 1}, \ldots, y_{\alpha p})'$, $\alpha = 1, \ldots, N_2$, from the second distribution, where the components characterizing the populations are dependent and

$$\bar{x} = \frac{1}{N_1} \sum_{\alpha=1}^{N_1} x^\alpha, \quad \bar{y} = \frac{1}{N_2} \sum_{\alpha=1}^{N_2} y^\alpha,$$

$$s = \sum_{\alpha=1}^{N_1} (x^\alpha - \bar{x})(x^\alpha - \bar{x})' + \sum_{\alpha=1}^{N_2} (y^\alpha - \bar{y})(y^\alpha - \bar{y})'.$$

The coefficient of racial likeness for the case of independent components was later modified by Morant (1928) and Mahalanobis (1927, 1930). Mahalanobis called his statistic the D^2-statistic. Subsequently Mahalanobis (1936) also modified his D^2-statistic for the case in which the components are dependent. This form is successfully applied to discrimination problems in anthropological and craniometric studies. For this problem Hotelling (1931) suggested the use of the T^2-statistic which is a constant multiple of Mahalanobis' D^2-statistic in the Studentized form, and obtained its null distribution. For a comprehensive review of this development the reader is referred to Das Gupta (1973).

Fisher (1936) was the first to suggest a linear function of variables representing different characters, hereafter called the linear discriminant function (discriminator) for classifying an individual into one of two populations. Its early applications led to several anthropometric discoveries such as sex differences in mandibles, the extraction from a dated series of the particular compound of cranial measurements showing secular trends and solutions of taxonomic problems in general. The motivation for the use of the linear discriminant function in multivariate populations came from Fisher's own idea in the univariate case.

For the univariate case he suggested a rule which classifies an observation x into the ith univariate population if

$$|x - \bar{x}_i| = \min(|x - \bar{x}_1|, |x - \bar{x}_2|), \quad i = 1, 2,$$

where $\bar{x}_i$, is the sample mean based on a sample of size N_i from the ith population. For two p-variate populations π_1 and π_2 (with the same covariance matrix) Fisher replaced the vector random variable by an optimum linear combination of its components obtained by maximizing the ratio of the difference of the expected values of a linear combination under π_1 and π_2 to its standard deviation. He then

Discriminant Analysis

used his univariate discrimination method with this optimum linear combination of components as the random variable.

The next stage of development of discriminant analysis was influenced by Neyman and Pearson's fundamental works (1933, 1936) in the theory of statistical inference. Advancement proceeded with the development of decision theory. Welch (1939) derived the forms of Bayes rules and the minimax Bayes rules for discriminating between two known multivariate populations with the same covariance matrix. This case was also considered by Wald (1944) when the parameters were unknown; he suggested some heuristic rules, replacing the unknown parameters by their corresponding maximum likelihood estimates. Wald also studied the distribution problem of his proposed test statistic. Von Mises (1945) obtained the rule which maximizes the minimum probability of correct classification. The problem of discrimination into two univariate normal populations with different variances was studied by Cavalli (1945) and Penrose (1947). The multivariate analog of this was studied by Smith (1947).

Rao (1946, 1947a,b, 1948, 1949a,b, 1950) studied the problem of discrimination following the approaches of Neyman-Pearson and Wald. He suggested a measure of distance between two populations, and considered the possibility of withholding decision though doubtful regions and preferential decision. Theoretical results on discriminant analysis from the viewpoint of decision theory are given in the book by Wald (1950) and in the paper by Wald and Wolfowitz (1950). Bahadur and Anderson (1962) also considered the problem of discriminating between two unknown multivariate normal populations with different covariance matrices. They derived the minimax rule and characterized the minimal complete class after restricting to the class of discriminant rules based on linear discriminant functions. For a complete bibliography the reader is referred to Das Gupta (1973) and Cacoullos (1973).

9.1. EXAMPLES

The following are some examples in which discriminant analysis can be applied with success.

Example 9.1.1. Rao (1948) considered three populations, the Brahmin, Artisan, and Korwa castes of India. He assumed that each of the three populations could be characterized by four characters—stature (x_1), sitting height (x_2), nasal depth (x_3), and nasal height (x_4)—of each member of the population. On the basis of sample observations on these characters from these three populations the problem is to classify an individual with observation $x = (x_1, \ldots, x_4)'$ into one of the three populations. Rao used a linear discriminator to obtain the solution.

Example 9.1.2. On a patient with a diagnosis of myocardial infarction, observations on his systolic blood pressure (x_1), diastolic blood pressure (x_2), heart rate (x_3), stroke index (x_4), and mean arterial pressure (x_5) are taken. On the basis of these observations it is possible to predict whether or not the patient will survive.

Example 9.1.3. In developing a certain rural area a question arises regarding the best strategy for this area to follow in its development. This problem can be considered as one of the problems of discriminant analysis. For example, the area can be grouped as catering to recreation users or attractive to industry by means of variables such as distance to the nearest city (x_1), distance to the nearest major airport (x_2), percentage of land under lakes (x_3), and percentage of land under forests (x_4).

Example 9.1.4. Admission of students to the state-supported medical program on the basis of examination marks in mathematics (x_1), physics (x_2), chemistry (x_3), English (x_4), and bioscience (x_5) is another example of discriminant analysis.

9.2. FORMULATION OF THE PROBLEM OF DISCRIMINANT ANALYSIS

Suppose we have k distinct populations $\pi_1, \ldots, \pi_k$. We want to classify an individual with observation $x = (x_1, \ldots, x_p)'$ or a group of N individuals with observations $x^\alpha = (x_{\alpha 1}, \ldots, x_{\alpha p})'$, $\alpha = 1, \ldots, N$, on p different characters, characterizing the individual or the group, into one of $\pi_1, \ldots, \pi_k$. When considering the group of individuals we make the basic assumption that the group as a whole belongs to only one population among the k given. Furthermore, we shall assume that each of the π_i can be specified by means of the distribution function F_i (or its probability density function f_i with respect to a Lebesgue measure) of a random vector $X = (X_1, \ldots, X_p)'$, whose components represent random measurements on the p different characters. For convenience we shall treat only the case in which the distribution possesses a density function, although the case of discrete distributions can be treated in almost the same way.

We shall assume that the functional form of F_i, for each i, is known and that the F_i are different for different i. However, the parameters involved in F_i may be known or unknown. If they are unknown, supplementary information about these parameters is obtained through additional samples from these populations. These additional samples are generally called training samples by engineers.

Let us denote by E^p the entire p-dimensional space of values of X. We are interested here in prescribing a rule to divide E^p into k disjoint regions $R_1, \ldots, R_k$

Discriminant Analysis

such that if x (or x^α, $\alpha = 1, \ldots, N$) falls in R_i, we assign the individual (or the group) to π_i. Evidently in using such a classification rule we may make an error by misclassifying an individual to π_i when he really belongs to $\pi_j (i \neq j)$. As in the case of testing of statistical hypotheses ($k = 2$), in prescribing a rule we should look for one that controls these errors of misclassification. Let the cost (penalty) of misclassifying an individual to π_j when he actually belongs to π_i be denoted by $C(j|i)$. Generally the $C(j|i)$ are not all equal, and depend on the relative importance of these errors. For example, the error of misclassifying a patient with myocardial infarction to survive is less serious than the error of misclassifying a patient to die. Furthermore, we shall assume throughout that there is no reward (negative penalty) for correct classification. In other words $C(i|i) = 0$ for all i. Let us first consider the case of classifying a single individual with observation x to one of the $\Pi_i (i = 1, \ldots, k)$. Let $R = (R_1, \ldots, R_k)$. We shall denote a classification rule which divides the space E into disjoint and exhaustive regions $R_1, \ldots, R_k$ by R. The probability of misclassifying an individual with observation x, from π_i, as coming from π_j (with the rule R) is

$$P(j|i, R) = \int_{R_j} f_i(x) dx \tag{9.1}$$

where $dx = \Pi_{i=1}^{p} dx_i$.

The expected cost of misclassifying an observation from π_i (using the rule R) is given by

$$r_i(R) = \sum_{j=1, j \neq i}^{k} C(j|i) P(j|i, R), \quad i = 1, \ldots, k. \tag{9.2}$$

In defining an optimum classification rule we now need to compare the cost vectors $r(R) = (r_1(R), \ldots, r_k(R))$ for different R.

Definition 9.2.1. Given any two classification rules R, R^* we say that R is as good as R^* if $r_i(R) \leq r_i(R^*)$ for all i and R is better than R^* if at least one inequality is strict.

Definition 9.2.2. *Admissible rule.* A classification rule R is said to be admissible if there does not exist a classification rule R^* which is better than R.

Definition 9.2.3. *Complete class.* A class of classification rules is said to be complete if for any rule R^* outside this class, we can find a rule R inside the class which is better than R^*.

Obviously the criterion of admissibility, in general, does not lead to a unique classification rule. Only in those circumstances in which $r(R)$ for different R can be ordered can one expect to arrive at a unique classification rule by using this criterion.

Definition 9.2.4. *Minimax rule.* A classification rule R^* is said to be minimax among the class of all rules R if

$$\max_i r_i(R^*) = \min_R \max_i r_i(R) \qquad (9.3)$$

This criterion leads to a unique classification rule whenever it exists and it minimizes the maximum expected loss (cost). Thus from a conservative viewpoint this may be considered as an optimum classification rule.

Let p_i denote the proportion of π_i in the population (of which the individual is a member), $i = 1, \ldots, k$. If the p_i are known, we can define the average cost of misclassifying an individual using the classification rule R. Since the probability of drawing an observation from π_i is p_i, the probability of drawing an observation from π_i and correctly classifying it to π_i with the help of the rule R is given by $p_i P(i|i, R)$, $i = 1, \ldots, k$. Similarly the probability of drawing an observation π_i, and misclassifying it to $\pi_j (i \neq j)$ is $p_i P(j|i, R)$. Thus the quantity

$$\sum_{i=1}^{k} p_i \sum_{j=1, j \neq i}^{k} C(j|i) P(j|i, R) \qquad (9.4)$$

is the average cost of misclassification for the rule R with respect to the a priori probabilities $p = (p_1, \ldots, p_k)$.

Definition 9.2.5. *Bayes rule.* Given p, a classification rule R which minimizes the average cost of misclassification is called a Bayes rule with respect to p.

It may be remarked that a Bayes rule may result in a large probability of misclassification, and there have been several attempts to overcome this difficulty (see Anderson, 1969). In cases in which the a priori probabilities p_i are known, the Bayes rule is optimum in the sense that it minimizes the average expected cost. For further results and details about these decision theoretic criteria the reader is referred to Wald (1950), Blackwell and Girshik (1954), and Ferguson (1967).

We shall now evaluate the explicit forms of these rules in cases in which each π_i admits of a probability density function f_i, $i = 1, \ldots, k$. We shall assume that all the classification procedures considered are the same if they differ only on sets of probability measure 0.

Theorem 9.2.1. *Bayes rule.* If the a priori probabilities $p_i, i = 1, \ldots, k$, are known and if π_i admits of a probability density function f_i with respect to a

Discriminant Analysis

Lebesgue measure, then the Bayes classification rule $R^* = (R_1^*, \ldots, R_k^*)$ which minimizes the average expected cost is defined by assigning x to the region R_l^* if

$$\sum_{i=1, i \neq l}^{k} p_i f_i(x) C(l|i) < \sum_{i=1, i \neq j}^{k} p_i f_i(x) C(j|i), \quad j = 1, \ldots, k, j \neq l. \quad (9.5)$$

If the probability of equality between the right-hand side and the left-hand side of (9.5) is 0 for each l and j and for each π_i, then the Bayes classification rule is unique except for sets of probability measure 0.

Proof. Let

$$h_i(x) = \sum_{i=1 (i \neq j)}^{k} p_i f_i(x) C(j|i). \quad (9.6)$$

Then the average expected cost of a classification rule $R = (R_1, \ldots, R_k)$ with respect to the a priori probabilities $p_i, i = 1, \ldots, k$, is given by

$$\sum_{j=1}^{k} \int_{R_j} h_j(x) dx = \int h(x) dx \quad (9.7)$$

where

$$h(x) = h_j(x) \quad \text{if} \quad x \in R_j. \quad (9.8)$$

For the Bayes classification rule R^*, $h(x)$ is equal to

$$h^*(x) = \min_j h_j(x). \quad (9.9)$$

In other words, $h^*(x) = h_j(x) = \min_i h_i(x)$ for $x \in R_j^*$. The difference between the average expected costs for any classification rules R and R^* is

$$\int [h(x) - h^*(x)] dx = \sum_j \int_{R_j} [h_j(x) - \min_i h_i(x)] dx \geq 0,$$

and the equality holds if $h_j(x) = \min_i h_i(x)$ for x in R_j (for all j). Q.E.D.

Remarks

(i) If (9.5) holds for all $j (\neq l)$ except for h indices, for which the inequality is replaced by equality, then x can be assigned to any one of these $(h+1)\pi_i$ terms.

(ii) If $C(i|j) = C(\neq 0)$ for all (i,j), $i \neq j$, then in R_l^* we obtain from (9.5)

$$\sum_{i'=1, i\neq l}^{k} p_i f_i(x) < \sum_{i=1, i\neq j}^{k} p_i f_i(x), \quad j = 1, \ldots, k, j \neq l,$$

which implies in R_l^*

$$p_j f_j(x) < p_l f_l(x), \quad j = 1, \ldots, k, j \neq l.$$

In other words, the point x is in R_l^* if l is the index for which $p_i f_i(x)$ is a maximum. If two different indices give the same maximum, it is irrelevant as to which index is selected.

Example 9.2.1. Suppose that

$$f_i(x) = \begin{cases} \beta_i^{-1} \exp\left(\dfrac{-x}{\beta_i}\right) & 0 < x < \infty \\ 0 & \text{otherwise,} \end{cases}$$

$i = 1, \ldots, k$, and $\beta_1 < \cdots < \beta_k$ are unknown parameters, and let $p_i = 1/k$, $i = 1, \ldots, k$. If x is observed, the Bayes rule with equal $C(i|j)$ requires us to classify x to π_i if

$$p_i f_i(x) \geq \max_{j(\neq i)} p_j f_j(x),$$

in other words, for $i < j$ if

$$\beta_i^{-1} \exp\left(\frac{-x}{\beta_i}\right) \geq \beta_j^{-1} \exp\left(\frac{-x}{\beta_j}\right),$$

which holds if and only if

$$x \leq \frac{\beta_i \beta_j}{\beta_j - \beta_i} (\log \beta_j - \log \beta_i).$$

It is easy to show that this is an increasing function of β_j for fixed $\beta_i < \beta_j$ and is an increasing function of β_i for fixed $\beta_j < \beta_i$. Since $f_i(x)$ is decreasing in x for $x > 0$, it implies that we classify x to π_i if

$$x_{i-1} \leq x < x_i$$

where

$$x_0 = 0, x_k = \infty, \quad \text{and} \quad x_i = \frac{\beta_i \beta_{i+1}}{\beta_{i+1} - \beta_i} (\log \beta_{i+1} - \log \beta_i).$$

Discriminant Analysis

It is interesting to note that if p_i is proportional to β_i, then the Bayes rule consists of making no observation on the individual and always classifying him to π_k.

Example 9.2.2 Let

$$f_i(x) = \begin{cases} \dfrac{1}{(2\pi)^{1/2}} \exp\left\{-\dfrac{1}{2}(x-\mu_i)^2\right\} & -\infty < x < \infty \\ 0 & \text{otherwise,} \end{cases}$$

where the μ_i are unknown parameters, and let $p_i = 1/k$, $i = 1, \ldots, k$. The Bayes rule with equal $C(i|j)$ requires us to classify an observed x to π_j if

$$(x-\mu_j)^2 < \max_{i, i \neq j}\{(x-\mu_i)^2\}. \tag{9.10}$$

For the particular case $k = 2$, the Bayes classification rule against the a priori (p_1, p_2) is given by

$$\text{Assign } x \text{ to } \begin{cases} \pi_1 & \text{if } \dfrac{f_1(x)}{f_2(x)} > \dfrac{C(1|2)p_2}{C(2|1)p_1} \\ \pi_2 & \text{if } \dfrac{f_1(x)}{f_2(x)} < \dfrac{C(1|2)p_2}{C(2|1)p_1} \\ \text{one of } \pi_1 \text{ and } \pi_2 & \text{if } \dfrac{f_1(x)}{f_2(x)} = \dfrac{C(1|2)p_2}{C(2|1)p_1}. \end{cases} \tag{9.11}$$

However, if under π_i, $i = 1, 2$,

$$P\left\{\dfrac{f_1(x)}{f_2(x)} = \dfrac{C(1|2)p_2}{C(2|1)p_1} \Big| \pi_i\right\} = 0, \tag{9.12}$$

Then the Bayes classification rule is unique except for sets of probability measure 0.

Some Heuristic Classification Rules

A likelihood ratio classification rule $R = (R_1, \ldots, R_k)$ is defined by

$$R_j : C_j f_j(x) > \max_{i, i \neq j} C_i f_i(x) \tag{9.13}$$

for positive constants $C_1, \ldots, C_k$. In particular, if the C_i are all equal, the classification rule is called a maximum likelihood rule.

If the distribution F_i is not completely known, supplementary information on it or on the parameters involved in it is obtained through a training sample from the corresponding population. Then assuming complete knowledge of the F_i, a good classification rule $R = (R_1, \ldots, R_k)$ (i.e., Bayes, minimax, likelihood ratio rule) is chosen. A plug-in classification rule R^* is obtained from R by replacing the F_i or the parameters involved in the definition of R by their corresponding estimates from the training samples.

For other heuristic rules based on the Mahalanobis distance the reader is referred to Das Gupta (1973), who also gives some results in this case and relevant references.

In concluding this section we state without proof some decision theoretic results of the classification rules. For a proof of these results see, for example, Wald (1950, Section 5.1.1), Ferguson (1967), and Anderson (1958).

Theorem 9.2.2. *Every admissible classification rule is a Bayes classification rule with respect to certain a priori probabilities on $\pi_1, \ldots, \pi_k$.*

Theorem 9.2.3. *The class of all admissible classification rules is complete.*

Theorem 9.2.4. *For every set of a priori probabilities $p = (p_1, \ldots, p_k)$ on $\pi = (\pi_1, \ldots, \pi_k)$, there exists an admissible Bayes classification rule.*

Theorem 9.2.5. *For $k = 2$, there exists a unique minimax classification rule R for which $r_1(R) = r_2(R)$.*

Theorem 9.2.6. *Suppose that $C(j|i) = C > 0$ for all $i \neq j$ and that the distribution functions $F_1, \ldots, F_k$ characterizing the populations $\pi_1, \ldots, \pi_k$ are absolutely continuous. Then there exists a unique minimax classification rule R for which*

$$r_1(R) = \cdots = r_k(R). \quad (9.14)$$

It may be cautioned that if either of these two conditions is violated, then (9.14) may not hold.

9.3. CLASSIFICATION INTO ONE OF TWO MULTIVARIATE NORMALS

Consider the problem of classifying an individual, with observation x on him, into one of two-known p-variate normal population with means μ_1 and μ_2, respectively, and the same positive definite covariance matrix Σ. Here

$$f_i(x) = (2\pi)^{-p/2}(\det \Sigma)^{-1/2} \exp\left\{-\frac{1}{2}(x - \mu_i)'\Sigma^{-1}(x - \mu_i)\right\}, \quad i = 1, 2. \quad (9.15)$$

Discriminant Analysis

The ratio of the densities is

$$\frac{f_1(x)}{f_2(x)} = \exp\left\{-\frac{1}{2}(x-\mu_1)'\Sigma^{-1}(x-\mu_1) + \frac{1}{2}(x-\mu_2)'\Sigma^{-1}(x-\mu_2)\right\}$$

$$= \exp\left\{x'\Sigma^{-1}(\mu_1-\mu_2) - \frac{1}{2}(\mu_1+\mu_2)'\Sigma^{-1}(\mu_1-\mu_2)\right\}. \quad (9.16)$$

The Bayes classification rule $R = (R_1, R_2)$ against the a priori probabilities (p_1, p_2) is given by

$$R_1 : \left(x - \frac{1}{2}(\mu_1+\mu_2)\right)'\Sigma^{-1}(\mu_1-\mu_2) \geq k,$$

$$R_2 : \left(x - \frac{1}{2}(\mu_1+\mu_2)\right)'\Sigma^{-1}(\mu_1-\mu_2) < k, \quad (9.17)$$

where $k = \log(p_2 C(1|2))/(p_1 C(2|1))$. For simplicity we have assigned the boundary to the region R_1, though we can equally assign it to R_2 also. The linear function $(x - \frac{1}{2}(\mu_1+\mu_2))'\Sigma^{-1}(\mu_1-\mu_2)$ of the components of the observation vector x is called the discriminant function, and the components of $\Sigma^{-1}(\mu_1-\mu_2)$ are called discriminant coefficients. It may be noted that if $p_1 = p_2 = 1/2$ and $C(1|2) = C(2|1)$, then $k = 0$. Now suppose that we do not have a priori probabilities for the π_i. In this case we cannot use the Bayes technique to obtain the Bayes classification rule given in (9.17). However, we can find the minimax classification rule by finding k such that the Bayes rule in (9.17) with unknown k satisfies

$$C(2|1)P(2|1, R) = C(1|2)P(1|2, R). \quad (9.18)$$

According to Ferguson (1967) such a classification rule is called an equalizer rule.

Let X be the random vector corresponding to the observed x and let

$$U = \left(X - \frac{1}{2}(\mu_1+\mu_2)\right)'\Sigma^{-1}(\mu_1-\mu_2). \quad (9.19)$$

On the assumption that X is distributed according to π_1, U is normally distributed with mean and variance

$$E_1(U) = \frac{1}{2}(\mu_1-\mu_2)'\Sigma^{-1}(\mu_1-\mu_2) = \frac{1}{2}\alpha$$

$$\text{var}(U) = E\{(\mu_1-\mu_2)'\Sigma^{-1}(X-\mu_1)(X-\mu_1)'\Sigma^{-1}(\mu_1-\mu_2)\} \quad (9.20)$$

$$= (\mu_1-\mu_2)'\Sigma^{-1}(\mu_1-\mu_2) = \alpha.$$

If X is distributed according to π_2, then U is normally distributed with mean and variance

$$E_2(U) = -\frac{1}{2}\alpha, \quad \text{var}(U) = \alpha. \tag{19.21}$$

The quantity α is called the Mahalanobis distance between two normal populations with the same covariance matrix. Now the minimax classification rule R is given by, writing $u = U(x)$,

$$R_1 : u \geq k, \quad R_2 : u < k, \tag{9.22}$$

where the constant k is given by

$$C(2|1) \int_{-\infty}^{k} \frac{1}{(2\pi\alpha)^{1/2}} \exp\left\{-\frac{1}{2\alpha}\left(\frac{u-\alpha}{2}\right)^2\right\} du$$

$$= C(1|2) \int_{k}^{\infty} \frac{1}{(2\pi\alpha)^{1/2}} \exp\left\{-\frac{1}{2\alpha}\left(\frac{u+\alpha}{2}\right)^2\right\} du$$

or, equivalently, by

$$C(2|1)\phi\left(\frac{k - \alpha/2}{\sqrt{\alpha}}\right) = C(1|2)\left(1 - \phi\left(\frac{k + \alpha/2}{\sqrt{\alpha}}\right)\right) \tag{9.23}$$

where $\phi(z) = \int_{-\infty}^{z} (2\pi)^{-1} \exp\{-1/2t^2\} dt$.

Suppose we have a group of N individuals, with observations x^α, $\alpha = 1, \ldots, N$, to be classified as a whole to one of the π_i, $i = 1, 2$. Since, writing $\bar{x} = (1/N)\Sigma_1^N x^\alpha$,

$$\prod_{\alpha=1}^{N} \frac{f_1(x^\alpha)}{f_2(x^\alpha)} = \exp\left\{N\left(\bar{x} - \frac{1}{2}(\mu_1 + \mu_2)\right)' \Sigma^{-1} (\mu_1 - \mu_2)\right\} \tag{9.24}$$

and $N(\bar{x} - 1/2(\mu_1 + \mu_2))'\Sigma^{-1}(\mu_1 - \mu_2)$ is normally distributed with means $N\alpha/2$, $-N\alpha/2$ and the same variance $N\alpha$ under π_1 and π_2, respectively, the Bayes classification rule $R = (R_1, R_2)$ against the a priori probabilities (p_1, p_2) is given by

$$R_1 : N\left(\bar{x} - \frac{1}{2}(\mu_1 + \mu_2)\right)' \Sigma^{-1}(\mu_1 - \mu_2) \geq k,$$

$$R_2 : N\left(\bar{x} - \frac{1}{2}(\mu_1 + \mu_2)\right)' \Sigma^{-1}(\mu_1 - \mu_2) < k. \tag{9.25}$$

Discriminant Analysis

The minimax classification rule $R = (R_1, R_2)$ is given by (9.25), where k is determined by

$$C(2|1)\phi\left(\frac{k - N\alpha/2}{(N\alpha)^{1/2}}\right) = C(1|2)\left(1 - \phi\left(\frac{k + N\alpha/2}{(N\alpha)^{1/2}}\right)\right). \tag{9.26}$$

If the parameters are unknown, estimates of these parameters are obtained from independent random samples of sizes N_1 and N_2 from π_1 and π_2, respectively. Let $x_\alpha^{(1)} = (x_{\alpha 1}^1, \ldots, x_{\alpha p}^1)'$, $\alpha = 1, \ldots, N_1$, $x_\alpha^{(2)} = (x_{\alpha 1}^2, \ldots, x_{\alpha p}^2)'$, $\alpha = 1, \ldots, N_2$, be the sample observations (independent) from π_1, π_2, respectively, and let

$$\bar{x}^{(i)} = \frac{1}{N_i}\sum_{\alpha=1}^{N_i} x_\alpha^{(i)}, \quad i = 1, 2$$

$$(N_1 + N_2 - 2)s = \sum_{i=1}^{2}\sum_{\alpha=1}^{N_i}(x_\alpha^{(i)} - \bar{x}^{(i)})(x_\alpha^{(i)} - \bar{x}^{(i)})'. \tag{9.27}$$

We substitute these estimates for the unknown parameters in the expression for U to obtain the sample discriminant function $[v(x)]$

$$v = \left(x - \frac{1}{2}(\bar{x}^{(1)} + \bar{x}^{(2)})\right)' s^{-1}(\bar{x}^{(1)} - \bar{x}^{(2)}), \tag{9.28}$$

which is used in the same way as U in the case of known parameters to define the classification rule R. When classifying a group of N individuals instead of a single one we can further improve the estimate of Σ by taking its estimate as s, defined by

$$(N_1 + N_2 + N - 3)s = \sum_{i=1}^{2}\sum_{\alpha=1}^{N_i}(x_\alpha^{(i)} - \bar{x}^{(i)})(x_\alpha^{(i)} - \bar{x}^{(i)})'$$

$$+ \sum_{\alpha=1}^{N}(x^\alpha - \bar{x})(x^\alpha - \bar{x})'.$$

The sample discriminant function in this case is

$$v = N\left(\bar{x} - \frac{1}{2}(\bar{x}^{(1)} + \bar{x}^{(2)})\right)' s^{-1}(\bar{x}^{(1)} - \bar{x}^{(2)}). \tag{9.29}$$

The classification rule based on v is a plug-in rule. To find the cutoff point k it is necessary to find the distribution of V. The distribution of V has been studied by Wald (1944), Anderson (1951), Sitgreaves (1952), Bowker (1960), Kabe (1963),

and Sinha and Giri (1975). Okamoto (1963) gave an asymptotic expression for the distribution of V.
Write

$$Z = X - \frac{1}{2}(\bar{X}^{(1)} + \bar{X}^{(2)}), Y = \bar{X}^{(1)} - \bar{X}^{(2)},$$

$$(N_1 + N_2 - 2)S = \sum_{i=1}^{2}\sum_{\alpha=1}^{N_i}(X_\alpha^{(i)} - \bar{X}^{(i)})(X_\alpha^{(i)} - \bar{X}^{(i)})'.$$

(9.30)

Obviously both Y and Z are distributed as p-variate normal with

$$E(Y) = \mu_1 - \mu_2, \operatorname{cov}(Y) = \left(\frac{1}{N_1} + \frac{1}{N_2}\right)\Sigma,$$

$$E_1(Y) = \frac{1}{2}(\mu_1 - \mu_2), E_2(Z) = \frac{1}{2}(\mu_2 - \mu_1),$$

(9.31)

$$\operatorname{cov}(Z) = \left(1 + \frac{1}{4N_1} + \frac{1}{4N_2}\right)\Sigma, \operatorname{cov}(Y, Z) = \left(\frac{1}{2N_2} - \frac{1}{2N_1}\right)\Sigma,$$

and $(N_1 + N_2 - 2)S$ is distributed independently of Z, Y as Wishart $W_p(N_1 + N_2 - 2, \Sigma)$ when $N_i > p$, $i = 1, 2$. If $N_1 = N_2$, Y and Z are independent. Wald (1944) and Anderson (1951) obtained the distribution of V when Z, Y are independent. Sitgreaves (1952) obtained the distribution of $Z'S^{-1}Y$ where Z, Y are independently distributed normal vectors whose means are proportional and S is distributed as Wishart, independently of (Z, Y). It may be remarked that the distribution of V is a particular case of this statistic. Sinha and Giri (1975) obtained the distribution of $Z'S^{-1}Y$ when the means of Z and Y are arbitrary vectors and Z, Y are not independent. However, all these distributions are too complicated for practical use.

It is easy to verify that if $N_1 = N_2$, the distribution of V if X comes from π_1 is the same as that of $-V$ if X comes from π_2. A similar result holds for V depending on X. If $v \geq 0$ is the region R_1 and $v < 0$ is the region R_2 (v is an observed value of V), then the probability of misclassifying x when it is actually from π_1 is equal to the probability of misclassifying it when it is from π_2. Furthermore, given $\bar{X}^{(i)} = \bar{x}^{(i)}$, $i = 1, 2$, $S = s$, the conditional distribution of V is normal with means

Discriminant Analysis

and variance

$$E_1(V) = \left(\mu_1 - \frac{1}{2}(\bar{x}^{(1)} + \bar{x}^{(2)})\right)' s^{-1}(\bar{x}^{(1)} - \bar{x}^{(2)})$$

$$E_2(V) = \left(\mu_2 - \frac{1}{2}(\bar{x}^{(1)} + \bar{x}^{(2)})\right)' s^{-1}(\bar{x}^{(1)} - \bar{x}^{(2)}) \qquad (9.32)$$

$$\text{var}(V) = (\bar{x}^{(1)} - \bar{x}^{(2)})' s^{-1}(\bar{x}^{(1)} - \bar{x}^{(2)}).$$

However, the unconditional distribution of V is not normal.

9.3.1. Evaluation of the Probability of Misclassification Based on V

As indicated earlier if $N_1 = N_2$, then the classification rule $R = (R_1, R_2)$ where $v > 0$ is the region R_1, $v < 0$ is the region R_2, has equal probabilities of misclassification. Various attempts have been made to evaluate these two probabilities of misclassification for the rule R in the general case $N_1 \neq N_2$. This classification rule is sometimes referred to as Anderson's rule in literature. As pointed out earlier in this section, the distribution of V, though known, is too complicated to be of any practical help in evaluating these probabilities. Let

$$P_1 = P(2|1, R), \quad P_2 = P(1|2, R). \qquad (9.33)$$

We shall now discuss several methods for estimating P_1, P_2. Let us recall that when the parameters are known these probabilities are given by [taking $k = 0$ in (9.23)]

$$P_1 = \phi\left(-\frac{1}{2}\sqrt{\alpha}\right), \quad P_2 = 1 - \phi\left(\frac{1}{2}\sqrt{\alpha}\right). \qquad (9.34)$$

Method 1 This method uses the sample observations $x_\alpha^{(1)}$, $\alpha = 1, \ldots, N_1$, from π_1, $x_\alpha^{(2)}$, $\alpha = 1, \ldots, N_2$, from π_2, used to estimate the unknown parameters, to assess the performance of R based on V. Each of these $N_1 + N_2$ observations $x_\alpha^{(i)}$, $\alpha = 1, \ldots, N_i$, is substituted in V and the proportions of misclassified observations from among these, using the rule R, are noted. These proportions are taken as the estimates of P_1, P_2. This method, which is sometimes called the resubstitution method, was suggested by Smith (1947). It is obviously very crude and often gives estimates of P_1 and P_2 that are too optimistic, as the same observations are used to compute the value of V and also to evaluate its performance.

Method 2 When the population parameters are known, using the analog statistic U, we have observed that the probabilities of misclassification are given by (9.34). Thus one way of estimating P_1 and P_2 is to replace α by its estimates from the samples $x_\alpha^{(i)}$, $\alpha = 1, \ldots, N_i$, $i = 1, 2$,

$$\hat{\alpha} = (\bar{x}^{(1)} - \bar{x}^{(2)})' s^{-1} (\bar{x}^{(1)} - \bar{x}^{(2)}), \quad (9.35)$$

as is done to obtain the sample discriminant function V from U. It follows from Theorem 6.8.1 that

$$E(\hat{\alpha}) = \frac{N_1 + N_2 - 2}{N_1 + N_2 - p + 1} \left(\alpha + \frac{pN_1 N_2}{N_1 + N_2} \right). \quad (9.36)$$

Thus $\phi(\frac{1}{2}\sqrt{\hat{\alpha}})$ is an underestimate of $\phi(\frac{1}{2}\sqrt{\alpha})$. A modification of this method will be to use an unbiased estimate of α, which is given by

$$\tilde{\alpha} = \frac{N_1 + N_2 - p + 1}{N_1 + N_2 - 2} \hat{\alpha} - \frac{pN_1 N_2}{N_1 + N_2}. \quad (9.37)$$

Method 3 This method is similar to the "jackknife technique" used in statistics (see Quenouille, 1956; Tukey, 1958; Schucany et al., 1971). Let $x_\alpha^{(i)}$, $\alpha = 1, \ldots, N_i$, $i = 1, 2$, be samples of sizes N_1, N_2 from π_1, π_2, respectively. In this method one observation is omitted from either $x_\alpha^{(1)}$ or $x_\alpha^{(2)}$, and v is computed by using the omitted observation as x and estimating the parameters from the remaining $N_1 + N_2 - 1$ observations in the samples. Since the estimates of the parameters are obtained without using the omitted observation, we can now classify the omitted observation which we correctly know to be from π_1 or π_2, using the statistic V and the rule R, and note if it is correctly or incorrectly classified. To estimate P_1 we repeat this procedure, omitting each $x_\alpha^{(1)}$, $\alpha = 1, \ldots, N_1$. Let m_1 be the number of $x_\alpha^{(1)}$ that are misclassified. Then m_1/N_1 is an estimate of P_1. To estimate P_2 the same procedure is repeated with respect to $x_\alpha^{(2)}$, $\alpha = 1, \ldots, N_2$. Intuitively it is felt that this method is not sensitive to the assumption of normality.

Method 4 This method is due to Lachenbruch and Mickey (1968). Let $v(x_\alpha^{(i)})$ be the value of V obtained from $x_\alpha^{(i)}$, $\alpha = 1, \ldots, N_i$, $i = 1, 2$, by omitting $x_\alpha^{(i)}$ as in Method 3, and let

$$u_1 = \frac{1}{N_1} \sum_{\alpha=1}^{N_1} v(x_\alpha^{(1)}), \quad u_2 = \frac{1}{N_2} \sum_{\alpha=1}^{N_2} v(x_\alpha^{(2)}),$$

$$(N_1 - 1) s_1^2 = \sum_{\alpha=1}^{N_1} (v(x_\alpha^{(1)}) - u_1)^2, \quad (9.38)$$

$$(N_2 - 1) s_2^2 = \sum_{\alpha=1}^{N_2} (v(x_\alpha^{(2)}) - u_2)^2.$$

Discriminant Analysis

Lachenbruch and Mickey propose $\phi(-u_1/s_1)$ as the estimate of P_1 and $\phi(u_2/s_2)$ as the estimate of P_2.

When the parameters are known, the probabilities of misclassifications for the classification rule $R = (R_1, R_2)$ where $R_1 : u \geq 0$, $R_2 : u < 0$ are given by

$$P_i = \phi\left((-1)^i \frac{E_i(U)}{(V(U))^{1/2}}\right), \quad i = 1, 2. \tag{9.39}$$

In case the parameters $E_i(U)$, $V(U)$ are unknown, for estimating $E_1(U)$ and $V(U)$, we can take $v(x_\alpha^{(1)})$, $\alpha = 1, \ldots, N_1$, as a sample of N_1 observations on U. So u_1, s_1^2 are appropriate estimates of $E_1(U)$ and $V(U)$. In other words, an appropriate estimate of P_1 is $\phi(-u_1/s_1)$. Similarly, $\phi(u_2/s_2)$ will be an appropriate estimate of P_2.

It may be added here that since U has the same variance irrespective of whether X comes from π_1 or π_2, a better estimate of $V(U)$ is

$$\frac{(N_1 - 1)s_1^2 + (N_2 - 1)s_2^2}{N_1 + N_2 - 2}$$

It is worth investigating the effect of replacing $V(U)$ by such an estimate in P_i.

Method 5 *Asymptotic case.* Let

$$\bar{X}^{(i)} = \frac{1}{N_i} \sum_{\alpha=1}^{N_i} X_\alpha^{(i)}, \quad i = 1, 2,$$

$$(N_1 + N_2 - 2)S = \sum_{i=1}^{\alpha} \sum_{\alpha=1}^{N_i} (X_\alpha^{(i)} - \bar{X}^{(i)})(X_\alpha^{(i)} - \bar{X}^{(i)})'$$

where $X_\alpha^{(i)}, \alpha = 1, \ldots, N_1$, and $X_\alpha^{(2)}, \alpha = 1, \ldots, N_2$, are independent random samples from π_1 and π_2, respectively. Since $\bar{X}^{(i)}$ is the mean of a random sample of size N_1 from a normal distribution with mean $\mu_{(i)}$ and covariance matrix Σ, then as shown in Chapter 6 $\bar{X}^{(i)}$ converges to μ_i in probability as $N_i \to \infty$, $i = 1, 2$. As also shown S converges to Σ in probability as both N_1 and N_2 tend to ∞. Hence it follows that $S^{-1}(\bar{X}^{(1)} - \bar{X}^{(2)})$ converges to $\Sigma^{-1}(\mu_1 - \mu_2)$ and $(\bar{X}_{(1)} + \bar{X}_{(2)})'S^{-1}(\bar{X}^{(1)} - \bar{X}^{(2)})$ converges to $(\mu_1 + \mu_2)'\Sigma^{-1}(\mu_1 - \mu_2)$ in probability as both $N_1, N_2 \to \infty$. Thus as $N_1, N_2 \to \infty$ the limiting distribution of V is normal with

$$E_1(V) = \frac{1}{2}\alpha, \quad E_2(V) = -\frac{1}{2}\alpha, \quad \text{and} \quad \text{var}(V) = \alpha.$$

If the dimension p is small, the sample sizes N_1, N_2 occurring in practice will probably be large enough to apply this result. However, if p is not small, we will probably require extremely large sample sizes to make this result relevant for our purpose. In this case one can achieve a better approximation of the probabilities of misclassifications by using the asymptotic results of Okamoto (1963). Okamoto obtained

$$P_1 = \phi\left(-\frac{1}{2}\alpha\right) + \frac{a_1}{N_1} + \frac{a_2}{N_2} + \frac{a_3}{N_1 + N_2 - 2} + \frac{b_{11}}{N_1^2} + \frac{b_{22}}{N_2^2} + \frac{b_{12}}{N_1 N_2}$$
$$+ \frac{b_{13}}{N_1(N_1 + N_2 - 2)} + \frac{b_{23}}{N_2(N_1 + N_2 - 2)} + \frac{b_{33}}{(N_1 + N_2 - 2)^2} + O_3 \quad (9.40)$$

where O_3 is $O(1/N_i^3)$, and he gave a similar expression for P_2. He gave the values of the a and b in terms of the parameters μ_1, μ_2, and Σ and tabulated the values of the a and b terms for some specific cases. To evaluate P_1 and P_2, α is to be replaced by its unbiased estimate as in (9.37) and the a and b are to be estimated by replacing the parameters by their corresponding estimates.

Lachenbruch and Mickey (1968) made a comparative study of all these methods on the basis of a series of Monte Carlo experiments. They concluded that Methods 1 and 2 give relatively poor results. Methods 3–5 do fairly well overall. If approximate normality can be assumed, Methods 4 and 5 are good. Cochran (1968), while commenting on this study, also reached the conclusion that Method 5 rank first, with Methods 3 and 4 not far behind. Obviously Method 5 needs sample sizes to be large and cannot be applied for small sample sizes. Methods 3 and 4 can be used for all sample sizes, but perform better for large sample sizes.

For the case of the equal covariance matrix, Kiefer and Schwartz (1965) indicated a method for obtaining a broad class of Bayes classification rules that are admissible. In particular, these authors showed that the likelihood ratio classification rules are admissible Bayes when Σ is unknown.

Rao (1954) derived an optimal classification rule in the class of rules for which P_1, P_2 depend only on α (the Mahalanobis distance) using the following criteria: (i) to minimize a linear combination of derivatives of P_1, P_2 with respect to α at $\alpha = 0$, subject to the condition that P_1, P_2 at $\alpha = 0$ leave a given ratio; (ii) the first criterion with the additional restriction that the derivatives of P_1, P_2 at $\alpha = 0$ bear a given ratio. See Kudo (1959, 1960) also for the minimax and the most stringent properties of the maximum likelihood classification rules.

9.3.2. Penrose's Shape and Size Factors

Let us assume that the common covariance matrix Σ of two p-variate normal populations with mean vectors $\mu_1 = (\mu_{11}, \ldots, \mu_{1p})'$, $\mu_2 = (\mu_{21}, \ldots, u_{2p})'$ has

Discriminant Analysis

the particular form

$$\Sigma = \begin{pmatrix} 1 & \rho & \cdots & \rho \\ \rho & 1 & \cdots & \rho \\ \vdots & \vdots & & \vdots \\ \rho & \rho & \cdots & 1 \end{pmatrix}, \tag{9.41}$$

since

$$x'\Sigma^{-1}(\mu_1 - \mu_2) = (\mu_1 - \mu_2)'\Sigma^{-1}x$$

$$= \frac{\sum_{i=1}^{p}(\mu_{1i} - \mu_{2i})}{p(1-\rho)} \left\{ b'x + \frac{1-\rho}{1+(p-1)\rho} \sum_{i=1}^{b} x_i \right\}, \tag{9.42}$$

where

$$b'x = \frac{p(\mu_1 - \mu_2)'x}{\sum_{i=1}^{p}(\mu_{1i} - \mu_{2i})} - \sum_{i=1}^{p} x_i. \tag{9.43}$$

Hence the discriminant function depends on two factors, $b'x$ and $\sum_{i=1}^{p} x_i$. Penrose (1947) called $\sum_{i=1}^{p} x_i$ the size factor, since it measures the total size, and $b'x$ the shape factor. This terminology is more appropriate for biological organs where Σ is of the form just given and $\sum_{i=1}^{p} x_i$, $b'x$ measure the size and the shape of an organ. It can be verified that

$$E_i\left(\sum_{j=1}^{p} X_j\right) = \sum_{j=1}^{p} \mu_{ij}, \quad E_i(b'X) = b'\mu_i, \quad i = 1, 2,$$

$$\text{cov}\left(b'X, \sum_{i=1}^{p} X_i\right) = 0, \quad \text{var}\left(\sum_{i=1}^{p} X_i\right) = p(1 + p\rho - \rho) \tag{9.44}$$

$$\text{var}(b'X) = p(1-\rho)\left[\frac{p(\mu_1 - \mu_2)'(\mu_1 - \mu_2)}{\sum_{i=1}^{p}(\mu_{1i} - \mu_{2i})^2} - 1\right].$$

Thus the random variables corresponding to the size and the shape factors are independently normally distributed with the means and variances just given. If the covariance matrix has this special form, the discriminant analysis can be performed with the help of two factors only. If Σ does not have this special form, it can sometimes be approximated to this form by first standardizing the variates to have unit variance for each component X_i and then replacing the correlation ρ_{ij} between the components X_i, X_j of X by ρ, the average correlation among all pairs (i, j). No doubt the discriminant analysis carried out in this fashion is not as efficient as with the true covariance matrix but it is certainly economical.

However, if ρ_{ij} for different (j, j) do not differ greatly, such an approximation may be quite adequate.

9.3.3. Unequal Covariance Matrices

The equal covariance assumption is rarely satisfied although in some cases the two covariance matrices are so close that it makes little or no difference in the results to assume equality. When they are quite different we obtain

$$\frac{f_1(x)}{f_2(x)} = \left(\frac{\det(\Sigma_2)}{\det(\Sigma_1)}\right)^{1/2} \exp\left\{-\frac{1}{2}x'(\Sigma_1^{-1} - \Sigma_2^{-1})x + x'(\Sigma_1^{-1}\mu_1 - \Sigma_2^{-1}\mu_2)\right.$$
$$\left. - \frac{1}{2}(\mu_1'\Sigma_1^{-1}\mu_1 - \mu_2'\Sigma_2^{-1}\mu_2)\right\}.$$

The Bayes classification rule $R = (R_1, R_2)$ against the prior probabilities (p_1, p_2) is given by

$$R_1 : \frac{1}{2}\log\left(\frac{\det \Sigma_2}{\det \Sigma_1}\right) - \frac{1}{2}\mu_1'\Sigma_1^{-1}\mu_1 + \frac{1}{2}\mu_2'\Sigma_2^{-1}\mu_2$$
$$- \frac{1}{2}(x'(\Sigma_1^{-1} - \Sigma_2^{-1})x - 2x'(\Sigma_1^{-1}\mu_1 - \Sigma_2^{-1}\mu_2)) \geq k,$$

where $k = \log(p_2 C(1|2)/p_1 C(2|1))$. The quantity

$$x'(\Sigma_1^{-1} - \Sigma_2^{-1})x - 2x'(\Sigma_1^{-1}\mu_1 - \Sigma_2^{-1}\mu_2) \qquad (9.45)$$

is called the quadratic discriminant function, and in the case of unequal covariance matrices one has to use a quadratic discriminant function since $\Sigma_1^{-1} - \Sigma_2^{-1}$ does not vanish. For the minimax classification rule R one has to find k such that (9.18) is satisfied. Typically this involves the finding of the distribution of the quadratic discriminant function when x comes from $\pi_i, i = 1, 2$. It may be remarked that the quadratic discriminant function is also the statistic involved in the likelihood ratio classification rule for this problem. The distribution of this quadratic function is very complicated. It was studied by Cavalli (1945) for the special case $p = 1$; by Smith (1947), Cooper (1963, 1965), and Bunke (1964); by Okamoto (1963) for the special case $\mu_1 = \mu_2$; by Bartlett and Please (1963) for the special case $\mu_1 = \mu_2 = 0$ and

$$\Sigma_i = \begin{pmatrix} 1 & \rho_i & \cdots & \rho_i \\ \rho_i & 1 & \cdots & \rho_i \\ \vdots & \vdots & & \vdots \\ \rho_i & \rho_i & \cdots & 1 \end{pmatrix}; \qquad (9.46)$$

Discriminant Analysis

and by Han (1968, 1969, 1970) for different special forms of Σ_i. Okamoto (1963) derived the minimax classification rule and the form of a Bayes classification rule when the parameters are known. He also studied some properties of Bayes classification risk function and suggested a method of choosing components. Okamoto also treated the case when the Σ_i are unknown and the common value of μ_i may be known or unknown. The asymptotic distribution of the sample quadratic discriminant function (plug-in-log likelihood statistic) was also obtained by him. Bunke (1964) showed that the plug-in minimax rule is consistent. Following the method of Kiefer and Schwartz (1965), Nishida (1971) obtained a class of admissible Bayes classification rules when the parameters are unknown. Since these results are not very elegant for presentation we shall not discuss them here. The reader is referred to the original references for these results. However, we shall discuss a solution of this problem by Bahadur and Anderson (1962), based on linear discriminant functions only.

Let $b (\neq 0)$ be a p-column vector and c a scalar. An observation x on an individual is classified as from π_1 if $b'x \leq c$ and as from π_2 if $b'x > c$. The probabilities of misclassification with this classification rule can be easily evaluated from the fact that $b'x$ is normally distributed with mean $b'\mu_1$ and variance $b'\Sigma_1 b$ if X comes from π_1, and with mean $b'\mu_2$ and variance $b'\Sigma_2 b$ if X comes from π_2, and are given by

$$P_1 = P(2|1, R) = 1 - \phi(z_1), \quad P_2 = P(1|2, R) = 1 - \phi(z_2), \quad (9.47)$$

where

$$z_1 = \frac{c - b'\mu_1}{(b'\Sigma_1 b)^{1/2}}, \quad z_2 = \frac{b'\mu_2 - c}{(b'\Sigma_2 b)^{1/2}}. \quad (9.48)$$

We shall assume in this treatment that $C(1|2) = C(2|1)$. Hence each procedure (obtained by varying b) can be evaluated in terms of the two probabilities of misclassification P_1, P_2. Since the transformation by the normal cumulative distribution $\phi(z)$ is strictly monotonic, comparisons of different linear procedures can just as well be made in terms of the arguments z_1, z_2 given in (9.48). For a given z_2, eliminating c, we obtain from (9.48)

$$z_1 = \frac{b'\delta - z_2 (b'\Sigma_2 b)^{1/2}}{(b'\Sigma_1 b)^{1/2}},$$

where $\delta = (\mu_2 - \mu_1)$. Since z_1 is homogeneous in b of degree 0, we can restrict b to lie on an ellipse, say $b'\Sigma_1 b = $ const, and on this bounded closed domain z_1 is continuous and hence has a maximum. Thus among the linear procedures with a specified z_2 coordinate (equivalently, with a specified P_2) there is at least one procedure which maximizes the z_1 coordinate (equivalently, minimizes P_1).

Lemma 9.3.1. *The maximum z_1 coordinate is a decreasing function of z_2.*

Proof. Let $z_2^* > z_2$ and let b^* be a vector maximizing z_1^* for given z_2^*. Then

$$\max_b z_1 = \max_b \frac{b'\delta - z_2(b'\Sigma_2 b)^{1/2}}{(b'\Sigma_1 b)^{1/2}} \geq \frac{b^{*'}\delta - z_2(b^{*'}\Sigma_2 b^*)^{1/2}}{(b^{*'}\Sigma_1 b^*)^{1/2}} \quad (9.49)$$

$$> \frac{b^*\delta - z_2^*(b^{*'}\Sigma_2 b^*)^{1/2}}{(b^{*'}\Sigma_1 b^*)^{1/2}} = \max z_1^*.$$

The set of z_2 with corresponding maximum z_1 is thus a curve in the (z_1, z_2) plane running downward and to the right. Since $\delta \neq 0$, the curve lies above and to the right of the origin. Q.E.D.

Theorem 9.3.1. *A linear classification rule R with $P_1 = 1 - \phi(z_1)$, $P_2 = 1 - \phi(z_2)$, where z_1 is maximized with respect to b for a given z_2, is admissible.*

Proof. Suppose R is not admissible. Then there is a linear classification rule $R^* = (R_1^*, R_2^*)$ with arguments (z_1^*, z_2^*) such that $z_1^* \geq z_1, z_2^* \geq z_2$ with at least one inequality being strict. If $z_2^* = z_2$, then $z_1^* > z_1$, which contradicts the fact that z_1 is a maximum. If $z_2^* > z_2$, the maximum coordinate corresponding to z_2^* must be less than z_1, which contradicts $z_1^* \geq z_1$. Q.E.D.

Furthermore, it can be verified that the set of admissible linear classification rules is complete in the sense that for any linear classification rule outside this set there is a better one in the set.

We now want to characterize analytically the admissible linear classification rules. To achieve this the following lemma will be quite helpful.

Lemma 9.3.2. *If a point (α_1, α_2) with $\alpha_i > 0, i = 1, 2$, is admissible, then there exists $t_i > 0, i = 1, 2$, such that the corresponding linear classification rule is defined by*

$$b = (t_1 \Sigma_1 + t_2 \Sigma_2)^{-1} \delta \quad (9.50)$$

$$c = b'\mu_1 + t_1 b'\Sigma_1 b = b'\mu_2 - t_2 b'\Sigma_2 b. \quad (9.51)$$

Proof. Let the admissible linear classification rule be defined by the vector β and the scalar γ. The line

$$z_1 = \frac{s - \beta'\mu_1}{(\beta'\Sigma_1\beta)^{1/2}}, \quad z_2 = \frac{\beta'\mu_2 - s}{(\beta'\Sigma_2\beta)^{1/2}}, \quad (9.52)$$

Discriminant Analysis

with s as parameter, has negative slope with the point (α_1, α_2) on it. Hence there exist positive numbers t_1, t_2 such that the line (9.52) is tangent to the ellipse

$$\frac{z_1^2}{t_1} + \frac{z_2^2}{t_2} = k \tag{9.53}$$

at the point (α_1, α_2). Consider the line defined by an arbitrary vector b and all scalars c. This line is tangent to an ellipse similar or concentric to (9.53) at the point (z_1, z_2) if c in (9.48) is chosen so that $-z_1 t_2 / z_2 t_1$ is equal to the slope of this line. For a given b, the values of c and the resulting z_1, z_2 are

$$c = \frac{t_1 b' \Sigma_1 b b' \mu_2 + t_2 b' \Sigma_2 b b' \mu_1}{t_1 b' \Sigma_1 b + t_2 b' \Sigma_2 b}, \quad z_1 = \frac{t_1 (b' \Sigma_1 b)^{1/2} b' \delta}{t_1 b' \Sigma_1 b + t_2 b' \Sigma_2 b}$$

$$z_2 = \frac{t_2 (b' \Sigma_2 b)^{1/2} b' \delta}{t_1 b' \Sigma_1 b + t_2 b' \Sigma_2 b} \tag{9.54}$$

This point (z_1, z_2) is on the ellipse

$$\frac{z_1^2}{t_1} + \frac{z_2^2}{t_2} = \frac{(b' \delta)^2}{b'(t_1 \Sigma_1 + t_2 \Sigma_2) b}. \tag{9.55}$$

The maximum of the right side of (9.55) with respect to b occurs when b is given by (9.50). However, the maximum must correspond to the admissible procedure, for if there were a b such that the constant in (9.55) were larger than k, the point (α_1, α_2) would be within the ellipse with the constant in (9.55) and would be nearer the origin than the line tangent at (z_1, z_2). Then some points on this line (corresponding to procedures with b and scalar c) would be better. The expressions for the value of c in (9.54) and (9.51) are the same if we use the value of b as given in (9.50). Q.E.D.

Remark. Since Σ_1, Σ_2 are positive definite and $t_i > 0$, $i = 1, 2$, $t_1 \Sigma_1 + t_2 \Sigma_2$ is positive definite, any multiples of (9.50) and (9.51) are equivalent solutions. When b in (9.50) is normalized so that

$$b' \delta = b'(t_1 \Sigma_1 + t_2 \Sigma_2)^{-1} b = \delta'(t_1 \Sigma_1 + t_2 \Sigma_2)^{-1} \delta, \tag{9.56}$$

then from (9.54) we get

$$z_1 = t_1 (b' \Sigma_1 b)^{1/2}, \quad z_2 = t_2 (b' \Sigma_2 b)^{1/2}. \tag{9.57}$$

Since these are homogeneous of degree 0 in t_1 and t_2 for b given by (9.50) we shall find it convenient to take $t_1 + t_2 = 1$ when $t_i > 0$, $i = 1, 2$, $t_1 - t_2 = 1$ when $t_1 > 0, t_2 < 0$, and $t_2 - t_1 = 1$ when $t_2 > 0, t_1 < 0$.

Theorem 9.3.2. *A linear classification rule with*

$$b = (t_1\Sigma_1 + t_2\Sigma_2)^{-1}\delta, \tag{9.58}$$

$$c = b'\mu_1 + t_1 b'\Sigma_1 b = b'\mu_2 - t_2 b'\Sigma_2 b \tag{9.59}$$

for any t_1, t_2 such that $t_1\Sigma_1 + t_2\Sigma_2$ is positive definite is admissible.

Proof. If $t_i > 0, i = 1, 2$, the corresponding z_1, z_2 are also positive. If this linear classification rule is not admissible, there would be a linear admissible classification rule that would be better (as the set of all linear admissible classification rules is complete) and both arguments for this rule would also be positive. By Lemma 9.3.2 the rule would be defined by

$$\beta = (\tau_1\Sigma_1 + \tau_2\Sigma_2)^{-1}\delta$$

for $\tau_i > 0, i = 1, 2$, such that $\tau_1 + \tau_2 = 1$. However, by the monotonicity properties of z_1, z_2 as functions of t_1, one of the coordinates corresponding to τ_1 would have to be less than one of the coordinates corresponding to t_1. This shows that the linear classification rule corresponding to β is not better than the rule defined by b. Hence the theorem is proved for $t_i > 0, i = 1, 2$.

If $t_1 = 0$, then $z_1 = 0, b = \Sigma_1^{-1}\delta, z_2 = (\delta'\Sigma_2^{-1}\delta)^{1/2}$. However, for any b if $z_1 = 0$, then $z_2 = b'\delta(b'\Sigma_2 b)^{-1/2}$, and z_2 is maximized if $b = \Sigma_2^{-1}\delta$. Similarly if $t_2 = 0$, the solution assumed in the theorem is optimum.

Now consider $t_1 > 0, t_2 < 0$, and $t_1 - t_2 = 1$. Any hyperbola

$$\frac{z_1^2}{t_1} + \frac{z_2^2}{t_2} = k \tag{9.60}$$

for $k > 0$ cuts the z_1 axis at $\pm(t_1 k)^{1/2}$. The rule assumed in the theorem has $z_1 > 0$ and $z_2 < 0$. From (9.48) we get

$$\frac{(c - b'\mu_1)^2}{t_1 b'\Sigma_1 b} + \frac{(b'\mu_2 - c)^2}{t_2 b'\Sigma_2 b} = k. \tag{9.61}$$

The maximum of this expression with respect to c for given b is attained for c as given in (9.54). Then z_1, z_2 are of the form (9.54), and (9.61) reduces to (9.55). The maximum of (9.61) is then given by $b = (t_1\Sigma_1 + t_2\Sigma_2)^{-1}\delta$. It is easy to argue that this point is admissible because otherwise there would be a better point which would lie on a hyperbola with greater k.

The case $t_1 < 0, t_2 > 0$ can be similarly treated. Q.E.D.

Discriminant Analysis

Given t_1, t_2 so that $t_1\Sigma_1 + t_2\Sigma_2$ is positive definite, one would compute the optimum b such that

$$(t_1\Sigma_1 + t_2\Sigma_2)b = \delta \tag{9.62}$$

and then compute c as given in (9.51). Usually t_1, t_2 are not given. A desired solution can be obtained as follows. For another solution the reader is referred to Bahadur and Anderson (1962).

Minimization of One Probability of Misclassification Given the Other

Suppose z_2 is given and let $z_2 > 0$. Then if the maximum $z_1 > 0$, we want to find $t_2 = 1 - t_1$ such that $z_2 = t_2(b'\Sigma_2 b)^{1/2}$ with b given by (9.62). The solution can be approximated by trial an error. For $t_2 = 0, z_2 = 0$ and for $t_2 = 1, z_2 = (b'\Sigma_2 b)^{1/2} = (b'\delta)^{1/2} = (\delta'\Sigma_2^{-1}\delta)^{1/2}$, where $\Sigma_2 b = \delta$. One could try other values of t_2 successively by solving (9.62) and inserting the solution in $b'\Sigma_2 b$ until $t_2(b'\Sigma_2 b)^{1/2}$ agrees closely enough with the desired z_2.

For $t_2 > 0, t_1 < 0$, and $t_2 - t_1 = 1, z_2$ is a decreasing function of $t_2 (t_2 \leq 1)$ and at $t_2 = 1, z_2 = (\delta'\Sigma_2^{-1}\delta)^{1/2}$. If the given z_2 is greater than $(\delta'\Sigma_2\delta)^{1/2}$, then $z_1 < 0$ and we look for a value of t_2 such that $z_2 = t_2(b'\Sigma_2 b)^{1/2}$. We require that t_2 be large enough so that $t_1\Sigma_1 + t_2\Sigma_2 = (t_2 - 1)\Sigma_1 + t_2\Sigma_2$ is positive definite.

The Minimax Classification

The minimax linear classification rule is the admissible rule with $z_1 = z_2$. Obviously in this case $z_1 = z_2 > 0$ and $t_i > 0, i = 1, 2$. Hence we want to find $t_1 = 1 - t_2$ such that

$$0 = z_1^2 - z_2^2 = b'(t_1^2\Sigma_1 - (1 - t_1)^2\Sigma_2)b. \tag{9.63}$$

The values of b and t_1 satisfying (9.63) and (9.62) are obtained by the trial and error method.

Since Σ_1, Σ_2 are positive definite there exists a nonsingular matrix C such that $\Sigma_1 = C'\Delta C, \Sigma_2 = C'C$ where Δ is a diagonal matrix with diagonal elements $\lambda_1, \ldots, \lambda_p$, the roots of $\det(\Sigma_1 - \lambda\Sigma_2) = 0$.

Let $b^* = (b_1^*, \ldots, b_p^*)' = Cb$. Then (9.63) can be written as

$$\sum_{i=1}^{p}(\lambda_i - \theta)b_i^{*2} = 0 \tag{9.64}$$

where $\theta = (1 - t_1^2)/t_1^2$. If $\lambda_i - \theta$ are all positive or all negative, (9.64) will not have a solution for b^*. To obtain a solution θ must lie between the minimum and the maximum of $\lambda_1, \ldots, \lambda_p$. This treatment is due to Banerjee and Marcus

9.3.4. Test Concerning Discriminant Coefficients

As we have observed earlier, for discriminating between two multivariate normal populations with means μ_1, μ_2 and the same positive definite covariance matrix Σ, the optimum classification rule depends on the linear discriminant function $x'\Sigma^{-1}(\mu_1 - \mu_2) - 1/2(\mu_1 + \mu_2)'\Sigma^{-1}(\mu_1 - \mu_2)$. The elements of $\Sigma^{-1}(\mu_1 - \mu_2)$ are called discriminant coefficients. In the case in which Σ, μ_1, μ_2 are unknown we can consider estimation and testing problems concerning these coefficients on the basis of sample observations $x_\alpha^{(1)}, \alpha = 1, \ldots, N_1$, from π_1, and $x_\alpha^{(2)}$, $\alpha = 1, \ldots, N_2$, from π_2. We have already tackled the problem of estimating these coefficients; here we will consider testing problems concerning them.

For testing hypotheses about these coefficients, the sufficiency consideration leads us to restrict our attention to the set of sufficient statistics $(\bar{X}^{(1)}, \bar{X}^{(2)}, S)$ as given in (9.27), where $\bar{X}^{(1)}, \bar{X}^{(2)}$ are independently distributed p-dimensional normal random vectors and $(N_1 + N_2 - 2)S$ is distributed independently of $(\bar{X}^{(1)}, \bar{X}^{(2)})$ as a Wishart random matrix with parameter Σ and $N_1 + N_2 - 2$ degrees of freedom. Further, invariance and sufficiency considerations permit us to consider the statistics $(\bar{X}^{(1)} - \bar{X}^{(2)}, S)$ instead of the random samples (independent) $X_\alpha^{(1)}, \alpha = 1, \ldots, N_1$, from π_1, and $X_\alpha^{(2)}, \alpha = 1, \ldots, N_2$, from π_2. Since $(1/N_1 + 1/N_2)^{-1/2}(\bar{X}^{(1)} - \bar{X}^{(2)})$ is distributed as a p-dimensional normal random vector with mean $(1/N_1 + 1/N_2)^{-1/2}(\mu_1 - \mu_2)$ and positive definite covariance matrix Σ, by relabeling variables we can consider the following canonical form where X is distributed as a p-dimensional normal random vector with mean $\mu = (\mu_1, \ldots, \mu_p)'$ and positive definite covariance matrix Σ, and S is distributed (independent of X) as Wishart with parameter Σ, and consider testing problems concerning the components of $\Gamma = \Sigma^{-1}\mu$. Equivalently this problem can be stated as follows: Let $X^\alpha = (X_{\alpha 1}, \ldots, X_{\alpha p})', \alpha = 1, \ldots, N$, be a random sample of size $N(> p)$ from a p-dimensional normal population with mean μ and covariance matrix Σ. Write

$$\bar{X} = \frac{1}{N}\sum_{\alpha=1}^{N} X^\alpha, \quad S = \sum_{\alpha=1}^{N}(X^\alpha - \bar{X})(X^\alpha - \bar{X})'.$$

(Note that we have changed the definition of S to be consistent with the notation of Chapter 7.) Let $\Gamma = (\Gamma_1, \ldots, \Gamma_p)' = \Sigma^{-1}\mu$. We shall now consider the following testing problems concerning Γ, using the notation of Section 7.2.2. We refer to Giri (1964, 1965) for further details.

A. To test the null hypothesis $H_0 : \Gamma = 0$ against the alternatives $H_1 : \Gamma \neq 0$ when μ, Σ are unknown. Since Σ is nonsingular this problem is equivalent to

Discriminant Analysis

testing $H_0 : \mu = 0$ against the alternatives $H_1 : \mu \neq 0$, which we have discussed in Chapter 7. This case does not seem to be of much interest in the context of linear discriminant functions but is included for completeness.

B. Let $\Gamma = (\Gamma_{(1)}, \Gamma_{(2)})'$, where the $\Gamma_{(i)}$ are subvectors of dimension $p_i \times 1$, $i = 1, 2$, with $p_1 + p_2 = p$. We are interested in testing the null hypothesis $H_0 : \Gamma_{(1)} = 0$ against the alternatives $H_1 : \Gamma_{(1)} \neq 0$ when it is given that $\Gamma_{(2)} = 0$ and μ, Σ are unknown. Let $S^* = S + N\bar{X}\bar{X}'$, and let $S^*, S, \bar{X}, \mu$, and Σ be partitioned as in (7.21) and (7.22) with $k = 2$. Let Ω be the parametric space of $((\Gamma_{(1)}, 0), \Sigma)$ and $\omega = (0, \Sigma)$ be the subspace of Ω when H_0 is true. The likelihood of the observations x^α on X^α, $\alpha = 1, \ldots, N$, is

$$L(\Gamma_{(1)}, \Sigma) = (2\pi)^{-Np/2}(\det \Sigma)^{-N/2}$$

$$\times \exp\left\{-\frac{1}{2}\mathrm{tr}(\Sigma^{-1}s^* - 2N\Gamma_{(1)}\bar{x}'_{(1)} + N\Sigma_{(11)}\Gamma_{(1)}\Gamma'_{(1)})\right\}.$$

Lemma 9.3.3.

$$\max_\Omega L(\Gamma_{(1)}, \Sigma) = (2N\pi)^{-Np/2}(\det s^*)^{-N/2}$$

$$\times (1 - N\bar{x}'_{(1)}(s_{(11)} + N\bar{x}_{(1)}\bar{x}'_{(1)})^{-1}\bar{x}_{(1)})^{-N/2} \exp\left\{-\frac{1}{2}Np\right\}.$$

Proof.

$$\max_\Omega L(\Gamma_{(1)}, \Sigma) = \max_{\Sigma, \Gamma_{(1)}}(2\pi)^{-Np/2}(\det \Sigma)^{-N/2}$$

$$\times \exp\left\{-\frac{1}{2}\mathrm{tr}(\Sigma^{-1}s^* + N\Sigma_{(11)}^{-1}(\bar{x}_{(1)} - \Sigma_{(11)}\Gamma_{(1)})\right.$$

$$\left.\times (\bar{x}_{(1)} - \Sigma_{(11)}\Gamma_{(1)})' - N\Sigma_{(11)}^{-1}\bar{x}_{(1)}\bar{x}'_{(1)})\right\}$$

$$= \max_\Sigma (2\pi)^{-Np/2}(\det \Sigma)^{-N/2} \exp\left\{-\frac{1}{2}\mathrm{tr}(\Sigma^{-1}s^* - N\Sigma_{(11)}^{-1}\bar{x}_{(1)}\bar{x}'_{(1)})\right\}.$$

(9.65)

Since Σ and s^* are positive definite there exist nonsingular upper triangular matrices K and T such that

$$\Sigma = KK', \quad s^* = TT'.$$

Partition K and T as

$$K = \begin{pmatrix} K_{(11)} & K_{(12)} \\ 0 & K_{(22)} \end{pmatrix}, \quad T = \begin{pmatrix} T_{(11)} & T_{(12)} \\ 0 & T_{(22)} \end{pmatrix}$$

where $K_{(11)}, T_{(11)}$ are (upper triangular) submatrices of K, T, respectively of dimension $p_1 \times p_1$. Now

$$K^{-1} = \begin{pmatrix} K_{(11)}^{-1} & -(K_{(11)}^{-1} K_{(12)} K_{(22)}^{-1}) \\ 0 & K_{(22)}^{-1} \end{pmatrix}, \quad T^{-1} = \begin{pmatrix} T_{(11)}^{-1} & -(T_{(11)}^{-1} T_{(12)} T_{(22)}^{-1}) \\ 0 & T_{(22)}^{-1} \end{pmatrix}$$

and $\Sigma_{(11)} = K'_{(11)} K_{(11)}$, $s^*_{(11)} = T'_{(11)} T_{(11)}$. Let $K = LT$ and $\Sigma^* = L'L$. Let L, Σ^* be partitioned in the same way as K into submatrices $L_{(ij)}, \Sigma_{(ij)}$, respectively.

Obviously $K_{(11)} = T_{(11)} L_{(11)}$. Writing $z'_{(1)} = x'_{(1)} T_{(11)}^{-1}$, from (9.65) we obtain

$$\max_{\Omega} L(\Gamma_{(1)}, \Sigma)$$

$$= \max_K (2\pi)^{-Np/2} (\det K)^{-N}$$

$$\times \exp\left\{-\frac{1}{2} \mathrm{tr}(K^{-1}(K')^{-1} T'T - N K_{(11)}^{-1} (K'_{(11)})^{-1} \bar{x}_{(1)} \bar{x}'_{(1)})\right\}$$

$$= \max_K (2\pi)^{-Np/2} (\det s^*)^{-N/2} (\det \Sigma^*)^{-N/2}$$

$$\times \exp\left\{-\frac{1}{2} \mathrm{tr}(\Sigma^{*-1} - N \Sigma^{*-1}_{(11)} z_{(1)} z'_{(1)})\right\}$$

$$= \max_\Lambda (2\pi)^{-Np/2} (\det s^*)^{-N/2} (\det \Lambda_{(22)})^{N/2} (\det(\Lambda_{(11)} - \Lambda_{(12)} \Lambda_{(22)}^{-1} \Lambda_{(21)}))^{N/2}$$

$$\times \exp\left\{-\frac{1}{2} \mathrm{tr}(\Lambda_{(11)} + \Lambda_{(22)} - (\Lambda_{(11)} - \Lambda_{(12)} \Lambda_{(22)}^{-1} \Lambda_{(21)})(N z_{(1)} z'_{(1)}))\right\}$$

$$= (2\pi/N)^{-Np/2} (\det s^*)^{-N/2} (\det(I - N z_{(1)} z'_{(1)}))^{-N/2} \exp\left\{-\frac{1}{2} Np\right\}$$

$$= (2\pi/N)^{-Np/2} (\det s^*)^{-N/2} (1 - N\bar{x}'_{(1)} (s_{(11)} + N\bar{x}'_{(1)})^{-1} \bar{x}_{(1)})^{-N/2}$$

$$\times \exp\left\{-\frac{1}{2} Np\right\}, \tag{9.66}$$

where $(\Sigma^*)^{-1} = \Lambda$ and Λ is partitioned into submatrices $\Lambda_{(ij)}$ similar to those of Σ^*. The next to last step in (9.66) follows from the fact that the maximum likelihood estimates of $\Lambda_{(22)}, \Lambda_{(11)}$ are $I/N, (I - N z_{(1)} z'_{(1)})/N$ (see Lemma 5.1.1) and that of $\Lambda_{(12)}$ is 0. Q.E.D.

Discriminant Analysis

Since

$$\max_{\omega} L(\Gamma_{(1)}, \Sigma) = (2\pi N)^{-Np/2}(\det s^*)^{-N/2} \exp\left\{-\frac{1}{2}Np\right\},$$

the likelihood ratio criterion for testing H_0 is given by

$$\lambda = \frac{\max_{\omega} L(\Gamma_{(1)}, \Sigma)}{\max_{\Omega} L(\Gamma_{(1)}, \Sigma)} = (1 - N\bar{x}'_{(1)}(s_{(11)} + N\bar{x}_{(1)}\bar{x}'_{(1)})^{-1}\bar{x}_{(1)})^{N/2}$$

$$= (1 - r_1)^{N/2}, \tag{9.67}$$

where r_1 is given in Section 7.2.2. (We have used the same notation for the classification regions R and the statistic R.) Thus the likelihood ratio test of H_0 rejects H_0 whenever

$$r_1 \geq C, \tag{9.68}$$

where the constant C depends on the level of significance α of the test. From Chapter 6 the probability density function of R_1 under H_1 is given by

$$f_{R_1}(r_1|\delta_1^2) = \frac{\Gamma(\frac{1}{2}N)}{\Gamma(\frac{1}{2}p_1)\Gamma(1/2(N-p_1))} r_1^{p_1/2-1}(1-r_1)^{(N-p_1)/2-1}$$

$$\times \exp\left\{-\frac{1}{2}\delta_1^2\right\} \phi\left(\frac{1}{2}N, \frac{1}{2}p_1; \frac{1}{2}(r_1\delta_1^2)\right) \tag{9.69}$$

provided $r_1 \geq 0$ and is zero elsewhere, where $\delta_1^2 = N\Gamma'_{(1)}\Sigma_{(11)}\Gamma_{(1)}$. Obviously under H_0, $\delta_1^2 = 0$ and R_1 is distributed as central beta with parameter $(\frac{1}{2}p_1, \frac{1}{2}(N-p_1))$.

Let G_{BT} (as defined in Section 7.2.2 with $k=2$) be the multiplicative group of lower triangular matrices

$$g = \begin{pmatrix} g_{(11)} & 0 \\ g_{(21)} & g_{(22)} \end{pmatrix}$$

of dimension $p \times p$. The problem of testing H_0 against H_1 remains invariant under G_{BT} with $k=2$ operating as $X_\alpha \to gX_\alpha$, $\alpha = 1, \ldots, N$, $g \in G_{BT}$. The induced transformation in the space of $(\bar{X}, S)$ is given by $(\bar{X}, S) \to (g\bar{X}, gSg')$ and in the space of (μ, Σ) is given by $(\mu, \Sigma) \to (g\mu, g\Sigma g')$. A set of maximal invariants in the space of $(\bar{X}, S)$ under G_{BT} is (R_1, R_2) as defined in (6.63) with $k=2$. A corresponding maximal invariant in the parametric space of (μ, Σ) is given by (δ_1^2, δ_2^2), where

$$\delta_1^2 = N(\Sigma_{(11)}\Gamma_{(1)} + \Sigma_{(12)}\Gamma_{(2)})'\Sigma_{(11)}^{-1}(\Sigma_{(11)}\Gamma_{(1)} + \Sigma_{(12)}\Gamma_{(2)})$$

$$\delta_1^2 + \delta_2^2 = N\Gamma'\Sigma\Gamma. \tag{9.70}$$

Since $\Gamma_{(2)} = 0$ in this case, we get $\delta_2^2 = 0$ and $\delta_1^2 = N\Gamma'_{(1)}\Sigma_{(11)}\Gamma_{(1)}$. Hence under $H_0 : \delta_1^2 = 0$ and under $H_1 : \delta_1^2 > 0$, the joint probability density function of (R_1, R_2) under H_1 is given by (6.73). The ratio of the density of (R_1, R_2) under H_1 to its density under H_0 is given by

$$\exp\left\{-\frac{1}{2}\delta_1^2\right\} \sum_{j=0}^{\infty} \left(\frac{r_1\delta_1^2}{2}\right)^j \frac{\Gamma(\frac{1}{2}N+j)\Gamma(\frac{1}{2}p_1)}{j!\Gamma(\frac{1}{2}p_1+j)\Gamma(\frac{1}{2}N)}. \quad (9.71)$$

Hence we have the following theorem.

Theorem 9.3.3. *For testing H_0 against H_1, the likelihood ratio test which rejects H_0 for large values of R_1 is uniformly most powerful invariant.*

C. To test the null hypothesis $H_0 : \Gamma_{(2)} = 0$ against the alternatives $H_1 : \Gamma_{(2)} \neq 0$ when μ and Σ are unknown and $\Gamma_{(1)}, \Gamma_{(2)}$ are defined as in case B. The likelihood of the observations x_α on $X_\alpha, \alpha = 1, \ldots, N$, is

$$L(\Gamma, \Sigma) = (2\pi)^{-Np/2}(\det \Sigma)^{-N/2}$$

$$\times \exp\left\{-\frac{1}{2}\text{tr}(\Sigma^{-1}s^* - 2N\Gamma'\bar{x} + N\Sigma\Gamma\Gamma')\right\}. \quad (9.72)$$

Proceeding exactly in the same way as in Lemma 9.3.3, we obtain

$$\max_{\Omega}(\Gamma, \Sigma) = (2\pi N)^{-Np/2}(\det s^*)^{-N/2}(1 - N\bar{x}'(s + N\bar{x}\bar{x}')^{-1}\bar{x})^{-N/2}$$

$$\times \exp\left\{-\frac{1}{2}Np\right\}, \quad (9.73)$$

where $\Omega = \{(\Gamma, \Sigma)\}$. From Lemma 9.3.3 and (9.73) the likelihood ratio criterion for testing H_0 is given by

$$\lambda = \frac{\max_\omega L(\Gamma, \Sigma)}{\max_\Omega L(\Gamma, \Sigma)} = \left(\frac{1 - r_1 - r_2}{1 - r_1}\right)^{N/2}, \quad (9.74)$$

where $\omega = ((\Gamma_{(1)}, 0), \Sigma)$ and r_1, r_2 are as defined in case B. Thus the likelihood ratio test for testing H_0 rejects H_0 whenever

$$z = \frac{1 - r_1 - r_2}{1 - r_1} \leq C, \quad (9.75)$$

Discriminant Analysis

where the constant C depends on the level of significance α of the test. From (6.73) the joint probability density function of Z and R_1 under H_0 is given by

$$\exp\left\{-\frac{1}{2}\delta_1^2\right\} \sum_{j=0}^{\infty} \Gamma\left(\frac{1}{2}N+j\right)\left(\frac{1}{2}r_1\delta_1^2\right)^j (r_1)^{p_1/2-1}$$

$$\times \frac{(1-r_1)^{(N-p_1)/2-1} z^{(N-p_1)/2-1}(1-z)^{(p-p_1)/2-1}}{j!\Gamma(\frac{1}{2}p_1+j)\Gamma(\frac{1}{2}(N-p))\Gamma(\frac{1}{2}(p-p_1)))}. \quad (9.76)$$

From this it follows that under H_0, Z is distributed as a central beta random variable with parameter $(\frac{1}{2}(N-p_1), \frac{1}{2}p_2)$ and is independent of R_1.

The problem of testing H_0 against H_1 remains invariant under the group of transformations G_{BT} with $k=2$, operating as $X^\alpha \to gX^\alpha$, $g \in G_{BT}$, $\alpha = 1, \ldots, N$. A set of maximal invariants in the space of $(\bar{X}, S)$ under G_{BT} is (R_1, R_2) of case B and the corresponding maximal invariant in the parametric space of (μ, Σ) is (δ_1^2, δ_2^2) of (9.70). Under H_0, $\delta_2^2 = 0$ and under H_1, $\delta_2^2 > 0$ (δ_1^2 is unknown). The joint probability density function of (R_1, R_2) is given by (6.73). From this we conclude that R_1 is sufficient for δ_1^2 when H_0 is true, and the marginal probability density function of R_1 when H_0 is true is given by (9.69). This is also the probability density function of R_1 when H_1 is true.

Lemma 9.3.4. *The family of probability density functions $\{f_{R_1}(r_1|\delta_1^2), \delta_1^2 \geq 0\}$ is boundedly complete.*

Proof. Let $\Psi(r_1)$ be any real valued function of r_1. Then

$$E_{\delta_1^2}(\Psi(R_1)) = \exp\left\{-\frac{1}{2}\delta_1^2\right\} \sum_{j=0}^{\infty} \left(\frac{1}{2}\delta_1^2\right)^j a_j \int_0^1 \Psi(r_1) r_1^{p_1/2+j-1}(1-r_1)^{(N-p_1)/2-1} dr_1$$

$$= \exp\left\{-\frac{1}{2}\delta_1^2\right\} \sum_{j=0}^{\infty} \left(\frac{1}{2}\delta_1^2\right)^j a_j \int_0^1 \Psi^*(r_1) r_1^j dr_1,$$

where

$$a_j = \frac{\Gamma(1/2N+j)}{j!\Gamma(\frac{1}{2}(N-p_1))\Gamma(\frac{1}{2}p_1+j)}, \quad \Psi^*(r_1) = r_1^{p_1/2-1}(1-r_1)^{(N-p_1)/2-1}\psi(r_1).$$

Hence $E_{\delta_1^2}(\Psi(R_1)) = 0$ identically in δ_1^2 implies that

$$\sum_{j=0}^{\infty} \left(\frac{1}{2}\delta_1^2\right)^j a_j \int_0^1 \Psi^*(r_1) r_1^j dr_1 = 0 \quad (9.77)$$

identically in δ_1^2. Since the left-hand side of (9.77) is a polynomial in δ_1^2, all its coefficients must be zero. In other words,

$$\int_0^1 \Psi^*(r_1) r_1^j dr_1 = 0, \quad j = 1, 2, \ldots, \tag{9.78}$$

which implies that $\Psi^{*+}(r_1) = \Psi^{*-}(r_1)$ for all r_1, except possibly for a set of values of r_1 of probability measure 0. Hence $\Psi^*(r_1) = 0$ almost everywhere, which implies that $\Psi(r_1) = 0$ almost everywhere. Q.E.D.

Theorem 9.3.4. *The likelihood ratio test of $H_0 : \Gamma_{(2)} = 0$ when μ, Σ are unknown is uniformly most powerful invariant similar against the alternatives $H_1 : \Gamma_{(2)} \neq 0$.*

Proof. Since R_1 is sufficient for δ_1^2 when H_0 is true and the distribution of R_1 is boundedly complete, it is well known that (see, e.g., Lehmann, 1959, p. 134) any level α invariant test $\phi(r_1, r_2)$ has Neyman structure with respect to R_1, i.e.,

$$E_{\delta_1^2}(\phi(R_1, R_2)|R_1 = r_1) = \alpha. \tag{9.79}$$

Now to find the uniformly most powerful test among all similar invariant tests we need the ratio of the conditional probability density function of R_2 given $R_1 = r_1$ under H_1 to that under H_0, and this ratio is given by

$$\exp\left\{-\frac{1}{2}\delta_1^2(1 - r_1)\right\} \sum_{j=0}^{\infty} \frac{(\frac{1}{2} r_2 \delta_2^2)^j \Gamma(\frac{1}{2}(N - p_1) + j)\Gamma(\frac{1}{2} p_2)}{j! \Gamma(\frac{1}{2} p_2 + j)\Gamma(\frac{1}{2}(N - p_1))}. \tag{9.80}$$

Since the distribution of R_2 on each surface $R_1 = r_1$ is independent of δ_1^2, condition (9.79) reduces the problem to that of testing a simple hypothesis $\delta_2^2 = 0$ against the alternatives $\delta_2^2 > 0$ on each surface $R_1 = r_1$. In this conditional situation, by Neyman and Pearson's fundamental lemma, the uniformly most powerful level α invariant test of $\delta_2^2 = 0$ against the alternatives $\delta_2^2 > 0$ [from (9.80)] rejects H_0 whenever

$$\sum_{j=0}^{\infty} \frac{(\frac{1}{2} r_2 \delta_2^2)^j \Gamma(\frac{1}{2}(N - p_1) + j)\Gamma(\frac{1}{2} p_2)}{j! \Gamma(\frac{1}{2} p_2 + j)\Gamma(\frac{1}{2}(N - p_1))} \geq C(r_1), \tag{9.81}$$

where $C(r_1)$ is a constant such that the test has level α on each surface $R_1 = r_1$. Since the left-hand side of (9.81) is an increasing function of r_2 and $r_2 = (1 - r_1)(1 - z)$, this reduces to rejecting H_0 on each surface $R_1 = r_1$ whenever $z \leq C$, where the constant C is chosen such that the test has level α. Since, under H_0, Z is independent of R_1, the constant C does not depend on r_1. Hence the theorem. Q.E.D.

Discriminant Analysis

D. Let $\Gamma = (\Gamma'_{(1)}, \Gamma'_{(2)}, \Gamma'_{(3)})$, where $\Gamma_{(i)}$ is $p_i \times 1$, $i = 1, 2, 3$, and $\Sigma_1^3 p_i = p$. We are interested in testing the null hypothesis $H_0 : \Gamma_{(2)} = 0$ against the alternatives $H_1 : \Gamma_{(2)} \neq 0$ when it is given that $\Gamma_{(3)} = 0$ and $\Gamma_{(1)}$ is unknown. Here

$$\Omega = \{(\Gamma_{(1)}, \Gamma_{(2)}, 0), \Sigma\}, \quad \omega = \{(\Gamma_{(1)}, 0, 0), \Sigma\}.$$

Let $S^*, S, \bar{X}, \mu$, and Σ be partitioned as in (7.21) and (7.22) with $k = 3$. Using Lemma 9.3.3 we get from (9.72)

$$\frac{\max_\omega L(\Gamma, \Sigma)}{\max_\Omega L(\Gamma, \Sigma)} = \left(\frac{1 - r_1 - r_2}{1 - r_1}\right)^{N/2}, \tag{9.82}$$

where r_1, r_2, r_3 are given in Section 7.2.2 with $k = 3$. The likelihood ratio test of H_0 rejects H_0 whenever

$$z = \frac{1 - r_1 - r_2}{1 - r_1} \leq C, \tag{9.83}$$

where C is a constant such that the test has size α. The joint probability density function of R_1, R_2, R_3 (under H_1) is given in (6.73) with $k = 3$, where

$$\delta_1^2 = N(\Sigma_{(11)}\Gamma_{(1)} + \Sigma_{(12)}\Gamma_{(2)})'\Sigma_{(11)}^{-1}(\Sigma_{(11)}\Gamma_{(1)} + \Sigma_{(12)}\Gamma_{(2)})$$

$$\delta_1^2 + \delta_2^2 = N\begin{pmatrix}\Sigma_{(11)}\Gamma_{(1)} + \Sigma_{(12)}\Gamma_{(2)} \\ \Sigma_{(21)}\Gamma_{(1)} + \Sigma_{(22)}\Gamma_{(2)}\end{pmatrix}' \begin{pmatrix}\Sigma_{(11)} & \Sigma_{(12)} \\ \Sigma_{(21)} & \Sigma_{(22)}\end{pmatrix}^{-1}$$

$$\times \begin{pmatrix}\Sigma_{(11)}\Gamma_{(1)} + \Sigma_{(12)}\Gamma_{(2)} \\ \Sigma_{(21)}\Gamma_{(1)} + \Sigma_{(22)}\Gamma_{(2)}\end{pmatrix}$$

$$\delta_3^2 = N\Gamma'_{(3)}\left(\Sigma_{(33)} - \begin{pmatrix}\Sigma_{(13)} \\ \Sigma_{(23)}\end{pmatrix}' \begin{pmatrix}\Sigma_{(11)} & \Sigma_{(12)} \\ \Sigma_{(21)} & \Sigma_{(22)}\end{pmatrix}^{-1} \begin{pmatrix}\Sigma_{(13)} \\ \Sigma_{(23)}\end{pmatrix}\right)\Gamma_{(3)} = 0,$$

$$\tag{9.84}$$

and under H_0, $\delta_2^2 = 0$. From this it follows that the joint probability density function of Z and R_1 under H_0 is given by (9.75) with p replaced by $p_1 + p_2$. Hence under H_0, Z is distributed as central beta with parameters $(\frac{1}{2}(N - p_1), 1/2p_2)$ and is independent of R_1.

The problem of testing H_0 against H_1 remains invariant under G_{BT} with $k = 3$ operating as $X_\alpha \to gX_\alpha$, $g \in G_{BT}$, $\alpha = 1, \ldots, N$. A set of maximal invariants in the space of $(\bar{X}, S)$ under G_{BT} with $k = 3$ is (R_1, R_2, R_3), and the corresponding maximal invariants in the parametric space is $(\delta_1^2, \delta_2^2, \delta_3^2)$ as given in (9.83). Under H_0, $\delta_2^2 = 0$ and under H_1, $\delta_2^2 > 0$, and it is given that $\delta_3^2 = 0$. As we have proved in case C, R_1 is sufficient for δ_1^2 under H_0 and the distribution of R_1 is boundedly complete. Now arguing in the same way as in case C we prove the following theorem.

Theorem 9.3.5. *For testing $H_0 : \Gamma_{(2)} = 0$ the likelihood ratio test which rejects H_0 whenever $z \leq C$, C depending on the level α of the test, is uniformly most powerful invariant similar against $H_1 : \Gamma_{(2)} \neq 0$ when it is given that $\Gamma_{(3)} = 0$.*

Tests depending on the Mahalanobis distance statistic are also used for testing hypotheses concerning discriminant coefficients. The reader is referred to Rao (1965) or Kshirsagar (1972) for an account of this. Recently Sinha and Giri (1975) have studied the optimum properties of the likelihood ratio tests of these problems from the point of view of Isaacson's type D and type E property (see Isaacson, 1951).

9.4. CLASSIFICATION INTO MORE THAN TWO MULTIVARIATE NORMALS

As pointed out in connection with Theorem 9.2.1 if $C(i|j) = C$ for all $i \neq j$, then the Bayes classification rule $R^* = (R_1^*, \ldots, R_k^*)$ against the a priori probabilities $(p_1, \ldots, p_k)$ classifies an observation x to R_l^* if

$$\frac{f_l(x)}{f_j(x)} \geq \frac{p_j}{p_l} \quad \text{for} \quad j = 1, \ldots, k, j \neq l. \tag{9.85}$$

In this section we shall assume that $f_i(x)$ is the probability density function of a p-variate normal random vector with mean μ_i and the same positive definite covariance matrix Σ. Most known results in this area are straightforward extensions of the results for the case $k = 2$. In this case the Bayes classification rule $R^* = (R_1^*, \ldots, R_k^*)$ classifies x to R_l^* whenever

$$u_{lj} = \log \frac{f_l(x)}{f_j(x)} = \left(x - \frac{1}{2}(\mu_l + \mu_j) \right)' \Sigma^{-1}(\mu_l - \mu_j) \geq \log \frac{p_j}{p_l}. \tag{9.86}$$

Each u_{lj} is the linear discriminant function related to the jth and the lth populations and obviously $u_{lj} = -u_{jl}$.

In the case in which the a priori probabilities are unknown the minimax classification rule $R = (R_1, \ldots, R_k)$ classifies x to R_l if

$$u_{lj} \geq C_l - C_j, \quad j = 1, \ldots, k, j \neq l, \tag{9.87}$$

where the C_j are nonnegative constants and are determined in such a way that all $P(i|i, R)$ are equal. Let us now evaluate $P(i|i, R)$. First observe that random variable

$$U_{ij} = \left(X - \frac{1}{2}(\mu_i + \mu_j) \right)' \Sigma^{-1}(\mu_i - \mu_j) \tag{9.88}$$

Discriminant Analysis

satisfies $U_{ij} = -U_{ji}$. Thus we use $k(k-1)/2$ linear discriminant functions U_{ij} if the mean vectors μ_i span a $(k-1)$-dimensional hyperplane. Now the U_{ij} are normally distributed with

$$E_i(U_{ij}) = \frac{1}{2}(\mu_i - \mu_j)'\Sigma^{-1}(\mu_i - \mu_j),$$

$$E_j(U_{ij}) = -\frac{1}{2}(\mu_i - \mu_j)'\Sigma^{-1}(\mu_i - \mu_j) \tag{9.89}$$

$$\text{var}(U_{ij}) = (\mu_i - \mu_j)'\Sigma^{-1}(\mu_i - \mu_j)$$

$$\text{cov}(U_{ij}, U_{ij'}) = (\mu_i - \mu_j)'\Sigma^{-1}(\mu_i - \mu_{j'}), \quad j \neq j',$$

where $E_i(U_{ij})$ denotes the expectation of U_{ij} when X comes from π_i. For a given j let us denote the joint probability density function of $U_{ji}, i = 1, \ldots, k; i \neq j$, by p_j. Then

$$P(j|j, R) = \int_{C_j - C_k}^{\infty} \cdots \int_{C_j - C_1}^{\infty} p_j \Pi_{i \neq j} du_{ji}.$$

Note that the sets of regions given by (9.87) form an admissible class.

If the parameters are unknown, they are replaced by their appropriate estimates from training samples from these populations to obtain sample discriminant functions as discussed in the case of two populations. We discussed earlier the problems associated with the distribution of sample discriminant functions and different methods of evaluating the probabilities of misclassification. For some relevant results the reader is referred to Das Gupta (1973) and the references therein.

The problem of unequal covariance matrices can be similarly resolved by using the results presented earlier for the case of two multivariate normal populations with unequal covariance matrices. For further discussions in this case the reader is referred to Fisher (1938), Brown (1947), Rao (1952, 1963), and Cacoullos (1965). Das Gupta (1962) considered the problems where $\mu_1, \ldots, \mu_k$ are linearly restricted and showed that the maximum likelihood classification rule is admissible Bayes when the common covariance matrix Σ is known. Following Kiefer and Schwartz (1965), Srivastava (1964) obtained similar results when Σ is unknown.

Example 9.4.1. Consider two populations π_1 and π_2 of plants of two distinct varieties of wheat. The measurements for each member of these two populations are

x_1 plant height (cm),
x_2 number of effective tillers,
x_3 length of ear (cm),
x_4 number of fertile spikelets per 10 ears,
x_5 number of grains per 10 ears,
x_6 weight of grains per 10 ears (gm).

Assuming that these are six-dimensional normal populations with different unknown mean vectors μ_1, μ_2 and with the same unknown covariance matrix Σ we shall consider here the problem of classifying an individual with observation $x = (x_1, \ldots, x_6)'$ on him to one of these populations. Since the parameters are unknown we obtained two training samples (Table 9.1) (of size 27 each) from them (these data were collected from the Indian Agricultural Research Institute, New Delhi, India). The sample mean vectors and the sample covariance matrix are given in Table 9.2 [see Eq. (9.27) for the notation]. Using the sample discriminant function

$$v = \left(x - \frac{1}{2}(\bar{x}^{(1)} + \bar{x}^{(2)})\right)' s^{-1}(\bar{x}^{(1)} - \bar{x}^{(2)}),$$

writing

$$d_1(x) = x's^{-1}\bar{x}^{(1)} - \frac{1}{2}\bar{x}^{(1)'}s^{-1}\bar{x}^{(1)}, \quad d_2(x) = x's^{-1}\bar{x}^{(2)} - \frac{1}{2}\bar{x}^{(2)'}s^{-1}\bar{x}^{(2)},$$

we classify x

to π_1 if $d_1(x) \geq d_2(x)$
to π_2 if $d_1(x) < d_2(x)$

Sample Covariance Matrix s

$$\begin{pmatrix} 3.13548 & & & & & \\ 2.61154 & 41.76262 & & & & \\ 0.37533 & 11.89829 & 0.82986 & & & \\ 0.75635 & 18.28440 & 1.27375 & 3.13548 & & \\ 18.28440 & 1214.74359 & 51.04744 & 90.44476 & 41.76282 & \\ 1.27375 & 51.04744 & 3.73134 & 6.54646 & 0.37415 & 0.82986 \end{pmatrix}$$

Discriminant Analysis

Table 9.1. Samples From Populations

	π_1						π_2					
Observation	x_1	x_2	x_3	x_4	x_5	x_6	x_1	x_2	x_3	x_4	x_5	x_6
1	77.60	136	9.65	12.6	322	14.7	65.55	166	9.29	11.3	323	13.1
2	83.45	177	9.76	13.1	321	14.5	67.10	132	9.52	11.7	319	13.6
3	76.20	164	10.52	13.9	384	17.1	66.25	173	9.88	12.1	319	13.6
4	80.30	185	9.76	12.5	259	15.4	80.45	155	11.19	13.8	394	17.6
5	82.30	187	9.77	13.4	314	14.4	78.30	202	10.78	13.3	376	16.7
6	86.00	171	9.25	13.0	278	13.0	77.80	155	10.86	14.0	401	18.2
7	90.50	211	9.75	12.9	308	13.6	79.20	161	10.68	14.3	417	17.8
8	81.50	158	10.38	13.6	258	14.8	82.65	158	10.64	12.2	382	17.4
9	79.75	176	9.31	12.0	307	13.2	79.85	156	10.83	13.7	366	16.1
10	86.85	175	10.23	14.2	330	14.6	67.30	157	9.98	11.8	354	14.0
11	72.90	139	10.29	12.9	346	15.5	70.65	173	9.97	12.2	310	12.5
12	73.50	124	9.68	12.0	308	14.1	67.15	159	9.99	12.3	325	11.9
13	86.85	149	10.33	13.5	337	15.1	80.85	160	10.47	12.7	358	15.5
14	89.15	224	9.70	13.0	317	12.4	81.80	162	10.87	13.9	403	18.3
15	78.05	149	9.63	12.6	285	12.5	81.15	178	11.07	13.8	401	16.2
16	81.95	200	9.28	12.8	272	12.5	82.95	177	11.04	13.5	366	16.6
17	81.70	187	9.46	12.6	276	12.3	81.20	172	11.14	14.1	412	19.3
18	89.65	200	9.58	11.1	285	12.5	83.85	192	11.24	14.1	372	17.2
19	79.90	152	9.49	13.2	275	11.7	67.60	164	10.07	11.9	305	11.8
20	71.15	144	9.55	12.0	292	11.9	64.35	170	9.34	11.0	303	11.6
21	83.05	147	10.30	13.3	326	14.2	66.40	158	9.71	11.9	326	12.9
22	87.25	231	10.32	13.1	332	14.7	79.10	162	10.49	12.9	395	17.0
23	78.65	183	9.90	14.1	324	14.6	81.65	171	11.31	14.1	403	17.2
24	79.95	165	9.34	12.5	290	12.1	79.35	162	10.43	12.6	390	15.9
25	86.65	198	10.07	12.7	293	12.3	78.90	166	11.14	14.0	432	18.4
26	92.05	212	9.81	13.1	304	13.9	80.45	172	11.32	14.3	306	18.7
27	76.80	193	9.80	13.1	288	13.4	83.75	202	10.38	13.4	343	13.8

Table 9.2. Sample Means

	π_1	π_2
x_1	81.98704	76.13333
x_2	175.44444	167.22222
x_3	9.81148	10.49741
x_4	12.91852	12.99630
x_5	305.22222	363.00000
x_6	13.82222	15.66296

Now

$$d_1(x) = 0.10070x_1 + 0.20551x_2 + 75.13581x_3 + 1.69460x_4 + 0.16121x_5 \\ - 15.98724x_6 - 315.81156$$

$$d_2(x) = -0.49307x_1 + 0.28011x_2 + 84.84069x_3 - 1.88664x_4 \\ + 0.22783x_5 - 16.30691x_6 - 351.33860.$$

To verify the efficacy of this plug-in classification rule we now classify the observed sample observations using the proposed criterion. The results are given in Table 9.3.

Table 9.3. Evaluations of the Classification Rule for Sample Observations

Observation	Population π_1 Classified to:	Population π_2 Classified to:
1	π_1	π_2
2	π_1	π_2
3	π_2	π_2
4	π_1	π_2
5	π_1	π_2
6	π_1	π_2
7	π_1	π_2
8	π_1	π_2
9	π_1	π_2
10	π_1	π_2
11	π_2	π_2
12	π_1	π_2
13	π_1	π_2
14	π_1	π_2
15	π_1	π_2
16	π_1	π_2
17	π_1	π_2
18	π_1	π_2
19	π_1	π_2
20	π_1	π_2
21	π_1	π_2
22	π_2	π_2
23	π_1	π_2
24	π_1	π_2
25	π_1	π_2
26	π_1	π_2
27	π_1	π_2

Discriminant Analysis

9.5. CONCLUDING REMARKS

We have limited our discussions mainly to the case of multivariate normal distributions. The cases of nonnormal and discrete distributions are equally important in practice and have been studied by various workers. For multinomial distributions the works of Matusita (1956), Chernoff (1956), Cochran and Hopkins (1961), Bunke (1966), and Glick (1969) are worth mentioning. For multivariate Bernouilli distributions we refer to Bahadur (1961), Solomon (1960, 1961), Hills (1966), Martin and Bradly (1972), Cooper (1963, 1965), Bhattacharya and Das Gupta (1964), and Anderson (1972). The works of Kendall (1966) and Marshall and Olkin (1968) are equally important for related results in connection with discrete distributions. The reader is also referred to the book edited by Cacoullos (1973) for an up-to-date account of research work in the area of discriminant analysis. Rukhin (1991) has shown that the natural estimator of the discriminant coefficient vector Γ is admissible under quadratic loss function when $\Sigma = \sigma^2 I$. Khatri and Bhavsar (1990) have treated the problem of the estimation of discriminant coefficients in the family of complex elliptically symmetric distributions. They have derived the asymptotic confidence bounds of the discriminatory values for the linear Fisher's discrimination for the future complex observation from this family.

9.6. DISCRIMINANT ANALYSIS AND CLUSTER ANALYSIS

Cluster analysis is distinct from the discriminant analysis. The discriminant analysis pertains to a known number of groups and the objective is to assign new observations to one of these groups. In cluster analysis no assumption is made about the number of groups (clusters) or their structure and it involves the search through the observations that are similar enough to each other to be identified as part of a common cluster. The clusters consist of observations that are close together and that the clusters themselves are clearly separated. If each observation is associated with one and only one cluster, the clusters constitute a partition of the data which is useful for statistical purposes.

The cluster analysis involves a search of the data of observations that are similar enough to each other to be identified as part of a common cluster. Better results are achieved by taking into account the cluster structure before attempting to estimate any of the relationship that may be present. It is not easy to find the cluster structure except in small problems.

Numerous algorithms have evolved for finding clusters in a reasonably efficient way. This development of algorithms has, for the most part, come out of applications-oriented disciplines, such as biology, psychology, medicine, education, business, etc.

Let X^α, $\alpha = 1, \ldots, N$ be a random sample from a population characterized by a probability distribution P. A clustering technique produces some clusters in the sample. A theoretical model generates some clusters in the population with the distribution P. We evaluate the technique by asking how well the sample clusters agree with the population clusters. For further study we refer to "Discriminant Analysis and Clustering", by the Panel on Discriminant Analysis, Classification and Clustering, published in Statistical Sciences, 1989, 4, 34–69, and the references included therein.

EXERCISES

1 Let π_1, π_2 be two p-variate normal populations with means μ_1, μ_2 and the same covariance matrix Σ. Let $X = (X_1, \ldots, X_p)'$ be a random vector distributed, according to π_1 or π_2 and let $b = (b_1, \ldots, b_p)'$ be a real vector. Show that

$$\frac{[E_1(b'X) - E_2(b'X)]^2}{\text{var}(b'X)}$$

is maximum for all choices of b whenever $b = \Sigma^{-1}(\mu_1 - \mu_2)$. [$E_i(b'X)$ is the expected value of $b'X$ under π_i.)

2 Let $x_\alpha^{(i)}$, $\alpha = 1, \ldots, N_i$, $i = 1, 2$. Define dummy variables $y_\alpha^{(i)}$

$$y_\alpha^{(i)} = \frac{N_i}{N_1 + N_2}, \quad \alpha = 1, \ldots, N_i, \quad i = 1, 2.$$

Find the regression on the variables $x_\alpha^{(i)}$ by choosing $b = (b_1, \ldots, b_p)'$ to minimize

$$\sum_{i=1}^{2} \sum_{\alpha=1}^{N_i} (y_\alpha^{(i)} - b'(x_\alpha^{(i)} - \bar{x}))^2,$$

where

$$\bar{x} = \frac{N_1 \bar{x}^{(1)} + N_2 \bar{x}^{(2)}}{N_1 + N_2}, \quad N_i \bar{x}^{(i)} = \sum_{\alpha=1}^{N_i} x_\alpha^{(i)}.$$

Show that the minimizing b is proportional to $s^{-1}(\bar{x}^{(1)} - \bar{x}^{(2)})$, where

$$(N_1 + N_2 - 2)s = \sum_{i=1}^{2} \sum_{\alpha=1}^{N_i} (x_\alpha^{(i)} - \bar{x}^{(i)})(x_\alpha^{(i)} - \bar{x}^{(i)})'.$$

3 (a) For discriminating between two p-dimensional normal distributions with unknown means μ_1, μ_2 and the same unknown covariance matrix Σ, show

Discriminant Analysis

that the sample discriminant function v can be obtained from

$$b'\left(x - \frac{1}{2}(\bar{x}^{(1)} + \bar{x}^{(2)})\right)$$

by finding b to maximize the ratio

$$\frac{[b'(\bar{x}^{(1)} - \bar{x}^{(2)})]^2}{(b'sb)}$$

where $\bar{x}^{(i)}$, s are given in (9.27).

(b) In the analysis of variance terminology (a) amounts to finding b to maximize the ratio of the between-population sum of squares to the within-population sum of squares. With this terminology show that the sample discriminant function obtained by finding b to maximize the ratio of the between-population sum of squares to the total sum of squares is proportional to v.

4. For discriminating between two-p-variate normal populations with known mean vectors μ_1, μ_2 and the same known positive definite covariance matrix Σ show that the linear discriminant function u is also good for any p-variate normal population with mean $a_1\mu_1 + a_2\mu_2$, where $a_1 + a_2 = 1$, and the same covariance Σ.

5. Prove Theorems 9.2.2 and 9.2.3.

6. Consider the problem of classifying an individual into one of two populations π_1, π_2 with probability density functions f_1, f_2, respectively.
   (a) Show that if $P(f_2(x) = 0|\pi_1) = 0$, $P(f_1(x) = 0|\pi_2) = 0$, then every Bayes classification rule is admissible.
   (b) Show that if $P(f_1(x)/f_2(x) = k|\pi_i) = 0$, $i = 1, 2$, $0 \leq k \leq \infty$, then every admissible classification rule is a Bayes classification rule.

7. Let $v = v(x)$ be defined as in (9.28). Show that for testing the equality of mean vectors of two p-variate normal populations with the same positive definite covariance matrix Σ, Hotelling's T^2-test on the basis of sample observations $x_\alpha^{(1)}$, $\alpha = 1, \ldots, N_1$, from the first population and $x_\alpha^{(2)}$, $\alpha = 1, \ldots, N_2$, from the second population, is proportional to $v(\bar{x}^{(1)})$ and $v(\bar{x}^{(2)})$.

8. Consider the problem of classifying an individual with observation $(x_1, \ldots, x_p)'$ between two p-dimensional normal populations with the same mean vector 0 and positive definite covariance matrices Σ_1, Σ_2.
   (a) Given $\Sigma_1 = \sigma_1^2 I$, $\Sigma_2 = \sigma_2^2 I$, where σ_1^2, σ_2^2 are known positive constants and $C(2|1) = C(1|2)$, find the minimax classification rule.

(b) (i) Let

$$\Sigma_1 = \begin{pmatrix} 1 & \rho_1 & \cdots & \rho_1 \\ \rho_1 & 1 & \cdots & \rho_1 \\ \vdots & \vdots & & \vdots \\ \rho_1 & \rho_1 & \cdots & 1 \end{pmatrix}, \quad \Sigma_2 = \sigma^2 \begin{pmatrix} 1 & \rho_2 & \cdots & \rho_2 \\ \rho_2 & 1 & \cdots & \rho_2 \\ \vdots & \vdots & & \vdots \\ \rho_2 & \rho_2 & \cdots & 1 \end{pmatrix}.$$

Show that the likelihood ratio classification rule leads to $aZ_1 - bZ_2 = C$ as the boundary separating the regions R_1, R_2 where

$$Z_1 = x'x, \quad Z_2 = (\Sigma_1^p x_i)^2$$

$$a = (1 - \rho_1)^{-1} - (\sigma^2(1 - \rho_2))^{-1},$$

$$b = \frac{\rho_1}{(1-\rho_1)(1+(p-1)\rho_1)} - \frac{\rho_2}{(1-\rho_2)\sigma^2(1+(p-1)\rho_2)}.$$

(ii) (Bartlett and Please, 1963). Suppose that $\rho_1 = \rho_2$ in (a). Then the classification rule reduces to: Classify x to π_1 if $u \geq c'$ and to π_2 if $u < c'$ where c' is a constant and

$$U = Z_1 - \frac{\rho}{1+(p-1)\rho} Z_2.$$

Show that the corresponding random variable U has a $((1-\rho)\sigma_i^2)\chi^2$ distribution with p degrees of freedom where $\sigma_i^2 = 1$ if X comes from π_1 and $\sigma_i^2 = \sigma^2$ is X comes from π_2.

9 Show that the likelihood ratio tests for cases C and D in Section 9.3 are uniformly most powerful similar among all tests whose power depends only on δ_1^2 and δ_2^2.

10 Giri (1973) Let $\xi = (\xi_1, \ldots, \xi_p)'$, $\eta = (\eta_1, \ldots, \eta_p)'$ be two p-dimensional independent complex Gaussian random vectors with complex means $E(\xi) = \alpha$, $E(\eta) = \beta$ and with the same Hermitian positive definite covariance matrix Σ.

(a) Find the likelihood ratio rule for classifying an observation into one of these two populations.

(b) Let ξ be distributed as a p-dimensional complex Gaussian random vector with mean $E(\xi) = \alpha$ and Hermitian positive definite covariance matrix Σ. Let $\Gamma = \Sigma^{-1}\alpha$. Find the likelihood ratio tests for problems analogous to B, C, and D in Section 9.3.

REFERENCES

Anderson, J. A. (1969). Discrimintion between k populations with constraints on the probabilities of misclassification. *J.R. Statist. Soc.* 31:123–139.

Anderson, J. A. (1972). Separate sample logistic discrimination. *Biometrika* 59:19–36.

Anderson, T. W. (1951). Classification by multivariate analysis. *Psychometrika* 16:631–650.

Anderson, T. W. (1958). *An Introduction to Multivariate Statistical Analysis*. New York: Wiley.

Bahadur, R. R. (1961). On classification based on response to N *dichotomus items*, In: Solomon, H. ed. *Studies in item analysis and prediction*. Stanford, California: Stanford Univ. Press. pp. 177–186.

Bahadur, R. R. and Anderson, T. W. (1962). Classification into two multivariate normal distributions with different covariance matrices. *Ann. Math. Statist.* 33:420–431.

Banerjee, K. S. and Marcus, L. F. (1965). Bounds in minimax classification procedures. *Biometrika* 52:153–154.

Bartlett, M. S. and Please, N. W. (1963). Discrimination in the case of zero mean differences. *Biometrika* 50:17–21.

Bhattacharya, P. K. and Das Gupta, S. (1964). Classification into exponential populations. *Sankhya* A26:17–24.

Blackwell, D. and Girshik, M. A. (1954). *Theory of Games and Statistical Decisions*. New York: Wiley.

Bowker, A. H. (1960). A representation of Hotelling's T^2- and Anderson's classification statistic, "Contribution to Probability and Statistics" (Hotelling's vol.). Stanford Univ. Press, Stanford, California.

Brown, G. R. (1947). Discriminant functions. *Ann. Math. Statist.* 18:514–528.

Bunke, O. (1964). Uber optimale verfahren der discriminazanalyse. *Abl. Deutsch. Akad. Wiss. Klasse. Math. Phys. Tech.* 4:35–41.

Bunke, O. (1966). Nichparametrische klassifikations verfahren für qualitative und quantitative Beobnachtunger. *Berlin Math. Naturwissensch. Reihe* 15:15–18.

Cacoullos, T. (1965). Comparing Mahalanobis distances, I and II. *Sankhya* A27:1–22, 27–32.

Cacoullos, T. (1973). *Discriminant Analysis and Applications.* New York: Academic Press.

Cavalli, L. L. (1945). Alumi problemi dela analyse biometrica di popolazioni naturali. *Mem. Inst. Indrobiol.* 2:301–323.

Chernoff, H. (1956). A Classification Problem. Tech. rep. no 33, Stanford Univ., Stanford, California.

Cochran, W. G. (1968). Commentary of estimation of error rates in discriminant analysis. *Technometrics* 10:204–210.

Cochran, W. G. and Hopkins, C. E. (1961). Some classification problems with multivariate quantitative data. *Biometrics* 17:10–32.

Cooper, D. W. (1963). Statistical classifications with quadratic forms. *Biometrika* 50:439–448.

Cooper, D. W. (1965). Quadratic discriminant function in pattern recognition. *IEEE Trans. Informat. II* 11:313–315.

Das Gupta, S. (1962). On the optimum properties of some classification rules. *Ann. Math. Statist.* 33:1504.

Das Gupta, S. (1973), Classification procedures, a review. In: Cacoullos, T. ed. *Discriminant Analysis and Applications.* New York: Academic Press.

Ferguson, T. S. (1967). *Mathematical Statistics.* Academic Press, New York.

Fisher, R. A. (1936). Use of multiple measurements in Taxonomic problems. *Ann. Eug.* 7:179–184.

Fisher, R. A. (1938). The statistical utilization of multiple measurements. *Ann. Eug.* 8:376–386.

Giri, N. (1964). On the likelihood ratio test of a normal multivariate testing problem. *Ann. Math. Statist.* 35:181–189.

Giri, N. (1965). On the likelihood ratio test of a normal multivariate testing problem, II. *Ann. Math. Statist.* 36:1061–1065.

Giri, N. (1973). On discriminant decision functions in complex Gaussian distributions. In: Behara, M., Krickeberg, K. and Wolfowits, J. *Probability and Information Theory.* Berlin and New York: Springer Verlag No. 296, pp. 139–148.

Glick, N. (1969). Estimating Unconditional Probabilities of Correct Classification. Stanford Univ., Dept. Statist. Tech. Rep. No. 3.

Han, Chien Pai (1968). A note on discrimination in the case of unequal covariance matrices. *Biometrika* 55:586–587.

Han, Chien Pai (1969). Distribution of discriminant function when covariance matrices are proportional. *Ann. Math. Statist.* 40:979–985.

Han, Chien Pai (1970). Distribution of discriminant function in Circular models. *Ann. Inst. Statist. Math.* 22:117–125.

Hills, M. (1966). Allocation rules and their error rates. *J.R. Statist. Soc. Ser. B* 28:1–31.

Hodges, J. L. (1950). Survey of discriminant analysis, USAF School of Aviation Medicine, rep. no. 1, Randolph field, Texas.

Hotelling, H. (1931). The generalization of Student's ratio. *Ann. Math. Statist.* 2:360–378.

Isaacson, S. L. (1951). On the theory of unbiased tests of simple statistical hypotheses specifying the values of two or more parameters. *Ann. Math. Statist.* 22:217–234.

Kabe, D. G. (1963). Some results on the distribution of two random matrices used in classification procedures. *Ann. Math. Statist.* 34:181–185.

Kendall, M. G. (1966). *Discrimination and classification*. Proc. Int. Symp. Multv. Anal. (P. R. Krishnaiah, ed.), pp. 165–185. Academic Press. New York.

Kiefer, J., and Schwartz, R. (1965). Admissible bayes character of T^2-. R^2-, and fully invariant tests for classical multivariate normal problem. *Ann. Math. Statist.* 36:747–770.

Khatri, C. G. and Bhavsar, C. D. (1990). Some asymptotic inferential problems connected with complex elliptical distribution. *Jour. Multi. Anal.*, 35:66–85.

Kshirsagar, A. M. (1972). *Multivariate Analysis*. Dekker, New York.

Kudo, A. (1959). The classification problem viewed as a two decision problem, I. *Mem. Fac. Sci. Kyushu Univ.* A13:96–125.

Kudo, A. (1960). The classification problem viewed as a two decision problem, II. *Mem. Fac. Sci. Kyushu Univ.* A14:63–83.

Lachenbruch, P. A. and Mickey, M. R. (1968). Estimation of error rates in discriminant analysis. *Technometries* 10:1–11.

Lehmann, E. (1959). *Testing Statistical Hypotheses*. Wiley, New York.

Mahalanobis, P. C. (1927). Analysis of race mixture in Bengal. *J. Proc. Asiatic Soc. Bengal* 23:3.

Mahalanobis, P. C. (1930). On tests and measurements of group divergence. *Proc. Asiatic Soc. Bengal* 26:541–589.

Mahalanobis, P. C. (1936). On the generalized distance in statistics. *Proc. Nat. Inst. Sci. India* 2:49–55.

Marshall, A. W. and Olkin, I. (1968). A general approach to some screening and classification problems. *J. R. Statist. Soc.* B30:407–435.

Martin, D. C. and Bradly, R. A. (1972). Probability models, estimation and classification for multivariate dichotomous populations. *Biometrika* 28:203–222.

Matusita, K. (1956). Decision rules based on the distance for the classification problem. *Ann. Inst. Statist. Math.* 8:67–77.

Morant, G. M. (1928). A preliminary classification of European races based on cranial measurements. *Biometrika*, 20:301–375.

Neyman, J., and Pearson, E. S. (1933). The problem of most efficient tests of statistical hypotheses. *Phil. Trans. R. Soc.* 231.

Neyman, J., and Pearson, E. S. (1936). Contribution to the theory of statistical hypotheses, I. *Statist. Res. Memo. I* 1–37.

Nishida, N. (1971). A note on the admissible tests and classification in multivariate analysis. *Hiroshima Math. J.* 1:427–434.

Okamoto, M. (1963). An asymptotic expansion for the distribution of linear discriminant function. *Ann. Math. Statist.* 34:1286–1301, correction vol. 39:1358–1359.

Pearson, K. (1926). On the coefficient of racial likeness. *Biometrika* 18:105–117.

Penrose, L. S. (1947). Some notes on discrimination. *Ann. Eug.* 13:228–237.

Quenouille, M. (1956). Notes on bias in estimation. *Biometrika* 43:353–360.

Rao, C. R. (1946). Tests with discriminant functions in multivariate analysis. *Sankhya* 7:407–413.

Rao, C. R. (1947a). The problem of classification and distance between two populations. *Nature* (London) 159:30–31.

Rao, C. R. (1947b). Statistical criterion to determine the group to which an individual belongs. *Nature* (London) 160:835–836.

Rao, C. R. (1948). The utilization of multiple measurements in problem of biological classification. *J.R. Statist. Soc. B* 10:159–203.

Rao, C. R. (1949a). On the distance between two populations. *Sankhya* 9: 246–248.

Rao, C. R. (1949b). On some problems arising out of discrimination with multiple characters. *Sankhya* 9:343–366.

Rao, C. R. (1950). Statistical inference applied to classification problems. *Sankhya* 10:229–256.

Rao, C. R. (1952). *Advanced Statistical Methods in Biometric Research*. New York: Wiley.

Rao, C. R. (1954). A general theory of discrimination when the information about alternative population is based on samples. *Ann. Math. Statist.* 25:651–670.

Rao, M. M. (1963). Discriminant analysis. *Ann. Inst. Statist. Math.* 15:15–24.

Rao, C. R. (1965). *Linear Statistical Inference and its Applications*. New York: Wiley.

Rukhin, A. L. (1991). Admissible estimators of discriminant coefficient. *Statistics and Decisions* 9:285–295.

Schucany, W. R. Gray, H. L. and Owen, D. B. (1971). On bias reduction in estimation. *Biometrika* 43:353.

Sinha, B. K. and Giri, N. (1975). On the distribution of a random matrix. *Commun. Statist.* 4:1057–1063.

Sinha, B. K. and Giri, N. (1976). On the optimality and non-optimality of some multivariate normal test procedures. *Sankhya* 38:244–249.

Sitgreaves, R. (1952). On the distribution of two random matrices used in classification procedures. *Ann. Math. Statist.* 23:263–270.

Smith, C. A. B. (1947). Some examples of discrimination. *Ann. Eug.* 13: 272–282.

Solomon, H. (1960). Classification procedures based on dichotomous response vectors, "Contributions to Probability and Statistics" (Hotelling's. volume), pp. 414–423. Stanford Univ. Press, Stanford, California.

Solomon, H. (1961). Classification procedures based on dichotomous response vectors. in "Studies in Item Analysis and Predictions", (H. Solomon, ed.), pp. 177–186: Stanford Univ. Press, Stanford, California.

Srivastava, M. S. (1964). Optimum procedures for Classification and Related Topics. Tech. rep. no. 11, Dept. Statist., Stanford Univ.

Tildesley, M. L. (1921). A first study of the Burmese skull. *Biometrika* 13: 247–251.

Tukey, J. W. (1958). Bias and confidence in not quite large samples. *Ann. Math. Statist.* 20:618.

Von Mises, R. (1945). On the classification of observation data into distinct groups. *Ann. Math. Statist.* 16:68–73.

Wald, A. (1944). On a statistical problem arising in the classification of an individual into one of two groups. *Ann. Math. Statist.* 15:145–162.

Wald, A. (1950). *Statistical Decision Function*. New York: Wiley.

Wald, A., and Wolfowitz, J. (1950). Characterization of minimum complete class of decision function when the number of decisions is finite. Proc. Berkeley Symp. Prob. Statist., 2nd, California.

Welch, B. L. (1939). Note on discriminant functions. *Biometrika* 31:218–220.

10
Principal Components

10.0. INTRODUCTION

In this and following chapters we will deal with covariance structures of multivariate distributions. Principal components, canonical correlations, and factor models are three interrelated concepts dealing with covariance structure. All these concepts aim at reducing the dimension of observable random variables. The principal components will be treated in this chapter. Canonical analysis and Factor analysis will be treated in Chapters 11 and 12 respectively. Though these concepts will be developed for any multivariate population, statistical inferences will be made under the assumption of normality. Proper references will be given for elliptical distributions.

10.1. PRINCIPAL COMPONENTS

Let $X = (X_1, \ldots, X_p)'$ be a random vector with

$$E(X) = \mu, \quad \text{cov}(X) = \Sigma = (\sigma_{ij}),$$

where μ is a real p-vector and Σ is a real positive semidefinite matrix. In multivariate analysis the dimension of X often causes problems in obtaining suitable statistical techniques to analyze a set of repeated observations (data) on X. For this reason it is natural to look for methods for rearranging the data so that with as little loss of information as possible, the dimension of the problem is considerably reduced. We have seen one such attempt in connection with discriminant analysis in Chapter 9.

This notion is motivated by the fact that in early stages of research interest was usually focused on those variables that tend to exhibit greatest variation from observation to observation. Since variables which do not change much from observation to observation can be treated as constants, by discarding low variance variables and centering attention on high variance variables, one can more conveniently study the problem of interest in a subspace of lower dimension. No doubt some information on the relationship among variables is lost by such a method; nevertheless, in many practical situations there is much more to gain than to lose by this approach.

The principal component approach was first introduced by Karl Pearson (1901) for nonstochastic variables. Hotelling (1933) generalized this concepts to random vectors. Principal components of X are normalized linear combinations of the components of X which have special properties in terms of variances. For example, the first principal components of X is the normalized linear combination

$$Z_1 = L'X, \quad L = (l_1, \ldots, l_p)' \in E^p,$$

where L is chosen so that $\text{var}(L'X)$ is maximum with respect of L. Obviously each weight l_i is a measure of the importance to be placed on the component X_i. We require the condition $L'L = 1$ in order to obtain a unique solution for the principal components. We shall assume that components of X are measured in the same units; otherwise the requirement $L'L = 1$ is not a sensible one. It will be seen that estimates of principal components are sensitive to units used in the analysis so that different sets of weights are obtained for different sets of units. Sometimes the sample correlation matrix is used instead of the sample covariance matrix to estimate these weights, thereby avoiding the problem of units, since the principal components are then invariant to changes in units of measurement. The use of the correlation matrix amounts to standardizing the variables to unit sample variance. However, since the new variables are not really standardized relative to the population, there is then introduced the problem of interpreting what has actually been computed. In practice such a technique is not recommended unless the sample size is large.

The second principal component is a linear combination that has maximum variance among all normalized linear combinations uncorrelated with Z_1 and so on up to the pth principal component of X. The original vector X can thus be transformed to the vector of its principal components by means of a rotation of the coordinate axes that has inherent statistical properties. The choosing of such a type of coordinate system is to be contrasted with previously treated problems where the coordinate system in which the original data are expressed is irrelevant.

Principal Components

The weights in the principal components associated with the random vector X are exactly the normalized characteristic vectors of the covariance matrix Σ of X, whereas the characteristic roots of Σ are the variances of the principal components, the largest root being the variance of the first principal component.

It may be cautioned that sample observations should not be indiscriminately subjected to principal component analysis merely to obtain fewer variables with which to work. Rather, principal component analysis should be used only if it complements the overall objective. For example, in problems in which correlation rather than variance is of primary interest or in which there are likely to be important nonlinear functions of observations that are of interest, most of the information about such relationships may be lost if all but the first few principal components are dropped.

10.2. POPULATION PRINCIPAL COMPONENTS

Let $X = (X_1, \ldots, X_p)'$ be a p-variate random vector with $E(X) = \mu$ and known covariance matrix Σ. We shall consider cases in which Σ is a positive semidefinite matrix or cases in which Σ has multiple roots. Since we shall only be concerned with variances and covariances of X we shall assume that $\mu = 0$. The first principal component of X is the normalized linear combination (say) $Z_1 = \alpha' X$, $\alpha = (\alpha_1, \ldots, \alpha_p)' \in E^p$ with $\alpha'\alpha = 1$ such that

$$\text{var}(\alpha' X) = \max_L \text{var}(L' X) \tag{10.1}$$

for all $L \in E^p$ satisfying $L'L = 1$. Now

$$\text{var}(L' X) = L' \Sigma L.$$

Thus to find the first principal component $\alpha' X$ we need to find the α that maximizes $L'\Sigma L$ for all choices of $L \in E^p$, subject to the restriction that $L'L = 1$. Using the Lagrange multiplier λ, we need to find the α that maximizes

$$\phi_1(L) = L'\Sigma L - \lambda(L'L - 1) \tag{10.2}$$

for all choices of L satisfying $L'L = 1$. Since $L'\Sigma L$ and $L'L$ have derivatives everywhere in a region containing $L'L = 1$, we conclude that the vector α which maximizes ϕ_1 must satisfy

$$2\Sigma\alpha - 2\lambda\alpha = 0, \tag{10.3}$$

or

$$(\Sigma - \lambda I)\alpha = 0. \tag{10.4}$$

Since $\alpha \neq 0$ (as a consequence of $\alpha'\alpha = 1$), Eq. (10.4) has a solution if

$$\det(\Sigma - \lambda I) = 0. \tag{10.5}$$

That is, λ is a characteristic root of Σ and α is the corresponding characteristic vector. Since Σ is of dimension $p \times p$, there are p values of λ which satisfy (10.5). Let

$$\lambda_1 \geq \lambda_2 \cdots \geq \lambda_p \tag{10.6}$$

denote the ordered characteristic roots of Σ and let

$$\alpha_1 = (\alpha_{11}, \ldots, \alpha_{1p})', \ldots, \alpha_p = (\alpha_{p1}, \ldots, \alpha_{pp})' \tag{10.7}$$

denote the corresponding characteristic vectors of Σ. Note that since Σ is positive semidefinite some of the characteristic roots may be zeros; in addition, some of the roots may have multiplicities greater than unity. From (10.4)

$$\alpha'\Sigma\alpha = \lambda\alpha'\alpha = \lambda. \tag{10.8}$$

Thus we conclude that if α with $\alpha'\alpha = 1$ satisfies (10.4), then

$$\text{var}(\alpha'X) = \alpha'\Sigma\alpha = \lambda, \tag{10.9}$$

where λ is the characteristic root of Σ corresponding to α. Thus to maximize $\text{var}(\alpha'X)$ we need to choose $\lambda = \lambda_1$, the largest characteristic root of Σ, and $\alpha = \alpha_1$ the characteristic vector of Σ corresponding to λ_1. If the rank of $\Sigma - \lambda_1 I$ is $p - 1$, then there is only one solution to

$$(\Sigma - \lambda_1 I)\alpha_1 = 0 \quad \text{with} \quad \alpha_1'\alpha_1 = 1.$$

Definition 10.2.1. *First principal component. The normalized linear function $\alpha_1'X = \Sigma_{i=1}^p \alpha_{1i}X_i$, where α_1 is the normalized characteristic vector of Σ corresponding to its largest characteristic root λ_1, is called the first principal component of X.*

We have assumed no distributional form for X. If X has a p-variate normal distribution with positive definite covariance matrix Σ, then the surfaces of constant probability density are concentric ellipsoids and $Z_1 = \alpha_1'X$ represents the major principal axis of these ellipsoids. In general under the assumption of normality of X, the principal components will represent a rotation of coordinate axes of its components to the principal axes of these ellipsoids. If there are multiple roots, the axes are not uniquely defined.

The second principal component is the normalized linear function $\alpha'X$ having maximum variance among all normalized linear functions $L'X$ that are uncorrelated with Z_1. If any normalized linear function $L'X$ is uncorrelated

Principal Components

with Z_1, then

$$E(L'XZ_1) = E(L'XZ'_1) = E(L'XX'\alpha_1)$$
$$= L'\Sigma\alpha_1 = L'\lambda_1\alpha_1 = \lambda_1 L'\alpha_1 = 0. \quad (10.10)$$

This implies that the vectors L and α_1 are orthogonal. We now want to find a linear combination $\alpha'X$ that has maximum variance among all normalized linear combinations $L'X, L \in E^p$, which are uncorrelated with Z_1. Using Lagrange multipliers λ, ν we want to find the α that maximizes

$$\phi_2(L) = L'\Sigma L - \lambda(L'L - 1) - 2\nu(L'\Sigma\alpha_1). \quad (10.11)$$

Since

$$\frac{\partial\phi_2}{\partial L} = 2\Sigma L - 2\lambda L - 2\nu\Sigma\alpha_1, \quad (10.12)$$

the maximizing α must satisfy

$$\alpha'_1\Sigma\alpha - \lambda\alpha'_1\alpha - \nu\alpha'_1\Sigma\alpha_1 = 0. \quad (10.13)$$

Since from (10.10) $\alpha'_1\Sigma\alpha = 0$ and $\alpha'_1\Sigma\alpha_1 = \lambda_1$, we get from (10.13),

$$\nu\lambda_1 = 0. \quad (10.14)$$

Since $\lambda_1 \neq 0$, we conclude that $\nu = 0$, and therefore from (10.12) we conclude that λ and α must satisfy (10.3) and (10.4). Thus it follows that the coefficients of the second principal component of X are the elements of the normalized characteristic vector α_2 of Σ, corresponding to its second largest characteristic root λ_2. The second principal component of Σ is

$$Z_2 = \alpha'_2 X.$$

This is continued to the rth ($r < p$) principal component Z_r. For the $(r+1)$th principal component we want to find a linear combination $\alpha'X$ that has maximum variance among all normalized linear combinations $L'X, L \in E^p$, which are uncorrelated with $Z_1, \ldots, Z_r$. So, with $Z_i = \alpha'_i X$,

$$\text{cov}(L'X, Z_i) = L'\Sigma\alpha_i = L'\lambda_i\alpha_i = \lambda_i L'\alpha_i = 0, \ i = 1, \ldots, r. \quad (10.15)$$

To find α we need to maximize

$$\phi_{r+1}(L) = L'\Sigma L - \lambda(L'L - 1) - 2\sum_{i=1}^{r}\nu_i L'\Sigma\alpha_i, \quad (10.16)$$

where $\lambda, \nu_1, \ldots, \nu_r$ are Lagrange multipliers. Setting the vector of partial derivatives

$$\frac{\partial \phi_{r+1}}{\partial L} = 0,$$

the vector α that maximizes $\phi_{r+1}(L)$ is given by

$$2\Sigma\alpha - 2\lambda\alpha - 2\sum_{i=1}^{r}\nu_i\Sigma\alpha_i = 0. \qquad (10.17)$$

Since from this

$$\alpha_i'\Sigma\alpha - \lambda\alpha_i'\alpha - \sum_{i\neq 1}\nu_i\lambda_i = 0 \qquad (10.18)$$

and $\alpha_i'\Sigma\alpha_i = \lambda_i$, we conclude from (10.17) and (10.18) that if $\lambda_i \neq 0$,

$$\nu_i\lambda_i = 0, \qquad (10.19)$$

that is, $\nu_i = 0$. If $\lambda_i = 0, \Sigma\alpha_i = \lambda_i\alpha_i = 0$, so that the factor $L'\Sigma\alpha_i$ in (10.16) vanishes. This argument holds for $i = 1, \ldots, r$, so we conclude from (10.17) that the maximizing α [satisfying (10.4)] is the characteristic vector of Σ, orthogonal to $\alpha_i, i = 1, \ldots, r$, corresponding to its characteristic root λ. If $\lambda_{r+1} \neq 0$, taking $\lambda = \lambda_{r+1}$ and α for the normalized characteristic vector α_{r+1}, corresponding to the $(r+1)$th largest characteristic root λ_{r+1}, the $(r+1)$th principal component is given by

$$Z_{r+1} = \alpha_{r+1}'X.$$

However, if $\lambda_{r+1} = 0$ and some $\lambda_i = 0$ for $1 \leq i \leq r$, then

$$\alpha_i'\Sigma\alpha_{r+1} = 0$$

does not imply that $\alpha_i'\alpha_{r+1} = 0$. In such cases replacing α_{r+1} by a linear combination of α_{r+1} and the α_i for which $\lambda_i = 0$, we can make the new α_{r+1} orthogonal to all $\alpha_i, i = 1, \ldots, r$. We continue in this way to the mth step such that at the $(m+1)$th step we cannot find a normalized vector α such that $\alpha'X$ is uncorrelated with all $Z_1, \ldots, Z_m$. Since Σ is of dimension $p \times p$, obviously $m = p$ or $m < p$. We now show that $m = p$ is the only solution. Assume $m < p$. There exist $p - m$ normalized orthogonal vectors $\beta_{m+1}, \ldots, \beta_p$ such that

$$\alpha_i'\beta_j = 0, i = 1, \ldots, m \; j = m+1, \ldots, p \qquad (10.20)$$

Write $B = (\beta_{m+1}, \ldots, \beta_p)$. Consider a root of $\det(B'\Sigma B - \lambda I) = 0$ and the corresponding $\beta = (\beta_{m+1}, \ldots, \beta_p)'$ satisfying

$$(B'\Sigma B - \lambda I)\beta = 0. \qquad (10.21)$$

Principal Components

Since

$$\alpha_i' \Sigma B\beta = \lambda_i \alpha_i' \sum_{j=m+1}^{p} \beta_j \beta_j' = \lambda_i \sum_{j=m+1}^{p} \beta_j' \alpha_i' \beta_j = 0,$$

the vector $\Sigma B\beta$ is orthogonal to $\alpha_i, i = 1, \ldots, r$. It is therefore a vector in the space spanned by $\beta_{m+1}, \ldots, \beta_p$, and can be written as

$$\Sigma B\beta = BC,$$

where C is a $(p-m)$-component vector. Now

$$B' \Sigma B\beta = B'BC = C.$$

Thus from (10.21)

$$\lambda \beta = C, \Sigma(B\beta) = \lambda B\beta.$$

Then $(B\beta)'X$ is uncorrelated with $\alpha_j'X, j = 1, \ldots, m$, and it leads to a new α_{m+1}. This contradicts the assumption that $m < p$, and we must have $m = p$. Let

$$A = (\alpha_1, \ldots, \alpha_p), \Lambda = \begin{pmatrix} \lambda_1 & 0 & \cdots & 0 \\ 0 & \lambda_2 & \cdots & 0 \\ \vdots & \vdots & & \vdots \\ 0 & 0 & \cdots & \lambda_p \end{pmatrix} \quad (10.22)$$

where $\lambda_1 \geq \lambda_2 \cdots \geq \lambda_p$ are the ordered characteristic roots of Σ and $\alpha_1, \ldots, \alpha_p$ are the corresponding normalized characteristic vectors. Since $AA' = I$ and $\Sigma A = A\Lambda$ we conclude that $A'\Sigma A = \Lambda$. Thus with $Z = (Z_1, \ldots, Z_p)'$ we have the following theorem.

Theorem 10.2.1. *There exists an orthogonal transformation*

$$Z = A'X$$

such that $\text{cov}(Z) = \Lambda$ *a diagonal matrix with diagonal elements* $\lambda_1 \geq \cdots \geq \lambda_p \geq 0$, *the ordered roots of* $\det(\Sigma - \lambda I) = 0$. *The ith column* α_i *of A satisfies* $(\Sigma - \lambda_i I)\alpha_i = 0$. *The components of Z are uncorrelated and* Z_i, *has maximum variance among all normalized linear combinations uncorrelated with* $Z_1, \ldots, Z_{i-1}$.

The vector Z is called the vector of principal components of X. In the case of multiple roots suppose that

$$\lambda_{r+1} = \cdots = \lambda_{r+m} = \lambda \text{ (say)}.$$

Then $(\Sigma - \lambda I)\alpha_i = 0, i = r+1, \ldots, r+m$. That is, $\alpha_i, i = r+1, \ldots, r+m$, are m linearly independent solutions of $(\Sigma - \lambda I)\alpha = 0$. They are the only linearly independent solutions. To show that there cannot be another linearly independent solution of

$$(\Sigma - \lambda I)\alpha = 0, \tag{10.23}$$

take $\sum_{i=1}^{p} a_i \alpha_i$, where the a_i are scalars. If it is a solution of (10.23), we must have

$$\lambda \sum_{i=1}^{p} a_i \alpha_i = \Sigma \left(\sum_{i=1}^{p} a_i \alpha_i \right) = \sum_{i=1}^{p} a_i \Sigma \alpha_i = \sum_{i=1}^{p} a_i \lambda_i \alpha_i.$$

Since $\lambda a_i = \lambda_i a_i$, we must have $a_i = 0$ unless $i = r+1, \ldots, r+m$. Thus the rank of $(\Sigma - \lambda I)\alpha$ is $p - m$. Obviously if $(\alpha_{r+1}, \ldots, \alpha_{r+m})$ is a solution of (10.23), then for any nonsingular matrix C,

$$(\alpha_{r+1}, \ldots, \alpha_{r+m})C$$

is also a solution of (10.23). But from the condition of orthonormality of $\alpha_{r+1}, \ldots, \alpha_{r+m}$, we easily conclude that C is an orthogonal matrix. Hence we have the following theorem.

Theorem 10.2.2. *If $\lambda_{r+1} = \cdots = \lambda_{r+m} = \lambda$, then $(\Sigma - \lambda I)$ is a matrix of rank $p - m$. Furthermore, the corresponding characteristic vector $(\alpha_{r+1}, \ldots, \alpha_{r+m})$ is uniquely determined except for multiplication from the right by an orthogonal matrix.*

From Theorem 10.2.1 it follows trivially that

$$\det \Sigma = \det \Lambda, \quad \operatorname{tr} \Sigma = \operatorname{tr} \Lambda, \tag{10.24}$$

and we conclude that the generalized variance of the vector X and its principal component vector Z are equal, and the same is true for the sum of variances of components of X and Z. Sometimes $\operatorname{tr} \Sigma$ is called the total system variance.

If the random vector X is distributed as $E_p(\mu, \Sigma)$, the contours of equal probability are ellipsoids and the principal components represent a rotation of the coordinate axes to the principal axes of the ellipsoid.

10.3. SAMPLE PRINCIPAL COMPONENTS

In practice the covariance matrix Σ is usually unknown. So the population principal components will be of no use and the decision as to which principal components have sufficiently small variances to be ignored must be made from sample observations on X. In the preceding discussion on population principal

Principal Components

components we do not need the specific form of the distribution of X. To deal with the problem of an unknown covariance matrix we shall assume that X has a p-variate normal distribution with mean μ and unknown positive definite covariance matrix Σ. In most applications of principal components all the characteristic roots of Σ are different, although the possibility of multiple roots cannot be entirely ruled out. For an interesting case in which Σ has only one root of multiplicity p see Exercise 10.1.

Let $x^\alpha = (x_{\alpha 1}, \ldots, x_{\alpha p})'$, $\alpha = 1, \ldots, N (N > p)$, be a sample of size N from the distribution of the random vector X which is assumed to be normal with unknown mean μ and unknown covariance matrix Σ. Let

$$\bar{x} = \frac{1}{N} \sum_{\alpha=1}^{N} x^\alpha, \quad s = \sum_{\alpha=1}^{N} (x^\alpha - \bar{x})(x^\alpha - \bar{x})'.$$

The maximum likelihood estimate of Σ is s/N and that of μ is $\bar{x}$.

Theorem 10.3.1. *The maximum likelihood estimates of the ordered characteristic roots $\lambda_1, \ldots, \lambda_p$ of Σ and the corresponding normalized characteristic vectors $\alpha_1, \ldots, \alpha_p$ of Σ are, respectively, the ordered characteristic roots $r_1, r_2, \ldots, r_p$, and the characteristic vectors $a_1, \ldots, a_p$ of s/N.*

Proof. Since the characteristic roots of Σ are all different, the normalized characteristic vectors $\alpha_1, \ldots, \alpha_p$ are uniquely determined except for multiplication by ± 1. To remove this arbitrariness we impose the condition that the first nonzero component of each α_i is positive. Now since (μ, Λ, A) is a single-valued function of μ, Σ, by Lemma 5.1.3, the maximum likelihood estimates of $\lambda_1, \ldots, \lambda_p$ are given by the ordered characteristic roots $r_1 > r_2 > \cdots > r_p$ of s/N, and that of α_i, is given by a_i, satisfying

$$(s/N - r_i I)a_i = 0, \quad a_i' a_i = 1, \tag{10.25}$$

with the added restriction that the first nonzero element of a_i is positive. Note that since $\det(\Sigma) \neq 0$ and $N > p$, the characteristic roots of S/N are all different with probability 1. Since $\Sigma = A\Lambda A'$, that is,

$$\Sigma = \sum_{i=1}^{p} \lambda_i \alpha_i \alpha_i', \tag{10.26}$$

we obtain

$$\frac{s}{N} = \sum_{i=1}^{p} r_i a_i a_i'. \qquad (10.27)$$

Obviously replacing a_i by $-a_i$ does not change this expression for s/N. Hence the maximum likelihood estimate of α_i is given by any solution of $(s/N - r_i I)a_i = 0$ with $a_i' a_i = 1$. Q.E.D.

The estimate of the total system variance is given by

$$\operatorname{tr}\left(\frac{s}{N}\right) = \sum_{i=1}^{p} r_i, \qquad (10.28)$$

and is called the total sample variance. The importance of the ith principal component is measured by

$$\frac{r_i}{\sum_{i=1}^{p} r_i} \qquad (10.29)$$

which, when expressed in percentage, will be called the percentage of contribution of the ith principal component to the total sample variance.

If the estimates of the principal components are obtained from the sample correlation matrix

$$r = (r_{ij}), \quad r_{ij} = \frac{s_{ij}}{(s_{ii} s_{jj})^{1/2}}, \qquad (10.30)$$

with $s = (s_{ij})$, then the estimate of the total sample variance will be $p = \operatorname{tr}(r)$.

If the first k principal components explain a large amount of total sample variance, they may be used in future investigations in place of the original vector X. For the computation of characteristic roots and vectors standard programs are now available.

10.4. EXAMPLE

Consider once again Example 9.1.1. We have two groups with 27 observations in each group. For group 1 the sample covariance matrix s/N and the sample

Principal Components

correlation matrix r are given by

$$\frac{s}{27} = \begin{pmatrix} 30.58 & & & & & \\ 108.70 & 781.8 & & & & \\ 0.1107 & -0.7453 & 0.1381 & & & \\ 0.4329 & 0.8684 & 0.1465 & 0.4600 & & \\ -10.72 & -98.22 & 6.302 & 7.992 & 840.2 & \\ -0.2647 & -1.910 & 0.3519 & 0.4715 & 25.76 & 1.761 \end{pmatrix}$$

$$r = \begin{pmatrix} 1.0000 & & & & & \\ 0.7025 & 1.0000 & & & & \\ 0.0539 & -0.0717 & 1.0000 & & & \\ 0.1154 & 0.0458 & 0.5812 & 1.0000 & & \\ -0.0669 & -0.1212 & 0.5851 & 0.4065 & 1.0000 & \\ -0.0361 & -0.0515 & 0.7137 & 0.5238 & 0.6698 & 1.0000 \end{pmatrix}$$

(i) The ordered characteristic roots of $s/27$ along with the corresponding percentages of contribution to the total sample variance (given within parentheses) are

920.312	717.984	15.1837	1.0756	0.3016	0.0533
(55.61%)	(43.39%)	(0.92%)	(0.06%)	(0.02%)	(0%)

(ii) The characteristic vectors a_i (column vectors) of $s/27$ are

1	2	3	4	5	6
0.0851	-0.1122	0.9898	-0.0029	0.0208	0.0100
0.6199	-0.7720	-0.1408	-0.0017	-0.0012	-0.0020
-0.0058	-0.0047	0.0126	0.1772	-0.1074	-0.9782
-0.0062	-0.0080	0.0186	0.3196	-0.9336	0.1607
-0.7797	-0.6253	-0.0040	-0.0332	-0.0007	0.0017
-0.0232	-0.0204	-0.0061	0.9302	0.3413	0.1312

(iii) The ordered characteristic roots of r along with the corresponding percentages of contribution to the total sample variance (given within parentheses) are

2.7578	1.7284	0.5892	0.3700	0.3277	0.2270
(45.96%)	(28.81%)	(9.82%)	(6.16%)	(5.46%)	(3.79%)

(iv) The characteristic vectors a_i (column vectors) of r are

	1	2	3	4	5	6
	0.0093	−0.7012	0.0619	−0.1850	0.5296	0.4356
	0.0628	−0.6933	0.1615	0.1894	−0.5158	−0.4329
	−0.5274	−0.0403	−0.0713	−0.6615	0.1488	0.5454
	−0.4436	−0.1455	−0.7751	0.4236	0.0237	0.0364
	−0.4861	0.0695	0.5602	0.5342	0.3616	−0.1700
	−0.5336	0.0035	0.2245	−0.1660	−0.5478	0.5807

For group 2 the sample covariance matrix and the sample correlation matrix are given by

$$\frac{s}{27} = \begin{pmatrix} 47.05 & & & & & \\ 35.21 & 214.3 & & & & \\ 3.831 & 2.719 & 0.3976 & & & \\ 5.838 & 4.355 & 0.6042 & 1.053 & & \\ 191.6 & 14.69 & 17.49 & 28.58 & 1598 & \\ 13.76 & 2.656 & 1.308 & 2.076 & 76.33 & 5.702 \end{pmatrix}$$

$$r = \begin{pmatrix} 1.0000 & & & & & \\ 0.3506 & 1.0000 & & & & \\ 0.8857 & 0.2945 & 1.0000 & & & \\ 0.8295 & 0.2899 & 0.9339 & 1.0000 & & \\ 0.7007 & 0.0252 & 0.6960 & 0.6987 & 1.0000 & \\ 0.8155 & 0.0761 & 0.8686 & 0.8474 & 0.8019 & 1.0000 \end{pmatrix}$$

(i) The ordered characteristic roots of $s/27$ along with the corresponding percentages of contribution to the total variance (given in parentheses) are

1617.46	219.829	19.0293	1.2743	0.2262	0.0283
(87.06%)	(11.83%)	(1.02%)	(0.07%)	(0.02%)	(0%)

(ii) The characteristic vectors a_i (column vectors) of $s/27$ are

	1	2	3	4	5	6
	0.1217	0.1641	0.9476	0.2412	0.0396	−0.0236
	0.0136	0.9855	−0.1672	−0.0225	0.0117	−0.0002
	0.0111	0.0124	0.0692	−0.1416	−0.3243	0.9326
	0.0181	0.0196	0.0924	−0.2749	−0.8873	−0.3576
	0.9911	−0.0347	−0.1268	0.0218	−0.0010	0.0011
	0.0480	0.0104	0.2114	−0.9194	0.3253	−0.0429

Principal Components

(iii) The ordered characteristic roots of r along with the corresponding percentages of contribution to the total sample variance are

4.3060	1.0439	0.3147	0.1685	0.1129	0.0529
(71.77%)	(17.40%)	(5.24%)	(2.81%)	(1.88%)	(0.90%)

(iv) The characteristic vectors a_i (column vectors) of r are

1	2	3	4	5	6
0.4477	0.1098	0.0619	−0.8278	−0.1882	0.2508
0.1418	0.9174	−0.3001	0.1309	0.1752	−0.0174
0.4624	0.0481	0.3534	0.0791	−0.1589	−0.7922
0.4543	0.0414	0.3311	0.5189	−0.3519	0.5378
0.3980	−0.3067	−0.8174	0.1376	−0.2282	−0.0914
0.4481	−0.2195	0.0588	0.0558	0.8560	0.1083

10.5. DISTRIBUTION OF CHARACTERISTIC ROOTS

We shall now investigate the distribution of the ordered characteristic roots $R_1, \ldots, R_p$ of the random (sample) covariance matrix S/N and the corresponding normalized characteristic vector A_i given by

$$(S/N - R_i I)A_i = 0, \quad i = 1, \ldots, p, \tag{10.31}$$

with $A_i' A_i = 1$. In Chapter 8 we derived the joint distribution of $R_1, \ldots, R_p$ when $\Sigma = I$ (identity matrix). We now give the large sample distribution of these statistics, the initial derivation of which was performed by Girshik (1936, 1939). Subsequently this was extended by Anderson (1965), Anderson (1951, 1963), Bartlett (1954), and Lawley (1956, 1963). In what follows we shall assume that the characteristic roots of Σ are different and N is large. These distribution results are given below without proof. Let

$$n = N - 1$$

$$U_i = \frac{N}{n} R_i, \, i = 1, \ldots, p$$

and let U, Λ be diagonal matrices with diagonal elements $U_1, \ldots, U_p$ and $\lambda_1, \ldots, \lambda_p$ respectively. From James (1960) the joint pdf of $U_1, \ldots, U_p$ is given by

$$\frac{(n/2)^{(1/2)np} \pi^{(1/2)p^2} (\det \Lambda)^{-n/2}}{\Gamma_p(n/2)\Gamma_p(p/2)} \prod_{i=1}^{p}(u_i)^{(n-p-1)/2}$$
$$\times \prod_{i<j}^{p}(u_i - u_j) {}_0F_0\left(-\tfrac{1}{2}nu, \Lambda^{-1}\right) \tag{10.32}$$

where the multivariate gamma function $\Gamma_p(a)$ is given by

$$\Gamma_p(a) = \pi^{(p(p-1))/4} \prod_{i=1}^{p} \Gamma\left(a - \tfrac{1}{2}(i-1)\right) \tag{10.33}$$

and for large N (Anderson, 1965)

$${}_0F_0\left(-\tfrac{1}{2}nu, \Lambda^{-1}\right) \simeq \frac{\Gamma_p(p/2)}{\pi^{p^2/2}} \exp\left\{-\frac{n}{2}\sum_{i=1}^{p}\frac{u_i}{\lambda_i}\right\} \prod_{i<j}^{p}\left(\frac{2\pi}{nc_{ij}}\right)^{1/2}$$

with $c_{ij} = (u_i - u_j)(\lambda_i - \lambda_j)/\lambda_i\lambda_j$. A large sample normal approximation is given in the following theorem:

Theorem 10.5.1. *(Girshik, 1939). If Σ is positive definite and all its characteristic roots are distinct so that $\lambda_1 > \lambda_2 > \cdots > \lambda_p > 0$, then*

(a) *as $N \to \infty$, the ordered characteristic roots $R_1, \ldots, R_p$ are independent, unbiased, and approximately normally distributed with*

$$E(R_i) = \lambda_i, \quad \text{var}(R_i) = 2\lambda_i^2/(N-1); \tag{10.34}$$

(b) *as $N \to \infty$, $(N-1)^{1/2}(A_i - \alpha_i)$ has a p-variate normal distribution with mean 0 and covariance matrix*

$$\lambda_i \sum_{j=1, j\neq i}^{p} \frac{\lambda_i}{(\lambda_j - \lambda_i)^2} \alpha_i \alpha_i'. \tag{10.35}$$

Roy's test (Roy, 1953) is based on the smallest or the largest characteristic root of a sample covariance matrix S. Their exact distribution are not easy to obtain. Theorem 10.5.1 is useful to obtain their asymptotic distributions.

Principal Components

Theorem 10.5.2. For $x \geq 0$

$$P(R_1 \leq x) \leq \prod_{i=1}^{p} P\left(\chi_n^2 \leq \frac{Nx}{\lambda_i}\right)$$

$$P(R_p \leq x) \leq \prod_{i=1}^{p} \left(\chi_n^2 \geq \frac{Nx}{\lambda_i}\right)$$

where χ_n^2 is the chi-square random variable with n degrees of freedom.

Proof. From Theorem 10.1.1 there exists a $p \times p$ orthogonal matrix A such that $A'\Sigma A = \Lambda$. Let $S^* = A'SA$. Then S^* is distributed as Wishart $W_p(n, \Lambda)$ and S/N. S^*/N have the same characteristic roots $R_1, \ldots, R_p$. Let $R_1 > \cdots > R_p$. Hence for $a \in E^p$ with $a'a = 1$

$$N^{-1}a'S^*a = a'0R0'a = b'Rb = \sum_{i=1}^{p} b_i^2 R_i$$

where $b = 0'a = (b_1, \ldots, b_p)'$ satisfying $b'b = a'00'a = a'a = 1$. Thus

$$N^{-1}a'S^*a \leq R_1\left(\sum_{1}^{p} b_i^2\right) = R_1,$$

$$N^{-1}a'S^*a \geq R_p\left(\sum_{1}^{p} b_i^2\right) = R_p.$$

(10.36)

Giving a the values $(1, 0, \ldots, 0)', (0, 1, 0, \ldots, 0)', \ldots, (0, \ldots 0, 1)'$, we conclude that

$$NR_1 \geq \max(S_{11}^*, \ldots, S_{pp}^*),$$

$$NR_p \leq \min(S_{11}^*, \ldots, S_{pp}^*)$$

(10.37)

where $S^* = (S_{ij}^*)$. Since NS_{ii}^*/λ_i, $i = 1, \ldots, p$ are independent chi-square random variables χ_n^2 with n degrees of freedom we conclude from (10.37) that

$$P(R_1 \leq x) \geq P(\max(S_{11}^*, \ldots, S_{pp}^*) \leq Nx)$$

$$= \prod_{i=1}^{p} P(S_{ii}^* \leq Nx)$$

$$= \prod_{i=1}^{p} P\left(\chi_n^2 \leq \frac{Nx}{\lambda_i}\right)$$

and

$$P(R_p \geq x) \leq P(\min(S_{11}^*, \ldots, S_{pp}^*) \geq Nx)$$
$$= \prod_{i=1}^{p} P(S_{ii}^* \geq Nx)$$
$$= \prod_{i=1}^{p} P\left(\chi_n^2 \geq \frac{Nx}{\lambda_i}\right).$$

Q.E.D.

10.6. TESTING IN PRINCIPAL COMPONENTS

The problem of testing the hypothesis $H_0: \lambda_1 = \cdots = \lambda_p = \sigma^2$ has been treated in Chapter 8 under the title sphericity test for $N_p(\mu, \Sigma)$ and $E_p(\mu, \Sigma)$. If H_0 is accepted we conclude that the principal components all have the same variance and hence contribute equally to the total variation. Thus no reduction in dimension can be achieved by transforming the variable to its principal components. On the other hand if H_0 is rejected it is natural to test the hypothesis that $\lambda_2 = \cdots = \lambda_p$ and so on. Theorem 10.6.1 gives the likelihood ratio test of $H_0: \lambda_{k+1} = \cdots = \lambda_p = \lambda$ where λ is unknown for $N_p(\mu, \Sigma)$. For elliptically symmetric distributions similar results can be obtained using Theorem 5.3.6.

Theorem 10.6.1. *On the basis of N observations x^α, $\alpha = 1, \ldots, N(N > p)$ from $N_p(\mu, \Sigma)$ the likelihood ratio test rejects $H_0: \lambda_{k+1} = \cdots = \lambda_p$ whenever*

$$\frac{\prod_{i=k+1}^{p} r_i}{[\sum_{k+1}^{p} r_i/(p-k)]^{p-k}} \leq C, \tag{10.38}$$

or equivalently

$$q = (p-k)(N-1)\log\left\{(p-k)^{-1}\sum_{k+1}^{p} r_i\right\} - N - 1)\sum_{k+1}^{p} r_i \geq K \tag{10.39}$$

where the constant C, K depend on the level α of the test.

Proof. The likelihood of x^α, $\alpha = 1, \ldots, N$ is

$$L(\mu, \Sigma) = (2\pi)^{-Np/2}(\det \Sigma)^{-N/2} \exp\{-\tfrac{1}{2}[\mathrm{tr}\Sigma^{-1}s + N(\bar{x} - \mu)'\Sigma^{-1}(\bar{x} - \mu)]\}.$$

Principal Components

Hence

$$\max_{\Omega} L(\mu, \Sigma) = \max_{\Sigma} L(\Sigma)$$

$$= (2\pi)^{-Np/2} \left[\det\left(\frac{S}{N}\right) \right]^{-N/2} \exp\{-Np/2\} \quad (10.40)$$

$$= (2\pi)^{-(Np/2)} \prod_{1}^{p} (r_i)^{-N/2} \exp\{-Np/2\}$$

where

$$L(\Sigma) = (2\pi)^{-Np/2} (\det \Sigma)^{-N/2} \exp\{-\tfrac{1}{2}\operatorname{tr}\Sigma^{-1}s\}. \quad (10.41)$$

To maximize $L(\Sigma)$ under H_0 we proceed as follows. Let $0(p)$ be the multiplicative group of $p \times p$ orthogonal matrices. Since S, Σ are both positive definite there exists $0_1, 0_2$ in $0(p)$ such that

$$\Sigma = 0_1 \Lambda 0_1', \quad S = N 0_2 R 0_2' \quad (10.42)$$

Λ, R are diagonal matrices with diagonal elements $\lambda_1, \ldots, \lambda_p (\lambda_1 \geq \cdots \geq \lambda_p)$ and $R_1, \ldots, R_p (R_1 > \cdots > R_p)$ respectively. Letting $0 = 0_2' 0_1$ we get $0 \in 0(p)$ and under H_0

$$\log L(\Sigma) = -\frac{Np}{2} \log 2\pi - \frac{N}{2} \Sigma_1^k \log \lambda_i - \frac{N(p-k)}{2} \log \lambda$$

$$- \frac{N}{2} \begin{pmatrix} \Lambda_1^{-1} & 0 \\ 0 & \lambda^{-1} I_k \end{pmatrix} 0' R 0 \quad (10.43)$$

where Λ_1 is the $k \times k$ diagonal matrix with diagonal elements $\lambda_1, \ldots, \lambda_k$. Write

$$0 = (0_{(1)}, 0_{(2)}) \quad (10.44)$$

where $0_{(1)}$ is $p \times k$. Since $00' = 0_{(1)} 0_{(1)}' + 0_{(2)} 0_{(2)}' = I$ we get

$$\operatorname{tr}\begin{pmatrix} \Lambda_1^{-1} & 0 \\ 0 & \lambda^{-1} I_k \end{pmatrix} 0' R 0 = \operatorname{tr} \Lambda_1^{-1} 0_{(1)}' R 0_{(1)} + \lambda^{-1} \operatorname{tr} R 0_{(2)} 0_{(2)}'$$

$$= \lambda^{-1} \sum_{1}^{p} R_i - \operatorname{tr}(\lambda^{-1} I_k - \Lambda_1^{-1}) 0_{(1)}' R 0_{(1)}.$$

Hence

$$\log L(\Sigma) = -\frac{Np}{2}\log 2\pi - \frac{N}{2}\sum_{1}^{k}\log \lambda_i - \frac{N(p-k)}{2}\log \lambda$$

$$-\frac{N}{2\lambda}\sum_{1}^{p}R_i + \text{tr}(\lambda^{-1}I_k - \Lambda_1^{-1})0_{(1)}^{-1}R0_{(1)}$$

(10.45)

It is straightforward to verify that $\log L(\Sigma)$ as a function of $0_{(1)}$ is maximum (see Exercise 4) when

$$0_{(1)} = \begin{pmatrix} 0_{(1)}^* \\ 0 \end{pmatrix}$$

where $0_{(1)}^*$ is a $k \times k$ matrix of the form

$$0_{(1)}^* = \begin{pmatrix} \pm 1, & 0, & 0, & \cdots, 0 \\ 0, & \pm 1, 0, & \cdots, & 0 \\ \cdot & \cdot & \cdot & \cdots \\ 0, & 0, & 0, & \cdots \pm 1 \end{pmatrix}$$

(10.46)

Thus

$$\max_{0 \in 0(p)} L(\Sigma) = -\frac{Np}{2}\log 2\pi - \frac{N}{2}\sum_{1}^{k}\log \lambda_i - \frac{N(p-k)}{2}\log \lambda$$

$$-\frac{N}{2\lambda}\sum_{k+1}^{p}R_i - \frac{N}{2}\sum_{1}^{k}\frac{R_i}{\lambda_i}$$

(10.47)

From (10.47) it follows that the maximum likelihood estimators $\hat{\lambda}_i, \hat{\lambda}$ of λ_i, λ are given by

$$\hat{\lambda}_i = R_i, \quad i = 1, \ldots, k.$$

$$\hat{\lambda} = \frac{\sum_{k+1}^{p}R_i}{p-k}$$

Hence

$$\max_{H_0} L(\mu, \Sigma) = (2\pi)^{-(Np/2)}\Pi^k(R_i)^{N/2}\left(\frac{\sum_{k+1}^{p}R_i}{p-k}\right)^{(N/2)(p-k)}$$

$$\times \exp\left\{-\frac{Np}{2}\right\}$$

(10.48)

Principal Components

From (10.40) and (10.48) we get

$$\lambda = \frac{\max_{H_0} L(\mu, \Sigma)}{\max_{\Omega} L(\mu, \Sigma)}$$

$$= \left[\frac{\prod_{k+1}^{p} R_i}{(p-k)^{-1} \sum_{k+1}^{p} R_i} \right]^{N/2},$$

where Ω is the parametric space of μ, Σ. Hence we prove the theorem.

Q.E.D.

Using Box (1949) the statistic Q with values q has approximately the central chi-square distribution with $\frac{1}{2}(p-k)(p-k+1) - 1$ degrees of freedom under H_0 as $N \to \infty$.

From Theorem 10.6.1 it is easy to conclude that for testing $H_0: \lambda_{i+1} = \cdots = \lambda_{i+k}$, $i+k < p$ the likelihood ratio test rejects H_0 whenever $q = k(N-1) \log(k^{-1} \Sigma_{j=1}^{k} R_j) - (N-1) \Sigma_{j=1}^{p} R_j \geq C$ where the constant C depends on the level α of the test and under H_0 the statistic Q with values q has the central chi-square distribution with $\frac{1}{2}k(k+1) - 1$ degrees of freedom as $N \to \infty$.

Partial least square (PLS) regression is often used in applied sciences and, in particular, in chemometrics. Using the redundancy index Lazzag and Cléroux (2000) wrote the PLS regression model in terms of the successive PLS components. These components are very similar to principle components and are used to explain or predict a set of dependent variables from a set of predictors particularly when the number of predictors in large but the number of observations is not so large. They studied their significance and build tests of hypothesis of this effect.

EXERCISES

1 Let $X = (X_1, \ldots, X_p)'$ be normally distributed with mean μ and covariance matrix Σ, and let Σ have one characteristic root λ_1 of multiplicity p.
   (a) On the basis of observations $x^\alpha = (x_{\alpha 1}, \ldots, x_{\alpha})'$, $\alpha = 1, \ldots, N$, show that the maximum likelihood estimate $\hat{\lambda}_1$ of λ_1 is given by

$$\hat{\lambda}_1 = \frac{1}{pN} \sum_{i=1}^{p} \sum_{\alpha=1}^{N} (x_{\alpha i} - \bar{x}_i)^2 \quad \text{where } \bar{x}_i = \frac{1}{N} \sum_{\alpha=1}^{N} x_{\alpha i}.$$

   (b) Show that the principal component of X is given by OX where O is any $p \times p$ orthogonal matrix.

2. Let $X = (X_1, \ldots, X_p)'$ be a random p-vector with covariance matrix

$$\Sigma = \sigma^2 \begin{pmatrix} 1 & \rho & \cdots & \rho \\ \rho & 1 & \cdots & \rho \\ \vdots & \vdots & & \vdots \\ \rho & \rho & \cdots & 1 \end{pmatrix}, \quad 0 < \rho \leq 1.$$

(a) Show that the largest characteristic root of Σ is

$$\lambda_1 = \sigma^2(1 + (p-1)\rho).$$

(b) Show that the first principal component of X is

$$Z_1 = \frac{1}{\sqrt{p}} \sum_{i=1}^{p} X_i.$$

3. Let $X = (X_1, \ldots, X_4)'$ be a random vector with covariance matrix

$$\Sigma = \begin{pmatrix} \sigma^2 & \sigma_{12} & \sigma_{13} & \sigma_{14} \\ & \sigma^2 & \sigma_{14} & \sigma_{13} \\ & & \sigma^2 & \sigma_{12} \\ & & & \sigma^2 \end{pmatrix}.$$

Show that the principal components of X are

$$Z_1 = \tfrac{1}{2}(X_1 + X_2 + X_3 + X_4), \quad Z_2 = \tfrac{1}{2}(X_1 + X_2 - X_3 - X_4),$$
$$Z_3 = \tfrac{1}{2}(X_1 - X_2 + X_3 - X_4), \quad Z_4 = \tfrac{1}{2}(X_1 - X_2 - X_3 + X_4).$$

4. Let R, Λ be as in Theorem 10.6.1. Show that $\operatorname{tr} \Lambda_1^{-1} 0_{(1)} R 0'_{(1)} \leq \Sigma_1^k (R_i / \lambda_i)$ and the equality holds if $0_{(1)}$ is of the from $\begin{pmatrix} 0^*_{(1)} \\ 0 \end{pmatrix}$ where $0^*_{(1)}$ is defined in Theorem 10.6.1.

5. For the data given in Example 5.3.1 find the the ordered characteristic roots and the normalized characteristic vectors.

REFERENCES

Anderson, G. A. (1965). An asymptotic expansion for the distribution of the latent roots of the estimated covariance matrix. *Ann. Math. Statist.* 36:1153–1173

Anderson, T. W. (1951). Classification by multivariate analysis. *Psychometrika* 16:31–50.

Principal Components

Anderson, T. W. (1963). Asymptotic theory for principal components analysis. *Ann. Math. Statist.* 3:122–148.

Bartlett, M. S. (1954). A note on the multiplying factors for various chi-square approximation. *J. R. Statist. Soc. B* 16:296–298.

Box, G. E. P. (1949). A general distribution theory for a class of likelihood ratio criteria. *Biometrika* 36:317–346.

Girshik, M. A. (1936). Principal components. *J. Am. Statist. Assoc*, 31:519–528.

Girshik, M. A. (1939). On the sampling theory of roots of determinantal equations. *Ann. Math. Statist.* 10:203–224.

Hotelling, H. (1933). Analysis of a complex of a statistical variables into principal components. *J. Educ. Psychol.* 24:417–441.

James, A. T. (1960). The distribution of the latent roots of the covariance matrix. *Ann. Math. Statist.* 31:151–158.

Lawley, D. N. (1956). Tests of significance for the latent roots of covariance and correlation matrices. *Biometrika* 43:128–136.

Lawley, D. N. (1963). On testing a set of correlation coefficients for equality. *Ann. Math. Statist.* 34:149–151.

Lazrag, A. and Cléroux, R. (2000). The pls multivariate regression model: Testing the significance of successive pls components. Univ. de Montreal (private communication).

Pearson, K. (1901). On lines and planes of closest fit to system of points in space. *Phil. Mag.* 2:559–572.

Roy, S. N. (1953). On a heuristic method of test construction and its use in multivariate analysis. *Ann. Math. Statist.* 24:220–238.

11
Canonical Correlations

11.0. INTRODUCTION

Suppose we have two sets of variates and we wish to study their interrelations. If the dimensions of both sets are large, one may wish to consider only a few linear combinations of each set and study those linear combinations which are highly correlated. The admission of students into a medical program is highly competitive. For an efficient selection one may wish to predict a linear combination of scores in the medical program for each candidate from certain linear combinations of scores obtained by the candidate in high school. Economists may find it useful to use a linear combination of easily available economic quantities to study the behavior of the prices of a group of stocks.

The canonical model was first developed by Hotelling (1936). It selects linear combinations of variables from each of the two sets, so that the correlations between the new variables in different sets are maximized subject to the restriction that the new variables in each set are uncorrelated with mean 0 and variance 1. In developing the concepts and the algebra we do not need a specific assumption of normality, though these will be necessary in making statistical inference. For more relevent materials on this topic we refer to Khirsagar (1972), Rao (1973), Mardia, Kent, and Biby (1979), Srivastava and Khatri (1979), Muirhead (1982) and Eaton (1983). Khatri and Bhavsar (1990) derived the asymptotic distribution of the canonical correlations for two sets of complex variates.

11.1. POPULATION CANONICAL CORRELATIONS

Consider a random vector $X = (X_1, \ldots, X_p)$ with mean μ and positive definite covariance matrix Σ. Since we shall be interested only in the covariances of the components of X, we shall take $\mu = 0$. Let

$$X = \begin{pmatrix} X_{(1)} \\ X_{(2)} \end{pmatrix},$$

where $X_{(1)}, X_{(2)}$ are subvectors of X of p_1, p_2 components, respectively. Assume that $p_1 \leq p_2$. Let Σ be similarly partitioned as

$$\Sigma = \begin{pmatrix} \Sigma_{(11)} & \Sigma_{(12)} \\ \Sigma_{(21)} & \Sigma_{(22)} \end{pmatrix},$$

where $\Sigma_{(ij)}$ is $p_i \times p_j$, $i, j = 1, 2$. Recall that if $p_1 = 1$, then the multiple correlation coefficient is the largest correlation attainable between $X_{(1)}$ and a linear combination of the components of $X_{(2)}$. For $p_1 > 1$, a natural generalization of the multiple correlation coefficient is the largest correlation coefficient ρ_1 (say), attainable between linear combinations of $X_{(1)}$ and linear combinations of $X_{(2)}$.

Consider arbitrary linear combinations

$$U_1 = \alpha' X_{(1)}, \quad V_1 = \beta' X_{(2)},$$

where $\alpha = (\alpha_1, \ldots, \alpha_{p_1})' \in E^{p_1}$, $\beta = (\beta_1, \ldots, \beta_{p_2})' \in E^{p_2}$. Since the coefficient of correlation between U_1 and V_1 remains invariant under affine transformations

$$U_1 \to aU_1 + b, \quad V_1 \to cV_1 + d,$$

where a, b, c, d are real constants and $a \neq 0$, $c \neq 0$, we can make an arbitrary normalization of α, β to study the correlation. We shall therefore require that

$$\text{var}(U_1) = \alpha' \Sigma_{(11)} \alpha = 1, \quad \text{var}(V_1) = \beta' \Sigma_{(22)} \beta = 1, \quad (11.1)$$

and maximize the coefficient of correlation between U_1 and V_1. Since $E(X) = 0$, using (11.1)

$$\rho(U_1, V_1) = \frac{E(U_1 V_1)}{(\text{var}(U_1)\text{var}(V_1))^{1/2}} = \frac{\alpha' \Sigma_{(12)} \beta}{((\alpha' \Sigma_{(11)} \alpha)(\beta' \Sigma_{(22)} \beta))^{1/2}}$$

$$= \alpha' \Sigma_{(12)} \beta = \text{cov}(U_1, V_1).$$

Thus we want to find α, β to maximize $\text{cov}(U_1, V_1)$ subject to (11.1). Let

$$\phi_1(\alpha, \beta) = \alpha' \Sigma_{(12)} \beta - \frac{1}{2} \rho(\alpha' \Sigma_{(11)} \alpha - 1) - \frac{1}{2} \nu(\beta' \Sigma_{(22)} \beta - 1) \quad (11.2)$$

where ρ, ν are Lagrange multipliers. Differentiating ϕ_1 with respect to the elements of α, β separately and setting the results equal to zero, we get

$$\frac{\partial \phi_1}{\partial \alpha} = \Sigma_{(12)}\beta - \rho\Sigma_{(11)}\alpha = 0, \quad \frac{\partial \phi_1}{\partial \beta} = \Sigma_{(21)}\alpha - \nu\Sigma_{(22)}\beta = 0. \tag{11.3}$$

From (11.1) and (11.3) we obtain

$$\rho = \nu = \alpha'\Sigma_{(12)}\beta, \ \begin{pmatrix} -\rho\Sigma_{(11)} & \Sigma_{(12)} \\ \Sigma_{(21)} & -\rho\Sigma_{(22)} \end{pmatrix} \begin{pmatrix} \alpha \\ \beta \end{pmatrix} = 0. \tag{11.4}$$

In order that there be a nontrivial solution of (11.4) it is necessary that

$$\det \begin{pmatrix} -\rho\Sigma_{(11)} & \Sigma_{(12)} \\ \Sigma_{(21)} & -\rho\Sigma_{(22)} \end{pmatrix} = 0. \tag{11.5}$$

The left-hand side of (11.5) is a polynomial of degree p in ρ and hence has p roots (say) $\rho_1 \geq \cdots \geq \rho_p$ and $\rho = \alpha'\Sigma_{(12)}\beta$ is the correlation between U_1 and V_1 subject to the restriction (11.1). From (11.4–11.5) we get

$$\det(\Sigma_{(12)}\Sigma_{(22)}^{-1}\Sigma_{(21)} - \rho^2\Sigma_{(11)}) = 0, \tag{11.6}$$

$$(\Sigma_{(12)}\Sigma_{(22)}^{-1}\Sigma_{(21)} - \rho^2\Sigma_{(11)})\alpha = 0, \tag{11.7}$$

which has p_1 solutions for ρ^2, $\rho_1^2 \geq \cdots \geq \rho_{p_1}^2$ (say), and p_1 solutions for α, and

$$\det(\Sigma_{(21)}\Sigma_{(11)}^{-1}\Sigma_{(12)} - \rho^2\Sigma_{(22)}) = 0, \tag{11.8}$$

$$(\Sigma_{(21)}\Sigma_{(11)}^{-1}\Sigma_{(12)} - \rho^2\Sigma_{(22)})\beta = 0, \tag{11.9}$$

which has p_2 solutions for ρ^2 and p_2 solutions for β. Now (11.6) implies that

$$\det(\Lambda\Lambda' - \rho^2 I) = 0, \quad \text{where } \Lambda = \Sigma_{(11)}^{-1/2}\Sigma_{(12)}\Sigma_{(22)}^{-1/2} \tag{11.10}$$

Since

$$\det(\Lambda\Lambda' - \rho^2 I) = \det(\Lambda\Lambda' - \rho^2 I) = \det(\Sigma_{(22)}^{-1}\Sigma_{(21)}\Sigma_{(11)}^{-1}\Sigma_{(12)}\Sigma_{(22)}^{(1/2)} - \rho^2 I),$$

we conclude that (11.6) and (11.7) have the same solutions. Thus (11.5) has p roots of which $p_2 - p_1$ are zeros, and the remaining $2p_1$ nonzero roots are of the form $\rho = \pm \rho_i, i = 1, \ldots, p_1$. The ordered p roots of (11.5) are thus $(\rho_1, \ldots, \rho_{p_1}, 0, \ldots, 0, -\rho_{p_1}, \ldots, -\rho_1)$. We shall show later that $\rho_i \geq 0$, $i = 1, \ldots, p_1$. To get the maximum correlation of U_1, V_1 we take $\rho = \rho_1$. Let $\alpha^{(1)}, \beta^{(1)}$ be the solution (11.4) when $\rho = \rho_1$. Thus $U_1 = \alpha^{(1)'}X_{(1)}$, $V_1 = \beta^{(1)'}X_{(2)}$ are normalized (with respect to variance) linear combinations of $X_{(1)}, X_{(2)}$, respectively, with maximum correlation ρ_1.

Definition 11.1.1. $U_1 = \alpha^{(1)'} X_{(1)}$, $V_1 = \beta^{(1)'} X_{(2)}$ are called the first canonical variates and ρ_1 is called the first canonical correlation between $X_{(1)}$ and $X_{(2)}$.

Next we define

$$U_2 = \alpha' X_{(1)}, \quad V_2 = \beta' X_{(2)},$$

$\alpha \in E^{p_1}$, $\beta \in E^{p_2}$ so that $\text{var}(U_2) = \text{var}(V_2) = 1$, U_2, V_2 are uncorrelated with U_1, V_1 respectively, and the coefficient of correlation $\rho(U_2, V_2)$ is as large as possible. It is now left as an exercise to establish that $\rho(U_2, V_2) = \rho_2$, the second largest root of (11.1). Let $\alpha^{(2)}$, $\beta^{(2)}$ be the solution of (11.5) where $\rho = \rho_2$.

Definition 11.1.2. $U_2 = \alpha^{(2)'} X_{(1)}$, $V_2 = \beta^{(2)'} X_{(2)}$ are called the second canonical variates and ρ_2 is called the second canonical correlation.

This procedure is continued and at each step we define canonical variates as normalized variates, which are uncorrelated with all previous canonical variates, having maximum correlation. Because of (11.6) and (11.7) the maximum number of pairs (U_i, V_i) of positively correlated canonical variates is p_1.
Let

$$U = (U_1, \ldots, U_{p_1})' = A' X_{(1)}, \quad A = (\alpha^{(1)}, \ldots, \alpha^{(p_1)}),$$
$$V_{(1)} = (V_1, \ldots, V_{p_1})' = B_1' X_{(2)}, \quad B_1 = (\beta^{(1)}, \ldots, \beta^{(p_1)}),$$
(11.11)

and let D be a diagonal matrix with diagonal elements $\rho_1, \ldots, \rho_{p_1}$. Since (U_i, V_i), $i = 1, \ldots, p_1$, are canonical variates,

$$\text{cov}(U) = A' \Sigma_{(11)} A = I, \quad \text{cov}(V_{(1)}) = B_1' \Sigma_{(22)} B_1 = I,$$
$$\text{cov}(U, V_{(1)}) = A' \Sigma_{(12)} B_1 = \Lambda.$$
(11.12)

Let $B_2 = (\beta^{(p_1+1)}, \ldots, \beta^{(p_2)})$ be a $p_2 \times (p_2 - p_1)$ matrix satisfying

$$B_2' \Sigma_{(22)} B_1 = 0, \quad B_2' \Sigma_{(22)} B_2 = I,$$

and formed one column at a time in the following way: $\beta^{(p_1+1)}$ is a vector orthogonal to $\Sigma_{(22)} B_1$ and $\beta^{(p_1+1)'} \Sigma_{(22)} \beta^{(p_1+1)} = 1$; $\beta^{(p_1+2)}$ is a vector orthogonal to $\Sigma_{(22)}(B_1, \beta^{(p_1+1)})$ and $\beta^{(p_1+2)'} \Sigma_{(22)} \beta^{(p_1+2)} = 1$; and so on. Let $B = (B_1, B_2)$.

Canonical Correlations

Since $B'\Sigma_{(22)}B = I$, we conclude that B is nonsingular. Now

$$\det\left[\begin{pmatrix} A' & 0 \\ 0 & B'_1 \\ 0 & B'_2 \end{pmatrix}\begin{pmatrix} -\rho\Sigma_{(11)} & \Sigma_{(12)} \\ \Sigma_{(21)} & -\rho\Sigma_{(22)} \end{pmatrix}\begin{pmatrix} A & 0 & 0 \\ 0 & B_1 & B_2 \end{pmatrix}\right]$$

$$= \det\begin{pmatrix} -\rho I & D & 0 \\ D & -\rho I & 0 \\ 0 & 0 & \rho I \end{pmatrix} = (-\rho)^{p_2-p_1}\det\begin{pmatrix} -\rho I & D \\ D & -\rho I \end{pmatrix} \quad (11.13)$$

$$= (-\rho)^{p_2-p_1}\det(\rho^2 I - DD)$$

$$= (-\rho)^{p_2-p_1}\prod_{i=1}^{p_1}(\rho^2 - \rho_i^2).$$

Hence the roots of the equation obtained by setting (11.13) equal to zero are the roots of (11.5). Observe that for $i = 1, \ldots, p_1$ [from (11.4)]

$$\Sigma_{(12)}\beta^{(i)} = -\rho_i \Sigma_{(11)}(-\alpha^{(i)}), \quad (11.14)$$

$$\Sigma_{(21)}(-\alpha^{(i)}) = -\rho_i \Sigma_{(22)}(\beta^{(i)}). \quad (11.15)$$

Thus, if ρ_i, $\alpha^{(i)}$, $\beta^{(i)}$ is a solution so is $-\rho_i$, $-\alpha^{(i)}$, $\beta^{(i)}$. Hence if the ρ_i were negative, then $-\rho_i$ would be nonnegative and $-\rho_i \geq \rho_i$. But since ρ_i was to be a maximum, we must have $\rho_i \geq -\rho_i$ and therefore $\rho_i \geq 0$.

The components of U are one set of canonical variates, the components of $(V_{(1)}, V_{(2)}) = B_2 X_{(2)}$ are other sets of canonical variates, and

$$\operatorname{cov}\begin{pmatrix} U \\ V_{(1)} \\ V_{(2)} \end{pmatrix} = \begin{pmatrix} I & \Lambda & 0 \\ \Lambda & I & 0 \\ 0 & 0 & I \end{pmatrix}.$$

Definition 11.1.3. The ith pair of canonical variates, $i = 1, \ldots, p_1$, is the pair of linear combinations $U_i = \alpha^{(i)'} X_{(1)}$, $V_i = \beta^{(i)'} X_{(2)}$, each of unit variance and uncorrelated with the first $(i-1)$ pairs of canonical variates (U_j, V_j), $j = 1, \ldots, i-1$, and having maximum correlation. The coefficient of correlation between U_i and V_i is called the ith canonical correlation. Hence we have the following theorem.

Theorem 11.1.1. *The ith canonical correlation between $X_{(1)}$ and $X_{(2)}$ is the ith largest root ρ_i of (11.5) and is positive. The coefficients $\alpha^{(i)}$, $\beta^{(i)}$ of the normalized ith canonical variates $U_i = \alpha^{(i)'} X_{(1)}$, $V_i = \beta^{(i)'} X_{(2)}$ satisfy (11.4) for $\rho = \rho_i$.*

In applications the first few pairs of canonical variates usually have appreciably large correlations, so that a large reduction in the dimension of two sets can be achieved by retaining these variates only.

11.2. SAMPLE CANONICAL CORRELATIONS

In practice μ, Σ are unknown. We need to estimate them on the basis of sample observations from the distribution of X. In what follows we shall assume that X has a p-variate normal distribution with mean μ and positive definite covariance matrix Σ (in the case of nonnormality see Rao, 1973). Let $x^\alpha = (x_{\alpha 1}, \ldots, x_{\alpha p})'$, $\alpha = 1, \ldots, N$, be a sample of N observations on X and let

$$\bar{x} = \frac{1}{N} \sum_{\alpha=1}^{N} x^\alpha, \quad s = \sum_{\alpha=1}^{N} (x^\alpha - \bar{x})(x^\alpha - \bar{x})'.$$

Partition s, similarly to Σ, as

$$s = \begin{pmatrix} s_{(11)} & s_{(12)} \\ s_{(21)} & s_{(22)} \end{pmatrix},$$

where $s_{(ij)}$ is $p_i \times p_j$, $i, j = 1, 2$. The maximum likelihood estimates of the $\Sigma_{(ij)}$ are $s_{(ij)}/N$. The maximum likelihood estimates $\hat{\alpha}^{(i)}, i = 1, \ldots, p_1, \hat{\beta}^{(j)}, j = 1, \ldots, p_2$ and $\hat{\rho}_i, i = 1, \ldots, p_1$, of $\alpha^{(i)}, \beta^{(j)}$, and ρ_i, respectively, are obtained from (11.4) and (11.5) by replacing $\Sigma_{(ij)}$ by $s_{(ij)}/N$. Standard programs are available for the computation of $\hat{\alpha}^{(i)}, \hat{\beta}^{(j)}, \hat{\rho}_i$, and we refer to Press (1971) for details. We define the squared sample canonical correlation R_i^2 (with values $r_i^2 = \hat{\rho}^2$) by the roots of

$$\det(S_{(12)} S_{(22)}^{-1} S_{(21)} - r^2 S_{(11)}) = 0, \quad (11.16)$$

which can be written as

$$\det(B - r^2(A + B)) = 0, \quad (11.17)$$

where $B = S_{(12)} S_{(22)}^{-1} S_{(21)}$, $A = S_{(11)} - S_{(12)} S_{22}^{-1} S_{(21)}$. From Theorem 6.4.1, A, B are independently distributed, A is distributed as Wishart

$$W_{p_1}(\Sigma_{(11)} - \Sigma_{(12)} \Sigma_{(22)}^{-1} \Sigma_{(21)}, N - 1 - p_2).$$

and the conditional distribution of $S_{(12)} S_{(22)}^{-1/2}$, given that $S_{(22)} = s_{(22)}$, is normal with mean $\Sigma_{(12)} \Sigma_{(22)}^{-1/2}$ and covariance matrix

$$(\Sigma_{(22)} - \Sigma_{(21)} \Sigma_{(11)}^{-1} \Sigma_{(12)}) \otimes s_{(22)}^{-1}$$

Canonical Correlations

Hence if $\Sigma_{(12)} = 0$, then A, B are independently distributed as $W_{p_1}(\Sigma_{(11)}, N - 1 - p_2)$, $W_{p_1}(\Sigma_{(11)}, p_2)$, respectively. Thus, in the case $\Sigma_{(12)} = 0$, the squared sample canonical correlation coefficients are the roots of the equation

$$\det(B - r^2(A + B)) = 0, \qquad (11.18)$$

where A, B are independent Wishart matrices with the same parameter $\Sigma_{(11)}$. The distribution of these ordered roots $R_1^2 > R_2^2 > \cdots > R_{p_1}^2$ (say) was derived in Chapter 8 and is given by

$$K \prod_{i=1}^{p_1} (r_i^2)^{(p_2 - p_1 - 1)/2} (1 - r_i^2)^{(N - 1 - p_1 - 1)/2} \prod_{i<j} (r_i^2 - r_j^2), \qquad (11.19)$$

where

$$K = \pi^{p/2} \left[\prod_{i=1}^{p_1} \Gamma(\tfrac{1}{2}(p_2 - i + 1)) \Gamma\left(\tfrac{i}{2}\right) \right]^{-1} \prod_{i=1}^{p_1} \frac{\Gamma(\tfrac{1}{2}(N - p_1 + p_2 - i))}{\Gamma(\tfrac{1}{2}(N - p_1 - i))}. \qquad (11.20)$$

These roots are maximal invariants in the space of the random Wishart matrix S under the transformations $S \to ASA'$ where

$$A = \begin{pmatrix} A_1 & 0 \\ 0 & A_2 \end{pmatrix} \quad \text{with} \quad A_i : p_i \times p_i, \quad i = 1, 2.$$

11.3. TESTS OF HYPOTHESES

Let us now consider the problem of testing the null hypothesis $H_{10} : \Sigma_{(12)} = 0$ against the alternatives $H_1 : \Sigma_{(12)} \neq 0$ on the basis of sample observations x^α, $\alpha = 1, \ldots, N (N \geq p)$. In other words, H_{10} is the hypothesis of joint nonsignificance of the first p_1 canonical correlations as a set. It can be easily calculated that the likelihood ratio test of H_{10} rejects H_{10} whenever

$$\lambda_1 = \frac{\det s}{\det(s_{(11)}) \det(s_{(22)})} \leq c,$$

where the constant c is chosen so that the test has level of significance α. Narain (1950) showed that the likelihood ratio test for testing $H_{(10)}$ is unbiased against H_1 (see Section 8.3). The exact distribution of

$$\lambda_1 = \frac{\det S}{\det(S_{(11)}) \det(S_{(22)})}$$

was studied by Hotelling (1936), Girshik (1939), and Anderson (1958, p. 237). These forms are quite complicated. Bartlett (1938, 1939, 1941) gave an approximate large sample distribution of λ_1. Since

$$\det(S) = \det(S_{(22)}) \det(S_{(11)} - S_{(12)} S_{(22)}^{-1} S_{(21)})$$

$$= \det S_{(22)} \det(S_{(11)}) \det(I - S_{(11)}^{-1} S_{(12)} S_{(22)}^{-1} S_{(21)}),$$

we can write λ_1 as

$$\lambda_1 = \det(I - S_{(11)}^{-1} S_{(12)} S_{(22)}^{-1} S_{(21)}) = \prod_{i=1}^{p_1}(1 - R_i^2).$$

Using Box (1949), as $N \to \infty$ and under H_{10}

$$P\{-\nu \log \lambda_1 \leq z\} = P\{\chi_f^2 \leq z\},$$

where $\nu = N - \frac{1}{2}(p_1 + p_2 + 1), f = p_1 p_2$.

Now suppose that H_{10} is rejected; that is, the likelihood ratio test accepts $H_1 : \Sigma_{(12)} \neq 0$. Bartlett (1941) suggested testing the hypothesis H_{20}: (the joint nonsignificance of $\rho_2, \ldots, \rho_{p_1}$ as a set), and proposed the test of rejecting H_{20} whenever

$$\lambda_2 = \prod_{i=2}^{p_1}(1 - r_i^2) \leq c,$$

where c depends on the level of significance α of the test, and under H_{20} for large N

$$P\{-\nu \log \lambda_2 \leq z\} = P\{\chi_{f_1}^2 \leq z\},$$

where $f_1 = (p_1 - 1)(p_2 - 1)$. That is, for large N, Bartlett suggested the possibility of testing the joint nonsignificance of $\rho_2, \ldots, \rho_{p_1}$. If H_{10} is rejected and H_{20} is accepted, then ρ_1 is the only significant canonical correlation. If H_{20} is also rejected, the procedure should be continued to test H_{30}: (the joint nonsignificance of $\rho_3, \ldots, \rho_{p_1}$ as a set), and then if necessary to test H_{40}, and so on. For H_{r0}: (the joint nonsignificance of $\rho_r, \ldots, \rho_{p_1}$ as a set), the tests rejects H_{r0} whenever

$$\lambda_r = \prod_{i=r}^{p_1}(1 - r_i^2) \leq c,$$

Canonical Correlations

Table 11.1. Bartlett's Test of Significance

i	Sample Canonical Correlations r_i	Likelihood Ratio λ_i	Chi-Square $-\nu \log \lambda_i$	Degrees of Freedom f_i
1	0.86018	0.09764	47.70060	36
2	0.64327	0.37527	20.09220	25
3	0.51725	0.64327	9.04426	16
4	0.30779	0.87824	2.66151	9
5	0.17273	0.97015	0.62126	4
6	0.00413	0.99998	0.00035	1

where the constant c depends on the level of significance α of the test, and for large N under H_{r0} (see Table 11.1)

$$P\{-\nu \log \lambda_r \leq z\} = P\{\chi^2_{f_r} \leq z\}$$

where $f_r = (p_1 - r)(p_2 - r)$.

EXAMPLE 11.3.1. Measurements on 12 different characters $x' = (x_1, \ldots, x_{12})$ for each of 27 randomly selected wheat plants of a particular variety grown at the Indian Agricultural Research Institute, New Delhi, are taken. The sample correlation matrix is given by

$$\begin{pmatrix} 1.0000 & & & & & & & & & & & \\ 0.7025 & 1.0000 & & & & & & & & & & \\ 0.0539 & -0.0717 & 1.0000 & & & & & & & & & \\ 0.1154 & 0.0458 & 0.5811 & 1.0000 & & & & & & & & \\ -0.0669 & -0.1212 & 0.5851 & 0.4065 & 1.0000 & & & & & & & \\ -0.0361 & -0.0515 & 0.7137 & 0.5238 & 0.6698 & 1.0000 & & & & & & \\ 0.4381 & 0.6109 & -0.2064 & -0.1113 & -0.4702 & -0.2029 & 1.0000 & & & & & \\ -0.1332 & 0.1667 & -0.0708 & -0.1186 & -0.0686 & -0.1693 & 0.3503 & 1.0000 & & & & \\ 0.4611 & 0.5927 & -0.2545 & -0.1213 & -0.4649 & -0.2284 & 0.8857 & 0.2945 & 1.0000 & & & \\ 0.5139 & 0.6633 & -0.3099 & -0.1602 & -0.3441 & -0.2141 & 0.8295 & 0.2899 & 0.9339 & 1.0000 & & \\ 0.4197 & 0.5148 & -0.1491 & -0.0216 & -0.3475 & -0.1929 & 0.7007 & 0.0252 & 0.6960 & 0.6987 & 1.0000 & \\ 0.6601 & 0.7129 & -0.1652 & -0.0121 & -0.3632 & -0.1119 & 0.8155 & 0.0761 & 0.8686 & 0.8474 & 0.0815 & 1.0000 \end{pmatrix}$$

We are interested in finding the canonical correlations between the set of first six characters and the set of the remaining six characters. Ordered sample canonical correlations r_i^2 and the corresponding normalized coefficient $\alpha^{(i)}, \beta^{(i)}$ of the canonical variates are given by:

$$r_1^2 = 0.86018$$

$$\alpha^{(1)} = (0.32592, 0.24328, -0.20063, -0.02249, -0.18167, 0.27907)'$$

$$\beta^{(1)} = (0.20072, -0.10910, -0.49652, 0.39825, -0.26955, 0.68558)'$$

$$r_2^2 = 0.64546$$

$$\alpha^{(2)} = (-0.09184, 0.02063, 0.22661, 0.05155, -0.36115, -0.01726)'$$

$$\beta^{(2)} = (0.17221, -0.01490, 0.66931, -0.69215, 0.12967, -0.16206)'$$

$$r_3^2 = 0.51725$$

$$\alpha^{(3)} = (-0.63826, 0.71640, 0.14862, -0.10546, 0.21983, -0.48038)'$$

$$\beta^{(3)} = (0.06472, 0.46436, -0.30851, 0.55171, 0.36140, -0.50000)'$$

$$r_4^2 = 0.30779$$

$$\alpha^{(4)} = (-0.04954, 0.17225, 0.35083, 0.13124, 0.12220, -0.22384)'$$

$$\beta^{(4)} = (-0.13684, 0.30706, -0.39213, -0.42461, 0.11164, 0.73523)'$$

$$r_5^2 = 0.17273$$

$$\alpha^{(5)} = (0.19760, -0.18106, -0.09297, 0.28382, 0.12898, -0.43494)'$$

$$\beta^{(5)} = (-0.68243, -0.01088, 0.51799, -0.03772, 0.48737, -0.16404)'$$

$$r_6^2 = 0.00413$$

$$\alpha^{(6)} = (0.19967, -0.11996, 0.32842, -0.33865, 0.04872, -0.19203)'$$

$$\beta^{(6)} = (0.51365, -0.17532, -0.67766, 0.27563, 0.28516, -0.29818)'$$

In the case of elliptically symmetric distributions we refer to Kariya and Sinha (1989) for the LBI tests and related results of this problem.

EXERCISES

1 Show that the canonical correlations remain unchanged when computed from $S/(N-1)$ instead of S/N but the canonical variables do not remain unchanged.

2 Find the canonical correlations and canonical variates between the variables (X_1, X_2, X_3) and (X_4, X_5, X_6) for 1971 and 1972 data in Example 5.3.1.

3 Let the covariance matrix of a random vector $X = (X_1, \ldots, X_p)'$ be given by

$$\begin{pmatrix} 1 & \rho & \rho^2 & \cdots & \rho^{p-1} \\ \rho & 1 & \rho & \cdots & \rho^{p-2} \\ \cdot & \cdot & \cdots & \cdots & \cdot \\ \rho^{p-1} & \rho^{p-2} & \cdot & \cdots & 1 \end{pmatrix}$$

Is it possible to replace X with a vector of lesser dimension for statistical inference.

4 The correlation between the jth and the kth component of a p-variate random vector is $1 - |j-k|/p$. Show that for $p = 4$ the latent roots are $1/4(2 \pm \sqrt{2}), 1/4(6 \pm \sqrt{26})$. Show that the system can not be represented in fewer than p dimensions.

REFERENCES

Anderson, T. W. (1958). *An Introduction to Multivariate Statistical Analysis*. New York: Wiley.

Bartlett, M. S. (1938). Further aspects of the theory of multiple regression. *Proc. Cambridge Phil. Soc.* 34:33–40.

Bartlett, M. S. (1939). A note on test of significance in multivariate analysis. *Proc. Cambridge Phil. Soc.* 35:180–185.

Bartlett, M. S. (1941). The statistical significance of canonical correlation. *Biometrika* 32:29–38.

Eaton, M. L. (1983). *Multivariate Statistics*. New York: Wiley.

Girshik, M. A. (1939). On the sampling theory of the roots of determinantal equation. *Ann. Math. Statist.* 10:203–224.

Hotelling, H. (1936). Relation between two sets of variates. *Biometrika* 28: 321–377.

Kariya, T. and Sinha, B. (1989). *Robustness of Statistical Tests*. New York: Academic Press.

Khatri, C. G. and Bhavsar, C. D. (1990). Some asymptotic inference problems connected with complex elliptical distribution. *Jour. Mult. Analysis* 35: 66–85.

Khirsagar, A. M. (1972). *Multivariate Analysis*. New York: Marcel Dekker.

Mardia, K. V., Kent, J. T., and Bibby, J. M. (1979). *Multivariate Analysis*. New York: Academic Press.

Muirhead, R. J. (1982). *Aspects of Multivariate Statistical Theory*. New York: Wiley.

Narain, R. D. (1950). On the completely unbiased character of tests of independence in multivariate normal system. *Ann. Math. Statist.* 21:293–298.

Press, J. (1971). *Applied Multivariate Analysis*. New York: Holt.

Rao, C. R. (1973) *Linear Statistical Inference and its Applications*. Second edition, New York: Wiley.

Srivastava, M. S. and Khatri, C. G. (1979). *An Introduction to Multivariate Statistics*. Amsterdam: North Holland.

12
Factor Analysis

12.0. INTRODUCTION

Factor analysis is a multivariate technique which attempts to account for the correlation pattern present in the distribution of an observable random vector $X = (X_1, \ldots, X_p)'$ in terms of a minimal number of unobservable random variables, called factors. In this approach each component X_i is examined to see if it could be generated by a linear function involving a minimum number of unobservable random variables, called common factor variates, and a single variable, called the specific factor variate.

The common factors will generate the covariance structure of X where the specific factor will account for the variance of the component X_i.

Though, in principle, the concept of latent factors seems to have been suggested by Galton (1888), the formulation and early development of factor analysis have their genesis in psychology and are generally attributed to Spearman (1904). He first hypothesized that the correlations among a set of intelligence test scores could be generated by linear functions of a single latent factor of general intellectual ability and a second set of specific factors representing the unique characteristics of individual tests. Thurston (1945) extended Spearman's model to include many latent factors and proposed a method, known as the centroid method, for estimating the coefficients of different factors (usually called factor loadings) in the linear model from a given correlation matrix. Lawley (1940), assuming normal distribution for the random vector X, estimated these factor loadings by using the method of maximum likelihood.

Factor analysis models are widely used in behavioral and social sciences. We refer to Armstrong (1967) for a complete exposition of factor analysis for an applied viewpoint, to Anderson and Rubin (1956) for a theoretical exposition, and to Thurston (1945) for a general treatment. We refer to Lawley (1949, 50, 53), Morrison (1967), Rao (1955) and Solomon (1960) for further relevant results in factor analysis.

12.1. ORTHOGONAL FACTOR MODEL

Let $X = (X_1, \ldots, X_p)'$ be an observable random vector with $E(X) = \mu$ and $\text{cov}(X) = \Sigma = (\sigma_{ij})$, a positive definite matrix. Assuming that each component X_i can be generated by a linear combination of $m (m < p)$ mutually uncorrelated (orthogonal) unobservable variables $Y_1, \ldots, Y_m$ upon which a set of errors may be superimposed, we write

$$X = \Lambda Y + \mu + U, \qquad (12.1)$$

where $Y = (Y_1, \ldots, Y_m)'$, $U = (U_1, \ldots, U_p)'$ denotes the error vector and $\Lambda = (\lambda_{(ij)})$ is a $p \times m$ matrix of unknown coefficients λ_{ij} which is usually called a factor loading matrix. The elements of Y are called common factors. We shall assume that U is distributed independently of Y with $E(U) = 0$ and $\text{cov}(U) = D$, a diagonal matrix with diagonal elements $\sigma_1^2, \ldots, \sigma_p^2$; $\text{var}(U_i) = \sigma_i^2$ is called the specific factor variance of X_i. The vector Y in some cases will be a random vector and in other cases will be an unknown parameter which varies from observation to observation. A component of U is made up of the error of measurement in the test plus specific factors representing the unique character of the individual test. The model (12.1) is similar to the multivariate regression model except that the independent variables Y in this case are not observable.

When Y is a random vector we shall assume that $E(Y) = 0$ and $\text{cov}(Y) = I$, the identity matrix. Since

$$E(X - \mu)(X - \mu)' = E(\Lambda Y + U)(\Lambda Y + U)' = \Lambda\Lambda' + D, \qquad (12.2)$$

we see that X has a p-variate normal distribution with mean μ and covariance matrix $\Sigma = \Lambda\Lambda' + D$, so that Σ is positive definite. Furthermore, since

$$E(XY') = E((\Lambda Y + U)Y') = \Lambda, \qquad (12.3)$$

the elements λ_{ij} of Λ are correlations of X_i, X_j. In behavioral science the term loading is used for correlation. The diagonal elements of $\Lambda\Lambda'$ are called communalities of the components. The purpose of factor analysis is the

Factor Analysis

determination of Λ with elements of D such that

$$\Sigma - D = \Lambda\Lambda'. \tag{12.4}$$

If the errors are small enough to be ignored, we can take $\Sigma = \Lambda\Lambda'$. From this point of view factor analysis is outwardly similar to finding the principal components of Σ since both procedures start with a linear model and end up with matrix factorization. However, the model for principal component analysis must be linear by the very fact that it refers to a rigid rotation of the original coordinate axes, whereas in the factor analysis model the linearity is as much a part of our hypothesis about the dependence structure as the choice of exactly m common factors. The linear model in factor analysis allows us to interpret λ_{ij} as correlation coefficients but if the covariances reproduced by the m-factor linear model fail to fit the linear model adequately, it is as proper to reject linearity as to advance the more usual finding that m common factors are inadequate to explain the correlation structure.

Existence. Since a necessary and sufficient condition that a $p \times p$ matrix A be expressed as BB', with B a $p \times m$ matrix, is that A is a positive semidefinite matrix of rank m, we see that the question of existence of a factor analysis model can be resolved if there exists a diagonal matrix D with nonnegative diagonal elements such that $\Sigma - D$ is a positive semidefinite matrix of rank m. So the question is how to tell if there exists such a diagonal matrix D, and we refer to Anderson and Rubin (1956) for answer to this question.

12.2. OBLIQUE FACTOR MODEL

This is obtained from the orthogonal factor model by replacing $\text{cov}(Y) = I$ by $\text{cov}(Y) = R$, where R is a positive definite correlation matrix; that is, all its diagonal elements are equal to unity. In other words, all factors in the oblique factor model are assumed to have mean 0 and variance 1 but are correlated. In this case $\Sigma = \Lambda R \Lambda' + D$.

12.3. ESTIMATION OF FACTOR LOADINGS

We shall assume that m is fixed beforehand and that X has the p-variate normal distribution with mean μ and covariance matrix Σ (positive definite). We are interested in the maximum likelihood estimates of these parameters. Let $x^\alpha = (x_{\alpha 1}, \ldots, x_{\alpha p})'$, $\alpha = 1, \ldots, N$, be a sample of size N on X. The maximum likelihood estimates of μ and Σ are given by

$$\hat{\mu} = \bar{x} = \frac{1}{N}\sum_{\alpha=1}^{N} x^{\alpha}, \quad \hat{\Sigma} = \frac{s}{N} = \sum_{\alpha=1}^{N}(x^{\alpha} - \bar{x})(x^{\alpha} - \bar{x})'/N.$$

Orthogonal Factor Model

Here $\Sigma = \Lambda\Lambda' + D$. The likelihood of x^{α}, $\alpha = 1, \ldots, N$, is given by

$$L(\Lambda, D, \mu) = (2\pi)^{-Np/2}[\det(\Lambda\Lambda' + D)]^{-N/2}$$
$$\times \exp\{-\tfrac{1}{2}\mathrm{tr}[(\Lambda\Lambda' + D)^{-1}$$
$$(s + N(\bar{x} - \mu)(\bar{x} - \mu)')]\}. \qquad (12.5)$$

Observe that changing Λ to ΛO, where O is an $m \times m$ orthogonal matrix, does not change $L(\Lambda, D, \mu)$. Thus if $\hat{\Lambda}$ is a maximum likelihood estimate of Λ, then $\hat{\Lambda}O$ is also a maximum likelihood estimate of Λ. To obtain uniqueness we impose the restriction that

$$\Lambda' D^{-1} \Lambda = \Gamma \qquad (12.6)$$

is a diagonal matrix with distinct diagonal elements $\gamma_1, \ldots, \gamma_p$. We are now interested in obtaining the maximum likelihood estimates $\hat{\mu}, \hat{\Lambda}, \hat{D}$ of μ, Λ, D respectively, subject to (12.6).

To maximize the likelihood function the term

$$\mathrm{tr}\{(\Lambda\Lambda' + D)^{-1} N(\bar{x} - \mu)(\bar{x} - \mu)'\}$$

may be put equal to zero in (12.5) since it vanishes when $\hat{\mu} = \bar{x}$. With this in mind let us find $\hat{\Lambda}, \hat{D}$.

Note. Λ will not depend on the units in which $Y_1, \ldots, Y_m$ are expressed. Suppose that Y has an m-dimensional normal distribution with mean 0 and covariance matrix $\theta\theta'$, where θ is a diagonal matrix with diagonal elements $\theta_1, \ldots, \theta_m$ such that $\theta_i^2 = \mathrm{var}(Y_i)$. Hence

$$\mathrm{cov}(X) = (\Lambda\theta)(\Lambda\theta)' + D = \Lambda^*\Lambda^{*'} + D,$$

where $\Lambda^* = \Lambda\theta$. Thus for the estimation of factor loadings, without any loss of generality we can assume that the Y_i have unit variance and $\mathrm{cov}(Y) = R$, a correlation matrix. For the orthogonal factor model $R = I$ and for the oblique factor model $R = R$.

Factor Analysis

Theorem 12.3.1. *The maximum likelihood estimates $\hat{\Lambda}, \hat{D}$ of Λ, D, respectively, in the orthogonal factor model are given by*

$$\text{diag}(\hat{D} + \hat{\Lambda}\hat{\Lambda}') = \text{diag}\left(\left(\frac{1}{N}\right)s\right), \tag{12.7}$$

$$\left(\frac{s}{N}\right)\hat{D}^{-1}\hat{\Lambda} = \hat{\Lambda}(I + \hat{\Lambda}'\hat{D}^{-1}\hat{\Lambda}). \tag{12.8a}$$

Proof. Let
$$L(\Lambda, D) = (2\pi)^{-Np/2}[\det(\Lambda\Lambda' + D)]^{-N/2} \exp\{-\tfrac{1}{2}\text{tr}(\Lambda\Lambda' + D)^{-1}s\} \tag{12.8b}$$

Then for $i = 1, \ldots, p$

$$\frac{\partial \log L(\Lambda, D)}{\partial \sigma_i^2} = -\tfrac{1}{2}N \frac{(\Lambda\Lambda' + D)_{ii}}{\det(\Lambda\Lambda' + D)}$$

$$+ \tfrac{1}{2}\text{tr}(\Lambda\Lambda' + D)^{-1}s(\Lambda\Lambda' + D)^{-1}\frac{\partial D}{\partial \sigma_i^2}, \tag{12.8c}$$

where $\partial D/\partial \sigma_i^2$ is the $p \times p$ matrix with unity in the ith diagonal position and zero elsewhere, and $(\Lambda\Lambda' + D)_{ii}$ is the cofactor of the ith diagonal element of $\Lambda\Lambda' + D$. Note that for any symmetric matrix $A = (a_{ij})$,

$$\frac{\partial \det A}{\partial a_{ii}} = A_{ii}, \quad \frac{\partial \det A}{\partial a_{ij}} = 2A_{ij}$$

where the A_{ij} are the cofactors of the elements a_{ij}. For $L(\Lambda, D)$ to be maximum it is necessary that each of the p derivatives in (12.8c) equal zero at $\Lambda = \hat{\Lambda}, D = \hat{D}$. This reduces to the condition that the diagonal elements of $(\hat{D} + \hat{\Lambda}\hat{\Lambda}')^{-1} \times [I - (s/N)(\hat{D} + \hat{\Lambda}\hat{\Lambda}')^{-1}]$ are zeros, that is,

$$\text{diag}\left\{(\hat{D} + \hat{\Lambda}\hat{\Lambda}')^{-1}\left[I - \left(\frac{s}{N}\right)(\hat{D} + \hat{\Lambda}\hat{\Lambda}')^{-1}\right]\right\} = 0. \tag{12.8}$$

Now differentiating $L(\Lambda, D)$ with respect to λ_{ij}, we get, with $\Lambda\Lambda' + D = \Sigma = (\sigma_{ij})$,

$$\frac{\partial \log L(\Lambda, D)}{\partial \lambda_{ij}} = -\frac{N}{2}\det(\Lambda\Lambda' + D)^{-1}\Sigma_{g,h=1}^{p}(\Lambda\Lambda' + D)_{gh}\frac{\partial \sigma_{gh}}{\partial \lambda_{ij}}$$

$$+ \frac{1}{2}\text{tr}(\Lambda\Lambda' + D)^{-1}\left(\frac{\partial \Sigma}{\partial \lambda_{ij}}\right)(\Lambda\Lambda' + D)^{-1}s$$

$$= -\frac{1}{2}N\text{tr}\Sigma^{-1}\left(\frac{\partial \Sigma}{\partial \lambda_{ij}}\right)$$

$$+ \frac{1}{2}\text{tr}(\Lambda\Lambda' + D)^{-1}\left(\frac{\partial \Sigma}{\partial \lambda_{ij}}\right)(\Lambda\Lambda' + D)^{-1}s. \tag{12.9}$$

Denoting $\Sigma^{-1} = (\sigma^{ij})$, from Exercise 1, we obtain

$$\operatorname{tr}\Sigma^{-1}\left(\frac{\partial \Sigma}{\partial \lambda_{ij}}\right) = 2(\sigma^j)'\lambda_j, \qquad (12.10)$$

where $(\sigma^j)'$ is the ith row of Σ^{-1} and λ_j is the jth column of Λ. Thus the first term in $\partial \log L(\Lambda, D)/\partial \Lambda$ is $-(1/2)N(\Lambda\Lambda' + D)^{-1}\Lambda$. Making two cyclic permutations of matrices within the trace symbol, we get

$$\operatorname{tr}(\Lambda\Lambda' + D)^{-1}\left(\frac{\partial \Sigma}{\partial \lambda_{ij}}\right)(\Lambda\Lambda' + D)^{-1}s$$

$$= \operatorname{tr}(\Lambda\Lambda' + D)^{-1}s(\Lambda\Lambda' + D)^{-1}\left(\frac{\partial \Sigma}{\partial \lambda_{ij}}\right). \qquad (12.11)$$

Write

$$Z = (\Lambda\Lambda' + D)^{-1}s(\Lambda\Lambda' + D)^{-1} = (Z_{ij})$$

and let the ith row of Z be Z'_i. From Exercise 1,

$$\operatorname{tr}Z\left(\frac{\partial \Sigma}{\partial \lambda_{ij}}\right) = 2Z'_i\lambda_j. \qquad (12.12)$$

Thus the second term in $\partial \log L(\Lambda, D)/\partial \Lambda$ is $Z\Lambda$. From (12.10–12.13) we get

$$[N(\hat{\Lambda}\hat{\Lambda}' + \hat{D})^{-1} - (\hat{\Lambda}\hat{\Lambda}' + \hat{D})^{-1}s(\hat{\Lambda}\hat{\Lambda}' + \hat{D})^{-1}]\hat{\Lambda} = 0$$

or, equivalently,

$$N\hat{\Lambda} = s(\hat{\Lambda}\hat{\Lambda}' + \hat{D})^{-1}\hat{\Lambda}. \qquad (12.13)$$

Since

$$(\hat{D} + \hat{\Lambda}\hat{\Lambda}')^{-1}\hat{\Lambda} = \hat{D}^{-1}\hat{\Lambda}(I + \hat{\Lambda}'\hat{D}^{-1}\hat{\Lambda})^{-1},$$

from (12.13) we get

$$N\hat{\Lambda} = s\hat{D}^{-1}\hat{\Lambda}(I + \hat{\Lambda}'\hat{D}^{-1}\hat{\Lambda})^{-1} \qquad (12.14)$$

or

$$\left(\frac{s}{N}\right)\hat{D}^{-1}\hat{\Lambda} = \hat{\Lambda}(I + \hat{\Lambda}'\hat{D}^{-1}\hat{\Lambda}),$$

which yields (12.7). From

$$(\hat{D} + \hat{\Lambda}\hat{\Lambda}')^{-1} = \hat{D}^{-1} - \hat{D}^{-1}\hat{\Lambda}\hat{\Lambda}'(\hat{D} + \hat{\Lambda}\hat{\Lambda}')^{-1},$$

$$(\hat{D} + \hat{\Lambda}\hat{\Lambda}')^{-1}\hat{D} = I - (\hat{D} + \hat{\Lambda}\hat{\Lambda}')^{-1}\hat{\Lambda}\hat{\Lambda}',$$

Factor Analysis

we get

$$\hat{D}(\hat{D} + \hat{\Lambda}\hat{\Lambda}')^{-1}\hat{D} = \hat{D} - \hat{\Lambda}\hat{\Lambda}'(\hat{D} + \hat{\Lambda}\hat{\Lambda}')^{-1}\hat{D}$$
$$= \hat{D} - \hat{\Lambda}\hat{\Lambda}' + \hat{\Lambda}\hat{\Lambda}'(\hat{D} + \hat{\Lambda}\hat{\Lambda}')^{-1}\hat{\Lambda}\hat{\Lambda}'. \quad (12.15)$$

Similarly,

$$\hat{D}(\hat{D} + \hat{\Lambda}\hat{\Lambda}')^{-1}\left(\frac{S}{N}\right)(\hat{D} + \hat{\Lambda}\hat{\Lambda}')^{-1}\hat{D} = \left(\frac{S}{N}\right) - \left(\frac{S}{N}\right)(\hat{D} + \hat{\Lambda}\hat{\Lambda}')^{-1}\hat{\Lambda}\hat{\Lambda}'$$
$$- \hat{\Lambda}\hat{\Lambda}'(\hat{D} + \hat{\Lambda}\hat{\Lambda}')^{-1}\left(\frac{S}{N}\right) + \hat{\Lambda}\hat{\Lambda}^{-1}(\hat{D} + \hat{\Lambda}\hat{\Lambda}')^{-1}\left(\frac{S}{N}\right)(\hat{D} + \hat{\Lambda}\hat{\Lambda}')^{-1}\hat{\Lambda}\hat{\Lambda}'.$$
$$(12.16)$$

Using (12.13) and (12.15–12.16), we get from (12.8),

$$\text{diag}(\hat{D} + \hat{\Lambda}\hat{\Lambda}') = \text{diag}\left(\frac{S}{N}\right),$$

which yields (12.7). It can be verified that these estimates yield a maximum for $L(\Lambda, D)$. Q.E.D.

Oblique Factor Model

Similarly for the oblique factor model with $\text{cov}(Y) = R$ (correlation matrix) we obtain the following theorem.

Theorem 12.3.2. *The maximum likelihood estimates $\hat{\Lambda}$, $\hat{R}$, $\hat{D}$ of Λ, R, D, respectively, for the oblique factor model are given by*

(1) $\hat{D} = \text{diag}(s/N - \hat{\Lambda}\hat{R}\hat{\Lambda}')$;
(2) $\hat{R}\hat{\Lambda}\hat{D}^{-1}\hat{\Lambda} + I = (\hat{\Lambda}\hat{D}^{-1}\hat{\Lambda}')^{-1}(\hat{\Lambda}'\hat{D}^{-1}(s/N)\hat{D}^{-1}\hat{\Lambda})$,
(3) $\hat{R}\hat{\Lambda}(\hat{\Lambda}\hat{\Lambda}' + \hat{D}^{-1}(I - (s/N)(\hat{\Lambda}\hat{\Lambda}' + \hat{D})^{-1})) = \hat{R}\hat{\Lambda}'[I - (\hat{\Lambda}\hat{\Lambda}' + \hat{D})^{-1}(s/N)]\hat{D}^{-1}$.

For numerical evaluation of these estimates, standard computer programs are now available (see Press, 1971). Anderson and Rubin (1956) have shown that as $N \to \infty$, $\sqrt{N}(\hat{\Lambda} - \Lambda)$ has mean 0 but the covariance matrix is extremely complicated.

Identification. For the orthogonal factor analysis model we want to represent the population covariance matrix as

$$\Sigma = \Lambda\Lambda' + D$$

For any orthogonal matrix 0 of dimension $p \times p$

$$\Sigma = \Lambda\Lambda' + D = \Lambda 00'\Lambda' + D = (\Lambda 0)(\Lambda 0)' + D = \Lambda^*\Lambda^{*'} + D.$$

Thus, regardless of the value of Λ used, it is always possible to transform Λ by an orthogonal matrix 0 to get a new Λ^* which gives the same representation for Λ. Furthermore, since Σ is symmetric, there are $p(p+1)/2$ distinct elements in Σ, and in the factor representation model there is generally a greater number, $p(m+1)$, of distinct parameters. So in general a unique estimate of λ is not possible and there remains the problem of identification in the factor analysis model. We refer to Anderson and Rubin (1956) for a detailed treatment of this topic.

12.4. TESTS OF HYPOTHESIS IN FACTOR MODELS

Let $x^\alpha = (x_{\alpha 1}, \ldots, x_{\alpha p})'$, $\alpha = 1, \ldots, N$, be a sample of size N from a p-variate normal population with positive definite covariance matrix Σ. On the basis of these observations we are interested in testing, with the orthogonal factor model, the null hypothesis $H_0 : \Sigma = \Lambda\Lambda' + D$ against the alternatives H_1 that Σ is a symmetric positive definite matrix. (The corresponding hypothesis in the oblique factor model is $H_0 : \Sigma = \Lambda R \Lambda^1 + D$.)

The likelihood of the observations x^α, $\alpha = 1, \ldots, N$, is

$$L(\Sigma, \mu) = (2\pi)^{-Np/2}(\det \Sigma)^{-N/2} \exp\left\{-\frac{1}{2}\text{tr}\Sigma^{-1}\left[\sum_{\alpha=1}^{N}(x^\alpha - \mu)(x^\alpha - \mu)'\right]\right\}$$

and hence

$$\max_{H_1} L(\Sigma, \mu) = (2\pi)^{-Np/2}(\det(s/N))^{-N/2} \exp\left\{-\frac{1}{2}Np\right\}.$$

Under H_0, $L(\Sigma, \mu)$ reduces to (for the orthogonal factor model)

$$L(\Lambda, D, \mu) = (2\pi)^{-Np/2}(\det(\Lambda\Lambda' + D))^{-N/2}$$

$$\times \exp\left\{-\frac{1}{2}\text{tr}(\Lambda\Lambda' + D)^{-1}\left[\sum_{\alpha=1}^{N}(x^\alpha - \mu)(x^\alpha - \mu)'\right]\right\}$$

and

$$\max_{H_0} L(\Lambda, D, \mu) = (2\pi)^{-Np/2}(\det(\hat\Lambda\hat\Lambda' + \hat D))^{-N/2} \exp\left\{-\frac{1}{2}\text{tr}(\hat\Lambda\hat\Lambda' + \hat D)^{-1}s\right\},$$

Factor Analysis

where $\hat{\Lambda}, \hat{D}$ are given in Theorem 12.3.1. Hence the modified likelihood ratio test of H_0 rejects H_0 whenever, with $N - 1 = n$ (say),

$$\lambda = \left[\frac{\det(s/N)}{\det(\hat{\Lambda}\hat{\Lambda}' + \hat{D})}\right]^{-n/2} \exp\left\{\frac{1}{2}\text{tr}(\hat{\Lambda}\hat{\Lambda}' + \hat{D})^{-1}s - \frac{1}{2}np\right\} \geq C, \quad (12.17)$$

where C depends on the level of significance α of the test. In large samples under H_0, using Box (1949),

$$P\{-2\log \lambda \leq z\} = P\{\chi_f^2 \leq z\},$$

where

$$f = \frac{1}{2}p(p+1) - [mp + p - \frac{1}{2}m(m+1) + m]. \quad (12.18)$$

The modification needed for the oblique factor model is obvious and the value of degrees of freedom f for the chi-square approximation in this case is

$$f = \frac{1}{2}p(p - 2m + 1). \quad (12.19)$$

Bartlett (1954) has pointed put that if $N - 1 = n$ is replaced by n_0, where

$$n_0 = n - \frac{1}{6}(2p + 5) - \frac{2}{3}m, \quad (12.20)$$

then under H_0, the convergence of $-2\log \lambda$ to chi-square distribution is more rapid.

12.5. TIMES SERIES

A time series is a sequence of observations usually ordered in time. The main distinguishing feature of time series analysis is its explicit recognition of the importance of the order in which the observations are made. Although in general statistical investigation the observations are independent, in a time series successive observations are dependent. Consider a stochastic process $X(t)$ as a random variable indexed by the continuous parameter t. Let t be a time scale and the process be observed at the particular p points $t_1 < t_2 \cdots < t_p$. The random vector

$$X(t) = (X(t_1), \ldots, X(t_p))'$$

is called a time series. It has a multidimensional distribution characterizing the process. In most situations we assume that $X(t)$ has a p-variate normal distribution specified by the mean $E(X(t))$ and covariance matrix with general elements

$$\text{cov}(X(t_i), X(t_j)) = \sigma_i \sigma_j \rho(t_i, t_j),$$

where $\text{var}(X(t_i)) = \sigma_i^2$, the term $\rho(t_i, t_j)$ is called the correlation function of the time series. The analysis of the time series data depends on the specific form of $\rho(t_i, t_j)$. A general model of time series can be written as

$$X(t) = f(t) + U(t),$$

where $\{f(t)\}$ is a completely determined sequence, often called the systematic part, and $\{U(t)\}$ is the random sequence having different probability laws. They are sometimes called signal and noise sequences, respectively. The sequence $\{f(t)\}$ may depend on unknown coefficients and known quantities depending on time. This model is analogous to the regression model discussed in Chapter 8. If $f(t)$ is a slowly moving function of t, for example a polynomial of lower degree, it is called a trend: if it is exemplified by a finite Fourier series, it is called cyclical. The effect of time may be present both in $f(t)$ (i.e., trend in time or cyclical) and in $U(t)$ as a stochastic process. When $f(t)$ has a given structure involving a finite number of parameters, we consider the problem of inference about these parameters. When the stochastic process is specified in terms of a finite number of parameters we want to estimate and test hypotheses about these parameters. We refer to Anderson (1971) for an explicit treatment of this topic.

EXERCISE

1 Consider the orthogonal factor analysis model of Section 12.1. Let $\Sigma = \Lambda\Lambda' + D$. Show that

$$\frac{\partial \Sigma}{\partial \lambda_{ij}} = \begin{pmatrix} 0 & \cdots & 0 & \lambda_{1,j} & 0 & \cdots & 0 \\ \vdots & & \vdots & \vdots & \vdots & & \vdots \\ 0 & \cdots & 0 & \lambda_{i-1,j} & 0 & \cdots & 0 \\ \lambda_{1j} & \cdots & \lambda_{i-1,j} & 2\lambda_{ij} & \lambda_{i+1,j} & \cdots & \lambda_{pj} \\ \vdots & & \vdots & \vdots & \vdots & & \vdots \\ 0 & \cdots & 0 & \lambda_{pj} & 0 & \cdots & 0 \end{pmatrix}$$

Hence show that

$$\text{tr}\Sigma^{-1} \frac{\partial \Sigma}{\partial \lambda_{ij}} = 2(\sigma^j)'\lambda_j.$$

REFERENCES

Anderson, T. W. (1971). *The Statistical Analysis of Time Series*. New York: Wiley.

Anderson, T. W. and Rubin, H. (1956). Statistical inference in factor analysis. Proc. Berkeley Symp. Math. Statist. Prob., 3rd 5, 110–150. Univ. of California, Berkeley, California.

Armstrong, J. S. (1967). Derivation of theory by means of factor analysis or Tom Swift and is electric factor analysis machine. *The American Statistician*, December, 17–21.

Bartlett, M. S. (1954). A note on the multiple factors for various chi-square approximation. *J. R. Statist. Soc. B* 16:296–298.

Box, G. E. P. (1949). A general distribution theory for a class of likelihood ratio criteria. *Biometrika* 36:317–346.

Galton, F. (1988). Co-relation and their measurements, chiefly from anthropometric data. *Proc. Roy. Soc.* 45:135–140.

Lawley, D. N. (1940). The estimation of factor loadings by the method of maximum likelihood. *Proc. Roy. Soc. Edinburgh, A* 60:64–82.

Lawley, D. N. (1949). Problems in factor analysis. *Proc. R. Soc. Edinbourgh* 62: 394–399.

Lawley, D. N. (1950). A further note on a problem in factor analysis. *Proc. R. Soc. Edinburgh* 63:93–94.

Lawley, D. N. (1953). *A modified method of estimation in factor analysis and some large sample results*. Uppsala Symp. Psycho Factor Analysis, 35–42, Almqvist and Wiksell, Sweden.

Morrison, D. F. (1967). *Multivariate Statistical Method*. New York: McGraw-Hill.

Press, J. (1971). *Applied Multivariate Analysis*. New York: Holt.

Rao, C. R. (1955). Estimation and tests of significance in factor analysis. *Psychometrika* 20:93–111.

Solomon, H. (1960). *A survey of mathematical models in factor analysis. Mathematical Thinking in the Measurement of Behaviour*. New York: Free Press, Glenco.

Spearman, C. (1904). General intelligence objectivity determined and measured. *Am. J. Psychol.* 15:201–293.

Thurston, L. (1945). *Multiple Factor Analysis*. Chicago, Illinois: Univ. of Chicago Press.

Bibliography of Related Recent Publications

1. Anderson, S. A. and Perlman, M. D. (1995). Unbiasedness of the likelihood ratio test for lattice conditional independence models. *J. Multivariate Anal.* 53:1–17.

2. Bai, Z. D., Rao, C. R., and Zhao, L. C. (1993). Manova type tests under a convex discrepancy function for the standard multivariate linear model. *Journal of Statistical Planning and Inference* 36:77–90.

3. Berger, J. O. (1993). The present and future of Bayesian multivariate analysis. In: Rao, C. R. ed. *Multivariate analysis—Future Direction*, Elsevier Science Publishers, pp. 25–53.

4. Bilodeau, M. (1994). Minimax estimators of mean vector in normal mixed models. *J. Multivariate Anal.* 52:73–82.

5. Brown, L. D., and Marden, J. I. (1992). Local admissibility and local unbiasedness in hypothesis testing problems. *Ann. Statist.*, 20:832–852.

6. Cohen, A., Kemperman, J. H. B., and Sackrowitz, H. B. (1993). Unbiased tests for normal order restricted hypotheses. *J. Multivariate Anal.* 46:139–153.

7. Eaton, M. L. (1989). Group Invariance Applications in Statistics. Regional Conference Series in Probability and Statistics, 1, Institute of Mathematical Statistics.

8. Fujisawa, H. (1995). A note on the maximum likelihood estimators for multivariate normal distributions with monotone data. *Communications in Statistics, Theory and Method* 24:1377–1382.

9. Guo, Y. Y., and Pal, N. (1993). A sequence of improvements over the James-Stein estimators. *J. Multivariate Anal.* 42:302–317.

10. Johannesson, B., and Giri, N. (1995). On approximation involving the Beta distributions. *Communications in Statistics, Simulation and Computation* 24:489–504.

11. Kono Yoshihiko, K. (1994). Estimation of normal covariance matrix with incomplete date under Stein's loss. *J. Multivariate Anal.* 52:325–337.

12. Kubokawa, T. (1991). An approach to improving the James-Stein estimator. *J. Multivariate Anal.* 36:121–136.

13. Marshall, A. W. and Olkin, I. (1995). Multivariate exponential and geometric distributions with limited memory. *J. Multivariate Anal.* 53:110–125.

14. Pal, N., Sinha, Bikas, K., Choudhuri, G., and Chang Ching-Hui (1995). Estimation of a multivariate normal mean vector and local improvements. *Statistics* 20:1–17.

15. Perron, F. (1992). Minimax estimators of a covariance matrix. *J. Multivariate Anal.* 43:6–28.

16. Pinelis, I. (1994). Extremal probabilistic problems and Hotelling's T^2 test under symmetry conditions. *Ann. Statist.* 22:357–368.

17. Rao, C. R. (Editor, 1993). *Multivariate Analysis-Future Directions*. B. V.: Elsevier Science Publishers.

18. Tracy, D. S., and Jinadasa, K. G. (1988). Patterned matrix derivatives. *Can. J. Statist.* 16:411–418.

19. Wijsman, R. A. (1990). *Invariant Measure on Groups and Their Use in Statistics*. IMS Lecture Notes-Mongraph Series, Institute of Mathematical Statistics.

20. Wong, Chi Song, and Liu Dongsheng (1994). Moments of generalized Wishart distributions. *J. Multivariate Anal.* 52:280–294.

21. Xiaomi, Hu and Wright, F. T. (1994). Monotonicity properties of the power functions of likelihood ratio tests for normal mean hypothesis constrained by a linear space and cone. *Ann. Statist.* 22:1547–1554.

Appendix A
Tables for the Chi-Square Adjustment Factor

$$C = C_{p,r,N-s}(\alpha)$$

$$= \frac{(N - s - \frac{1}{2}(p - r + 1))ln[u_{p,r,N-s}(\alpha)]}{\chi^2_{pr}(\alpha)}$$

for different values of p, r, $M = N - s - p + 1$. To obtain the required percentile point of

$$-(N - s) - \frac{1}{2}(p - r + 1)ln[u_{p,r,N-s}(\alpha)]$$

one multiplies the corresponding upper percentile point of χ^2_{pr} by the tabulated value of the adjustment factor. These tables are reproduced here with the kind permission of the Biometrika Trustees.

Table A.1 Tables of chi-square adjustments to Wilks's criterion U. Factor C for lower percentiles of U (upper percentiles of χ^2)

	$p = 3, r = 4$					$p = 4, r = 4$				
α \ M	0·100	0·050	0·025	0·010	0·005	0·100	0·050	0·025	0·010	0·005
1	1·379	1·422	1·463	1·514	1·550	1·405	1·451	1·494	1·550	1·589
2	1·159	1·174	1·188	1·207	1·220	1·178	1·194	1·209	1·229	1·243
3	1·091	1·099	1·107	1·116	1·123	1·105	1·114	1·122	1·132	1·139
4	1·060	1·065	1·070	1·076	1·080	1·071	1·076	1·081	1·088	1·092
5	1·043	1·046	1·050	1·054	1·057	1·051	1·055	1·058	1·063	1·066
6	1·032	1·035	1·037	1·040	1·042	1·039	1·042	1·044	1·048	1·050
7	1·025	1·027	1·029	1·031	1·033	1·031	1·033	1·035	1·037	1·039
8	1·020	1·022	1·023	1·025	1·026	1·025	1·027	1·028	1·030	1·032
9	1·017	1·018	1·019	1·021	1·022	1·020	1·022	1·023	1·025	1·026
10	1·014	1·015	1·016	1·017	1·018	1·017	1·018	1·019	1·021	1·022
12	1·010	1·011	1·012	1·012	1·013	1·013	1·014	1·014	1·015	1·016
14	1·008	1·008	1·009	1·010	1·010	1·010	1·010	1·011	1·012	1·012
16	1·006	1·007	1·007	1·007	1·008	1·008	1·008	1·009	1·009	1·010
18	1·005	1·005	1·006	1·006	1·006	1·006	1·007	1·007	1·008	1·008
20	1·004	1·004	1·005	1·005	1·005	1·005	1·006	1·006	1·006	1·007
24	1·003	1·003	1·003	1·004	1·004	1·004	1·004	1·004	1·004	1·005
30	1·002	1·002	1·002	1·002	1·002	1·002	1·003	1·003	1·003	1·003
40	1·001	1·001	1·001	1·001	1·001	1·001	1·002	1·002	1·002	1·002
60	1·000	1·001	1·001	1·001	1·001	1·001	1·001	1·001	1·001	1·001
120	1·000	1·000	1·000	1·000	1·000	1·000	1·000	1·000	1·000	1·000
∞	1·000	1·000	1·000	1·000	1·000	1·000	1·000	1·000	1·000	1·000
χ^2_{pq}	18·5494	21·0261	23·3367	26·2170	28·2995	23·5418	26·2962	28·8454	31·9999	34·2672

	$p = 5, r = 4$					$p = 6, r = 4$				
α \ M	0·100	0·050	0·025	0·010	0·005	0·100	0·050	0·025	0·010	0·005
1	1·435	1·483	1·530	1·589	1·632	1·466	1·517	1·566	1·628	1·674
2	1·199	1·216	1·233	1·253	1·269	1·222	1·240	1·257	1·279	1·295
3	1·121	1·130	1·139	1·150	1·158	1·138	1·148	1·157	1·168	1·177
4	1·083	1·089	1·094	1·101	1·106	1·096	1·102	1·108	1·115	1·121
5	1·061	1·065	1·069	1·074	1·077	1·071	1·076	1·080	1·085	1·089
6	1·047	1·050	1·053	1·056	1·059	1·055	1·059	1·062	1·066	1·068
7	1·037	1·040	1·042	1·044	1·046	1·044	1·047	1·049	1·052	1·055
8	1·030	1·032	1·034	1·036	1·038	1·036	1·038	1·040	1·043	1·045
9	1·025	1·027	1·028	1·030	1·031	1·030	1·032	1·034	1·036	1·037
10	1·021	1·023	1·024	1·025	1·026	1·026	1·027	1·029	1·030	1·032
12	1·016	1·017	1·018	1·019	1·020	1·019	1·020	1·021	1·023	1·024
14	1·012	1·013	1·014	1·014	1·015	1·015	1·016	1·017	1·018	1·018
16	1·010	1·010	1·011	1·012	1·012	1·012	1·013	1·013	1·014	1·015
18	1·008	1·008	1·009	1·009	1·010	1·010	1·010	1·011	1·012	1·012
20	1·007	1·007	1·007	1·008	1·008	1·008	1·009	1·009	1·010	1·010
24	1·005	1·005	1·005	1·006	1·006	1·006	1·006	1·007	1·007	1·007
30	1·003	1·003	1·004	1·004	1·004	1·004	1·004	1·004	1·005	1·005
40	1·002	1·002	1·002	1·002	1·002	1·002	1·002	1·003	1·003	1·003
60	1·001	1·001	1·001	1·001	1·001	1·001	1·001	1·001	1·001	1·001
120	1·000	1·000	1·000	1·000	1·000	1·000	1·000	1·000	1·000	1·000
∞	1·000	1·000	1·000	1·000	1·000	1·000	1·000	1·000	1·000	1·000
χ^2_{pq}	28·4120	31·4104	34·1696	37·5662	39·9968	33·1963	36·4151	39·3641	42·9798	45·5585

Tables of chi-square adjustments

Table A.1 (*cont.*)

$M \backslash \alpha$	\multicolumn{5}{c	}{$p = 7, r = 4$}	\multicolumn{5}{c}{$p = 8, r = 4$}
	0·100	0·050	0·025	0·010	0·005	0·100	0·050	0·025	0·010	0·005
1	1·497	1·550	1·601	1·667	1·715	1·528	1·583	1·636	1·704	1·754
2	1·244	1·263	1·281	1·305	1·322	1·266	1·286	1·305	1·330	1·348
3	1·155	1·165	1·175	1·188	1·197	1·172	1·183	1·193	1·207	1·216
4	1·109	1·116	1·122	1·130	1·136	1·123	1·130	1·137	1·146	1·152
5	1·082	1·087	1·092	1·097	1·101	1·093	1·099	1·103	1·109	1·114
6	1·064	1·068	1·071	1·076	1·079	1·074	1·078	1·081	1·086	1·089
7	1·052	1·055	1·057	1·061	1·063	1·060	1·063	1·066	1·070	1·072
8	1·043	1·045	1·047	1·050	1·052	1·050	1·052	1·055	1·058	1·060
9	1·036	1·038	1·040	1·042	1·044	1·042	1·044	1·046	1·048	1·050
10	1·031	1·032	1·034	1·036	1·037	1·036	1·038	1·039	1·041	1·043
12	1·023	1·024	1·026	1·027	1·028	1·027	1·029	1·030	1·031	1·033
14	1·018	1·019	1·020	1·021	1·022	1·021	1·023	1·023	1·025	1·026
16	1·015	1·015	1·016	1·017	1·017	1·017	1·018	1·019	1·020	1·021
18	1·012	1·013	1·013	1·014	1·014	1·014	1·015	1·016	1·016	1·017
20	1·010	1·011	1·011	1·012	1·012	1·012	1 013	1·013	1·014	1·014
24	1·007	1·008	1·008	1·008	1·009	1·009	1·009	1·010	1·010	1·010
30	1·005	1·005	1·005	1·006	1·006	1·006	1·006	1·007	1·007	1·007
40	1·003	1·003	1·003	1·003	1·004	1·003	1·004	1·004	1·004	1·004
60	1·001	1·001	1·002	1·002	1·002	1·002	1·002	1·002	1·002	1·002
120	1·000	1·000	1·000	1·000	1·000	1·000	1·000	1·000	1·001	1·001
∞	1·000	1·000	1·000	1·000	1·000	1·000	1·000	1·000	1·000	1·000
χ^2_{pq}	37·9159	41·3372	44·4607	48·2782	50·9933	42·5847	46·1943	49·4804	53·4858	56·3281

$M \backslash \alpha$	\multicolumn{5}{c	}{$p = 9, r = 4$}	\multicolumn{5}{c}{$p = 10, r = 4$}
	0·100	0·050	0·025	0·010	0·005	0·100	0·050	0·025	0·010	0·005
1	1·557	1·614	1·669	1·740	1·792	1·585	1·644	1·701	1·774	1·828
2	1·288	1·309	1·329	1·355	1·373	1·309	1·331	1·352	1·379	1·398
3	1·189	1·201	1·212	1·226	1·236	1·206	1·218	1·230	1·244	1·255
4	1·137	1·144	1·152	1·161	1·167	1·150	1·159	1·166	1·176	1·183
5	1·105	1·110	1·115	1·122	1·127	1·116	1·122	1·128	1·134	1·139
6	1·083	1·088	1·091	1·096	1·100	1·093	1·097	1·102	1·107	1·111
7	1·068	1·071	1·075	1·078	1·081	1·076	1·080	1·083	1·088	1·090
8	1·057	1·060	1·062	1·065	1·068	1·064	1·067	1·070	1·073	1·076
9	1·048	1·050	1·053	1·055	1·057	1·054	1·057	1·059	1·062	1·064
10	1·041	1·043	1·045	1·047	1·049	1·047	1·049	1·051	1·054	1·055
12	1·032	1·033	1·034	1·036	1·037	1·036	1·038	1·039	1·041	1·042
14	1·025	1·026	1·027	1·029	1·029	1·029	1·030	1·031	1·033	1·034
16	1·020	1·021	1·022	1·023	1·024	1·023	1·024	1·025	1·026	1·027
18	1·017	1·018	1·018	1·019	1·020	1·019	1·020	1·021	1·022	1·023
20	1·014	1·015	1·015	1·016	1·017	1·016	1·017	1·018	1·019	1·019
24	1·010	1·011	1·011	1·012	1·012	1·012	1·013	1·013	1·014	1·014
30	1·007	1·007	1·008	1·008	1·008	1·008	1·009	1·009	1·009	1·010
40	1·004	1·004	1·005	1·005	1·005	1·005	1·005	1·005	1·006	1·006
60	1·002	1·002	1·002	1·002	1·002	1·002	1·003	1·003	1·003	1·003
120	1·001	1·001	1·001	1·001	1·001	1·001	1·001	1·001	1·001	1·001
∞	1·000	1·000	1·000	1·000	1·000	1·000	1·000	1·000	1·000	1·000
χ^2_{pq}	47·2122	50·9985	54·4373	58·6192	61·5812	51·8050	55·7585	59·3417	63·6907	66·7659

Table A.1 (*cont.*)

	$p = 3, r = 6$				
M \ α	0·100	0·050	0·025	0·010	0·005
1	1·482	1·535	1·586	1·649	1·694
2	1·222	1·241	1·259	1·282	1·298
3	1·135	1·145	1·155	1·167	1·176
4	1·092	1·099	1·105	1·113	1·119
5	1·068	1·072	1·077	1·082	1·086
6	1·052	1·056	1·059	1·063	1·066
7	1·041	1·044	1·047	1·050	1·052
8	1·034	1·036	1·038	1·041	1·042
9	1·028	1·030	1·032	1·034	1·035
10	1·024	1·025	1·027	1·028	1·030
12	1·018	1·019	1·020	1·021	1·022
14	1·014	1·014	1·015	1·016	1·017
16	1·011	1·012	1·012	1·013	1·014
18	1·009	1·009	1·010	1·011	1·011
20	1·007	1·008	1·008	1·009	1·009
24	1·005	1·006	1·006	1·006	1·007
30	1·004	1·004	1·004	1·004	1·004
40	1·002	1·002	1·002	1·002	1·003
60	1·001	1·001	1·001	1·001	1·001
120	1·000	1·000	1·000	1·000	1·000
∞	1·000	1·000	1·000	1·000	1·000
χ^2_{pq}	25·9894	28·8693	31·5264	34·8053	37·1564

$p = 4, r = 6$

The values for C are obtained by entering the table on page 532 for $p = 6$, $r = 4$ with $M = n - 3$

	$p = 5, r = 6$					$p = 6, r = 6$				
M \ α	0·100	0·050	0·025	0·010	0·005	0·100	0·050	0·025	0·010	0·005
1	1·465	1·514	1·563	1·625	1·671	1·471	1·520	1·568	1·631	1·677
2	1·228	1·245	1·262	1·284	1·300	1·237	1·255	1·272	1·294	1·310
3	1·144	1·154	1·163	1·175	1·183	1·153	1·163	1·172	1·183	1·192
4	1·102	1·108	1·114	1·121	1·127	1·109	1·116	1·122	1·129	1·134
5	1·077	1·081	1·085	1·090	1·094	1·083	1·088	1·092	1·097	1·101
6	1·060	1·063	1·066	1·070	1·073	1·066	1·069	1·072	1·076	1·079
7	1·048	1·051	1·053	1·056	1·059	1·053	1·056	1·058	1·061	1·064
8	1·040	1·042	1·044	1·046	1·048	1·044	1·046	1·048	1·051	1·053
9	1·034	1·035	1·037	1·039	1·040	1·037	1·039	1·041	1·043	1·044
10	1·029	1·030	1·031	1·033	1·034	1·032	1·034	1·035	1·037	1·038
12	1·022	1·023	1·024	1·025	1·026	1·024	1·025	1·026	1·028	1·029
14	1·017	1·018	1·019	1·019	1·020	1·019	1·020	1·021	1·022	1·022
16	1·014	1·014	1·015	1·016	1·016	1·015	1·016	1·017	1·018	1·018
18	1·011	1·012	1·012	1·013	1·013	1·013	1·013	1·014	1·014	1·015
20	1·009	1·010	1·010	1·011	1·011	1·011	1·011	1·012	1·012	1·013
24	1·007	1·007	1·007	1·008	1·008	1·008	1·008	1·009	1·009	1·009
30	1·005	1·005	1·005	1·005	1·005	1·005	1·006	1·006	1·006	1·006
40	1·003	1·003	1·003	1·003	1·003	1·003	1·003	1·003	1·004	1·004
60	1·001	1·001	1·001	1·001	1·002	1·002	1·002	1·002	1·002	1·002
120	1·000	1·000	1·000	1·000	1·000	1·000	1·000	1·000	1·000	1·000
∞	1·000	1·000	1·000	1·000	1·000	1·000	1·000	1·000	1·000	1·000
χ^2_{pq}	40·2560	43·7730	46·9792	50·8922	53·6720	47·2122	50·9985	54·4373	58·6192	61·5812

Tables of chi-square adjustments

Table A.1 (*cont.*)

M \ α	\multicolumn{5}{c}{$p = 7, r = 6$}	\multicolumn{5}{c}{$p = 8, r = 6$}
	0·100	0·050	0·025	0·010	0·005	0·100	0·050	0·025	0·010	0·005
1	1·481	1·530	1·579	1·642	1·688	1·494	1·543	1·592	1·656	1·703
2	1·249	1·266	1·284	1·306	1·322	1·261	1·279	1·297	1·319	1·336
3	1·163	1·173	1·182	1·194	1·203	1·174	1·184	1·194	1·205	1·214
4	1·118	1·124	1·131	1·138	1·144	1·127	1·134	1·140	1·148	1·153
5	1·090	1·095	1·099	1·105	1·109	1·098	1·103	1·108	1·113	1·117
6	1·072	1·075	1·079	1·083	1·086	1·079	1·082	1·086	1·090	1·093
7	1·059	1·062	1·064	1·067	1·070	1·065	1·068	1·070	1·074	1·076
8	1·049	1·051	1·053	1·056	1·058	1·054	1·057	1·059	1·062	1·063
9	1·042	1·043	1·045	1·047	1·049	1·046	1·048	1·050	1·052	1·054
10	1·036	1·037	1·039	1·041	1·042	1·040	1·042	1·043	1·045	1·046
12	1·027	1·029	1·030	1·031	1·032	1·031	1·032	1·033	1·035	1·036
14	1·022	1·023	1·023	1·024	1·025	1·024	1·025	1·026	1·027	1·028
16	1·018	1·018	1·019	1·020	1·020	1·020	1·021	1·021	1·022	1·022
18	1·014	1·015	1·016	1·016	1 017	1·016	1·017	1·018	1·018	1·019
20	1·012	1·013	1·013	1·014	1·014	1·014	1·014	1·015	1·015	1·016
24	1·009	1·009	1·010	1·010	1·010	1·010	1·011	1·011	1·011	1·012
30	1·006	1·006	1·007	1·007	1·007	1·007	1·007	1·008	1·008	1·008
40	1·004	1·004	1·004	1·004	1·004	1·004	1·004	1·005	1·005	1·005
60	1·002	1·002	1·002	1·002	1·002	1·002	1·002	1·002	1·002	1·002
120	1·000	1·000	1·000	1·001	1·001	1·001	1·001	1·001	1·001	1·001
∞	1·000	1·000	1·000	1·000	1·000	1·000	1·000	1·000	1·000	1·000
χ^2_{pq}	54·0902	58·1240	61·7768	66·2062	69·3360	60·9066	65·1708	69·0226	73·6826	76·9688

M \ α	\multicolumn{5}{c}{$p = 9, r = 6$}	\multicolumn{5}{c}{$p = 10, r = 6$}
	0·100	0·050	0·025	0·010	0·005	0·100	0·050	0·025	0·010	0·005
1	1·508	1·558	1·607	1·671	1·719	1·523	1·573	1·623	1·687	1·736
2	1·275	1·293	1·311	1·333	1·350	1·288	1·307	1·325	1·348	1·365
3	1·185	1·196	1·205	1·218	1·227	1·197	1·208	1·218	1·230	1·239
4	1·137	1·144	1·150	1·158	1·164	1·147	1·154	1·161	1·169	1·175
5	1·107	1·112	1·116	1·122	1·126	1·115	1·120	1·125	1·131	1·135
6	1·086	1·090	1·093	1·098	1·101	1·093	1·097	1·101	1·106	1·109
7	1·071	1·074	1·077	1·080	1·083	1·078	1·081	1·084	1·087	1·090
8	1·060	1·062	1·065	1·067	1·069	1·066	1·068	1·071	1·074	1·076
9	1·051	1·053	1·055	1·058	1·059	1·056	1·059	1·061	1·063	1·065
10	1·044	1·046	1·048	1·050	1·051	1·049	1·051	1·053	1·055	1·056
12	1·034	1·035	1·036	1·037	1·037	1·038	1·040	1·041	1·042	1·043
14	1·027	1·028	1·029	1·031	1·031	1·031	1·032	1·033	1·034	1·034
16	1·022	1·023	1·024	1·024	1·025	1·025	1·026	1·026	1·027	1·028
18	1·019	1·019	1·020	1·020	1·021	1·021	1·021	1·022	1·023	1·023
20	1·016	1·016	1·017	1·017	1·018	1·018	1·018	1·019	1·019	1·020
24	1·012	1·012	1·013	1·013	1·013	1·013	1·014	1·014	1·015	1·015
30	1·008	1·008	1·009	1·009	1·009	1·009	1·010	1·010	1·010	1·010
40	1·005	1·005	1·005	1·005	1·006	1·006	1·006	1·006	1·006	1·006
60	1·002	1·002	1·003	1·003	1·003	1·003	1·003	1·003	1·003	1·003
120	1·001	1·001	1·001	1·001	1·002	1·001	1·001	1·001	1·001	1·001
∞	1·000	1·000	1·000	1·000	1·000	1·000	1·000	1·000	1·000	1·000
χ^2_{pq}	67·6728	72·1532	76·1921	81·0688	84·5016	74·3970	79·0819	83·2976	88·3794	91·9517

Table A.1 (*cont.*)

	$p = 3, r = 8$					$p = 5, r = 8$				
α \\ M	0·100	0·050	0·025	0·010	0·005	0·100	0·050	0·025	0·010	0·005
1	1·572	1·632	1·690	1·763	1·816	1·505	1·556	1·607	1·672	1·721
2	1·280	1·302	1·324	1·350	1·370	1·261	1·280	1·298	1·321	1·338
3	1·177	1·190	1·201	1·216	1·227	1·171	1·182	1·192	1·204	1·213
4	1·125	1·133	1·141	1·150	1·157	1·124	1·131	1·137	1·145	1·151
5	1·094	1·100	1·105	1·112	1·117	1·095	1·100	1·105	1·110	1·114
6	1·073	1·078	1·082	1·087	1·091	1·076	1·079	1·083	1·087	1·090
7	1·059	1·063	1·066	1·070	1·073	1·062	1·065	1·068	1·071	1·073
8	1·049	1·052	1·054	1·058	1·060	1·052	1·054	1·056	1·059	1·061
9	1·041	1·043	1·046	1·048	1·050	1·044	1·046	1·048	1·050	1·052
10	1·035	1·037	1·039	1·041	1·043	1·038	1·039	1·041	1·043	1·044
12	1·026	1·028	1·029	1·031	1·032	1·029	1·030	1·031	1·033	1·034
14	1·021	1·022	1·023	1·024	1·025	1·023	1·024	1·025	1·026	1·027
16	1·017	1·018	1·018	1·019	1·020	1·018	1·019	1·020	1·021	1·022
18	1·014	1·014	1·015	1·016	1·017	1·015	1·016	1·017	1·017	1·018
20	1·011	1·012	1·013	1·013	1·014	1·013	1·013	1·014	1·015	1·015
24	1·008	1·009	1·009	1·010	1·010	1·009	1·010	1·010	1·011	1·011
30	1·006	1·006	1·006	1·007	1·007	1·006	1·007	1·007	1·007	1·008
40	1·003	1·004	1·004	1·004	1·004	1·004	1·004	1·004	1·004	1·005
60	1·002	1·002	1·002	1·002	1·002	1·002	1·002	1·002	1·002	1·002
120	1·000	1·000	1·000	1·000	1·001	1·001	1·001	1·001	1·001	1·001
∞	1·000	1·000	1·000	1·000	1·000	1·000	1·000	1·000	1·000	1·000
χ^2_{pq}	33·1963	36·4151	39·3641	42·9798	45·5585	51·8050	55·7585	59·3417	63·6907	66·7659

	$p = 7, r = 8$					$p = 8, r = 8$				
α \\ M	0·100	0·050	0·025	0·010	0·005	0·100	0·050	0·025	0·010	0·005
1	1·490	1·538	1·586	1·648	1·694	1·491	1·538	1·585	1·646	1·692
2	1·265	1·282	1·299	1·321	1·337	1·270	1·288	1·305	1·326	1·342
3	1·179	1·189	1·198	1·210	1·218	1·185	1·195	1·204	1·215	1·224
4	1·132	1·139	1·145	1·152	1·158	1·138	1·144	1·150	1·158	1·163
5	1·103	1·108	1·112	1·117	1·121	1·108	1·113	1·117	1·123	1·126
6	1·083	1·086	1·090	1·094	1·097	1·088	1·091	1·095	1·099	1·102
7	1·068	1·071	1·074	1·077	1·080	1·073	1·076	1·078	1·082	1·084
8	1·058	1·060	1·062	1·065	1·067	1·061	1·064	1·066	1·069	1·071
9	1·049	1·051	1·053	1·055	1·057	1·053	1·055	1·057	1·059	1·060
10	1·043	1·044	1·046	1·048	1·049	1·046	1·048	1·049	1·051	1·052
12	1·033	1·034	1·035	1·037	1·038	1·036	1·038	1·039	1·040	1·041
14	1·026	1·027	1·028	1·028	1·029	1·028	1·030	1·031	1·031	1·032
16	1·021	1·022	1·023	1·023	1·024	1·023	1·025	1·026	1·026	1·027
18	1·018	1·018	1·019	1·020	1·020	1·020	1·021	1·022	1·022	1·023
20	1·015	1·016	1·016	1·017	1·017	1·017	1·017	1·018	1·018	1·018
24	1·011	1·012	1·012	1·012	1·013	1·012	1·013	1·013	1·014	1·014
30	1·008	1·008	1·008	1·009	1·009	1·009	1·009	1·009	1·009	1·010
40	1·005	1·005	1·005	1·005	1·005	1·005	1·005	1·006	1·006	1·006
60	1·002	1·002	1·002	1·003	1·003	1·003	1·003	1·003	1·003	1·003
120	1·001	1·001	1·001	1·001	1·001	1·001	1·001	1·001	1·001	1·001
∞	1·000	1·000	1·000	1·000	1·000	1·000	1·000	1·000	1·000	1·000
χ^2_{pq}	69·9185	74·4683	78·5672	83·5134	86·9938	78·8596	83·6753	88·0041	93·2169	96·8781

Tables of chi-square adjustments

Table A.1 (*cont.*)

	$p = 3, r = 10$					$p = 5, r = 10$				
α \\ M	0·100	0·050	0·025	0·010	0·005	0·100	0·050	0·025	0·010	0·005
1	1·650	1·716	1·781	1·862	1·921	1·547	1·600	1·653	1·721	1·772
2	1·333	1·359	1·383	1·413	1·435	1·295	1·315	1·334	1·359	1·377
3	1·218	1·232	1·245	1·262	1·274	1·199	1·211	1·221	1·235	1·244
4	1·157	1·167	1·175	1·187	1·195	1·147	1·155	1·162	1·171	1·177
5	1·120	1·127	1·133	1·141	1·147	1·115	1·120	1·125	1·131	1·136
6	1·095	1·101	1·106	1·112	1·116	1·092	1·097	1·101	1·105	1·109
7	1·078	1·082	1·086	1·091	1·094	1·076	1·080	1·083	1·087	1·089
8	1·065	1·068	1·072	1·075	1·078	1·064	1·067	1·070	1·073	1·075
9	1·055	1·058	1·061	1·064	1·066	1·055	1·057	1·059	1·062	1·064
10	1·047	1·050	1·052	1·055	1·057	1·048	1·050	1·051	1·054	1·055
12	1·036	1·038	1·040	1·042	1·043	1·037	1·038	1·040	1·041	1·043
14	1·029	1·030	1·031	1·033	1·034	1·029	1·031	1·032	1·033	1·034
16	1·023	1·024	1·025	1·027	1·028	1·024	1·025	1·026	1·027	1·028
18	1·019	1·020	1·021	1·022	1·023	1·020	1·021	1·022	1·022	1·023
20	1·016	1·017	1·018	1·019	1·019	1·017	1·018	1·018	1·019	1·019
24	1·012	1·012	1·013	1·014	1·014	1·013	1·013	1·014	1·014	1·014
30	1·008	1·009	1·009	1·009	1·010	1·009	1·009	1·009	1·010	1·010
40	1·005	1·005	1·005	1·006	1·006	1·005	1·005	1·006	1·006	1·006
60	1·002	1·002	1·003	1·003	1·003	1·003	1·003	1·003	1·003	1·003
120	1·001	1·001	1·001	1·001	1·001	1·001	1·001	1·001	1·002	1·002
∞	1·000	1·000	1·000	1·000	1·000	1·000	1·000	1·000	1·000	1·000
χ^2_{pq}	40·2560	43·7730	46·9792	50·8922	53·6720	63·1671	67·5048	71·4202	76·1539	79·4900

	$p = 7, r = 10$				
α \\ M	0·100	0·050	0·025	0·010	0·005
1	1·509	1·557	1·604	1·666	1·713
2	1·285	1·303	1·320	1·342	1·359
3	1·197	1·208	1·217	1·229	1·238
4	1·148	1·155	1·162	1·169	1·175
5	1·117	1·122	1·127	1·132	1·136
6	1·096	1·099	1·103	1·107	1·110
7	1·080	1·083	1·086	1·089	1·091
8	1·068	1·070	1·073	1·075	1·077
9	1·058	1·060	1·062	1·065	1·066
10	1·051	1·053	1·054	1·056	1·058
12	1·040	1·042	1·042	1·044	1·045
14	1·032	1·034	1·034	1·036	1·036
16	1·026	1·028	1·028	1·029	1·029
18	1·022	1·023	1·023	1·024	1·024
20	1·019	1·019	1·020	1·020	1·021
24	1·014	1·014	1·015	1·015	1·016
30	1·010	1·010	1·010	1·011	1·011
40	1·006	1·006	1·006	1·007	1·007
60	1·003	1·003	1·003	1·003	1·003
120	1·001	1·001	1·001	1·001	1·001
∞	1·000	1·000	1·000	1·000	1·000
χ^2_{pq}	85·5271	90·5312	95·0231	100·4250	104·2150

For $p = 4, r = 10$ enter the table on p. 533 for $p = 10, r = 4$ with $M = n-3$.
For $p = 6, r = 10$ enter the table on p. 535 for $p = 10, r = 6$ with $M = n-5$.

Table A.2 Chi-square adjustments to Wilks's criterion U. Factor C for lower percentiles of U (upper percentiles of χ^2), $p = 3$

	$p = 3, r = 12$					$p = 3, r = 14$				
α \ M	0·100	0·050	0·025	0·010	0·005	0·100	0·050	0·025	0·010	0·005
1	1·718	1·791	1·860	1·949	2·013	1·780	1·857	1·931	2·026	2·095
2	1·382	1·410	1·437	1·470	1·495	1·427	1·458	1·486	1·523	1·549
3	1·256	1·272	1·287	1·306	1·319	1·292	1·309	1·326	1·346	1·361
4	1·188	1·199	1·209	1·221	1·230	1·217	1·229	1·240	1·254	1·264
5	1·146	1·154	1·161	1·170	1·176	1·171	1·179	1·188	1·198	1·205
6	1·117	1·123	1·129	1·136	1·141	1·138	1·145	1·152	1·159	1·165
7	1·097	1·101	1·106	1·111	1·115	1·115	1·121	1·126	1·132	1·136
8	1·081	1·085	1·089	1·093	1·097	1·097	1·102	1·106	1·111	1·115
9	1·069	1·073	1·076	1·080	1·082	1·084	1·088	1·091	1·095	1·099
10	1·060	1·063	1·066	1·069	1·071	1·073	1·076	1·079	1·082	1·085
12	1·046	1·048	1·050	1·053	1·054	1·057	1·059	1·061	1·064	1·066
14	1·037	1·039	1·040	1·042	1·043	1·046	1·048	1·049	1·052	1·053
16	1·030	1·032	1·033	1·034	1·035	1·037	1·039	1·041	1·042	1·044
18	1·025	1·026	1·027	1·029	1·029	1·031	1·033	1·034	1·035	1·036
20	1·021	1·022	1·023	1·024	1·025	1·027	1·028	1·029	1·030	1·031
24	1·016	1·017	1·017	1·018	1·019	1·020	1·021	1·022	1·023	1·023
30	1·011	1·011	1·012	1·012	1·013	1·014	1·015	1·015	1·016	1·016
40	1·007	1·007	1·007	1·008	1·008	1·009	1·009	1·009	1·010	1·010
60	1·003	1·003	1·004	1·004	1·004	1·004	1·004	1·005	1·005	1·005
120	1·001	1·001	1·001	1·001	1·001	1·001	1·001	1·001	1·001	1·001
∞	1·000	1·000	1·000	1·000	1·000	1·000	1·000	1·000	1·000	1·000
$\chi^2_{pf_2}$	47·2122	50·9985	54·4373	58·6192	61·5812	54·0902	58·1240	61·7768	66·2062	69·3360

	$r = 16$					$p = 3, r = 18$				
α \ M	0·100	0·050	0·025	0·010	0·005	0·100	0·050	0·025	0·010	0·005
1	1·835	1·916	1·995	2·095	2·169	1·886	1·971	2·053	2·158	2·235
2	1·469	1·501	1·532	1·571	1·599	1·508	1·542	1·575	1·616	1·646
3	1·325	1·344	1·362	1·384	1·400	1·357	1·377	1·396	1·420	1·437
4	1·245	1·258	1·271	1·285	1·296	1·272	1·286	1·299	1·315	1·327
5	1·195	1·204	1·213	1·224	1·232	1·218	1·228	1·238	1·249	1·258
6	1·159	1·167	1·174	1·182	1·188	1·179	1·188	1·195	1·204	1·211
7	1·133	1·139	1·145	1·152	1·157	1·151	1·158	1·164	1·171	1·177
8	1·114	1·119	1·123	1·129	1·133	1·129	1·135	1·140	1·146	1·151
9	1·098	1·102	1·106	1·111	1·115	1·112	1·117	1·121	1·127	1·130
10	1·085	1·089	1·092	1·097	1·099	1·099	1·103	1·107	1·111	1·114
12	1·067	1·070	1·073	1·076	1·078	1·078	1·081	1·084	1·087	1·090
14	1·054	1·057	1·059	1·061	1·063	1·064	1·066	1·068	1·071	1·073
16	1·045	1·047	1·049	1·051	1·052	1·053	1·055	1·057	1·059	1·061
18	1·038	1·039	1·041	1·043	1·044	1·045	1·046	1·048	1·050	1·051
20	1·032	1·034	1·035	1·036	1·037	1·038	1·040	1·041	1·043	1·044
24	1·025	1·026	1·026	1·027	1·028	1·029	1·030	1·031	1·032	1·033
30	1·017	1·018	1·018	1·019	1·020	1·021	1·021	1·022	1·023	1·023
40	1·011	1·011	1·011	1·012	1·012	1·013	1·013	1·014	1·014	1·015
60	1·005	1·006	1·006	1·006	1·006	1·006	1·007	1·007	1·007	1·007
120	1·002	1·002	1·002	1·002	1·002	1·002	1·002	1·002	1·002	1·002
∞	1·000	1·000	1·000	1·000	1·000	1·000	1·000	1·000	1·000	1·000
$\chi^2_{pf_2}$	60·9066	65·1708	69·0226	73·6826	76·9688	67·6728	72·1532	76·1920	81·0688	84·5019

Tables of chi-square adjustments

Table A.2 (*cont.*)

M \ α	$p=3, r=20$					$p=3, r=22$				
	0·100	0·050	0·025	0·010	0·005	0·100	0·050	0·025	0·010	0·005
1	1·932	2·021	2·106	2·216	2·297	1·975	2·067	2·156	2·269	2·353
2	1·544	1·580	1·614	1·657	1·689	1·578	1·616	1·651	1·696	1·729
3	1·387	1·408	1·428	1·453	1·472	1·415	1·438	1·459	1·485	1·504
4	1·298	1·313	1·327	1·344	1·356	1·322	1·338	1·353	1·371	1·384
5	1·240	1·251	1·261	1·274	1·283	1·261	1·273	1·284	1·297	1·307
6	1·199	1·208	1·216	1·226	1·233	1·218	1·227	1·236	1·246	1·254
7	1·168	1·176	1·182	1·190	1·196	1·185	1·193	1·200	1·209	1·215
8	1·145	1·151	1·157	1·163	1·168	1·160	1·167	1·173	1·180	1·185
9	1·127	1·132	1·136	1·142	1·146	1·142	1·147	1·151	1·157	1·161
10	1·112	1·116	1·120	1·125	1·128	1·124	1·129	1·133	1·139	1·141
12	1·089	1·092	1·095	1·099	1·102	1·099	1·103	1·106	1·110	1·113
14	1·073	1·075	1·078	1·081	1·083	1·082	1·085	1·087	1·091	1·093
16	1·061	1·063	1·065	1·067	1·069	1·069	1·071	1·073	1·076	1·078
18	1·052	1·053	1·055	1·057	1·059	1·059	1·061	1·063	1·065	1·066
20	1·044	1·046	1·048	1·049	1·050	1·051	1·052	1·054	1·056	1·057
24	1·034	1·035	1·036	1·038	1·039	1·039	1·040	1·041	1·043	1·044
30	1·024	1·025	1·026	1·027	1·027	1·028	1·029	1·030	1·031	1·031
40	1·015	1·016	1·016	1·017	1·017	1·018	1·018	1·019	1·020	1·020
60	1·008	1·008	1·008	1·009	1·009	1·009	1·009	1·010	1·010	1·010
120	1·002	1·002	1·002	1·002	1·003	1·003	1·003	1·003	1·003	1·003
∞	1·000	1·000	1·000	1·000	1·000	1·000	1·000	1·000	1·000	1·000
$\chi^2_{pf_2}$	74·3970	79·0819	83·2976	88·3794	91·9517	81·0855	85·9649	90·3489	95·6257	99·3304

Table A.2 (*cont.*)

	$p=4, r=12$					$p=4, r=14$				
α M	0·100	0·050	0·025	0·010	0·005	0·100	0·050	0·025	0·010	0·005
1	1·638	1·700	1·760	1·838	1·895	1·686	1·751	1·814	1·896	1·956
2	1·350	1·373	1·396	1·424	1·446	1·388	1·413	1·436	1·467	1·489
3	1·238	1·252	1·264	1·280	1·292	1·269	1·284	1·297	1·314	1·327
4	1·177	1·186	1·195	1·205	1·213	1·203	1·213	1·222	1·234	1·242
5	1·139	1·145	1·152	1·159	1·165	1·161	1·168	1·175	1·183	1·189
6	1·112	1·118	1·122	1·128	1·132	1·131	1·137	1·142	1·149	1·154
7	1·093	1·097	1·101	1·106	1·109	1·110	1·115	1·119	1·124	1·128
8	1·079	1·082	1·085	1·089	1·092	1·094	1·097	1·101	1·105	1·109
9	1·068	1·070	1·073	1·076	1·079	1·081	1·084	1·087	1·091	1·093
10	1·059	1·061	1·063	1·066	1·068	1·071	1·073	1·076	1·079	1·081
12	1·046	1·047	1·049	1·051	1·053	1·055	1·058	1·059	1·062	1·064
14	1·037	1·038	1·039	1·041	1·042	1·045	1·046	1·048	1·050	1·051
16	1·030	1·031	1·032	1·033	1·034	1·037	1·038	1·039	1·041	1·042
18	1·025	1·026	1·027	1·028	1·029	1·031	1·032	1·033	1·034	1·035
20	1·021	1·022	1·023	1·024	1·024	1·026	1·027	1·028	1·029	1·030
24	1·016	1·017	1·017	1·018	1·018	1·020	1·021	1·021	1·022	1·023
30	1·011	1·011	1·012	1·012	1·013	1·014	1·014	1·015	1·015	1·016
40	1·007	1·007	1·007	1·008	1·008	1·009	1·009	1·009	1·009	1·010
60	1·003	1·003	1·004	1·004	1·004	1·004	1·004	1·005	1·005	1·005
120	1·001	1·001	1·001	1·001	1·001	1·001	1·001	1·001	1·001	1·001
∞	1·000	1·000	1·000	1·000	1·000	1·000	1·000	1·000	1·000	1·000
$\chi^2_{pf_2}$	60·9066	65·1708	69·0226	73·6826	76·9688	69·9185	74·4683	78·5671	83·5134	86·9937

	$p=4, r=16$					$p=4, r=18$				
α M	0·100	0·050	0·025	0·010	0·005	0·100	0·050	0·025	0·010	0·005
1	1·731	1·799	1·864	1·949	2·012	1·773	1·843	1·911	1·999	2·065
2	1·423	1·450	1·475	1·507	1·531	1·457	1·485	1·511	1·545	1·570
3	1·299	1·314	1·329	1·347	1·360	1·327	1·343	1·359	1·378	1·392
4	1·228	1·239	1·249	1·261	1·270	1·252	1·264	1·274	1·287	1·297
5	1·182	1·190	1·198	1·207	1·213	1·203	1·212	1·220	1·230	1·237
6	1·150	1·157	1·163	1·169	1·174	1·169	1·176	1·182	1·189	1·195
7	1·127	1·132	1·136	1·142	1·146	1·143	1·149	1·154	1·160	1·164
8	1·108	1·113	1·117	1·121	1·125	1·123	1·128	1·132	1·137	1·141
9	1·094	1·098	1·101	1·105	1·108	1·107	1·111	1·115	1·119	1·122
10	1·083	1·086	1·089	1·092	1·094	1·095	1·098	1·101	1·105	1·108
12	1·065	1·068	1·070	1·073	1·074	1·075	1·078	1·080	1·083	1·085
14	1·053	1·055	1·056	1·058	1·060	1·061	1·063	1·065	1·068	1·069
16	1·044	1·045	1·047	1·049	1·050	1·051	1·053	1·054	1·056	1·058
18	1·037	1·038	1·040	1·041	1·042	1·044	1·045	1·046	1·048	1·049
20	1·032	1·033	1·034	1·035	1·036	1·037	1·039	1·040	1·041	1·042
24	1·024	1·025	1·026	1·027	1·027	1·029	1·030	1·030	1·031	1·032
30	1·017	1·018	1·018	1·019	1·019	1·020	1·021	1·022	1·022	1·023
40	1·011	1·011	1·011	1·012	1·012	1·013	1·013	1·014	1·014	1·014
60	1·005	1·005	1·006	1·006	1·006	1·006	1·007	1·007	1·007	1·007
120	1·001	1·002	1·002	1·002	1·002	1·002	1·002	1·002	1·002	1·002
∞	1·000	1·000	1·000	1·000	1·000	1·000	1·000	1·000	1·000	1·000
$\chi^2_{pf_2}$	78·8597	83·6753	88·0040	93·2168	96·8781	87·7431	92·8083	97·3531	102·816	106·648

Tables of chi-square adjustments

Table A.2 (*cont.*)

M \ α	$p=5,\ r=12$		$p=6,\ r=14$		$p=5,\ r=16$	
	0·050	0·010	0·050	0·010	0·050	0·010
1	1·643	1·768	1·683	1·813	1·722	1·855
2	1·350	1·396	1·383	1·431	1·415	1·465
3	1·240	1·265	1·267	1·294	1·294	1·323
4	1·179	1·196	1·203	1·221	1·226	1·245
5	1·141	1·153	1·161	1·174	1·181	1·196
6	1·114	1·124	1·132	1·143	1·150	1·161
7	1·095	1·103	1·111	1·119	1·127	1·136
8	1·081	1·087	1·095	1·102	1·109	1·116
9	1·070	1·075	1·082	1·088	1·095	1·101
10	1·061	1·065	1·072	1·077	1·083	1·089
12	1·047	1·051	1·057	1·060	1·066	1·070
14	1·038	1·040	1·045	1·048	1·053	1·057
16	1·031	1·033	1·038	1·040	1·045	1·047
18	1·026	1·028	1·032	1·034	1·038	1·040
20	1·022	1·024	1·027	1·029	1·033	1·034
24	1·017	1·018	1·021	1·022	1·025	1·026
30	1·012	1·012	1·014	1·015	1·018	1·019
40	1·007	1·008	1·009	1·010	1·011	1·012
60	1·004	1·004	1·004	1·005	1·006	1·006
120	1·001	1·001	1·001	1·001	1·002	1·002
∞	1·000	1·000	1·000	1·000	1·000	1·000
$\chi^2_{pf_2}$	79·0819	88·3794	90·5312	100·4250	101·8790	112·3290

M \ α	$p=6,\ r=11$		$p=6,\ r=12$		$p=6,\ r=13$	
	0·050	0·010	0·050	0·010	0·050	0·010
1	1·589	1·704	1·605	1·722	1·621	1·739
2	1·321	1·363	1·335	1·378	1·349	1·393
3	1·220	1·243	1·232	1·255	1·244	1·268
4	1·164	1·180	1·175	1·191	1·185	1·201
5	1·129	1·140	1·138	1·150	1·148	1·159
6	1·105	1·114	1·113	1·122	1·121	1·130
7	1·088	1·094	1·095	1·102	1·102	1·109
8	1·074	1·080	1·081	1·086	1·088	1·093
9	1·064	1·069	1·070	1·075	1·075	1·080
10	1·055	1·060	1·061	1·065	1·066	1·070
12	1·044	1·047	1·048	1·051	1·052	1·055
14	1·035	1·037	1·038	1·040	1·042	1·044
16	1·029	1·030	1·032	1·033	1·035	1·037
18	1·024	1·026	1·027	1·028	1·029	1·031
20	1·020	1·022	1·023	1·024	1·025	1·026
24	1·015	1·016	1·017	1·018	1·019	1·020
30	1·011	1·011	1·012	1·013	1·013	1·014
40	1·007	1·007	1·007	1·008	1·008	1·009
60	1·003	1·003	1·004	1·004	1·004	1·004
120	1·001	1·001	1·001	1·001	1·001	1·001
∞	1·000	1·000	1·000	1·000	1·000	1·000
$\chi^2_{pf_2}$	85·9649	95·6257	92·8083	102·8160	99·6169	109·9580

Table A.2 (*cont.*)

	$p=4,\ r=20$					$p=4,\ r=22$				
α M	0·100	0·050	0·025	0·010	0·005	0·100	0·050	0·025	0·010	0·005
1	1·812	1·884	1·954	2·045	2·113	1·848	1·922	1·994	2·088	2·158
2	1·488	1·518	1·545	1·580	1·606	1·518	1·549	1·577	1·614	1·641
3	1·353	1·371	1·387	1·408	1·422	1·379	1·397	1·414	1·436	1·451
4	1·275	1·288	1·299	1·313	1·323	1·298	1·310	1·322	1·337	1·347
5	1·224	1·233	1·241	1·252	1·259	1·243	1·253	1·262	1·273	1·281
6	1·187	1·194	1·201	1·208	1·215	1·204	1·212	1·219	1·228	1·234
7	1·159	1·165	1·170	1·177	1·182	1·175	1·181	1·187	1·194	1·199
8	1·138	1·143	1·147	1·153	1·157	1·152	1·157	1·162	1·168	1·172
9	1·121	1·125	1·129	1·133	1·137	1·134	1·138	1·142	1·147	1·151
10	1·107	1·110	1·114	1·118	1·121	1·119	1·123	1·126	1·130	1·134
12	1·086	1·088	1·091	1·094	1·096	1·095	1·098	1·101	1·104	1·107
14	1·070	1·072	1·074	1·077	1·078	1·079	1·081	1·083	1·086	1·088
16	1·059	1·061	1·062	1·064	1·066	1·066	1·068	1 070	1·072	1·074
18	1·050	1·052	1·053	1·055	1·056	1·057	1·058	1·060	1·062	1·063
20	1·043	1·045	1·046	1·047	1·048	1·049	1·051	1·052	1·053	1·055
24	1·033	1·034	1·035	1·036	1·037	1·038	1·039	1·040	1·041	1·042
30	1·024	1·024	1·025	1·026	1·026	1·027	1·028	1·029	1·030	1·030
40	1·015	1·016	1·016	1·016	1·017	1·017	1·018	1·018	1·019	1·019
60	1·008	1·008	1·008	1·008	1·008	1·009	1·009	1·009	1·010	1·010
120	1·002	1·002	1·002	1·002	1·002	1·003	1·003	1·003	1·003	1·003
∞	1·000	1·000	1·000	1·000	1·000	1·000	1·000	1·000	1·000	1·000
$\chi^2_{pf_2}$	96·5782	101·879	106·629	112·329	116·321	105·372	110·898	115·841	121·767	125·913

Appendix B
Publications of the Author

Books

1. Introduction to probability and statistics, part I: probability, 1974, Marcel Dekker, N.Y.
2. Introduction to probability and statistics, part II: statistics, 1975, Marcel Dekker, N.Y.
3. Invariance and minimax statistical tests, 1975, University press of Canada and Hindusthan publishing corporation.
4. Multivariate statistical inference, 1977, Academic press, N.Y.
5. Design and analysis of experiments (with M. N. Das), 1979, Wiley Eastern Limited.
6. Analysis of variance, 1986, South Asian Publisher and National Book Trust Of India.
7. Design and analysis of experiments (with M. N. Das), Second edition, 1986, Wiley Eastern Limited and John Wiley, N.Y. (Halsted Press).
8. Introduction to probability and statistics, second edition, 1993, revised and expanded, Marcel Dekker, N.Y.
9. Basic statistics (with S. R. Chakravorty), 1997, South Asian Publisher.
10. Group invariance in statistical inference, 1996, World Scientific.
11. Multivariate statistical analysis, 1996, Marcel Dekker, N.Y.

Research papers

1955

1. Application of non-parametric tests in industrial problems I.S.Q.C. Bulletin. pp. 15–21.
2. Family budget analysis of Jute growers. Proceedings Indian Science Congress, 5 pages.

1957

1. One row balance in PBIB designs, Journal of Indian Society of agricultural statistics. pp. 168–178.
2. On a reinforced PBIB designs. Journal of Indian Society of agricultural research statistics. pp. 41–51.

1961

1. On tests with likelihood ratio criteria in some problems of Multivariate analysis. Ph.D. thesis, Stanford University, California. (Thesis Adviser: Prof. Charles Stein).

1962

1. On a multivariate testing problem. Calcutta Statistical Assoc. Bulletin. pp. 55–60.

1963

1. A note on the combined analysis of Youlden squares and Latin square designs with some common treatments. Biometrics. pp. 20–27.
2. Minimax property of T^2-test in the simplest case (with J. Kiefer, Cornell Univ. and C. Stein, Stanford University). Annals of Mathematical Statistics. pp. 1525–1535.

1964

1. On the likelihood ratio tests of a normal multivariate testing problem. Annals of Mathematical statistics. pp. 181–189.
2. Local and asymptotic minimax properties of multivariate tests (with J. Kiefer). Annals of Mathematical Statistics. pp. 21–35.
3. Minimax property of R^2-test in the simplest case (with J. Kiefer). Annals of Mathematical Statistics. pp. 1475–1490.

1965

1. In the complex analogues of T^2 and R^2-tests. Annals of Mathematical Statistics. pp. 664–670.

2. On the likelihood ratio test of a normal multivariate testing problem II. Annals of Mathematical Statistics. pp. 1061–1065.
3. On the F-test in the intra-block analysis of a class of partially balanced incomplete block designs. Journal of American Statistical Association. pp. 285–293.

1967

1. On a multivariate likelihood ratio test. Seminar volume in statistics, Banarus Hindu University, India. pp. 12–24.

1968

1. Locally and asymptotically minimax tests of a multivariate problem. Annals of Mathematical Statistics. pp. 171–178.
2. On tests of the equality of two covariance matrices. Annals of Mathematical Statistics. pp. 275–277.

1970

1. On symmetrical designs for exploring response surfaces. Indian Institute of Technology, Kanpur, India, publication, 50 pages.
2. Use of Standard test scores for preparing merit lists. I.I.T. Kanpur publication, 30 pages
3. Bayesian Estimation of means of a one-way classification with random effect model. Statistiche Hefte, 10 pages.

1971

1. On the distribution of a multivariate statistic. Sankhya. pp. 207–210.
2. On the distribution of a multivariate complex statistic. Archiv der mathematik pp. 431–436.
3. On the optimum test of some multivariate problem (with M. Behara of Mc. Master Université). Archiv der mathematic. pp. 436–442.
4. On tests with discriminant coefficents (Invited paper). Silver Jubilee volume. Journal of the Indian Society of Agricultural Statistics. pp. 28–35.

1972

1. On testing problems concerning the mean of multivariate complex Gaussian distribution. Annals of the Institute of Statistical Mathematics. pp. 245–250.

1973

1. On discriminant decision function in complex Gaussian distribution. Seminar volume of McMaster University. Prob. and Information Theory II, Springer-Verlag. pub. pp. 139–148.

2. An integral—its evaluation and applications. Sankhya. pp. 334–340.
3. Oh a class of unbiased tests for the equality of K covariance matrices. Research Seminar Volume. Dalhousie University. pp. 57–61.
4. On a class of tests concerning covariance matrices of normal distributions (with S. Das Gupta, University of Minnesota). Annals of Mathematical Statistics, 5 pages.

1975

1. On the distribution of a random matrix (with B. Sinha, University of Maryland). Communication in Statistics. pp. 1053–1067.

1976

1. On the optimality and nonoptimality of some multivariate test procedures (with Prof. B, Sinha). Sankhya. pp. 244–349.

1977

1. Alternative derivation of some multivariate distributions (with S. Basu). University of North Carolina), Archiv der Mathematik. pp. 210–216.
2. Bayesian inference of Statistical models (with B. Clément of Polytechnique) statistiche Hefte. pp. 181–192.

1978

1. An alternative measure theoretic derivation of some multivariate complex distributions. Archiv der Mathematike. pp. 215–224
2. An algebric version of the Central limit Theorem (with W. V. Waldenfels, Heidelberg University). Zeitschrift fur Wahrscheinlichbleits theoric. pp. 129–134.
3. Effect of additional variates on the power functions of R^2-test (with B. Clément and B. Sinha) Sankhya B. pp. 74–82.

1979

1. Locally minimax tests for multiple correlations. The Canadian Journal of Statistics. pp. 53–60.
2. Locally minimax test for the equality of two covariance matrices (with S. Chakravorti, Ind. Stat. Inst.). Archiv der Mathematik. pp. 583–589.

1980

1. On D-, E-, D_A- and D_M- optimality of test procedures on hypotheses concerning the covariance matrix of a normal distribution (with P. Banerjee, Univ. of New Brunswick). Research seminar at Dalhousie University, North-Holland. pp. 11–19.

1981

1. Asymptotically minimax test of Manova (with S. Chakravorti). Sankhya A, 8 pages.
2. Tests for the mean vector under interclass covariance structure (with B. Clément, S. Chakravorti and B. Sinha). Journal of Statistical Computations and Simulations. pp. 237–245.
3. The asymptotic behaviour of the probability ratio of general linear hypothesis (with W.V. Waldenfels). Sankhya A. pp. 311–330.
4. Invariance concepts in statistics (Invited paper). Encyclopedia of Statistical Sciences, vol. 3. Ed. N. Johnson, K.S. Klotz, John Wiley. pp. 386–389.
5. The Hunt-Stein theorem (Invited paper). Encyclopedia of Statistical Sciences. Ed. N. Johnson and S. Klotz. Vol. 4, John Wiley. pp. 219–225.

1982

1. Numerical comparison of power functions of invariant tests for means with covariates (with O. Mouqadem. Maroc). Statistiche Hefte. pp. 115–121.
2. Critical values of the locally optimum combination of two independent test statistics (B. Johannesson). Journal of Statistical Computation and Simulation. pp. 1–35.

1983

1. Estimation of mixing proportion of two unknown distributions (with M. Ahmad, l"Univ. de Quebec à Montréal and B. Sinha). Sankhya A. pp. 357–371.
2. Generalized variance statistic in the testing of hypothesis in complex Gaussian distributions (with M. Behara). Archiv de Mathematik. pp. 538–543.
3. Optimum tests of means and discriminant coefficents with additional informations (S. Chaksavorti). Journal of Statistical Planning and Inference, 5 pages.
4. Comparison of power functions of some stepdown tests for means with additional observations (B. Johannesson). Journal of Statistical Computation and Simulation. pp. 1–30.

1985

1. Tests for means with additional informations (with B. Clément and B. Sinha). Communication in Statistics A., pp. 1427–1453.

1987

1. Robust tests of mean vector in Symmetrical multivariate distribution (with B. Sinha). Sankhya A. pp. 254–263.
2. On a locally best invariant and locally minimax test in symmetrical multivariate distribution. Pillai volume, Advances in Multivariate Statistical Analysis, D. Reidel. Pub. Co. pp. 63–83.
3. Robustness of t-test (with T. Kariya and B. Sinha). Journal of Statistical Society of Japan. pp. 165–173.

1988

1. Locally minimax tests in symmetrical distributions. Annals of Statistical Mathematics. pp. 381–394.
2. On robust test of the extended GMANOVA problem in Elliptically symmetrical distributions (with K. Das, University of Calcutta). Sankhya A. pp. 234–248.
3. Equivariant estimator of a mean vector μ of $N(\mu, \Sigma)$ with $\mu'\Sigma^{-1}\mu = 1$, $\Sigma^{-1/2}\mu = c$ or $\Sigma = \sigma^2(\mu'\mu)I$, (with T. Kariya and F. Perron). Journal of multivariate Analysis. pp. 270–283.

1989

1. Elliptically symmetrical distributions (Invited paper). La Gazette des sciences mathématiques du Québec). pp. 25–32.
2. Some robust tests of independence in Symmetrical Multivariate distributions, Canadian Journal of Statistics. pp. 419–428, 1988.

1990

1. On the best equivariant estimator of mean of a multivariate normal population (with F. Perron). J. of Multivariate Analysis. pp. 1–16.
2. Locally minimax tests of independence with additional observations (with K. Das). Sankhya B. pp. 14–22.
3. Inadmissibility of an estimator of the ratio of the variance components (with K. Das and Q. Meneghini). Statistics and Prob. Letters, vol. 10, 2, pp. 151–158.

1991

1. Best equivariant estimation in curved covariance model (with F. Perron). Jour. Mult. Analysis. pp. 46–55.
2. Improved estimators of variance components in balanced hierarchical mixed models (with Q. Meneghini and K. Das). Communication on Statistics, Theory & Methods. pp. 1653–1665.

Publications of the Author

1992

1. On an optimum test of the equality of two covariance matrices. Ann. Inst. Stat. Math., 357–362.
2. Optimum equivariant estimation in curved model (with E. Marchand). Selecta Statitica Canadiana, 8, pp. 37–57.
3. On the robustness of the locally minimax test of independence with additional observations (with P. Banerjee and M. Behara). Metrika. pp. 37–57.

1993

1. James-stein estimation with constraints on the norm (with E. Marchand). Communication in Statistics, Theory and Method, 22, pp. 2903–2923.

1994

1. Some distributions related to a noncentral Wishart (with S. Dahel). Comm. in Statistics, Theory and Method, 23, pp. 1229–1239.
2. Robustnesss of multivariate tests (with M. Behara). Selecta Statistica Canadiana, 9, pp. 105–140.
3. Locally minimax tests for a multivariate data problem (with S. Dahel and Y. Lepage). Metrika. pp. 363–376.
4. On approximations involving beta distribution (with B. Johannesson). Comm. in Statistics, Simulation. pp. 489–503.

1995

1. Designs through recording of varietal and level codes (with M. Ahmad and M. N. Das). Statistics and Probability letters. pp. 371–380.

2001

1. Note on Manova in oblique axes system. Statistics and applications, 3, pp. 129–132.

2002

1. Power comparison of some optimum invariant tests of multiple correlation with partial information (with R. Cléroux). Statistics and Application, 4, pp. 69–74.
2. Approximations and tables of beta distributions. Handbook of beta distribution and its applications, edited by A. K. Gupta and N. Saralees, Marcel Dekker, N.Y. (To appear).

Author Index

Amari, S., 188
Anderson, T. W., 231, 269, 331, 346, 374, 384, 385, 386, 437, 444, 447, 448, 455, 473, 495, 512, 518, 519, 523, 524, 526
Anderson, S. A., 61
Armstrong, J. S., 518

Bahadur, R. R., 131, 141, 437, 455, 473
Banerjee, K. S., 460
Bantegui, C. G., 375
Baranchik, A. J., 168
Bartholomeu, D. J., 308
Bartlett, M. S., 25, 382, 454, 495, 512
Basu, D., 82, 83, 132
Bennett, B. M., 299
Berger, J. O., 132, 152, 159, 173
Berry, P. J., 185
Besilevsy, A., 1
Bhattacharya, P. K., 473
Bhargava, R. P., 414
Bhavsar, C. D., 473, 505
Bibby, J. M., 505
Bickel, P. J., 185
Biloudeaux, M., 178

Birkhoff, G., 1
Blackwell, D., 48
Bonder, J. V., 61
Bose, R. C., 250, 275
Bowker, A. H., 234, 447
Box, G. E. P., 70, 346, 374, 385, 501, 512
Bradly, R. A., 473
Briden, C., 195
Bradwein, A. C., 159
Brown, G. R., 469
Bunke, O., 454, 455, 473

Cacoullas, J., 437, 469, 473
Cambanis, S., 120
Cavalli, L., 437, 454
Chacko, V. J., 303
Chernoff, H., 473
Chu, K'ai-Ching, 91
Clement, B., 416
Cléroux, R., 120, 501
Cochran, W. G., 218, 452, 473
Cohen, A., 303
Constantine, A. G., 231, 369
Consul, P. C., 339

551

Cooper, D. W., 454, 473
Cox, D. R., 184
Craig, A. T., 218
Cramer, H., 82, 110

Darmois, G., 83
Davis, A. W., 375
DasGupta, A., 185
DasGupta, S., 120, 329, 331, 332, 341, 386, 436, 437, 444, 468, 469
Dykstra, R. L., 136

Eaton, M., 41, 120, 136, 220, 269, 303, 414, 505
Efron, B., 184, 193
Elfving, G., 224

Farrell, R. H., 269
Ferguson, T. S., 41, 152, 159, 444, 445
Fisher, R. A., 131, 184, 224, 436, 468
Foster, R. G., 385
Fourdrinier, D., 184
Frechet, M., 80
Fujikashi, Y., 375, 386

Gabriel, K. R., 385
Galton, F., 113, 517
Ganadesikan, R., 250, 385
Ghosh, J. K., 55
Ghosh, M. N., 57, 386
Giri, N. C., 1, 41, 49, 54, 57, 58, 61, 79, 120, 142, 143, 151, 159, 184, 193, 194, 202, 205, 211, 245, 250, 269, 287, 303, 317, 319, 332, 353, 367, 369, 386, 416, 448, 460, 468
Girshik, M. A., 495, 496
Glic, N., 473
Goodman, N. R., 178, 205, 307
Gray, H. L., 450
Graybill, F. A., 1, 218
Gupta, A. K., 375

Haff, L. R., 171
Hall, W. J., 55
Halmos, P., 55, 142, 178
Han, C. P., 455
Heck, D. L., 385
Hills, M., 473

Hinkley, D. V., 184, 188
Hodges, J. L., 435
Hogg, R. V., 218
Hopkins, C. E., 473
Hotelling, H., 273, 375, 436, 484, 505, 512
Hsu, P. L., 275, 372, 374
Huges, J. B., 375, 385
Huang, S., 120

Ingham, D. E., 224
Isaacson, S. L., 468
Ito, K., 375

James, A. T., 231
James, W., 162, 170, 171, 173, 495
Jayachandran, K., 386
John, S., 386

Kabe, D. G., 250, 447
Kagan, A., 82
Kariya, T., 120, 184, 269, 303, 363, 414
Karlin, S., 250
Kelkar, D., 120
Kendall, M. G., 473
Kent, J. T., 505
Khatri, C. G., 224, 250, 339, 375, 473, 505
Khirsagar, A. M., 220, 250, 468, 505
Kiefer, J., 57, 58, 132, 173, 245, 250, 282, 284, 287, 335, 341, 353, 386, 416, 452, 455, 469
Koehn, U., 61
Kolmogorov, A. N., 43
Krishnaiah, P. R., 375, 413
Kubokawa, T., 171, 178
Kudo, A., 303, 452

Lachenbruch, P. A., 450, 452
Lawley, D. N., 375, 495, 517, 518
Lazrag, A., 120, 501
Le Cam, L., 131
Lee, Y. S., 375, 386
Lehmann, E., 41, 44, 48, 50, 57, 141, 281, 284, 329, 352
Lehmar, E., 275
Linnik, Ju.Vo., 57, 82, 287

Author Index

MacLane, S., 1
Mahalanobis, P. C., 250, 436
Marcus, L. F., 460
Marcus, M., 1
Marchand, E., 184
Marden, J. I., 303
Marshall, A. W., 473
Mardia, K. V., 195, 505
Martin, D. C., 473
Matusita, K., 473
Mauchly, J. W., 338
Mauldan, J. G., 224
McDermott, M., 303
McGraw, D. K., 91
Mickey, M. R., 450, 452
Mijares, T. A., 375
Mikhail, N. N., 386
Mikail, W. F., 386
Mine, H., 1
Mitra, S. K., 109
Morant, G. M., 436
Morrison, D. F., 518
Morris, C., 193
Morrow, D. J., 375, 385
Mudholkar, G. S., 386
Muirhead, R. J., 269, 375, 505

Nachbin, L., 34, 53, 220
Nagao, H., 329, 340
Nagasenker, B. N., 328, 339
Nandi, H. K., 269, 386
Narain, R. D., 224, 386
Neyman, J., 132, 437
Nishida, N., 455
Nüesch, P. E., 303, 304

Ogawa, J., 218, 224
Okamoto, M., 454, 455
Olkin, I., 184
Owen, D. B., 450

Pearlman, M., 120, 136, 303, 386
Pearson, E. S., 437
Pearson, K., 138, 434, 435
Penrose, L. S., 453
Perlis, S., 1
Perron, F., 184, 193, 194, 202

Pillai, K. C. S., 328, 339, 375, 385, 386
Pitman, E. J. G., 57
Please, N. W., 454
Pliss, V. A., 57
Press, J., 523

Quenouille, M., 450

Raiffa, H., 152
Rao, C. R., 34, 82, 109, 146, 218, 269, 413, 437, 452, 468, 505, 518
Rasch, G., 224
Rees, D. D., 385
Robert, C. P., 171
Roy, S. N., 34, 224, 250, 269, 275, 369, 375, 386, 496
Rubin, H., 250, 518, 519, 523, 524
Rukhin, A. L., 473

Salaevskii, O. V., 57, 287
Sampson, P., 375
Savage, L. J., 120, 142, 178
Saw, J. G., 375
Schatzoff, M., 375, 386
Scheffé, H., 141, 299
Schlaifer, R., 152
Schoenberg, I. J., 109
Schucany, W. R., 450
Schwartz, R., 61, 282, 335, 341, 364, 386, 416, 452, 455, 469
Scott, E., 132
Seillah, J. B., 171
Simaika, J. B., 280, 349
Sharack, G. R., 303
Sigiura, N., 340
Sinha, B. K., 120, 269, 317, 369, 416, 448, 468
Siotani, M., 375
Sitgreaves, R., 447, 448
Skitovic, V. P., 83
Smith, C. A. B., 385
Sobel, M., 120
Solomon, H., 473, 518
Spearman, C., 517
Srivastava, M. S., 171, 178, 339, 469, 505

Stein, C., 29, 57, 59, 60, 136, 159, 160, 162, 167, 170, 171, 172, 173, 250, 352, 353, 360, 414, 416
Strawderman, W. E., 159, 170
Sugiura, N., 329
Sverdrupe, E., 224

Tang, P. C., 275
Tiao, G. C., 70
Thurston, L., 517, 518
Tildsley, M. L., 435

Tukey, 450

Wald, A., 346, 437, 444, 447, 448
Wang, Y., 303
Welch, B. L., 437
Wijsman, R. A., 41, 55, 60, 61, 250
Wilks, S. S., 372, 374, 382
Wishart, J., 224, 250
Wolfowitz, J., 224, 250

Zellner, L. C., 91

Subject Index

Abelian groups, 30
Additive groups, 30
Adjoint, 84
Admissible classification rules, 439
Admissible estimator of mean, 159
Admissibility
  of R^2-test, 349
  of T^2-test, 281
  of the test of independence, 349
Affine groups, 31
Almost invariance and invariance, 49
Almost invariance, 50
Analysis of variance, 65
Ancillary statistic, 185
Application of T^2-test, 293
Asymptotically minimax property of T^2-test, 194
Asymptotically minimax tests, 289

Bartlett decomposition, 153
Basis of a vector space, 3
Bayes classification rule, 440
Bayes
  estimator, 151

[Bayes]
  extended, 155
  generalized, 155
Behern–Fisher problem, 298
Best equivariant estimation of mean, 186
  in curved covariance model, 195
Bivariate complex normal, 87
Bivariate normal random variable, 75
Boundedly complete family, 54

Canonical correlation, 505
Cartan G-space, 60
Characteristic
  equation, 8
  function, 43
  roots, 8
  vector, 8
Characterization
  equivariant estimators of covariance matrix, 196
  multivariate normal, 82
  of regression, 199
Classical properties of MLES, 98

Classification
  into one of two multivariate normal, 434
  with unequal covariance matrices, 454
Cochran's Theorem, 217
Coefficient of determination, 80
Cofactor, 5
Completeness, 144
Complex analog of R^2-test, 406
Complex matrices, 24
Complex multivariate normal distribution, 84
Concave function, 39
Concentration ellipsoid and axes, 110
Confidence region of mean vector, 293
Confluent hypergeometric function, 237
Consistency, 142
Contaminated normal, 95
Convex function, 39
Coordinates of a vector, 3
Covariance matrix, 71
Cumulants, 119
Curved model, 185

Diagonal matrix, 5
Direct product, 32
Dirichlet distribution, 97
Discriminant analysis, 435
  and cluster analysis, 473
Distribution
  of characteristic roots, 495
  of Hotelling's T^2, 234
  of multiple correlation coefficient, 245
  of partial correlation coefficient, 245
  of quadratic forms, 213
  of U statistic, 384
  in complex normal, 248
  in symmetrical distributions, 250

Efficiency, 145
Elliptically symmetric distributions
  multivariate, 106
  univariate, 92
Equality of several covariance matrices, 389
Equivalent to invariant test, 50

Equivariant estimator, 49
Estimation of covariance matrix, 171
Estimation of factor loadings, 519
Estimation of mean, 159, 170
Estimation of parameters
  in complex normal, 176
  in symmetrical distributions, 178

Factor analysis, 517
Fisher–Neyman factorization theorem, 142
Full linear groups, 31
Functions
  bimeasurable, 38
  one-to-one, 38

General linear hypothesis, 65
Generalized inverse, 109
Generalized Rao–Cramer inequality, 146
Generalized Variance, 232
Gram–Schmidt orthogonalization process, 75
Groups, 29
  acts topologically, 60

Hermitian matrix, 24
Hermitian positive definite matrix, 24
Hermitian semidefinite matrix, 24
Heuristic classification rules, 443
Homomorphism, 32
Hotelling's T^2-test, 273
Hunt–Stein theorem, 57
Hypothesis: a covariance matrix unknown, 328

Idempotent matrix, 26
Identity matrix, 5
Independence of two subvectors, 360
Induced transformation, 44
Invariance
  function, 48
  and optimum tests, 57
  of parametric space, 45
  of statistical problems, 46
  in statistical testing, 44
Inverse matrix, 6
Inverted Wishart distribution, 205, 231

Subject Index

Isomorphism, 32

Jacobian, 33
James–Stein estimator, 159, 163
  positive part of, 165

Kurtosis, 118

Lawley–Hotelling's test, 361, 375
Linear hypotheses, 65
Linear space, 39
Locally best invariant test, 58, 60
Locally minimax tests, 288

Mahalanobis distance, 446
Matrix, 4
Matrix derivatives, 21
Maximal invariant, 48
Maximum likelihood estimator, 132
  of multiple and partial correlation coefficients, 140
  of redundancy index, 140
  of regression, 138
Minimax
  character of R^2-test, 353
  classification rule, 440, 459
  estimator, 155
  property of T^2-test, 281
Minor, 5
Most stringent test, 58
  and invariance, 58
Multinomial distribution, 124
Multiple correlation, 114, 241
  with partial information, 366
Multiplicative groups, 30
Multivariate distributions, 41
Multivariate beta (Dirichlet) distributions, 126
Multivariate elliptically symmetric distribution, 106
Multivariate exponential power distribution, 127
Multivariate general linear hypothesis, 369
Multivariate log-normal distribution, 125
Multivariate complex normal, 84
Multivariate normal distribution, 70, 80

Multivariate t-distribution, 93, 96
Multivariate classification
  one-way, 387
  two-way, 387, 388
Multivariate regression model, 378

Negative definite, 8
Negative semidefinite, 8
Noncentral chi-square distribution, 211
Noncentral F-distribution, 213
Noncentral student's t, 212
Noncentral Wishart distribution, 231
Nonsingular matrix, 6
Normal subgroup, 31

Oblique factor model, 519
Optimum invariant properties of T^2-test, 275
Orthogonal factor model, 518
Orthogonal matrix, 5
Orthogonal vectors, 2

Paired T^2-test, 298
Partial correlation coefficient, 115, 241
Partition matrix, 16
Penrose shape and size factors, 452
Permulation groups, 31
Pillai's test, 361, 375
Population canonical correlation, 506
Population principal components, 485
Positive definite matrix, 8
Principal components, 483
Problem of symmetry, 295
Profile analysis, 318
Projection of a vector, 2
Proper action, 60

Quadratic forms, 7
Quotient groups, 32

R^2-test, 342, 347
Randomized block design, 297
Rank
  of a matrix, 7
  of a vector, 4
Ratio of distributions, 59
Rectangular coordinates, 153
Regression surface, 113

Redundancy index, 120
Relatively left invariant measure, 59
Roy's test, 361, 375

Sample canonical correlation, 510
Sample principal components, 490
Simultaneous confidence interval, 293
Singular symmetrical distributions, 109
Skew matrix, 23
Smoother shrinkage estimation of mean, 168
Sphericity test, 337, 408
Statistical test, 279
Stein's theorem, 60
Subgroups, 30
Sufficiency, 141
  and invariance, 55
Symmetric distributions, 91

Tensor product, 107
Test
  of equality of several multivariate normal distributions, 404
  of equality of two mean vectors, 294
  with missing data,
    mean vector, 412
    covariance matrix, 416
Test of hypotheses
  about canonical correlation, 511
  concerning discriminant coefficients, 460
  of independence, 342

[Test of hypotheses]
  of mean against one sided alternatives, 303
  of mean in symmetrical distributions, 309
  of mean vectors, 269, 310
  of mean vector in complex normal, 307
  of scale matrices in $E_p(\mu, \Sigma)$, 407
  of significance of contrasts, 295
  of symmetry of biological organs, 318
  of subvectors of mean, 299, 316
Tests of means
  known covariances, 270
  unknown covariances, 272
Testing principal components, 498
Time series, 525
Trace of a matrix, 7
Translation groups, 31
Triangular matrices, 5

Unbiasedness, 141
  and invariance, 56
Uniformly most powerful invariant test, 58
Unimodular group, 31
Union-Intersection principle, 318

Vectors, 1
Vector space, 3

Wilks' criterion, 374, 385
Wishart distribution, 218
  inverted, 231
  square root of, 261